Environmental Radioactivity

From Natural, Industrial,
and Military Sources

FOURTH EDITION

Environmental Radioactivity

From Natural, Industrial, and Military Sources

FOURTH EDITION

Merril Eisenbud

Professor Emeritus
New York University Medical Center
Nelson Institute of Environmental Medicine
New York, New York

Thomas Gesell

Health Physics Program
Department of Physics
Idaho State University
Pocatello, Idaho

ACADEMIC PRESS

San Diego · London · Boston · New York · Sydney · Tokyo · Toronto

This book is printed on acid-free paper. ∞

Copyright © 1997 by ACADEMIC PRESS

Academic Press
a division of Harcourt Brace & Company
525 B Street, Suite 1900, San Diego, California 92101-4495, USA
http://www.apnet.com

Academic Press Limited
24-28 Oval Road, London NW1 7DX, UK
http://www.hbuk.co.uk/ap/

Library of Congress Cataloging-in-Publication Data

Environmental radioactivity / edited by Merril Eisenbud, Thomas F.
    Gesell. -- 4th ed.
        p.   cm.
    Includes bibliographical references and index.
    ISBN 0-12-235154-1 (alk. paper)
    1. Radioactive pollution. I. Eisenbud, Merril. II. Gesell,
Thomas F.
TD196.R3E597   1997
363.1'79--dc20   96-41834
                                                       CIP

PRINTED IN THE UNITED STATES OF AMERICA
97 98 99 00 01 02 QW 9 8 7 6 5 4 3 2 1

Contents

Preface to the Fourth Edition   xi

Preface to the Third Edition   xv

Preface to the Second Edition   xix

Preface to the First Edition   xxi

Acronyms   xxiii

CHAPTER 1

Introduction

The Early History of Radioactivity   2

The Nuclear Energy Industry   5

Early Studies of Radioactive Contamination of the Environment   7

CHAPTER 2

The Biological Basis of Radiation Protection

Early Knowledge of Radiation Effects   10

Summary of Present Knowledge of Radiation Effects on Humans   13

CHAPTER 3

Radiation Protection Standards

Evolution of Radiation Dose Standards   40
Changes in Concepts of Standards Setting   55
Organizations Involved in Establishing and Implementing
    Radiation Protection Standards   58

CHAPTER 4

Atmospheric Pathways

Properties of the Atmosphere   63
Deposition and Resuspension   82
Tropospheric and Stratospheric Behavior   85

CHAPTER 5

Terrestrial and Aquatic Pathways

The Food Chain from Soil to Humans   89
Transport in Groundwater Systems   113
Transport in Surface Water Systems   115

CHAPTER 6

Natural Radioactivity

Naturally Occurring Radioactive Substances   135
Natural Radioactivity in Phosphate Fertilizers   172
Natural Radioactivity in Building Materials   174
Natural Radioactivity in Fossil Fuels   175
Induced Radionuclides   180
Natural Sources of External Ionizing Radiation   182
Technological Developments That Increase Human Exposure   190
Areas Having Unusually High Natural Radioactivity   191
Summary of Human Exposures to Natural Ionizing Radiation   200

CHAPTER 7

Production and Reprocessing of Nuclear Fuels

Uranium	201
Thorium	217
Average Doses from the Production and Reprocessing of Nuclear Fuels	218

CHAPTER 8

Power Reactors

Some Physical Aspects of Reactor Design and Operation	222
Types of Reactors	229
Low-Level Discharges from Light-Water Reactors	232
Reactor Accidents	244

CHAPTER 9

Nuclear Weapons

Physical Aspects of Nuclear Explosions	271
Short-Term Radiological Effects of Nuclear War	285
Some Problems of Recovery from Nuclear Attack	290
Worldwide Fallout from Nuclear Weapons Tests	291
Behavior of Individual Radionuclides from Fallout	297
Lung Dose Due to Inhalation of Dust from Weapons Tests	314
External Radiation	314

CHAPTER 10

Various Other Sources of Exposure

The Early History of Radium	317
Exposure from Radioluminescent Paints	318
Uranium and Thorium in Ceramics and Glass	319
Depleted Uranium Projectiles	320
Thorium in Gas Mantles	321

Thorium in Welding Rods   321
Use of Specific Radionuclides ("Isotopes") in Research and Industry   322
Americium-241 in Smoke Detectors   323
Radionuclides as Sources of Power   325
Transportation of Radioactive Substances   328
Summary of Doses Received from Various Sources   333

CHAPTER 11

Radioactive Waste Management

Low-Level Wastes   336
High-Level Wastes   345
Proposed Repositories in the United States   366
The Special Problems of Gaseous or Highly Soluble Long-
     Lived Radionuclides   373

CHAPTER 12

Experience with Radioactive Contamination Due to Accidents

Fallout from the Thermonuclear Weapons Test of March 1, 1954   379
The Accident at the Windscale Reactor Number One, October 1957   387
The Houston Incident of March 1957   393
The Oak Ridge Plutonium Release of November 1959   395
The Army Stationary Low-Power Reactor (SL-1), January 1961   398
Abortive Reentry of the SNAP 9A, April 1964   399
Reentry of the Satellite Cosmos 954, January 1978   401
The Accident at Three Mile Island Unit 2, March 1979   402
The Chernobyl Accident, April 1986   409
Other Contaminating Events That Occurred in the Former
     Soviet Union   425
Accidents Involving Military Aircraft Carrying Nuclear Weapons   429
Accidents Involving Abandoned, Lost, or Stolen
     Gamma-Radiation Sources   433

CHAPTER 13

Methods of Environmental Surveillance

Design Phase   442
Operational Phase   443
Environmental Surveillance under Emergency Conditions   467

CHAPTER 14

Radiological Assessment and Its Application to Dose Reconstruction

Radiological Assessment   471
Dose Reconstruction   475
Events Leading to the Need for Dose Reconstructions   479
Summary of Dose Reconstruction Projects at U.S. Nuclear
    Weapons Plants   483

CHAPTER 15

Remediation of Contaminated Sites

Examples of U.S. Standards for Cleanup   495
Risk Assessment, Risk Management, and the Public   497
The Remediation Process   498
Remediation Measures and Technology   500
Sites Predating World War II   502
Sites Related to the Development of Nuclear Energy and
    Nuclear Weapons   504
Some Concluding Remarks   521

CHAPTER 16

Radiation Exposure and Risks: Some Contemporary Social Aspects

The Occupational Safety Record of the U.S. Nuclear Energy Program   524
Sources of Exposure of the General Population   527

The Disparity between Actual and Perceived Risk   530
Evolution of the Modern Environmental Movement   531
Public Perception of Technology   541
The Positive Side of Popular Environmentalism   544
What about the Future of Nuclear Energy?   545

Appendix: The Properties of Certain Radionuclides   547

References   569

Index   641

Preface to the Fourth Edition

It is symptomatic of the rapid rate of change in the subject of this book that each of its four editions has been published in a distinctly different technical and sociopolitical era.

In 1963, the year when the first edition was published, President Kennedy and Premier Khrushchev signed an agreement that ended the frenetic pace of nuclear weapons testing in the atmosphere, and the first order for a privately funded nuclear power reactor was placed in the United States. The subject of *Environmental Radioactivity* was still primarily concerned with the behavior of weapons testing fallout, but there was great optimism about the prospects for future developments in the field of nuclear power. Incidentally, 1963 was also the year in which people were reading Rachel Carson's best-seller, *Silent Spring*. That book, more than any other single event, initiated the environmental movement that has swept much of the world and influenced the subject of this book.

The second edition, published in 1977, appeared at a time of great optimism about the future of nuclear power in the United States. There was good reason for this. In 1963 there were only four nuclear power plants in the United States, with an installed generating capacity of 641 megawatts electric (MWe). By 1977 this had grown to 13,440 MWe, with another 116,000 MWe in the planning stage, but there was much public and political controversy about the safety of nuclear reactors and the significance of their radioactive emissions on the environment. Although fallout from nuclear weapons testing had all but ceased, the Cold War continued as the numbers of nuclear weapons in the stockpiles of the superpowers grew. It was a period when the public was becoming increasingly obsessed with the dangers of traces of radioactive and chemical substances in the environment, while remaining almost oblivious to the great danger imposed by those stockpiles.

The third edition was published in 1987, when the subject was influenced greatly by the nuclear power plant accidents at Three Mile Island and Chernobyl. The Three Mile accident caused no injuries to either the workers or the general public, but it was financially costly and was one of several factors that resulted in a cessation of new orders for nuclear power plants in the United States. In contrast, the Chernobyl accident was of such magnitude as to have major implications for public health.

As this fourth edition goes to press, there are again major new developments. It is a source of great relief that the Cold War has ended and that the superpowers are now in the process of dismantling nearly 50,000 nuclear weapons that have accumulated in their stockpiles. An all-out nuclear war that seemed so probable when the first edition of this book was published is far less of a threat, although for many years to come we will live with the lesser danger that weapons may be diverted to malevolent hands. Moreover, developments at home require that this edition deal with two new facets of our subject. First, we must discuss what we know about how to clean up (remediate) the nuclear weapons plants and laboratories and their environs contaminated with radioactive and chemical materials. Second, this edition, being published at a time of unprecedented legal activity and class-action suits that claim lack of prudence in the way the weapons production program was conducted, has required that new methods of retrospective dose assessment be developed to estimate the doses received by radiation workers and people living in the vicinity of the weapons plants.

The first three editions of this book were written by myself, but it was inevitable that the time would come when I would seek to share the burden with a coauthor. That time is here, and I was greatly pleased when Thomas F. Gesell, Professor of Health Physics at Idaho State University, agreed to serve as coauthor of this new edition. I hope he will assume an increasingly important role in the preparation of other editions that may be needed in the decades ahead.

Preparation of the earlier editions was greatly assisted by the staff and students of the Nelson Institute of Environmental Medicine at the New York University Medical Center. For this edition Dr. Gesell and I have been assisted by the faculty, staff, and students of the Idaho State University Health Physics Program, especially Bruce Busby, who diligently updated the appendix on properties of radionuclides with the most recent information. We also acknowledge our many colleagues, who generously made materials available for this book, and our families, who encouraged us during its preparation.

Merril Eisenbud

Preface to the Third Edition

The first edition of this book was published in 1963, at a time when interest in environmental radioactivity was focused on fallout from testing nuclear weapons. An agreement to ban testing in the atmosphere was signed in that year by the major nuclear powers, and would have resulted in diminished interest in the subject except that 1963 was also when the first commercial order for a privately owned nuclear power plant was placed by a utility company in the United States. Attention thus shifted from the effects of nuclear weapons testing on the environment to the effects of nuclear power plants, an unfortunate coupling that may explain some of the apprehensions of the public about the dangers from the civilian uses of atomic energy.

When the second edition was published in 1973, the nuclear power industry was in a phase of rapid growth that was sustained until the late 1970s, when orders for new nuclear power plants ceased in the United States and many existing orders were canceled. This was due to many factors, including reduced demand for energy as a result of economic recession, greater energy conservation, growing public opposition to nuclear power, inflation of construction costs, and finally the 1979 accident that destroyed Reactor 2 at the nuclear power plant at Three Mile Island, Pennsylvania. Information about the catastrophic accident at Chernobyl in the U.S.S.R. came to the world's attention in the final stages of preparation of this manuscript. This disaster is certain to further complicate the future of nuclear power.

This edition is being written nearly twenty-five years after the first one was published, and it is going to press at a time of great uncertainty with respect to the future of both the military and civilian uses of nuclear energy. Above all is the continuing danger of nuclear war. The stockpiles of nuclear weapons continue to increase, and delivery systems of greater sophistication are being developed. Public anxiety about the products of nuclear technology is to be expected as long as the major powers have the capacity to cover much of the world with life-threatening amounts of radioactivity.

The civilian nuclear energy industry has established an impressive safety record, but it is beset with economic, regulatory, and political problems that will not be resolved easily. This is even true of such benign applications as those in medical research but the problems are most acute in the electric utility industry. However, the wisdom of continued dependence on fossil fuels is also being questioned increasingly because of potential serious long-term environmental effects. Since substitution of renewable resources such as solar energy does not seem feasible for the foreseeable future, it is to be hoped that the present hiatus in the development of nuclear power will be used to design reactors that will be more acceptable to the public. There may be no other choice if the world is to meet its future energy needs.

There is one feature of this text that requires a word of explanation. After considerable thought and many discussions with my colleagues, I made the decision to avoid the use of SI units. I realized that in this interim period when the new units are coming rapidly into general use, it would be preferable to use both systems of units, with one in parentheses. This was in fact done in an early stage of manuscript preparation, but it caused many of the pages, and particularly those with complex tables, to become so cluttered that I decided to use only one system of units.

Many individuals have assisted me in the preparation of this work and its previous editions. In the preparation of this edition, I am especially grateful for the many suggestions I have received from John Auxier, Beverly Cohen, Norman Cohen, Thomas Gesell, Frank Gifford, Edward Hardy, Catherine Klusek, Paul Linsalata, Frank Parker, Norman Rasmussen, Charles Roessler, Keith Schiager, Lauriston Taylor, Arthur Upton, Herbert Volchok, and Robert Watters. As was true in the last edition, Eleanor Clemm labored diligently in my behalf, not only as an excellent typist, but also as editor, bibliographic assistant, and general coordinator.

I wish to express my appreciation to the Office of Energy and Health Research of the U.S. Department of Energy for a grant awarded me to support the cost of manuscript preparation. I am also indebted to the National Institute of Environmental Health Sciences for the many years of support I received under Center Grant No. ES 00260. Finally, it is with pleasure and pride that I record my indebtedness to the Institute of Environ-mental Medicine of the New York University Medical Center, where I recently completed a long and pleasant association with stimulating members of the faculty, staff, and student body.

Merril Eisenbud

Preface to the Second Edition

The first edition of this book was published early in 1963 (McGraw-Hill, New York) at a time when worldwide concern existed because of radioactive fallout from testing nuclear weapons. In that year the principal nuclear powers agreed to ban open-air testing, and the environmental levels of radioactivity from that source have accordingly diminished considerably since then despite occasional tests by countries that did not participate in the test-ban agreement. However, interest in the subject of environmental radioactivity has not diminished correspondingly, but has been more than sustained by the rapid growth of the nuclear power industry and the use of radioactive materials in medicine, research, space exploration, and industry.

When the original edition was published, there were only four commercially operated power reactors in the United States, with an installed generating capacity of about 641 megawatts. At this writing, the generating capacity of power plants already in operation in the United States is about 13,400 megawatts and an additional 116,000 megawatts are in the planning stage. Nuclear power will represent an increasing fraction of the electrical generating capacity in many countries, and the potential environmental impact of this new industry is receiving wide attention.

As this second edition goes to press, the United States continues to be embroiled in public and political controversy over the safety of nuclear reactors and the significance of radioactive emissions to the environment. To the best of my ability, I have attempted to provide an objective account of the relevant biological and physical information that has been accumulated in the thirty years since control over the fission process was first achieved. Because this book is intended to be a coherent technical summary, I have dealt only incidentally with the status of the public controversy and the confusion that has been caused by the impact of extreme statements that have been widely publicized in recent years.

The book was written as a reference source for the scientist, engineer, or administrator with a professional interest in the subject, but it may also be of value to the reader who wishes to understand the technical facts behind the public debate. Hopefully the book will contribute to public understanding of what is fundamentally a complex subject.

The subject of environmental radioactivity is one of vast dimensions, and I cannot claim that this book is an exhaustive treatise on the subject. However, it does represent what I hope is a useful review of what is now known, though perhaps presented with some degree of imbalance due to the fact that my experience is primarily with atomic energy programs in the United States.

I am indebted in many ways to my present and past associates and students for the many thousands of hours of dialogue and research that have made it possible to produce the two editions of this book. My students in particular have proved to be my most effective teachers.

In addition to those to whom I acknowledged assistance in the preparation of the first edition, I feel particularly obliged to mention the help given to me by McDonald E. Wrenn, Norman Cohen, Peter Freudenthal, and Steven Jinks in the preparation of this edition. My work was enormously facilitated also by the assistance of Eleanor Clemm, to whom I am greatly indebted.

Merril Eisenbud

Preface to the First Edition

It is now more than half a century since the phenomenon of natural radioactivity was first discovered by Becquerel and the Curies, and about a quarter of a century since the discovery of nuclear fission by Hahn and Strassmann. Enormous amounts of energy that have been locked in the nucleus of the atom since the beginning of time can now be released at will, and the greatest challenge in his history confronts man as he decides whether to use this new source of energy to reach ever higher levels of social accomplishment or to destroy what mankind has created.

Whether the fission process is exploited in peace or in war, an unavoidable result will be the production of radionuclides in enormous amounts. If these by-products of the fission process should be produced in a major nuclear war, radioactivity will become one of the dominant features of the environment and one of many major obstacles to survival that will face those who have the good fortune to escape prompt death by blast, radiation, or fire. If, on the other hand, mankind does find a peaceful solution to its political problems, and our future civilian industries are supplied with nuclear power, the radionuclides will be accumulated in such great quantities as to require constant vigilance and great wisdom to prevent the effluents of the nuclear industry from contaminating the environment to a serious extent.

Ever since early in World War II, extensive research has been conducted to understand the physical and chemical properties of radioactive sub-

stances, the manner in which they are transported physically through the
environment, and the way in which some of them enter into man's food
supplies, the water he drinks, and the air he breathes. In 1959, when I
accepted an opportunity to develop a graduate teaching program in the
general field of radiological hygiene, I was impressed with the need to
consolidate the vast amount of information that had been developed on the
subject of environmental radioactivity in order that the subject could be
presented to students and others in a comprehensive and systematic
manner.

The subject of environmental radioactivity has aspects of vast dimensions, and the task of bringing together the pertinent information in so
many diverse disciplines proved to be not without its difficulties. There was
first the question of what to include, and in this regard I decided that
the text should be concerned primarily with the behavior of radioactive
substances when they enter the environment. The important and elaborate
technology by which passage of radioactive materials to the environment
may be prevented and the equally important field of health physics that is
concerned with protecting the atomic energy worker were thus placed
beyond the bounds of this work, although it has been necessary frequently
to deal briefly with both subjects in the present text.

I am greatly indebted to my many associates, past and present, who
assisted me in the preparation of this work. It is not possible to acknowledge
all the assistance I have received, but certain of my colleagues have been
particularly helpful in the review of early drafts of certain chapters. In this
regard I am particularly indebted for the help of Norton Nelson, Roy E.
Albert, Abraham S. Goldin, Bernard S. Pasternack, Gerard R. Laurer, Harold
H. Rossi, and Ben Davidson. The index was prepared by Stephen F. Cleary,
and every word of every one of the three to five drafts that ultimately
evolved into the seventeen chapters of this book was typed with remarkable
proficiency and patience by Patricia S. Richtmann. Finally, as is customary
for reasons that can only be understood fully by authors, and authors'
families, I wish to express my appreciation for the help and encouragement
provided by my wife, Irma, to whom this book is dedicated.

Merril Eisenbud

Acronyms

AEC	Atomic Energy Commission
ALARA	As low as reasonably achievable
ALI	Annual limit of intake
BRC	Below regulatory concern
BWR	Boiling-water reactor
CERCLA	Comprehensive Environmental Response, Compensation, and Liability Act
CFR	Code of Federal Regulations
COGEMA	Compagnie General des Matieres Nucleaires (France)
DAC	Derived air concentration (for radionuclides)
DNB	Departure from nucleate boiling
DOE	Department of Energy
DOT	Department of Transportation
EDE	Effective dose equivalent
EPA	Environmental Protection Agency
EPRI	Electric Power Research Institute
FAO	Food and Agriculture Organization
FDA	Food and Drug Administration
FRC	Federal Radiation Council
FUSRAP	Formerly Utilized Sites Remedial Action Program
HM	Heavy metal, usually referring to fissionable material
HTRG	High-temperature gas-cooled reactor

IAEA	International Atomic Energy Agency
ICRP	International Commission on Radiological Protection
INEL	Idaho National Engineering Laboratory
LD_{50}	Radiation dose that is lethal to 50% of experimental animals
LWR	Light-water reactor
MCL	Maximum contaminant level
MED	Manhattan Engineering District, agency for the World War II nuclear program
MESODIF	Mesoscale Diffusion Code (atmospheric model)
MPC	Maximum permissible concentration
MRC	Medical Research Council (United Kingdom)
MWe	Megawatts (electric) in reference to nuclear power plant capacity
MWt	Megawatts (thermal) in reference to nuclear power plant capacity
NAS	National Academy of Sciences
NAS/NRC	National Research Council of the National Academy of Sciences
NCRP	National Council on Radiation Protection and Measurements
NEPA	National Environmental Policy Act
NESHAP	National Emission Standards for Hazardous Air Pollutants
NOAA	National Oceanic and Atmospheric Administration
NRC	Nuclear Regulatory Commission
NRPB	National Radiological Protection Board (United Kingdom)
OECD	Organization for Economic Cooperation and Development
ORNL	Oak Ridge National Laboratory
PWR	Pressurized-water reactor
RBMK	A pressurized-tube, graphite-moderated reactor of the type at Chernobyl
RCRA	Resource Conservation and Recovery Act
RTG	Radioisotope thermal generator
SNAP	Satellite Nuclear Auxiliary Power
TRAC	Terrain Responsive Atmospheric Code (atmospheric model)
TRU	Transuranic, referring to elements with atomic weights greater than uranium (92)
UMTRAP	Uranium Mill Tailings Remedial Action Program
UMTRCA	Uranium Mill Tailings Radiation Control Act
UNSCEAR	United Nations Scientific Committee on the Effects of Atomic Radiation
USDA	U.S. Department of Agriculture

USGPO	U.S. Government Printing Office
WHO	World Health Organization
WIPP	Waste Isolation Pilot Plant
WLM	Working level month
WSSRAP	Weldon Spring Site Remedial Action Program

CHAPTER 1

Introduction

The discovery in 1939 of methods by which the energy contained within the atomic nucleus can be released has led to major advances in our knowledge of the physical world and to far-reaching advances in technology. It so happened that the discovery of nuclear fission coincided with the outbreak of World War II, and the first application of nuclear energy was therefore for military purposes. The dramatic announcement of the destruction of Hiroshima by a nuclear weapon created an image of "the atom" that has been imprinted indelibly on the consciousness of the world's citizens. The linkage of nuclear energy and nuclear war is certainly a major factor in the widespread public fears about even the most benign applications of this form of energy and its by-products.

The most obvious long-range benefit from the fission process is the potential to provide a source of power that would assure a higher standard of living in those countries that do not have adequate reserves of fossil fuel. As the energy requirements of the world increase and as the reserves of fossil fuel become smaller, nuclear energy is bound to play an increasingly important role in civilian economies. Political instabilities in the Middle East have been a major reason for the use of nuclear power in many countries. Its use may be encouraged in the years ahead because, unlike energy derived from fossil fuels, it does not add to the atmospheric burden of "greenhouse gases" that threaten to cause global warming.

Nuclear reactors are already producing a significant fraction of the electricity used in many countries of the world. In 1994, 26 nations were using nuclear power. France employed it to the largest extent, generating 78% of its electric energy. In the United States, 21% of the electricity used was derived from nuclear fuels. However, public fears about the effects of nuclear power began to grow with the 1979 accident at Three Mile Island in Pennsylvania, the 1986 accident at Chernobyl in the Ukraine, and rising concerns about the disposal of nuclear waste (see Chapters 11 and 12). In addition, a worldwide recession that started in the mid-1970s, and an unforeseen rise in the cost of capital required for building nuclear power plants, resulted in a cessation of orders for new power plants in the United States and many other countries.

A less conspicuous but highly important contribution of atomic energy to humankind has been the copious quantities of radionuclides that have become available. In the fields of medical and biological research, the use of radioactive labels is now so commonplace that we take for granted the research that has become possible because of these useful substances. The use of radionuclides as a research tool has grown rapidly and yet with comparative unobtrusiveness, and many discoveries in the biomedical sciences would not have been possible if not for the ready availability of radionuclides.

Regrettably, we cannot forget the negative side to nuclear energy—the possibility of nuclear war. As with many technological advances, society possesses the wisdom to use new knowledge constructively if it wishes to do so, but is also capable of great folly. Only time will tell if, on balance, nuclear energy has been used to bring blessings to humankind or to hasten its social destruction. Happily, this edition is being prepared at a time when major reductions in nuclear weaponry are being implemented by the countries of the former USSR and the United States. The probability of a major nuclear war seems to be greatly reduced, but this may be offset by the acquisition of nuclear weapons by lesser powers, some of which have been notoriously unstable. As this edition goes to press, the most grave problem in nuclear safety is how to keep weapons-grade nuclear material out of the hands of terrorist states. The worldwide increase in well-financed terrorism also raises the possibility that nuclear weapons may be diverted for such use from existing stockpiles of weapons that are now being reduced.

THE EARLY HISTORY OF RADIOACTIVITY

Experience with the dangers of radioactive materials preceded by many years the discovery of the phenomenon of radioactivity. As will be seen in Chapter 2, the atmospheres of mines in Central Europe that had been exploited for their heavy metals since medieval times were unknowingly so radioactive that the miners developed a fatal lung disease that was later diagnosed as lung cancer (Lorenz, 1944), caused by the presence of radon. This was nearly 400 years after the German physician Georgius Agricola wrote his classic book "De Re Metallica" (Agricola, 1556), in which he described the high mortality among the miners. Some radioactive substances were used even before it was known that they were radioactive. The Welsbach gas mantle, which was developed in 1885, utilized the incandescent properties of thorium oxide to greatly increase the luminosity of gaslight in many parts of the world, and uranium oxide has long been used to provide a vivid orange color in ceramic glazes. Other oxides of uranium and thorium have also been used as glazes and for tinting glass (see Chapter 10).

Since the turn of the century and continuing up to the present, in some parts of the world natural radioactivity has been exploited for its supposed benefit to health (see Chapter 6). There is no fully satisfactory explanation as to how this custom originated, but it is known that the popularity of mineral waters in spas around the world led to their establishment as health resorts as long ago as Roman times. It is also known that the laxative properties of spring waters having high mineral content were highly prized, and extensive resorts grew up at places such as Saratoga Springs in New York State and the spas of Europe, Japan, and South America. When the phenomenon of radioactivity was discovered, tests of these mineral waters showed some of them to contain abnormally high concentrations of natural radioactive elements, and it is this fact that may have given rise to the idea that the newly discovered and mysterious property of matter was the reason why the mineral waters seemed to possess curative powers.

The discovery that the radiations from radium could destroy cancerous tissue possibly abetted development of the fad, and the spas of the world soon began to advertise the radioactivity of their waters. To this day the labels on bottled mineral waters in many countries contain measurements of the radioactivity of the spring from which the water was obtained. We will see later that the radioactive sand of Brazilian beaches and the high radon concentrations in the air of old mines in Austria and the United States have attracted tourists who believe that exposure to natural radioactivity can cure arthritis, general debility, and a variety of other diseases (Lewis, 1955; Scheminzky, 1961).

In the early 1920s and continuing up to about 1940, radioactive substances, particularly radium and radon, also found a place in the medical faddism of the period. Radium was injected intravenously for a variety of ills but, far from being cured, many of the patients later developed bone cancer or other malignant disease. Devices that were sold for home use made it possible to add radon to drinking water, and radioactive poultices were prescribed for arthritic joints (Fig. 1-1).

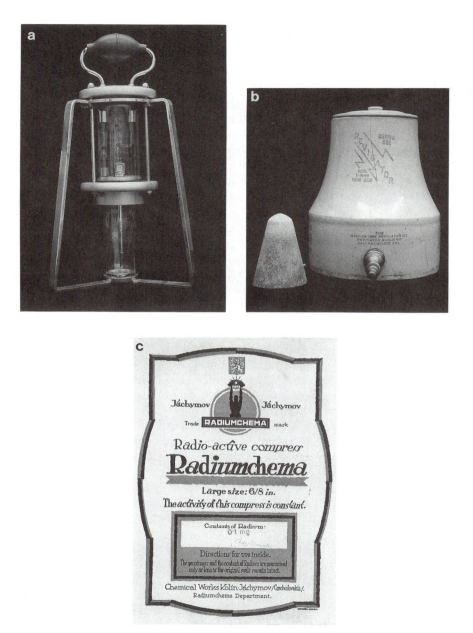

FIGURE 1-1 Early uses of radioactivity in health fads. (a) The Radiumator: air was pumped by hand bulb through a small radium source, entraining radon, which was bubbled through the glass of drinking water. (b) The Revigator (patented 1912): the cone is a mildly radioactive "ore" that was placed in the drinking water crock. (c) Radioactive compress used for miscellaneous aches and pains. The compress contained 0.1 mg of ^{226}Ra and was certified by the Radium Institute of the Faculte des Sciences de Paris.

Finally, a word must be said about the early use of radium in luminous paints. It had been discovered that a slight amount of radium mixed with the fluorescent material zinc sulfide produced a luminous material. The dials of thousands of timepieces, compasses, and other devices were painted with such paints during and immediately after World War I, with no precautions to protect the employees. Many cases of aplastic anemia, leukemia, and bone cancer developed among the factory workers engaged in applying these luminous paints prior to about 1940, when hygienic practices were developed that proved practical and prevented further injuries from occurring. We will see in Chapter 2 that the information derived from studies of the early radium cases has contributed in an important way to the excellent safety record of the modern atomic energy program. The use of radioluminescent materials will be discussed more fully in Chapter 10.

THE NUCLEAR ENERGY INDUSTRY

This introductory chapter is a convenient place in which to provide a bird's-eye view of the nuclear industry, the major components of which will be discussed in greater detail in later chapters.

One way to visualize the main operations of the nuclear energy industry is by the flow diagram in Fig. 1-2. After being mined and concentrated (see

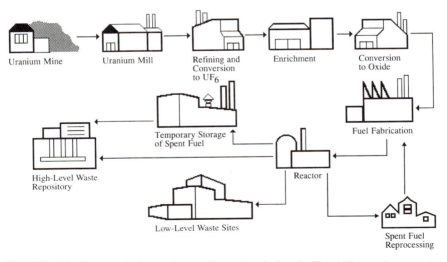

FIGURE 1-2   The principal operations in the nuclear fuel cycle. The civilian nuclear energy industry in the United States has not processed spent fuel since the mid-1970s (see Chapter 7).

Chapter 7), the uranium is shipped to refineries for conversion to a uranium metal or oxide of sufficient purity to be used in reactors (see Chapter 7). The refinery products may be shipped directly to fuel-element fabrication plants or the uranium may be converted to UF_4, a green salt, which is then converted to UF_6, a volatile corrosive compound. To produce uranium that is suitable for use in reactors, it is necessary to increase its ^{235}U content. Since the midcentury, this has been accomplished in the United States by the gaseous diffusion process in enormous plants at Oak Ridge, Tennessee; Paducah, Kentucky; and Portsmouth, Ohio. Depending on the purpose for which the uranium is intended, the degree of enrichment may vary from a few tenths of 1% to more than 90%. It is possible that the gaseous diffusion process in the United States may be replaced eventually by a process that employs lasers. Some European countries are now using centrifuges for this purpose (Tait, 1983). The uranium, in either natural or enriched form as the metal, oxide, or other compounds or alloys, is then transported to the fuel-element fabrication plants. The exact shape to be taken by the uranium and the manner in which it will be clad with stainless steel, zirconium, or various alloys depend on the reactor design. The fabricated fuel elements are then shipped to reactors (see Chapter 8), some of which have been used in the past for production of ^{239}Pu for military purposes.

Other fuel elements are used in power reactors operated by the civilian power industry and in naval vessels. In addition, some of the fuel will be used by research reactors located in laboratories and industrial plants. The products of these reactors may be heat, plutonium, tritium, other radioisotopes, or radiations for research or industrial purposes. The spent fuel can be either treated as a solid waste and placed in a radioactive waste repository or transported to reprocessing plants in which the fuel is dissolved and the unused uranium and possibly plutonium are recovered. The fission products can be processed into a form convenient for waste storage and disposal (see Chapter 11), but in some cases radioactive isotopes may be separated from these fission products for research, medical, or industrial applications. United States policy has not encouraged the reprocessing of civilian nuclear fuels since the late 1970s, but reprocessing is practiced in other countries.

In addition to the radioactive by-products from the spent-fuel-reprocessing plants, certain radionuclides are produced in research reactors by neutron irradiation. For example, the naturally occurring stable isotope ^{59}Co may be placed in a reactor to produce ^{60}Co, which is an important source of gamma radiation for medical and industrial purposes. Many of the more common radioactive isotopes, such as ^{14}C, ^{131}I, and ^{32}P, are produced by neutron bombardment of the appropriate parent nuclide.

EARLY STUDIES OF RADIOACTIVE CONTAMINATION OF THE ENVIRONMENT

A limited amount of information about environmental radioactivity was already available prior to World War II, but there was little diffusion of this knowledge beyond the relatively few highly specialized laboratories that were then equipped to make measurements of radioactivity. The world inventory of radioactive materials was confined to those separated from natural sources, with the insignificant exception of a relatively few millicuries of artificial radioactivity that were produced in cyclotrons during the late 1930s.

During World War II, the construction of large water-cooled plutonium-producing reactors at the Hanford Reservation, near Richland, Washington, and the associated operations for extracting the plutonium from the irradiated uranium resulted in the first possibilities of major contamination of the environment by radioactivity. This was also true to a lesser extent at two other major nuclear research and production centers at Oak Ridge, Tennessee, and Los Alamos, New Mexico. The early Hanford studies on the behavior of various radionuclides in the environment are classics in the field and have served to demonstrate, on the one hand, the caution one must adopt in discharging radioactive substances to the environment and, on the other hand, the fact that substantial quantities of radioactivity can be discharged safely if the properties of the individual radionuclides and their behavior in the environment are well understood. With cessation of nuclear weapons production in the United States, a great effort is being made to decontaminate these and other facilities (NAS/NRC, 1989).

The policies laid down by the Manhattan Engineering District (MED) of the Corps of Engineers (Groves, 1962), which was the World War II military organization responsible for the atomic energy program, placed a high priority on the importance of operating in such a way as to keep environmental contamination at a minimum. When the Atomic Energy Commission (AEC), a civilian organization, succeeded the MED in 1946 (Hewlett and Anderson, 1962), the cautious policies toward release of radioactive materials to the environment from industrial and research activities continued, but by the mid-1970s these policies were overtaken by new policies that applied to toxic chemicals with which the Manhattan District and AEC had paid less attention. The AEC at first did not believe it was subject to the rules and regulations of the U.S. Environmental Protection Agency (EPA), but this was reversed by the courts in 1984 in a decision that would saddle the Department of Energy (DOE), AEC's successor, with hundreds of billions of dollars in cleanup costs.

Starting in the late 1940s and continuing at an accelerated rate until 1962, there began a series of nuclear weapons tests in the atmosphere, first by the United States and then by the Soviet Union, the United Kingdom, France, India, and China, that discharged into the environment amounts of radioactivity that were enormous in relation to the prohibitions self-imposed by the AEC in the operation of its research and industrial facilities. The radionuclides produced in the nuclear explosions in various parts of the world soon permeated the atmosphere, the soils, and the food chains to such an extent that widespread apprehension began to develop, first in certain scientific circles and later among the general public throughout the world. Responding to this concern, the Congress of the United States held a number of hearings on the subject of fallout from weapons testing and also on radioactive waste-disposal practices. At about the same time, the National Academy of Sciences in the United States and the Medical Research Council in Great Britain undertook to assess the state of knowledge on the effects of small doses of radioactivity (NAS/NRC, 1956; Medical Research Council, 1956).

In 1955, the United Nations appointed a committee, consisting originally of scientific representatives of 15 nations (but later expanded), to investigate the effects of radiation on humans. The United Nations Scientific Committee on the Effects of Atomic Radiation has published a series of authoritative reports that are classics in international scientific collaboration (UNSCEAR, 1958–1994).

The widespread interest in the subject of environmental radioactivity resulted in acceleration of research into the behavior of trace substances in the environment. Many branches of the biological and physical sciences, including genetics, inorganic chemistry, trace-element metabolism, micrometeorology, upper-atmosphere meteorology, and oceanography, have made great progress because of the need for a better understanding of environmental radioactivity. Large, well-equipped ecological laboratories were established at the major atomic energy production and research centers in the United States and other countries, and funds and equipment for ecological research were supplied to individual investigators at many universities. The concern that began to pervade the scientific community about contamination by toxic chemicals in the mid-1960s was to some extent the result of ecological knowledge that was obtained from studies of fallout from nuclear weapons tests.

Those studies raised many difficult questions that at first seemed unique to the subject of environmental radioactivity. What are the ecological pathways by which these substances reach humans? Do they accumulate in such a way that they can result in unforeseen ecological injury? Are there synergistic effects with other environmental pollutants? By the late 1960s, it was apparent that the same questions could be asked about insecticides, food additives, fossil fuel combustion products, trace metals, and other nonradioactive pollutants of the environment. In many respects, the pioneering studies of the environmental effects of radioactivity provided the tools by which more general problems of environmental pollution could be understood.

The plants and laboratories of the weapons production complex were built half a century ago, and increasing conservatism has developed concerning exposure of the public to radioactive and chemical pollution. Waste disposal and operating practices that were acceptable during the period from 1945 until about 1960 are no longer tolerated by the public or the regulatory agencies. Public intolerance toward risks from nuclear facilities has developed despite the fact that, with few exceptions, exposures have been maintained well below the standards in force today. Similar objections have developed in the manufacture and use of many chemical products. Widespread apprehensions have grown despite the fact that health effects can rarely be identified even with the best of clinical or epidemiological methods. This public concern is a great impediment to the development of new technology. Difficulties have been encountered not only with environmental radioactivity, but with electronic products, agricultural chemicals, and pharmaceutical products. This will be further discussed in Chapter 16.

CHAPTER 2

The Biological Basis of Radiation Protection

This chapter is intended to permit the general reader to appreciate the relative scale of risks associated with various levels of exposure and to understand the biological basis on which levels of maximum permissible exposure are established. The subject is treated elsewhere in a number of more comprehensive reviews (UNSCEAR, 1993, 1994; NAS/NRC, 1990). Although much remains to be learned, more is known about the effects of ionizing radiation exposure than about the effects of any other of the many noxious agents that have been introduced artificially into the environment. This is because of the large amount of research that has been performed worldwide since 1942, when the first nuclear reactor went into operation. In the United States, government funds for radiation research have been far greater than the expenditures for studies of the effects of the many human-produced chemical pollutants of air, water, and food.

EARLY KNOWLEDGE OF RADIATION EFFECTS

Reports of radiation injury began to appear in the literature with astonishing rapidity after the announcement on November 8, 1895, of Wilhelm Roentgen's discovery of X rays. The first volume of the *American X-ray Journal,* published in 1897, included a compilation (Scott, 1897) of 69 casese of X-ray injuries reported from laboratories and clinics in many countries. The reason why so many injuries were reported so quickly is related to the way X rays were discovered and the kind of research that had been under way for many years even before the existence of X rays was known.

X-ray effects seem to have been first observed in 1859, nearly 36 years before Roentgen's announcement, when Julius Plücker recorded the fact that an apple green fluorescence was seen on the inner wall of a vacuum tube within which a current was flowing under high voltage (Grübbe, 1933). Plücker's work was followed by a number of experiments in other laboratories, and in 1875 Sir William Crookes developed the first high-vacuum tube (which thereafter bore his name) and discovered that the apple green fluorescence reported by Plücker originated from a discharge at the negative electrode of the tube. Thus was inaugurated the term "cathode rays." Roentgen's contribution was the startling observation that he could see the shadow created by the bones of his hand when it was placed between a Crookes tube and a screen covered with fluorescent chemical. To this penetrating radiation he gave the name X rays.

The relevant point is that research with Crookes tubes was occurring in many parts of the world prior to Roentgen's discovery and that, unknown to those investigators, X rays were being generated by many of the tubes they were using. Emil Grübbe, who manufactured Crookes tubes in Chicago and was studying the fluorescence of chemicals at the time of Roentgen's announcement, began immediately to experiment with the newly named X rays and promptly developed an acute dermatitis that was followed by skin desquamation and eventually cancer. An early description of his injury was presented at a clinical conference at the Hahnemann Medical College in Chicago on January 27, 1896, at which the interesting observation was made that "any physical agent capable of doing so much damage to normal cells and tissues might offer possibilities, if used as a therapeutic agent, in the treatment of a pathological condition in which pronounced irritative blistering or even destructive effects might be desirable." Two days later, less than three months after publication of Roentgen's discovery, a patient was referred to Grübbe for treatment of a breast carcinoma. In his interesting recital of the facts, Grübbe (1933) claimed that he was the first person to be injured by X rays, that he was the first person to apply X rays to pathological lesions for therapeutic purposes, and, incidentally, that he was the first to use sheet lead for protection against X-ray effects. More recently, Hodges (1964) and Brecker and Brecker (1969) have cast doubt on the authenticity of the Grübbe claims because of inconsistencies in the chronology. However, Hodges does conclude that Grübbe was the first to treat cancer with X rays. Although the question of priorities of radiological discoveries may never be fully satisfied in the historical sense, the 1933 Grübbe paper and the subsequent work by others do provide insight into the early history of radiation injury.

An indirect result of Roentgen's announcement was Antoine Henri Becquerel's accidental discovery of radioactivity in the same year as the discovery of X rays. By 1900 the radioactive constituents of ore containing radium had been sufficiently concentrated so that burns were produced on the skins of the pioneers in radioactivity research (Becquerel and Curie, 1901). The 1897 report by N. S. Scott of injuries produced by X rays was followed by that of E. A. Codman (1902) in which he reported that the number of cases of radiation injury in the literature had increased to at least 147. Interestingly, Codman noted that the number of cases had begun to diminish, which he attributed to "the bitter teaching of experience and the fact that the introduction of better apparatus has done away with long exposures and the close approximation of the tube." Codman's observation, though justified at the time, has regrettably not been supported by history. The total number of people who have been injured and killed by the use and misuse of X ray and radium prior to the development of proper standards of radiation hygiene will probably never be known, but certainly numbers in the many hundreds.

The early interest of the experimental biologists in the effects of ionizing radiation is illustrated by the discovery in 1897 of the ability of ionizing radiation to cause lenticular cataracts in exposed animals (Clapp, 1934). The production of genetic mutations by exposure to X rays was first reported by H. J. Muller (1927) and opened an era of research on the hereditary effects of radiation.

During and immediately following World War I, the use of radium in luminous paints was attended by hazards arising out of ignorance of the effects of this radioactive element when inhaled or ingested (Martland, 1925, 1951; Evans *et al.*, 1969). Radium-226 and radium-228 were also used in the practice of medicine. Among a total population of about 6600 exposed to either radium isotope in the United States, a total of 85 cases of bone cancer are known to have occurred (Rowland, 1994). These luminous dial painters, mostly women, ingested radium because of the practice of using the lips to point the paint brushes (NAS/NRC, 1990). The dial-painting cases have been studied thoroughly by a number of investigators, but the main credit belongs to Evans for having worked out the basic biophysical principles of radium injury in sufficient detail so that safe practices could be adopted. These practices proved effective for protection not only against radium but also against many of the hazards that developed later in the atomic energy industry, where the information gained with radium was utilized to great advantage. In the first 40 years of this century, only about 1 kg of radium was extracted from the earth's crust and at least 100 people are known to have died from various misuses of this material. In contrast, since 1942 the atomic energy programs have produced the radioactive equivalent of many tons of radium and have done so with an excellent record of safety, except in the mining of uranium. The safety record will be discussed further in Chapter 16.

The uranium mining experience in the United States, in which several hundred workers died of lung cancer caused by their exposure to radon, has been a tragedy that should have been avoided. It had been known for centuries that men who worked in the eastern European metal mines of Schneeberg and Joachimsthal (Hartung and Hesse, 1879) were prone to a quickly fatal lung disease, but only toward the latter part of the nineteenth century was it learned that the disease was bronchiogenic carcinoma. In the twentieth century, these mines became a source of pitchblende. When it was realized that cancer could result from internal irradiation, it was suggested that the high incidence of lung cancer among the miners might be explained by the presence of radioactive substances in the mine atmospheres. Studies of the mine air revealed the presence of high concentrations of radon, and this radioactive gas came to be regarded as the likely etiologic agent in the high incidence of lung cancer (Hueper, 1942; Lorenz, 1944). A standard for protection against the effects of radon inhalation already existed in the United States by World War II, having been published by the U.S. Committee on the Safe Handling of X Rays and Radium (NCRP, 1941), which was then housed in the Bureau of Standards. Had that standard been enforced in the uranium mining industry, hundreds of lives would have been saved.

Experience early in this century also demonstrated that ionizing radiations in a sufficient dose could produce sterility, damage to blood-forming tissues, and, in the case of high levels of exposure, a complex of symptoms that came to be known as the acute radiation syndrome. However, not all of this information was obtained from the misuse of ionizing radiations. The use of X rays and radium for treatment of cancer created the need to understand the effects of large doses on healthy as well as cancerous tissue to permit the radiologist to limit side effects. Since the radiotherapists pushed the doses as high as possible in treating the life-threatening cancers, they needed to understand the reactions of the noncancerous tissues to the high doses received.

SUMMARY OF PRESENT KNOWLEDGE OF RADIATION EFFECTS ON HUMANS

A great amount of research conducted since 1942 has been directed toward understanding both the mechanisms of radiation injury and the ecological relationships that exist in an environment contaminated with radioactive material. Knowledge about the effects of radiation exposure comes from many sources. The greatest volume of literature comes from work with experimental animals. These studies have taught us much about the mechanisms of radiation injury and the relationships between dose and response in various species. However, the dose–response relationships developed in experimental animals are not necessarily transferable to humans. Although the basic mechanisms of radiation injury may be the same in all species, there are major differences among species in their susceptibility to radiation injury. Thus, information about the size of the dose required to produce a given effect in a species of laboratory animal cannot be applied directly to humans.

Much of what we have learned has come from human experience, for example, from the early misuses of radium and X rays, as discussed earlier. The most useful information has come from studies of the delayed effects of radiation among the survivors of the bombings of Hiroshima and Nagasaki. Studies of radon exposure of underground miners and exposures from accidents have also been important sources of information.

Any general discussion of radiation effects should begin with certain dichotomies:

1. Whether the source of radiation is external to the body (as in the case of exposure to medical X rays) or is an internally deposited radionuclide (as in the case of radioiodine in the thyroid).
2. Whether the dose was from a relatively massive exposure delivered in a short period of time (less than a few days) or was delivered in small bits over longer periods of time, which may extend to many years.
3. Whether the effects appear soon after exposure ("acute effects") or are delayed for months or years ("delayed effects").

THE PROMPT EFFECTS OF EXPOSURE

When a massive dose of whole-body radiation is received instantaneously or when the exposure is received predominantly in the first few days, as in the case of the external radiation from fresh fission products, symptoms of acute radiation injury may be seen as early as a few hours after exposure and will follow a course dependent on the size of the dose received (Glasstone and Dolan, 1977). Table 2-1, in which the expected effects of massive exposure to external radiation are summarized, shows that relatively minor effects would occur at doses less than 1 Gy (gray, or 100 rad), but that about 50% fatalities would be expected in the range of 4 to 5 Gy (400 to 500 rad). As the whole-body dose approaches 10 Gy (1000 rad), the fatalities would reach 100%.

Nausea occurs in increasing frequency above 1 Gy (100 rad) and will be seen in almost all exposures above about 3 Gy (300 rad). The nausea may occur within hours after exposure and be followed by an asymptotic period of as much as 2 weeks after a dose of 1 to 2.5 Gy (100 to 250 rad), but less than 1 day when the dose is greater than 7 Gy (700 rad). The signs and symptoms that then develop may include epilation, sore throat, hemorrhage, and diarrhea.

Among the most striking changes due to doses above 1 Gy (100 rad) are those from injury to the blood-forming organs, which reduce the rate at which the component elements of blood are produced and have a dramatic effect on the composition of circulating blood (Wald *et al.*, 1962). At whole-body doses in excess of 10 Gy (1000 rad), the predominant symptoms may be due to injury to the gastrointestinal tract and central nervous system. Disorientation within a matter of minutes owing to central nervous system injury was a conspicuous feature of at least one case (Shipman, 1961) after accidental exposure to more than 10 Gy (1000 rad).

The radioiodines may be the only internal emitters capable of producing prompt effects. These nuclides are short-lived, are readily absorbed into the body, and concentrate in the thyroid gland, which weighs only a few grams. For these reasons, among the internal emitters, the radioiodines are uniquely capable of delivering a high dose over a short period of time. Exposure to heavy doses of radioactive iodine can reduce thyroid function, and doses of several thousand rad (several tens of Gy), such as are used in treatment of thyroid cancer, can result in destruction of the gland (NCRP, 1985a). It is likely that the external radiation dose from generalized fission-product contamination of the environment would be so high as to be over-whelming relative to the internal dose from radionuclides other than the radioiodines. It will be seen in Chapter 12 that doses to the thyroid of up to 1800 rad (18 Gy), superimposed on whole-body doses of about 175 rad (1.75 Gy) following exposure to heavy fallout from weapons testing, resulted in impaired thyroid function (a relatively prompt thyroid effect) in addition to causing thyroid cancer after many years.

DELAYED EFFECTS

The delayed effects of radiation may not appear for several decades after exposure, and can result either from massive doses that have caused prompt effects or from relatively small exposures repeated over an extended period of time. The effects that develop in the exposued individual are referred to as somatic effects to differentiate them from genetic effects, which occur in the progeny of the exposed person and are the result of changes transmitted by hereditary mechanisms. Radiation injury can also occur in the devel-

TABLE 2-1

Summary of Clinical Effects of Acute Ionizing Radiation Doses[a]

	0 to 100 rads (0–1 Gy): Subclinical range	100 to 1000 rads (1–10 Gy): Therapeutic range			Over 1000 rads (10 Gy): Lethal range	
		100 to 200 rads (1–2 Gy)	200 to 600 rads (2–6 Gy)	600 to 1000 rads (6–10 Gy)	1000 to 5000 rads (10–50 Gy)	Over 5000 rads (50 Gy)
		Clinical surveillance	Therapy effective	Therapy promising	Therapy palliative	
Incidence of vomiting	None	100 rads: infrequent 200 rads: common	300 rads: 100%	100%	100%	
Initial phase						
Onset	—	3 to 6 hours	½ to 6 hours	¼ to ½ hour	5 to 30 minutes	Almost immediately[c]
Duration	—	≤1 day	1 to 2 days	≤2 days	≤1 day	
Latent phase						
Onset	—	≤1 day	1 to 2 days	≤2 days	≤1 day[b]	Almost immediately[c]
Duration	—	≤2 weeks	1 to 4 weeks	5 to 10 days	0 to 7 days[b]	
Final phase						
Onset	—	10 to 14 days	1 to 4 weeks	5 to 10 days	0 to 10 days	Almost immediately[c]
Duration	—	4 weeks	1 to 8 weeks	1 to 4 weeks	2 to 10 days	

16

Leading organ	Hematopoietic tissue			Gastrointestinal tract	Central nervous system
Characteristic signs	None below 50 rads	Moderate leukopenia	Severe leukopenia; purpura; hemorrhage; infection; epilation above 300 rads	Diarrhea; fever; disturbance of electrolyte balance	Convulsions; tremor; ataxia; lethargy
Critical period postexposure	—	—	1 to 6 weeks	2 to 14 days	1 to 48 hours
Therapy	Reassurance	Reassurance; hematologic surveillance	Blood transfusion; antibiotics	Maintenance of electrolyte balance	Sedatives
Prognosis	Excellent	Excellent	Guarded	Guarded	Hopeless
Convalescent period	None	Several weeks	1 to 12 months	Long	—
Incidence of death	None	None	0 to 90%	90 to 100%	100%
Death occurs within	—	—	2 to 12 weeks	1 to 6 weeks	<1 day to 2 days
Cause of death	—	—	Hemorrhage; infection	Circulatory collapse	Respiratory failure; brain edema

[a] Adapted from Glasstone and Dolan (1977).

[b] At the higher doses within this range there may be no latent phase.

[c] Initial phase merges into final phase, death usually occurring from a few hours to about 2 days; this chronology is possibly interrupted by a very short latent phase.

oping fetus, but this is a teratogenic effect that is considered somatic rather than genetic, since it does not involve the germ plasm.

Until the early 1960s, the genetic consequences were thought to be the most important delayed effect of radiation exposure. A major publication prepared for the AEC in 1958 on the biological effects of radiation placed great emphasis on the genetic effects and included hardly any information on cancer (Claus, 1958). During the 1950s, there was already evidence that the dose–response relationship for induction of genetic effects was linear, and that it was probably without a threshold. This was in contrast to cancer induction, for which a threshold was then believed to exist and for which linearity of response was not generally accepted. It was believed that the dose–response relationship for cancer was sigmoidal in form, and that a threshold existed below which cancer was not produced. This is the classical dose–response relationship that is widely applicable in toxicology. The sigmoid response is shown in Fig. 2-1a, in which the effects per unit of dose gradually increase until a dose level is reached at which the rate of increase begins to diminish and eventually reaches a plateau when all susceptible

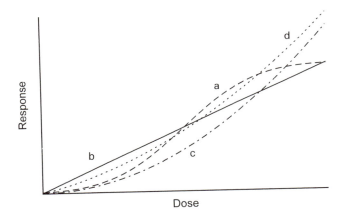

FIGURE 2-1   Types of dose–response curves. (a) The classic sigmoid curve, no longer considered applicable for stochastic effects, such as cancer, in radiobiology. At the low end, the response may be asymptotic with dose, or a true threshold may exist but will be obscured by statistical uncertainties inherent in the data. (b) The linear dose–response relationship. Note that (a) and (b) are close together over such a considerable range that it may not be possible to distinguish them if the experimental or epidemiological data fall between or near the two curves. (c) The quadratic dose–response curve, in which the response takes the form of KD^2 at higher doses. (d) The so-called linear–quadratic curve, which takes the form of $aD + bD^2 + c$. The slope of the linear portion may vary, depending on dose rate. Most of the levels to which the public is exposed are assumed to lie within the linear portion of the linear–quadratic curve.

members of the population have been affected. As noted in Fig. 2-1a, there may be a threshold below which no effects occur, or the question as to whether a threshold exists may be indeterminate because of statistical limitations in the data. Figure 2-1b illustrates the linear–no threshold response, which, because of statistical limitations, cannot always be distinguished from other dose–response relationships.

However, research with animals and improved understanding of the underlying mechanisms involved in cancer induction raised doubts about the existence of a threshold and suggested that the carcinogenic response might also be linear, as in the case of genetic effects. Moreover, studies of the Hiroshima and Nagasaki survivors demonstrated during the 1960s that the cancers produced by their exposure to radiation were occurring with greater frequency than had been expected, and that the kinds of cancers were more varied. Until then, leukemia had been the only cancer seen in excess among the survivors, but other cancers began to appear, and it was soon concluded that there would be at least five "solid" tumors for each case of leukemia (NAS/NRC, 1990). At the same time, experiments with rats indicated that the genetic effects of radiation were less than had been believed (Russell, 1968), and the studies in Japan could find no genetic effects among the offspring of survivors of the atomic bombings. For all of these reasons, a shift in thinking took place in the early 1960s and the risk of cancer induction became the main concern on the part of those involved with the health effects of ionizing radiation.

Further laboratory experimentation began to support the hypothesis that both the genetic and carcinogenic effects of ionizing radiation were inherently stochastic, which implies that no threshold exists, that the frequency with which effects are seen depends on probabilistic mechanisms, and that the probability of an effect is a continuous function of dose. For stochastic effects the severity of the effect is independent of dose, but the probability that an effect will occur is proportional to the dose received.

A hypothesis that competes with the linear dose–response relationship is the quadratic form (Fig. 2-1c), in which the number of effects produced per unit dose increases progressively with increasing dose. In recent years, the so-called linear–quadratic relationship (Fig. 2-1d) has been favored by both radiobiologists and epidemiologists. In this relationship the effects are linear at low doses but become curvilinear at higher doses according to the relationship

$$I = aD + bD^2 + c$$

in which I is incidence at dose D, c is the incidence at zero dose, and a and b are constants determined empirically (Upton, 1977; NAS/NRC, 1990).

The most important source of information on which the standards for protection against the delayed effects of radiation are based are the data gathered during the past five decades in Hiroshima and Nagasaki. The risk coefficients used to predict the number of cancers that can be expected in a population exposed to a given dose come largely from those data. Although the medical history of the surviving population is well known, there are uncertainties in the doses received. This is particularly true of the relative contribution of neutrons to the total dose received by the survivors of the Hiroshima bombing. This question will continue to cast doubt on the risk coefficients used by UNSCEAR, the International Commission on Radiological Protection (ICRP), and the National Council on Radiation Protection and Measurements (NCRP). The question was believed to have been resolved in a thorough reassessment by a joint Japanese–United States team in 1987 (Roesch, 1987), but new questions continue to arise. We may expect more precision in the dose estimates than is possible under the circumstances. This should be taken into consideration concerning the risk coefficients to be discussed later in this book. The uncertainty in the gamma radiation dose estimates is probably valid within a factor of two, which, all things considered, is quite remarkable.

Excess cancers are clearly demonstrated at the high doses associated with the survivors of the atomic bombs in Japan and in a number of other high-dose studies discussed later in this chapter. However, epidemiological studies have not shown statistically significant effects in studies of populations exposed at elevated levels of natural radiation (NAS/NRC, 1990) or in the recent, large international study of nuclear industry workers (Cardis *et al.*, 1995). An inherent difficulty with studies at the relatively low doses associated with environmental and well-controlled occupational exposures is that biological variability and confounding factors can mask effects, even if they are present. For this reason, a common conclusion drawn from these studies is that they provide assurance that the risk rates developed at higher doses do not underestimate risk at lower doses, but that they do not contradict the risk rates developed at higher doses. Alternatively it may be argued from the lack of demonstrable effects in large studies that low-dose effects, if they exist at all, are not worthy of concern.

Except under highly unusual conditions, such as the accidents described in Chapter 12, environmental radioactivity results in doses on the order of natural background (about 300 mrem y^{-1} or 3 mSv y^{-1}) or less, where the effects are assumed, for radiation protection purposes, to be governed by a linear relationship.

There is evidence from experiments with laboratory animals that the effects of a given dose are reduced when it is received gradually over an extended period of time. Protracting the period in which the dose is received has the effect of reducing the slope of the linear portion of the linear–quadratic response shown in Fig. 2-1d. The risk coefficients derived by extrapolation from high dose and high dose rate exposure to low doses delivered at low dose rates are thus too high for the more usual case in which the dose delivery is protracted in time. Because of the paucity of data, there is no agreement on the magnitude of the dose rate reduction factor, beyond the conclusion that it lies between 2 and 10 for X-ray and gamma radiation. A dose rate reduction factor of two to three seems to be a reasonable assumption for X-ray and gamma radiation (UNSCEAR, 1993). However, contrary evidence also exists. The effect per unit dose of alpha radiation on the incidence of lung cancer has been observed to be independent of dose and dose rate in both experimental animals and epidemiological studies (Cuddihy, 1982; Schlenker, 1982, NAS/NRC, 1988). The frequency of breast cancer among women irradiated with X rays has also been shown to be linear with dose and independent of the manner in which the dose was fractionated in delivery (NAS/NRC, 1990).

Another factor that may affect the dose–response relationship is the effect of progressively smaller doses on the time it takes for the cancer to develop. Radiation-induced cancers (other than leukemia) rarely appear sooner than 10 years postexposure and can occur longer than 30 years after exposure. It has been shown by Evans (1967) that the latency period for the development of bone cancer in luminous-dial workers is inversely correlated with the radium body burden. This caused Evans to postulate that the question of whether a threshold exists may be moot because there is a finite bone dose below which the time to tumor appearance is so long as to exceed the life span of the individual. Evans coined the term "practical threshold" to explain this feature of dose–response relationships. The concept of a practical threshold, at least for certain radiation-induced cancers, has been greatly strengthened by the thorough analysis of Raabe (1994).

An inverse relationship between dose and leukemia latency has been reported both in experimental animals (Upton, 1977) and among the Japanese survivors of the World War II atomic bombings (Ishimaru *et al.*, 1982). Albert and Altshuler (1973) examined this question both theoretically and epidemiologically and concluded that there may be a basis for assuming that there is an inverse relationship between dose to chemical carcinogens and tumor latency.

The assumptions of a linear dose–response relationship and absence of a threshold have important implications for risk assessment and formulation of public policy. The absence of a threshold implies that there is no such thing as an absolutely safe level of exposure. Every increment of dose above zero, however small, is calculated to result in an increment of risk as well, but the risk becomes smaller as the dose diminishes. It will be seen in Chapter 3 that this complicates the problem of setting standards of permissible exposure.

It has been proposed that the question of whether a specific cancer is causally related to a given dose of radiation should be answered in probabilistic terms. For this purpose, tables of probability of causation have been prepared that give estimates of the likelihood that a particular cancer might have resulted from a given dose. These tables were prepared in the hope that legal decisions regarding liability for cancer induction might be based on the probability that it was caused by radiation, taking into consideration the magnitude of dose, age, and other relevant factors (NAS/NRC, 1984; Department of Health and Human Services, 1985).

A dilemma that also arises from the assumptions of linearity and the absence of a threshold is that the risk to individuals can be very small, but a finite number of cancers can result if a sufficiently large population is exposed. A lifetime risk of 10^{-6} is negligible to an individual, but in the world population of 5×10^9 people, that level of risk would result in 5000 radiation-induced cancers. Thus, for purposes of risk assessment, the use of "population" or "collective" dose is useful (Lindell, 1985). The number of effects produced depends on the mean dose multiplied by the number of persons exposed. The effect on the population, expressed as the number of cancers produced, will be the same whether N persons receive a dose D, or $0.1N$ receive a dose of $10D$.

Somatic Effects

With the foregoing as background, we can now proceed to a brief discussion of each of the somatic health effects seen in irradiated human populations (NCRP, 1993c). The main focus of this discussion will be cancer from low doses of X or gamma radiation. For purposes of this discussion, low doses will be defined as those usually less than the exposure to the typical natural background of 300 mrem (3 mGy) per year. In round numbers, the no-threshold assumption leads to a lifetime risk to the general population of developing fatal cancer from low-level radiation exposure at low dose rates of about 5×10^{-2} Sv^{-1} (5×10^{-4} rem^{-1}) (ICRP, 1991; NCRP, 1993c). This estimate includes a dose rate effectiveness factor of two. The cancers will include at least 5 solid tumors for every case of leukemia, but leukemia is of special interest in epidemiological studies because the cases occur well in advance of the solid tumors, and excess cases of leukemia can be more readily detected because of the low normal incidence of the disease compared to cancers of most other radiosensitive organs. The latter include breast, lung, stomach, thyroid, and digestive tract. Nonfatal cancers are estimated to occur at a rate of about 20% of the fatal cancers. Since the information collected in Japan provides the best basis for this risk coefficient, and about half of the irradiated population is still alive, the coefficient may be increased by additional excess cancers that are likely to develop in the future. By 1987 a total of about 510 excess cancers were reported, and the number was slowly increasing (UNSCEAR, 1994).

It has frequently been suggested that the risks of low-level radiation might be determined by studies of people who live in areas of the world in which the radioactive background is elevated above the normal range of values. Such places exist in Brazil, India, and China (see Chapter 6). The most meaningful study thus far completed has been in China, where no effects were found in a study of about 80,000 people who have lived for generations in an area in which the soil is enriched in thorium, and the average gamma radiation dose is about 330 mrem y^{-1} (3.3 mSv y^{-1}). In fact, cancer mortality adjusted for age was found to be lower among residents of the high-background area than in the control population (Wei *et al.*, 1990; NAS/NRC, 1990). No differences were found in the prevalences of genetic disease or congenital defects.

Leukemia

Leukemia is a relatively rare disease that is known to have occurred in increased frequency among Japanese survivors at Hiroshima and Nagasaki (NAS/NRC, 1990), among children irradiated in infancy for thymic enlargement (Shore *et al.*, 1986; NCRP, 1985a), among patients irradiated for ankylosing spondylitis (Smith and Doll, 1982), among physicians exposed in the practice of radiology (Cronkite, 1961), and possibly among children who were irradiated *in utero* in the course of pelvic examination during pregnancy (Kneale and Stewart, 1976; Totter and McPherson, 1981; Monson and MacMahon, 1984). There is also evidence of a slight increase in the prevalence of leukemia among residents of Utah who were exposed as children to fallout from the open-air testing of nuclear weapons at the test site in Nevada (Stevens *et al.*, 1990). Of these groups, only the radiologists can be described as having been exposed to repeated small doses, in contrast to the other groups, which were subjected to either single or slightly fractionated exposures. The chronic lymphocytic form of leukemia, a disease seen in older persons, has not been associated with ionizing radiation exposure.

Among the Japanese atom bomb survivors who were exposed in 1945, the incidence of leukemia reached a peak in the early 1950s and returned to near-normal by about 1970, although a slight excess in the incidence of leukemia has persisted as late as 1987 (UNSCEAR, 1994). By 1987 a total of 231 deaths from leukemia were reported among a study sample of about 93,000 survivors who comprise the "life-span study." Of these cases, 156 would have been expected in an unexposed population, leaving an excess of 75 due to radiation induction (UNSCEAR, 1994). A more recent update places the number of excess leukemia deaths at 87 (Pierce *et al.*, 1996). The mean dose received by the survivors has been estimated to be 0.25 Sv (25 rem).

Several studies of the incidence of leukemia in children who were irradiated *in utero* show conflicting findings (NAS/NRC, 1990). MacMahon (1962) and Stewart and Kneale (1970) reported an apparent increase in the incidence of leukemia and other neoplasms in the first 10 years of life in children who were irradiated *in utero* in the course of maternal pelvic X-ray examination. This finding has not always been supported by other studies. Jablon and Kato (1970) studied the 25-year cancer experience of 1292 children who were *in utero* during the bombings of Hiroshima and Nagasaki, and who received very much higher doses than the children whose mothers were X-rayed during pregnancy. Although some of the Japanese children were estimated to have received as much as 250 rem (2.5 Sv) prenatally, no increase was at first found in the incidence of leukemia or other cancers during the first 25 years of life. This finding, among others, resulted in many years of controversy over the question of the sensitivity of the fetus. However, on the basis of more recent data from the Japanese studies, there was agreement by 1990 that an excess of leukemias, and possibly other cancers, is a consequence of prenatal irradiation (NAS/NRC, 1990).

There have been occasional reports of leukemia clusters near nuclear installations as well as other localities. The most persistent reports have been from England and Wales, with the major facility at Sellafield attracting the most attention. However, a review of the results of epidemiological and laboratory research failed to support the hypothesis that such an association exists (Little *et al.*, 1995).

From all the epidemiological evidence, it can be concluded that for the purposes of estimating the upper limit of risk to an exposed population, the risk coefficient for development of leukemia can be taken as about 10^{-2} Gy^{-1} (10^{-4} rem^{-1}) (NCRP, 1993c).

Solid Cancers

Bone Cancer   When radium or other radioelements that are chemically similar to calcium are ingested, they deposit in bone from which they are removed very slowly. As noted earlier, bone cancer developed among individuals who were exposed to radium both in the process of luminous dial manufacturing and as patients who were prescribed radium medicinally during the early years of this century.

Bone cancer among radium dial painters was first observed and called "radium jaw" in 1924 by Theodore Blum, a New York dentist. The cases originated from a luminous dial plant in northern New Jersey, and by the late 1920s it was already understood that the cases of bone cancer, mainly among young women, resulted from ingestion of radium as a consequence of lip-pointing brushes used to paint the numerals. The cases attracted the attention of a forensic pathologist, Harrison Martland, and a toxicologist, A. O. Getler, who undertook to study the first several cases (Martland, 1951).

Valuable epidemiological information has been obtained from these cases and also from studies of persons to whom radium was administered either intravenously or orally for medical reasons. Many thousands of patients were treated clinically in this way. One clinic administered more than 14,000 intravenous injections of radium, usually in doses of about 10 μCi [microcuries, or 370 kBq (kilobecquerels)]. In addition, a number of popular medicinal waters that contained radium were marketed in the early 1920s. One of the most popular of these nostrums was Radithor, a mixture of 1 μCi (37 kBq) each of ^{226}Ra and ^{228}Ra (Evans, 1966).

About 85 cases of cancer due to occupational ^{226}Ra and ^{228}Ra exposure have been reported (Rowland, 1994). One-fourth of these were carcinomas of the nasal sinus and mastoid air spaces, which are believed to have been caused by the accumulation of radon and its decay products within these cavities.

On the basis of his epidemiological studies prior to about 1942, Evans (1943) proposed that 0.1 μCi (3.7 kBq) of ^{226}Ra be adopted as the maximum permissible body burden. All the known cases of injury had occurred in individuals whose body burden was greater than 0.5 μCi (19 kBq) at the time of observation, which was usually many years after cessation of exposure. Evans later estimated that the maximum permissible body burden has a safety factor of about 15 (Evans, 1967; Evans *et al.*, 1969).

A much larger number of individuals have shown less severe bone pathology than cancer, ranging from small areas of osteoporosis to extensive necrosis. Advanced osteoporitic or necrotic changes were frequently associated with spontaneous fractures.

Bone cancers have also occurred among patients treated by administrations of ^{224}Ra for tuberculosis and ankylosing spondylitis. This short-lived ($T_{1/2} = 3.6$ days) isotope was used in this way in Europe for more than 40 years, and studies have been undertaken in Germany and elsewhere of the effects of these practices on about 3800 patients (Mays, 1978; Health Physics, 1983; Wick and Gossner, 1983; Mays *et al.*, 1978).

The ability of the various bone-seeking radioactive elements to produce osteogenic sarcoma varies because of differences in the kinds of radiations emitted and also because the different elements deposit in different parts of the bone structure. Raabe (1984) concludes, on the basis of laboratory studies with beagles, that the relative biological effectiveness compared to [226]Ra is 3 for [228]Ra, 9 for [239]Pu, 6.4 for [241]Am, 10.7 for [228]Th, and 15.5 for [238]Pu.

The question of whether a threshold exists for production of bone cancer from internal emitters is of considerable importance in relation to standards of permissible public exposure, because traces of bone-seeking nuclides such as ^{90}Sr and [239]Pu are present in the general environment from weapons testing and can be measured in the skeletons of the general population. Although the dose to individuals is very small, the number of persons exposed is very large, since everyone in the world is exposed to some extent. If the dose–response relationship is linear, a finite number of bone cancers may result, even when the dose to the individual is exceedingly small. As noted earlier, Evans (1967) found that the latency period for the development of bone cancer in the population of dial painters was inversely related to the [226]Ra body burden. However, this was not shown to be true in the [224]Ra series (NAS/NRC, 1988; Rowland, 1994).

Lung Cancer   The respiratory tract (Fig. 2-2) is designed to transport air deep into the lung to the alveoli, where oxygen and carbon dioxide exchange takes place between the blood and air. Most particles of dust contained in the inhaled air are removed by deposition within the upper respiratory tract before they reach the alveoli, the exact fraction depending primarily on the particle size and density of the particles (ICRP, 1966, 1980, 1994b). Deposition within the respiratory tract may result from inertial impaction, settling, or Brownian motion in the case of particles less than about 0.1 μm in diameter. In general, the larger particles tend to be removed between the nasal passageways and the lower bronchi, whereas the smaller particles have a higher probability of penetrating to the alveoli. If the dust is highly soluble, it will be quickly absorbed from the respiratory tract into the blood.

The respiratory tract above the terminal bronchioles is lined with ciliated epithelium that has the ability to move the deposited dust up and out of the respiratory tract, from which it enters the gastrointestinal tract with swallowed phlegm. Dust deposited on the bronchial epithelium in humans is removed from the lung in a matter of hours, whereas dust that deposits in the alveolar regions of the lung can remain within the lung for weeks or years (ICRP, 1975, 1994b).

Insoluble dust deposited on the alveolar walls is removed by phagocytes, which are motile white cells that have the ability to engulf the particles and transport them out of the alveolar spaces. The dust-laden phagocyte will either wander onto the ciliated epithelium from which it will pass to the

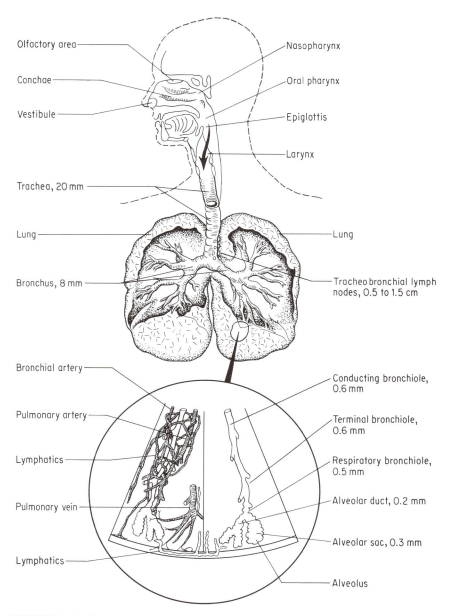

FIGURE 2-2 Principal anatomical features of the human respiratory tract. [From NAS/ NRC (1961).]

mouth by mucociliary action or pass into the lymphatic vessels and then to the lymph nodes.

The dose received by the different portions of the lung from inhaled radioactive dust thus depends on the concentration of the radionuclides in the inhaled air, respiration rate, physical properties of the radionuclide, the region of the lung in which the dust is deposited, and the rate at which it is removed. The latter factors depend on particle size and density, as well as on physiological and morphological factors that vary among individuals.

The tissues may not be irradiated uniformly by particles deposited in the lung. This is particularly true when the dust particles are alpha emitters. The basic question is whether the risk of developing cancer is increased because the energy is absorbed in a smaller volume of tissue. Both the ICRP and NCRP have taken the position that the hazard of radiation from particulate sources in the lung is probably less than if the same absorbed energy is distributed uniformly throughout the lung (ICRP, 1977; NCRP, 1975a).

There are several ways of computing the dose from alpha-emitting radioactive dust deposited in the lung. The simplest assumption is that the energy from the alpha particles is absorbed uniformly throughout the total lung mass (taken to be 1000 g in the adult human) and one can calculate the dose accordingly. However, since most lung cancers originate in the bronchial epithelial tissue, and since most of the radioactive dust is deposited on the bronchial epithelium, the dose from inhaled insoluble dust is calculated by assuming that the energy is absorbed in that tissue. For radon and its decay products, the depth of penetration in tissue is only a few tens of micrometers. Dose is calculated at the depth of the basal cell layer of the bronchial epithelium, assumed to be 22 μm (NAS/NRC, 1988). The dose calculated in this way is orders of magnitude higher than when it is assumed the energy is absorbed in the whole lung (Altshuler *et al.*, 1964; NCRP, 1984c).

Radon   Radon is an inert radioactive gas that is produced in the environment as a component of the ^{238}U decay series. More details about the occurrence and physical properties of radon will be discussed in Chapter 6. When radon is inhaled, it is usually not in equilibrium with its decay products, which include a series of short-lived nuclides that decay within hours to ^{210}Pb ($T_{1/2}$ = 22 y). Because of the difficulty in describing the mixture in the usual units (pCi L^{-1} or Bq m^{-3}), the term "working level" (WL) was devised (Holaday *et al.*, 1957) to express the concentration of radioactivity of the mine atmosphere. The WL is defined as that concentration of short-lived radon decay products that emits 1.3×10^5 MeV of alpha radiation per liter of air. Exposure to one WL for one month is referred to as the WLM and is a convenient way of expressing integrated exposure. Conversion of WLM, which is a measurement in air, to dose is complicated by the fact that the decay products do not deposit uniformly on the bronchial surfaces but, depending on particle size, deposit at various depths in the bronchial tree. Various estimates of dose per WLM have been made, but they vary widely depending on the assumptions made. The NCRP (1984b) found the dose to the bronchial epithelium to range from 0.7 to 140 mGy (0.07 to 14 rem) per WLM among 26 estimates reported by various investigators. There are a great many variables in calculations of this kind. These include the particle size of the dust to which the decay products are attached, the unattached fraction of decay products, assumed breathing rate, the level of the bronchial tree at which the dose is calculated, and the thickness of the bronchial mucosa. The National Academy of Sciences (NAS/NRC, 1988) has examined three well-refined models and concluded that the variation among them amounts to about a factor of three. The recommended conversion factor is about 1 rad (10 mGy) per WLM. Among U.S. uranium miners, there is evidence of an excess of lung cancer at cumulative exposures greater than about 100 WLM.

Studies of Underground Miners   The first evidence that radon could cause lung cancer came from studies of metal miners in eastern Europe (Hueper, 1942; Lorenz, 1944; Behounek, 1970). These mines had served for centuries as a source of a number of heavy metals and it had been known that the miners suffered from a lung disease that was not diagnosed as lung cancer until the latter part of the nineteenth century. Studies in the 1920s led to the conclusion that radon in the mine atmosphere was probably responsible. By the 1950s, based on work initiated by Harley (1952), it was apparent that the principal dose to the lung from inhalation of radon was not due to the radon itself but to decay within the lung of the short-lived decay products of radon attached to inert dust normally present in the atmosphere.

Beginning in the 1960s, an excess of lung cancers has also been reported among uranium miners in the United States (Archer, 1981; Holaday, 1969; NCRP, 1984b; NAS/NRC, 1988, 1994a). These miners, in the southwestern states, were known to have been exposed to concentrations of radon and radon decay products comparable with those reported earlier in the mines of eastern Europe. Excesses of lung cancer have also been reported among miners in nonuranium mines in which elevated concentrations of radon are present. These include Newfoundland fluorspar mines, where radon enters the mine via groundwater (deVilliers and Windish, 1964), and iron, zinc, and lead mines in Sweden (Axelson and Sundell, 1978). Lung cancer has also occurred in great numbers among miners in a large tin mine in southern Hunan Province of the People's Republic of China, where more than 1500 cases have been reported. Arsenic is known to be present in the mine air and may increase the effect of the radon decay products (Lubin *et al.,* 1990;

Xuan *et al.,* 1993). To date, the relationship between radon exposure and lung cancer has been studied in 11 separate cohorts comprising 68,000 underground miners, among whom 2780 cases of lung cancer have developed (NAS/NCR, 1994a). There is a linear relationship between exposure to the radon products and excess cancer. An important finding is that the effect per unit of dose is increased at lower exposure rates, which is the reverse of the negative dose rate effect noted earlier in this chapter. The 1994 report from the National Academy of Sciences concluded that the risk of developing lung cancer from radon exposure is increased by smoking, and that the interaction is somewhere between additive and multiplicative.

Exposure to Indoor Radon   Because radon can seep into buildings from the soil below, or enter by way of radon-laden tap water, the general public can be exposed to relatively high concentrations within buildings. This has led to estimates that radon exposure may be responsible for a significant fraction of the lung cancers that occur among nonsmokers (NCRP, 1984b). The mechanisms by which radon enters the home and methods for its alleviation will be discussed further in Chapter 6. It has been estimated that as many as 10% of the cases of lung cancer in the United States, about 13,000 deaths per year, may be due to indoor radon exposure. There is considerable uncertainty in this estimate, which ranges from 6000 to 36,000 deaths (Lubin, 1994). In 1986 the EPA estimated that the number could range from 5000 to 20,000 (EPA, 1986a), but now places the estimate at 7000 to 30,000 (EPA, 1992a). These estimates should be regarded as highly provisional because of the many uncertainties involved. EPA has recommended that remedial measures be taken to reduce the radon when levels greater than 4 pCi L^{-1} (150 Bq m^{-3}) are found.

Although there is evidence that indoor radon exposure is by far the source of the greatest dose received by the public—much more so than other sources of radiation exposure that have caused far greater public anxiety—people have been remarkably reluctant to take corrective actions. The placement of inexpensive charcoal canisters or other simple devices in the home for a short period of time can determine whether or not a problem exists, and corrective action may require no more than an improvement in basement ventilation (Moeller and Fujimoto, 1984; Mahar *et al.,* 1987).

The public inaction may be due to uncertainty in the validity of the estimates of health effects. The question arises of whether the risk estimates based on experience with uranium miners are applicable to the general population exposed to indoor ^{222}Rn. To some extent the conditions are different. The particle size of the dust to which the ^{222}Rn decay products are attached, differences in the unattached fractions of the ^{222}Rn decay products, the extent to which decay product equilibrium has been achieved, the radiosensitivity of children compared to adults, and even differences in breathing rates affect the dose estimates and the risk coefficients. These factors have been examined by Harley (1984), who concluded that the risk estimates based on miners are applicable to the general population. However, Lubin *et al.* (1995) caution that the exposures of miners to diesel smoke and arsenic, plus other differences between mines and dwellings, make generalizing the miner epidemiological findings to domestic radon exposures highly uncertain.

Whether the estimates of national mortality are realistic may be answerable by epidemiological studies, but the results have been inconclusive (Neuberger, 1992; Cole, 1993; Lubin, 1994). To the contrary, more than 15 studies have "failed to demonstrate consistent correlations between population lung cancer rates and measures of radon exposure" (Alavanja *et al.,* 1994; Lubin, 1994; Stidley and Samet, 1993). The 15 studies were "ecologic" in design in that they attempted to determine whether the variability of the rates of lung cancers in different populations can be explained by difference in the levels of radon exposure. (Figure 2-3 from Lubin (1994)

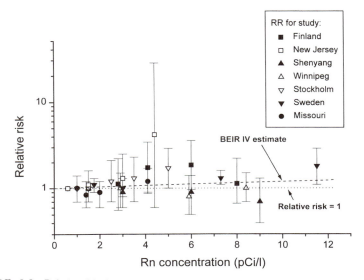

FIGURE 2-3 Relationship between lung cancer and indoor radon exposure from seven case–control epidemiological studies. The uncertainty limits of most of the studies encompass both the zero effect line and the risk predicted by downward extrapolation of the underground miner data (BEIR IV), illustrating the difficulty of drawing conclusions from indoor radon epidemiological data. [From Lubin (1994).]

illustrates results from seven epidemiological studies of persons exposed to indoor radon. It is seen that the uncertainty ranges of nearly all the data points encompass both the BEIR IV radon risk estimate and the zero relative risk line. A 1991 international workshop on the subject concluded that case–control studies are more likely to produce definitive information and that such studies rather than ecologic investigations should be favored in the future (Samet *et al.*, 1991). In 1993 the World Health Organization also concluded that direct epidemiological evidence of increased lung cancer due to radon exposures in dwellings was equivocal (WHO, 1993).

However, the recent and largest domestic indoor radon case–control study to date does report a positive association between radon levels and lung cancer (Pershagen *et al.*, 1993, 1994). This study was based on 1360 lung cancer cases and roughly twice as many controls and involved radon measurements over a period of more than 30 years. The lung cancer risk was shown to increase to a statistically significant degree with increasing radon exposure. In particular, the risk was elevated for an average radon concentration in the range 140–400 Bq m^{-3}. The risk estimates from this study are consistent with those from the miner studies. Moreover, there was statistically significant evidence that the joint effect of radon and smoking exceeded additivity and the data were consistent with a multiplicative effect. In view of the large number of studies showing no effect or even inverse correlations between radon exposure and lung cancer, this Swedish study should be regarded with caution until it is confirmed by additional studies.

Thyroid Cancer   Several radioactive isotopes of iodine are produced in fission, of which ^{131}I, with a half-life of 8 days, is of most interest. A number of other radionuclides of iodine are produced (^{132}I, ^{133}I, and ^{135}I), but they have half-lives measured in hours, which limits the dose they can deliver under most circumstances because they will have decayed either during transport in the atmosphere or in the course of moving through food chains. However, in situations in which the dose is received from the short-lived isotopes, the risk per unit of dose may be greater than for ^{131}I, because the dose is delivered in a shorter time.

That irradiation of the thyroid gland can cause cancer in the human thyroid has been shown by a number of studies (Hemplemann *et al.*, 1975; Ron *et al.*, 1995; NCRP, 1985a; Shore *et al.*, 1986; Shore, 1992). Among the cohorts studied have been survivors of the Japanese atomic bombings, patients irradiated in childhood for enlarged thymus glands, patients irradiated for ringworm of the scalp, and inhabitants of the Marshall Islands who were accidentally exposed to heavy fallout from a thermonuclear weapon test in 1954 (see Chapter 12). Evidence, limited by statistical uncertainty, has also been presented more recently of an increased prevalence of between 0 and 6 malignant thyroid tumors among residents exposed to fallout from nuclear weapons tests conducted in the southwestern United States (Kerber *et al.*, 1993), and an excess of about 500 thyroid cancers had been reported by 1995 in Belarus, presumably as a consequence of the Chernobyl accident (see Chapter 12). In addition, considerable useful information has come from the extensive use of radioactive iodine for both diagnostic and therapeutic purposes.

From all the evidence, it appears that the risk of thyroid cancer is greater by a factor of two when exposure occurs during the first two decades of life, that women are about twice as susceptible as men, and that the risk is higher in persons of Jewish descent than in others (Shore *et al.*, 1986).

The risk of thyroid cancer from radioiodine released to the environment by nuclear weapons or as a result of reactor accidents is of great practical concern. Two such accidents, in England and the former USSR, are described in Chapter 12.

Ron *et al.* (1995), from a pooled analysis of seven studies, concluded that "the thyroid gland in children has one of the highest risk coefficients of any organ and is the only tissue with convincing evidence for risk at about 0.1 Gy (10 rem)." Iodine-131 has been shown to be less effective in producing thyroid cancer than X rays, which may be due to the relatively low dose rate at which the dose from the radionuclide is delivered. The NCRP has concluded that the carcinogenicity of ^{131}I is no more than one-third that of external radiation (NCRP, 1985a).

Holm and associates (1988) have followed a cohort of 38,700 patients who were administered ^{131}I for diagnostic reasons between 1951 and 1969. The average dose received by the thyroids in these patients was 0.5 Sv (50 rem). The investigators concluded that their data "provide little proof that ^{131}I is carcinogenic in humans," but this should be accepted with caution. Studies in regions affected by the ^{131}I releases from the Chernobyl accident suggest that thyroid cancer is occurring in children at a rate equivalent to that seen in other studies from external radiation (Beral, 1996).

A group of about 4000 women who were treated for hypothyroidism has been followed by Hoffman (1984), who found no difference in the incidence of thyroid or other cancers between those treated by ^{131}I and others treated by surgery. In that series of cases the dose was much greater than 1000 rad (10 Gy).

An excess of 30 cases of thyroid cancer has been reported among Japanese survivors of the atomic bombings (UNSCEAR, 1994). The effect has been found to be proportional to radiation dose, and the increase has been greater in women than in men (Parker *et al.*, 1974).

In March 1954, a group of natives of the Marshall Islands were exposed to fallout from a thermonuclear explosion at Bikini (see Chapter 12). Of 67 people exposed on Rongelap, 34 developed thyroid abnormalities, including hypothyroidism, nodules, and seven malignant tumors, at the end of about 27 years. The thyroid doses from absorbed radioiodine were estimated to have ranged from 335 rem (3.35 Sv) in adults to 1800 rem (18 Sv) in children less than 10 years old. The whole-body dose was about 175 rem (1.75 Sv) (Conard et al., 1980; Conard, 1984). The thyroid doses are somewhat uncertain because they were reconstructed from urine analysis data several weeks postexposure. The fact that the dose received was from both radioiodine and external gamma radiation must be taken into consideration when interpreting this series of cases.

Using data from the various cohorts, the lifetime risk of developing a fatal thyroid cancer when a dose from ^{131}I within the range of 0.06 to 15 Gy (6 to 1500 rad) is delivered to the general population has been estimated to be about 7.5×10^{-4} Gy^{-1} (7.5×10^{-6} rad^{-1}). The risk of developing a nonfatal thyroid cancer would be about 10 times higher (NCRP, 1985a). The minimum latency period has been given as 5 years (Shore et al., 1984).

For purposes of calculating the thyroid dose in an individual who inhales or ingests radioiodine, it is assumed that the thyroid absorbs 30% of inhaled or ingested radioiodine (ICRP, 1960). Yet according to Dolphin (1968, 1971), thyroidal uptake can vary from 10 to 30% and may actually be a function of the thyroid mass, which is usually taken as 20 g, but has been shown by Mochizuki et al. (1963) to vary somewhat around a mean weight of 16.7 $\pm$ 6.9 g in New York City adults. The radioiodine uptake can be almost completely blocked by oral administration of potassium iodide in doses in the range of 50 to 200 mg at the start of exposure (NCRP, 1977b).

Cataracts   Cataracts are a deterministic effect on the lens of the eye from relatively high doses of X rays, gamma rays, beta particles, or neutrons (UNSCEAR, 1993). As noted earlier, cataract was reported in experimental animals exposed to X rays within 2 years after Roentgen's discovery. Cataracts in human beings were observed among survivors of the Japanese bombings (Nefzger et al., 1968), among patients whose eyes were treated by X, gamma, or beta rays for medical purposes (Merriam and Focht, 1957; NAS/NRC, 1961), and among physicists who were exposed to neutron radiation from cyclotrons. Although lens changes have been reported from doses as low as 200 rem (2 Sv), the minimum X-ray dose capable of causing clinically significant cataract is thought to lie between 600 and 1000 rad (6 to 10 Gy) in adults but may be less in children. Neutrons are thought to be 5 to 10 times more effective than X or gamma rays in the production of cataracts. Cataract has not been reported as a result of low-level chronic exposure.

Genetic Effects

As in the case of all higher animal forms, the essential characteristics of humans are passed from one generation to the next by means of chromosomes located in the nuclei of reproductive cells. Human cells normally contain 46 chromosomes of which one-half are derived from the mother and the other half from the father. The inheritable characteristics are communicated by genes, which are strung together in beadlike fashion to form tiny filaments that are the chromosomes.

The genes from the mother and father are united within the ovum at conception, and thereafter throughout life the chromosomes are almost always faithfully reproduced at each cell division. The 46 human chromosomes are believed to contain on the order of 10^5 genes, and it is this complex of genes and chromosomes that, when passed on to the next generation, will determine many of the physical and psychological characteristics of the individual.

Genes are large molecules that may undergo structural changes ("mutations") as the result of action by a number of agents, including heat, ionizing radiation, and mutagenic chemicals (NAS/NRC, 1990). Mutations can also result from breaks and rearrangements of the chromosomes themselves. It is possible for a mutation to be passed directly to progeny in whom the mutant characteristic may ultimately manifest itself. This occurs only for those mutations that are "dominant," that is, have the ability to express themselves in the child that inherits the mutation. It is estimated that 1% of all persons inherit at least one dominant mutation. Most mutations of the type produced by ionizing radiations are recessive and do not express themselves in the offspring unless a similar mutation is encountered in the chromosomes of the mate. However, the mutated gene will be reproduced and transmitted to successive generations of progeny. Because it is highly improbable that two similar mutations would encounter each other in the first postmutation generation, the mutated genes would simply diffuse through the genetic material of subsequent generations, and would accumulate in number until they became so numerous as to increase the probability that they will encounter each other in the reproductive process. In the offspring produced from such encounters, the mutations would express themselves as inherited changes in the characteristics of individuals.

The measure of potential damage from radiation exposure is the total collective dose delivered to the gonads of the breeding population or, stated another way, the per capita gonadal dose. In any given population, allowance must be made for age distribution and the probability of reproduction by the various members of the population. When the per capita gonadal dose is corrected in this way, it is known as the "genetically significant dose."

In a population subjected to ionizing radiation exposure, the total number of mutations produced will continue to increase until the rate at which new mutations are produced is exactly offset by the rate at which new carriers of the mutations die. It would take many generations for such an equilibrium to be established.

It is estimated that about 4% of all individuals inherit characteristics that result from recessive mutations due to natural factors in the environment. Many of these characteristics are so minor as to be of little or no concern, such as variation in eye or hair color. Other mutations can result in tragic consequences that may include serious deformity, crippling disease, and premature death. The genetically induced defects due to ionizing radiation exposure are generally of the same kind as those that occur spontaneously.

The genetic effects of radiation have not been seen in irradiated human populations, even among the two large populations of atom bomb survivors. This does not mean that genetic effects have not occurred, but rather that they may occur too infrequently to be detected (NAS/NRC, 1990).

Based on 45 years of study of the population that survived the Japanese atomic bombings, it is estimated that, at a minimum, a per capita dose between 1.7 and 2.2 Sv (170 and 220 rem), delivered in a single dose to an entire population, could cause the spontaneous mutation rate to double. The estimated minimum doubling dose for chronic radiation is twice as great (Neel *et al.,* 1990).

Radiation-Induced Chromosome Changes

It has been known for many years that aberrations in human chromosome structure occur more frequently in irradiated persons and can be observed in chromosomes from peripheral blood lymphocytes at doses as low as a few rads (a few tens of mGy) (UNSCEAR, 1977). The biological significance of such changes, particularly when seen in somatic cells, is not known. Considerable interest in the phenomenon derives from the possibility that the frequency and types of chromosomal abberations may provide a method of estimating past exposure to ionizing radiation (Pohl-Rüling and Fischer, 1979; Du Frain *et al.,* 1980). Soviet scientists who studied the aftermath of the 1986 reactor accident in Chernobyl claimed that cytogenetic studies made it possible to estimate the doses received by exposed individuals (USSR, 1986). The difficulties of using chromosomal abnormalities for dosimetric purposes have been summarized by Littlefield and Lushbaugh (1990), who studied 176 patients involved in radiation accidents and called attention to the differences due to partial body radiation and dose fractionation. Nevertheless, it has been shown that cytogenetic dosimetry can play an important role in dose reconstruction following accidental exposure at dose levels above 0.5–1.0 Gy (50–100 rad) (Nugis and Konchalovskii, 1993).

Effects on Growth and Development

It has been shown in animal experiments and epidemiological studies that prenatal irradiation may affect physical and mental development. This is particularly true when irradiation occurs during the period when the organs are developing rapidly.

Mental retardation is a major consideration. Human effects have been observed only among Japanese who were *in utero* at the time of the atomic bombings (NAS/NRC, 1990). Among a cohort of 1598 such individuals, severe mental retardation developed among 30 individuals who were exposed between the 8th and 26th week of gestation, with the maximum sensitivity occurring between 8 and 15 weeks of gestational age. Severe mental retardation occurred in 43% of those exposed to 1 Gy (100 rad). There is a suggestion of a threshold in the range of 0.05 Gy (5 rad), but the cases are so few that there is great uncertainty at such low doses (ICRP, 1991).

Among the same groups (between 8 and 26 weeks of gestation), there has also been a dose related diminution in intelligence test scores, estimated to be from 21 to 29 points per Gy (100 rad). These findings have resulted in adoption of special precautions to prevent fetal damage among female radiation workers (Chapter 3).

ESTIMATES OF DETRIMENT FROM RADIATION

The International Commission on Radiological Protection (1991) has estimated the probabilities for stochastic effects, including fatal cancers, nonfa-

TABLE 2-2

Nominal Probability Coefficients for Stochastic Effects Estimated by the ICRP[a]

| Exposed population | Detriment (10^{-2} Sv^{-1} or 10^{-4} rem^{-1}) | | | |
	Fatal cancer	Nonfatal cancer	Severe hereditary effects	Total
Adult workers	4.0	0.8	0.8	5.6
Whole population	5.0	1.0	1.3	7.3

[a] Adapted from ICRP (1991).

tal cancers, severe hereditary effects, and "detriment" from ionizing radiation. Detriment is the sum of the probability of attributable fatal cancer, the weighted probability of attributable nonfatal cancer, the weighted probability of severe hereditary effects, and the relative length of life lost. These probabilites are intended for use in radiation protection, with dose ranges where the probabilities are small, so they are treated as additive. The detriment for workers is estimated to be smaller than that for the whole population because the age distribution is limited to the working years. Table 2-2 gives the nominal probability factors for detriment per unit effective dose for adult workers and the whole population as developed by the ICRP.

CHAPTER 3

Radiation Protection Standards

Soon after the discovery of X rays and radium, their dangers were so well recognized and their uses were becoming so commonplace that there was a clear need to develop uniform standards of radiation protection. In 1928, the International Society of Radiology sponsored formation of the International Commission on Radiological Protection (ICRP) (Taylor, 1971, 1981). This commission still exists, comprising 13 members and 4 committees that have a total of about 70 experts from 20 countries. The ICRP has held an important position in the field of radiation protection and in recent years has received financial support from the World Health Organization and other international and national organizations.

The National Council on Radiation Protection and Measurements (NCRP; originally called the Advisory Committee on X-ray and Radium Protection) was founded in the United States in 1929, one year after the ICRP was formed. For many years, the NCRP was housed administratively within the Bureau of Standards, although it was not supported financially by it. In 1964, the NCRP was granted a congressional charter that gave it a broad mandate to continue and extend its activities in the radiation protection field (Taylor, 1971). Since that time it has operated as an independent organization financed by contributions from government, scientific societies, and manufacturing associations. The council has up to 75 members and about 700 individuals are associated with its scientific committees, which help develop the NCRP reports and recommendations (Sinclair, 1993).

EVOLUTION OF RADIATION DOSE STANDARDS

The first efforts to develop safe practices were handicapped because of primitive concepts of dose (Wyckoff, 1980). Until the roentgen was adopted in the early 1930s, radiation dose was related to the length of time it took for a given radiation flux to produce reddening of the skin. The maximum permissible exposure for radiation workers during any month was taken to be 1/1000th of the amount of radiation that would produce erythema in an acute exposure. Later, when the roentgen (R) was defined as the unit of radiation exposure, it was estimated (Failla, 1932) that the erythema dose of X radiation was approximately 600 R, and that the values recommended by various experts between 1925 and 1932 had ranged between 0.04 and 2.0 R per day. The ICRP and NCRP subsequently adopted 0.02 R and 0.01 R, respectively, as the daily "tolerance dose." One roentgen corresponds approximately to 1 rad or 0.01 Gy in air. The discrepancy of a factor of two is explained by the fact that the ICRP calibration was free in air whereas the NCRP calibration was on the body and included backscatter. The two recommendations were thus in approximate agreement.

LIMITS FOR OCCUPATIONAL EXPOSURE

When the first radiation protection standards were proposed, the total population at risk consisted of a few thousand X-ray and radium workers, so it was natural that the initial standards were designed to protect those exposed occupationally. Until that time the hazard was thought to be due entirely to external radiation, although, unknown to the scientists then concerned, a number of workers in industry had already absorbed radium in amounts that were to prove lethal to some of them. The fact that ionizing radiations could produce genetic mutations was not sufficiently appreciated, and the possibility that there might not be a threshold for the production of radiation-produced neoplasms had not yet been suggested. Thus, radiation protection practices were deemed to be sufficient if no apparent damage was being caused at a given level of radiation exposure. The clinical sign of radiation injury that was then considered to be particularly important was a change in the white cell count, and a periodic blood examination was the main procedure by which the health of radiation workers was monitored. A worker was believed to have been protected if he maintained a normal blood count.

The need to develop limits of permissible exposure to internal emitters began to arise in the late 1920s, when it was learned that bone cancer (osteogenic sarcoma) had developed among workers applying radium to luminous dials used in timepieces (Martland, 1925). In a group of about 30 former dial painters studied by Evans (1967), it was found that the lowest [226]Ra body burden of an individual who had developed bone cancer was 1.2 μCi (44 kBq). The NCRP (1941), on the basis of these data, proposed that the maximum permissible body burden of [226]Ra be limited to 0.1 μCi (3.7 kBq). In later years, the number of known radium workers increased manyfold, and X-ray evidence of bone disease other than cancer has been reported at body burdens as low as 0.5 μCi (19kBq). However, the NCRP recommendation that the permissible body burden be limited to 0.1 μCi (3.7 kBq) essentially remains in effect to the present time, more than 50 years after it was first proposed, although the limit is now expressed in units of dose rather than activity.

The two benchmarks adopted prior to World War II, namely, 0.1 R (~1 mGy) per day for whole-body external radiation and 0.1 μCi (3.7 kBq) as the maximum body burden of [226]Ra, were the standards on which the early atomic energy program was conducted, with an excellent safety record. The fact that these important standards became available just prior to the start of the atomic energy program in 1942 was a remarkable historical coincidence. If the relatively few cases of radium poisoning had gone unnoticed, or if they had not attracted the interest of such perceptive investigators as Martland, Evans, and Failla, the quantitative basis for the safe handling of bone-seeking internal emitters would not have been available by World War II. Similarly, had the uses of X rays developed one or two decades later than they did, the ICRP and NCRP would not have had the guidance that the early misadventures with X rays provided, and the maximum permissible dose might not have been available when needed during World War II. Thus the two all-important benchmarks were developed just in time to serve the needs of the early atomic energy program.

Much improvising was needed to fill the gaps that developed during World War II. The recommendations of the NCRP for protection against external radiation from X rays and radium were quickly adapted to the sudden need to protect people against a wider variety of ionizing radiations, many of which, like neutrons, had not been extensively studied by biologists. Failla and others devised the concept of relative biological effectiveness, and Parker invented two new units, the rep and the rem[1] (Cantril and Parker, 1945). This made it possible to extend to particulate radiation the use of the recommended daily limit of 0.1 R (~1 mGy) then used in the United States for gamma-ray and X-ray protection. Also, the maximum permissible body burden adopted for ^{226}Ra was used as the basis for computing the permissible body burdens of other alpha-emitting bone seekers, some of which, such as plutonium, did not exist before World War II.

The original daily "tolerance dose" of 0.1 R (~1 mGy) for radiation workers served well during the period of great expansion in the medical use of X ray, industrial radiography, and the early development of the atomic energy program. There was no evidence that the standard resulted in unsafe conditions for the workers, but development of the concept of stochastic dose–response relationships, and the accompanying statistical problems referred to in Chapter 2, suggested to some that the permissible dose should be reduced. In 1949, in a report that was not published formally until 1954, the NCRP recommended that the permissible dose from occupational exposure be reduced from 0.1 R per day (~1 mGy per day) to 0.3 rem per week (3 mSv per week) (NCRP, 1954).

In 1956 another change was made, this time by the ICRP, which limited the dose to the gonads and blood-forming organs to 5 rem (50 mSv) per year. The NCRP adopted this recommendation and further stipulated that the accumulated dose to individuals should be limited to $5(N - 18)$ rem [$50(N - 18)$ mSv], where N is the age of the individual in years. This formula also reflected the position that an individual should not be exposed to

---

[1]The quantities and units of radiation dose are inherently more complex than those used in toxicology or pharmacology, and additional complexity has resulted from frequent changes required by evolving concepts in radiation dosimetry. The rep (roentgen equivalent physical) was the unit of the quantity absorbed dose (D) having the magnitude 93 erg g^{-1} (0.0093 J kg^{-1}) of absorbing material. The rep was replaced by the rad, which has a magnitude of 100 erg g^{-1} (0.01 J kg^{-1}). The quantity dose equivalent (H) and its unit of rem was introduced to account for the different biological effects of the same physical dose from different types of radiations. H is the product of D, Q, and N at a point of interest in tissue, where D is the absorbed dose, Q is the quality factor, and N is the product of any other modifying factors. In recent recommendations, the use of N has been dropped and Système International (SI) units have been adopted. Dose is now expressed in grays (Gy) and equivalent dose (and associated quantities such as "effective dose") in sieverts (Sv). These units are equal to one joule per kilogram. In terms of the traditional units, 1 Gy = 100 rad and 1 Sv = 100 rem. SI units have been almost universally adopted internationally and in the U.S. scientific community but have not been embraced enthusiastically by the U.S. regulatory and engineering communities. In the medical radiation therapy community, where a serious mistake caused by confusion over units could be potentially fatal, the unit "centigray" is frequently used. This compromise unit incorporates the gray but is numerically equal to the familiar rad. The principal international authority on radiological quantities and units is the International Commission on Radiation Units and Measurements (ICRU), which maintains administrative offices within the NCRP headquarters in Bethesda, Maryland.

ionizing radiation before the age of 18. A further stipulation by both the IRCP and NCRP was that the dose received during a 13-week period should not exceed 3 rem (30 mSv).

The average dose over a working lifetime was thus reduced to about 5 rem per year (50 mSv per year), one-sixth of the limit used during World War II. The ICRP and NCRP recommendations were quickly added to the regulations of the governmental agencies.

Another important development took place in 1956, when reports by the Medical Research Council of the United Kingdom (MRC, 1956) and the U.S. National Academy of Sciences (NAS/NRC, 1956) provided the first quantitative insights into the genetic effects of population exposure. In 1960 the U.S. government established an interagency Federal Radiation Council (now inactive as such), which published Radiation Protection Guides that followed the $5(N - 18)$ formula of the NCRP and ICRP for occupational exposure.

The next major ICRP change took place in 1977 (ICRP, 1977), when it was recommended that exposure be limited to 5 rem (50 mSv) per year, but that the practice of allowing exposure accumulation by the formula $5(N - 18)$ rem be discontinued. The age proration formula was still permitted in the United States up to 1994 but was rarely called on in practice. The ICRP concluded that the risk to workers if exposure were limited annually to 5 rem (50 mSv) would be no greater than the risks taken by employees of other industries that have relatively good safety records.

Between 1960 and 1977, the system used by ICRP (1960) to specify the maximum permissible concentration of a radionuclide in air or water was based on the criteria that the dose to the "critical organ," that is, the organ that receives the highest dose from a radionuclide absorbed into the body, should not exceed a stated annual dose rate after 50 years of exposure. The annual limits were 5 rem (50 mSv) for the gonads and the 15 rem (0.15 Sv) for other organs, except that the limit for bone-seeking radionuclides was based on analogy with 0.1 μCi (3.7 kBq) of ^{226}Ra, which, with a quality factor of 10, delivers an annual dose of about 30 rem (0.3 Sv).

The 1977 ICRP report included a system of permissible limits that considered the differences in the radiosensitivity of the various organs and tissues based on risk coefficients estimated from epidemiological data. Under the revised ICRP system, the limits are based on the criteria that the risk of organ irradiation should not be greater than that from 5 rem (50 mSv) per year of whole-body radiation (Vennart, 1981). On the basis of the information available from epidemiological sources, the ICRP introduced the factor L, which is the dose equivalent to individual organs that results in the same risk as irradiation of the whole body at a rate of 5 rem (50 mSv) per year. The ICRP also introduced weighting factors to express the risk of organ irradiation relative to whole-body irradiation. The sum of the organ doses times their weighting factors was expressed as the effective dose equivalent (EDE). The major change that took place as a result of these ICRP reports is that the dose limits are now being stated by the ICRP on the basis of the risk that will be incurred (Peterson, 1984).

In 1987, the NCRP issued new basic radiation protection guidance (NCRP, 1987e). This report reaffirmed the annual limit on dose equivalent of 50 mSv (5 rem) but emphasized the upper boundary nature of the limit. The NCRP joined the ICRP in recommending that the cumulative limit, expressed as $5(N - 18)$ rem or $50(N - 18)$ mSv, be discontinued. However, a significant departure from the status quo was made in Report 91, which encouraged the community of radiation users to limit cumulative occupational dose equivalent, expressed in units of 10 mSv (1 rem), to the numerical value of the individual worker's age. This was done because evidence was accumulating, particularly from the continuing study of the atomic bomb survivors, that stochastic risks might be larger than previous evidence would indicate. Report 91 (NCRP, 1987e) also adopted the ICRP system of effective dose equivalent and utilized the same weighting factors to express the relative stochastic risk of various tissues and organs to radiation.

The ICRP issued its latest basic radiation protection guidance as Publication 60 (ICRP, 1991). This report introduces modified dose concepts as well as new dose limits. The probability of occurrence of a stochastic effect such as cancer is related not only to the absorbed dose but to the quality (kind and energy) of the radiation and to the organ or tissue irradiated. These radiation- and tissue-related factors are accounted for by the ICRP in a systematic way. Dose equivalent has been replaced by equivalent dose (H_T) and quality factor by the radiation weighting factor (w_R). Equivalent dose is expressed as

$$H_T = \sum_R w_R \cdot D_{T,R} \qquad (3\text{-}1)$$

where $D_{R,T}$ is the dose averaged over tissue T due to radiation R. H_T is distinguished from dose equivalent by the fact that it is an average over the organ or tissue of interest rather than a point quantity. The values assigned to w_R by ICRP Publication 60 are given in Table 3-1 (ICRP, 1991). The ICRP emphasizes that the weighting factors are rounded values intended strictly for radiation protection. The values of w_R are similar and in several cases identical to the previously used quality factors.

ICRP Publication 60 replaces effective dose equivalent with effective dose. Effective dose is a quantity intended to express the overall risk to humans when several tissues are irradiated by several qualities of radiation. It is especially useful for dose from internally deposited radionuclides and

TABLE 3-1

Radiation Weighting Factors for the Common Radiations[a]

Type and energy range	Radiation weighting factor, w_R
Photons, electrons, and muons, all energies	1
Neutrons with energy:	
<10 keV	5
10 to 100 keV	10
>100 keV to 2 MeV	20
>2 to 20 MeV	10
>20 MeV	5
Protons, other than recoil protons, energy > 2 MeV	5
Alpha particles, fission fragments, heavy nuclei	20

[a] Adapted from ICRP (1991).

in the case of highly nonuniform external radiation. It provides a means of summing an index of risk from uniform external (whole-body) radiation, nonuniform external radiation, and internal radiation and expressing it as a single number. Effective dose is defined as

$$E = \sum_T w_T \cdot H_T \qquad (3\text{-}2)$$

where H_T is determined from Eq. (3-1) and values for the w_T are given in Table 3-2. The values of w_T are somewhat different than previous weighting factors and are given for more tissues. They represent current estimates of the relative sensitivity for stochastic effects of various organs and tissues.

In addition to new dosimetric quantities, ICRP Publication 60 also recommends a new, lower limit for occupational dose (ICRP, 1991). Occupational effective dose is recommended to be limited to 20 mSv (2 rem) per year averaged over defined periods of 5 years, with the further provision that effective dose in any one year not exceed 50 mSv (5 rem). Less restrictive limits are recommended for lens of the eye, skin, and hands and feet. The limits are summarized in Table 3-3.

The newest NCRP guidance (NCRP, 1993a) adopts the new ICRP dosimetric quantities but not the ICRP occupational dose standard. Instead it elevates to a basic recommendation its 1987 "encouragement" to limit the numerical value of worker's dose in units of 10 mSv (1 rem) to the individual's

TABLE 3-2

Tissue Weighting Factors[a]

Tissue or organ	Tissue weighting factor, w_T
Gonads	0.20
Bone marrow (red)	0.12
Colon	0.12
Lung	0.12
Stomach	0.12
Bladder	0.05
Breast	0.05
Liver	0.05
Esophagus	0.05
Thyroid	0.05
Skin	0.01
Bone surface	0.01
Remainder	0.05

[a]From ICRP (1991).

age in years. This was done because the higher risk estimates that seemed likely in 1987 had become more firmly established with publications such as the "BEIR V" report (NAS/NRC, 1990).

LIMITS FOR EXPOSURE OF THE GENERAL PUBLIC

The question arose during the 1950s of how the standards should be modified to protect the general public. The subject assumed increased importance at that time because of concerns about worldwide exposure to fallout from testing nuclear weapons in the atmosphere. The tests coincided with increasing agreement within the scientific community that the assumption of a threshold should be abandoned. The absence of a threshold was first invoked for genetic effects, but was soon extended to cancer (Lewis, 1959), and resulted in an increasingly conservative attitude toward ionizing radiation exposure that has persisted to the present time and great caution in establishing levels of maximum permissible exposure for the population at large.

It is general practice in public health to set lower limits of exposure of the general public to pollutants than is permitted for those exposed occupationally. This practice has developed for at least three reasons: It seems reasonable to expect an occupationally exposed individual to accept some measure of risk as long as the risk is no greater than other on-the-

TABLE 3-3

Recommended Dose Limits[a]

Application	Dose limit	
	Occupational	Public
Effective dose	20 mSv (2 rem) per year averaged over defined periods of 5 years[b]	1 mSv (0.1 rem) in a year
Annual effective dose in		
Lens of the eye	150 mSv (15 rem)	15 mSv (1.5 rem)
Skin	500 mSv (50 rem)	50 mSv (5 rem)
Hands and feet	500 mSv (50 rem)	—

[a] From ICRP (1991).

[b] Not to exceed 50 mSv (5 rem) in any one year; additional restrictions apply to pregnant women.

job risks that are accepted as part of modern living. In most cases the individual accepts occupational risks knowingly. However, these arguments do not hold for members of the community. Members of the general population may be exposed prenatally or in childhood. Children and the fetus may be more susceptible to injury than adults. Exposure of the public may be complicated by ecological relationships that influence the passage of radioactive materials to humans. In occupational exposure, one is normally concerned only with exposure to external radiation and airborne contamination. However, control over exposure of the general population requires that the regulations also minimize internal exposure that may occur via complex environmental pathways (see Chapters 4 and 5).

The assumption that there is no threshold for either the genetic or carcinogenic effects of ionizing radiation exposure resulted in the basic precept that there should be no man-made exposure without the expectation of benefit [Federal Radiation Council (FRC), 1960]. In one way or another this principle has been stated in many publications of the ICRP, NCRP, and FRC. The concept can also be found in much earlier statements made by officials of the World War II atomic energy program (Auxier and Dickson, 1983). The concept was first stated by the NCRP in a report (NCRP, 1954) that emphasized that radiation criteria involved value judgments that must weigh both the benefits and risks of exposure. The 1954 NCRP report stated further that "exposure to radiation be kept at the lowest practicable level," a principle that is now strongly entrenched in radiation protection practice. It has come to be known by the acronym ALARA (as low as reasonably achievable).

Radiation Protection Standards versus Environmental Radiation Standards

Kocher (1991) makes a useful distinction between a "radiation protection standard for the public" and "environmental radiation standards." The former is usually applicable to the sum of all except natural and medical sources of radiation. In contrast, environmental radiation standards are promulgated for specific practices or sources. These latter standards include judgment of the reduction of levels of exposures that are reasonably achievable for particular practices or sources, in accordance with the principle of "as low as reasonably achievable." Depending on circumstances, environmental radiation standards may be expressed in terms of dose, concentration, total activity permitted to be released, or even site design and operation.

Radiation Protection Standards for the Public

The first recommendation for a public radiation protection standard (ICRP, 1955) was that public exposures be limited to one-tenth of occupational levels. Occupational limits at that time were 15 rem (0.15 Sv) per year, so the recommended public radiation standard amounted to 1.5 rem (15 mSv) per year. The NCRP (1957) recommended an annual limit of 0.5 rem (5 mSv), which was 10% of the 5 rem (50 mSv) per year occupational standard recommended by that time. In 1959 the ICRP recommended a limit of 5 rem (50 mSv) in 30 years for the average gonadal dose to populations (ICRP, 1960). In 1960 the Federal Radiation Council also recommended that no individual in the general population be exposed to more than 0.5 rem per year (5 mSv per year) and that the "average" dose to members of the general population not exceed 5 rem (50 mSv) in 30 years, which corresponds to an annual limit of 0.17 rem (1.7 mSv). This limit was based on genetic considerations (FRC, 1960).

On the basis of stochastic risk, the ICRP (1977, 1985) recommended that for prolonged exposures, public dose be limited to a lifetime annual average of 0.1 rem (1 mSv), but also permitted the earlier annual maximum public dose of 0.5 rem (5 mSv), provided the average was not exceeded. These same values were adopted by the NCRP (1987e) in Report 91. Although the ICRP subsequently reduced the recommended occupational limit in Report 60 (ICRP, 1991), it recommended no changes in the public radiation protection standard. The NCRP made no changes to its public dose standard in its 1993 basic recommendations (NCRP, 1993a). The widespread use of very restrictive environmental radiation standards virtually assures that the dose to the public from controlled sources will be much less than the public radiation protection standard.

The standards discussed here are generally applicable for practices that involve the intended release of radioactivity to the environment from installations. Both the NCRP (1993a) and the ICRP (1991) give different guidance for cases where intervention is the only possible remedy. Both organizations recognize that intervention to control radiation exposure from past practices or natural radiation sources should be justified in terms of net benefit. However, the ICRP explicitly states that neither its numerical limits for "practices" nor any other predetermined numerical limits should be used to justify intervention. Remedial action has to be judged by comparing the benefit of the reductions in dose with the costs and risks of the remedial work. The NCRP, on the other hand, recommends a total annual dose level from natural sources of 500 mrem as an action level. It further recommends that once action is decided upon, reduction to levels substantially below the remedial action level may be obtainable and appropriate. In contrast, the EPA and the Nuclear Regulatory Commission (NRC) have circulated draft documents that propose limiting the annual dose from contaminated sites to 15 mrem (0.15 mSv).

Environmental Radiation Standards

Environmental radiation standards for specific practices or sources are usually promulgated by regulatory agencies but may also be found in recommendations such as the NCRP (1987e, 1993a) standard for indoor radon or the Food and Agriculture Organization (FAO, 1995) guidelines for levels for radionuclides in food following accidental nuclear contamination. Emission of radioactivity to air is regulated in the United States by the EPA through its regulations entitled National Emission Standards for Hazardous Air Pollutants (NESHAP), first issued in 1985. In some cases, implementation of the EPA regulation may be delegated to another agency such as the NRC. These air regulations provide good examples of the diverse approaches to environmental radiation standards, depending on the source.

The authorizing statute for NESHAP is the Clean Air Act, enacted in 1970, and its subsequent reauthorizations and amendments, enacted in 1977. Radionuclides comprise only one category of air emissions that are regulated under NESHAP. It is interesting to note that after EPA listed radionuclides as hazardous air pollutants and drafted standards for radionuclides, it attempted to withdraw the draft standards for DOE and other federal facilities, NRC-regulated facilities, and elemental phosphorous plants on the grounds that existing emission control and operational practices provided an ample margin of safety in protecting the public health. However, the Sierra Club, an environmental organization, filed suit and the court ordered EPA to either "delist" radionuclides as hazardous air pollutants or publish standards. EPA believed that radionuclides could not be "delisted" because emissions from underground uranium mines appeared to reach levels that warranted regulation (Mills *et al.,* 1988). The required standards, a regulatory apparatus, and reporting requirements were established governing most practices and sources that emit radioactivity to the air. EPA has set limits that reflect current practices control technology, noting that the ALARA policy has generally resulted in low emissions.

Different approaches to controlling radioactive air emissions have been taken, depending on the source. Current requirements can be found in the latest revision of the Code of Federal Regulations, 40 CFR 61. Emissions from federal facilities, including DOE, and from NRC-licensed facilities, but excluding certain categories such as low-energy accelerators and sealed sources, are regulated on the basis of annual dose to any single member of the public. The original standard placed limits on dose equivalent of 25 mrem (0.25 mSv) for the whole body and 75 mrem (0.75 mSv) for the critical organ. The standards were subsequently revised, effective for 1990, to utilize the newer dose concepts and set at an effective dose equivalent of 10 mrem (0.1 mSv). The facilities demonstrate compliance by some combination of emissions monitoring or inventory, environmental modeling, and environmental surveillance. To ensure uniformity, EPA requires that all facilities use specific, relatively simple environmental models provided by the agency. Facilities with more sophisticated, site-specific environmental models are encouraged to report results from those models as well. Methods of environmental modeling are discussed in subsequent chapters.

Calciners and kilns at elemental phosphate plants emit natural radionuclides contained in the ore during the refining process. Polonium-210 is the most important nuclide from a dose standpoint and the only one regulated. The original standard was an annual limit of 21 Ci (0.8 TBq), but this was reduced in steps to an annual limit of 2 Ci (74 GBq) by 1991.

Underground uranium mines emit ^{222}Rn with the ventilation air. Because conventional technology for removal of radon from airstreams was deemed ineffective or prohibitively expensive, EPA chose to limit emissions by process control. Mine operators are required to install bulkheads in unused sections of mines to limit the amount of surface area available to contribute radon to the mine air. Emissions from uranium mill tailings are controlled in a similar fashion. Operators are required to limit the surface area of active tailings impoundments and to continuously dispose of tailings in such a way as to minimize release of ^{222}Rn to the atmosphere.

It is interesting to note that some classes of NRC licensees have recently been exempted from reporting their emissions and calculated doses to EPA, the NRC requirements having been deemed equivalent.

The reader is referred to Mills *et al.* (1988) for a comprehensive summary of 23 radiation protection regulations, standards, and guides in effect or proposed at that time. Because these regulations are in a constant state of flux, the reader should rely on the official, current publications of the various agencies for detailed information about the standards in effect at any given time. The current collection of U.S. radiation standards is not embraced uncritically. For example, Mills *et al.* (1988) conclude that:

> A cursory review of the legal and technical facts contained in many of the basic U.S. radiation protection standards suggests that the standards are numerous and complex, principally control activities that make relatively small contributions to the overall U.S. population dose, have become more restrictive over time, and follow no common rationale in achieving public health objectives.

A concise illustration of the wide range in the level of protection afforded by various environmental radiation standards is given in Table 3-4, which is an updated version of the original compilation of Kocher (1991). This table lists the relative risk to an individual exposed to the various levels permitted by standards and includes some natural exposure levels for comparison. The nominal value of effective dose equivalent attributable to natural background of 300 mrem (3 mSv) is assigned a risk of one, and consistent risk factors were used to calculate others. The risks associated with levels permitted by the various environmental standards vary by about a factor of 300,000. It would appear that the standards reflect the relative ease of control and society's level of concern about various sources of radiation far more than calculated risk.

How Modern Radiation Protection Limits Are Established

Regarding the assumption of stochastic dose–response relationships, the process of setting meaningful radiation standards cannot begin until sufficient experimental or epidemiological information exists to permit estimates of risk coefficient to be made. This was not always so. Prior to the 1950s it was thought that there were threshold doses below which radiation injury would not occur. However, once it was accepted that the genetic and carcinogenic effects of radiation are stochastic as well as nonthreshold, the problem of standards setting became more difficult. The question was no longer "what dose is safe" but "how safe is safe enough." This no longer involved purely scientific judgment, and is a question that could only be answered by making social judgments.

Of primary importance in the process of setting standards is the need to assemble the published scientific literature on the biological effects of

TABLE 3-4

Estimated Relative Risk from Chronic Exposures Associated with Selected
Radiation Protection Standards, Environmental Radiation Standards and Guides,
and Radiation Exposures[a]

Relative risk[b]	Standard, guide, or exposure
3.3	EPA guidance on Rn in dwellings of 4 pCi L^{-1} (148 Bq m^{-3}), which corresponds to ~1 rem (10 mSv) annually
1.7	Maximum annual dose equivalent for members of the public found in several radiation protection standards of 5 mSv (500 mrem)
1.0	Average annual dose equivalent due to natural background in the U.S. of 3 mSv (300 mrem)
0.67	Average annual dose equivalent due to radon in the U.S. of 2 mSv (200 mrem)
0.33	Annual effective dose equivalent limit for members of the public found in several radiation protection standards of 1 mSv (100 mrem)
0.083	Annual dose equivalent limit for members of the public found in several environmental radiation standards of 0.25 mSv (25 mrem)
0.05	Proposed EPA contaminated site cleanup standard of 0.15 mSv (15 mrem)
0.033	Annual effective dose equivalent limit for members of the public found in Clean Air Act regulations of 0.1 mSv (10 mrem)
0.017	Limit for ^{226}Ra plus ^{228}Ra in drinking water
0.013	Annual dose equivalent to the whole body of 0.04 mSv (4 mrem) in drinking water standards
0.0075	Annual dose equivalent to thyroid of 0.75 mSv (75 mrem) in several environmental radiation standards
0.0033	Negligible individual annual effective dose equivalent of 0.01 mSv (1 mrem) recommended by the NCRP
0.0025	Annual dose equivalent to bone of 0.25 mSv (25 mrem) in several environmental radiation standards
0.0012	Annual dose equivalent to bone of 0.04 mSv (4 mrem) from ^{90}Sr in drinking water standards
0.00040	Annual dose equivalent to thyroid of 0.04 mSv (4 mrem) from ^{129}I in drinking water standards
0.000012	Containment requirements for disposal of high-level waste (average risk in U.S. population)

[a]Adapted from Kocher (1991) and subsequently proposed or promulgated standards.
[b]Values are normalized to unit risk for an annual natural background radiation level of 3 mSv (300 mrem).

radiation so that there can be agreement regarding what is and is not known. This role has been performed effectively on an international scale since 1955 by the United Nations Scientific Committee on the Effects of Atomic

Radiation, which has published a series of reports that summarize the state of our basic knowledge with respect to biological effects as well as the behavior of radioactive material in the environment (UNSCEAR, 1958–1994). The publications of the ICRP and the NCRP, in addition to providing guidance for the establishment of radiation standards, also include much basic information on the biological effects of radiation (NCRP, 1993a, ICRP, 1991). Other important sources of information are the periodic reports of the National Academy of Sciences–National Research Council (NAS/NRC) Committee on the Biological Effects of Ionizing Radiation, of which five exist at this writing (e.g., NAS/NRC, 1990).

UNSCEAR and the NAS/NRC have limited their role to assembling the basic scientific information and, in recent years, to estimating the risk coefficients, that is, the increased probability of occurrence of genetic effects or cancer per unit of radiation dose per unit of population. Neither UNSCEAR nor NAS/NRC recommends limits of acceptable dose.

In the United States, the EPA has the primary responsibility for using the available scientific information, including the recommendations of the ICRP and NCRP, to develop the basic numerical limits of permissible exposure. EPA consults with a number of agencies, including the NRC, the Food and Drug Administration, and the Department of Transportation, in this process.

Having decided on a level of permissible dose, there remains the need to specify the design, construction, and operating practices that must be followed to assure that the limit will not be exceeded. For example, the risks due to transportation of radioactive materials can be more efficiently controlled by specifying the procedures to be followed in packaging and shipping the radioactive materials than by simply providing numerical limits of exposure to people along the route. More and more, as experience accumulates, the radiation protection programs will be implemented by specifying operating practices that preclude the possibility of contamination of the environment above acceptable limits.

UNDERLYING FACTORS IN STANDARDS SETTING

The promulgation and implementation of standards have become increasingly complex in recent years. This is not because past standards-setting techniques have proved unsatisfactory: there is every reason to believe that the standards that became available nearly two decades ago were adequate to provide at least the same margin of safety accepted for other standards applicable to toxic or carcinogenic substances. The changes that have taken place in radiation protection standards have been largely due to factors such as the ALARA principle, better control technology, and conservatism in regard to environmental standards setting generally. In addition, the risk coefficients have been increased based on new information (e.g., NAS/NRC, 1990).

The process is also complicated by a number of difficult public policy questions that arise out of the assumption that the dose–response curve is linear at low doses and that there is no threshold. This assumption makes it seemingly possible to quantify effects at the most minuscule level of exposure. Modern methods of measuring radioactivity can detect quantities of alpha-emitting radionuclides as low as 1×10^{-18} Ci (37 μBq), and result in dose estimates that become vanishingly small. Moreover, modern methods of environmental modeling now make it possible to predict levels of exposure far into the future, and at concentrations that are far below the measurement capabilities of even our most sophisticated instrumentation. For example, as a result of the accident that occurred in 1979, the mean per capita dose to persons living within 50 miles of Three Mile Island was estimated to be about 1 mrem (10 μSv). This mean dose was not measured—this is not possible even with the best available instrumentation—but was estimated by using environmental dosimetric models, such as those to be discussed in subsequent chapters.

The assumption of linearity and the use of risk coefficients make it possible to estimate the numbers of cancers that will result theoretically from exceedingly small doses applied to very large populations. The risk to individuals may be minute, but a finite number of cancers can be predicted if a sufficiently large population is exposed. The assumed absence of a threshold has led to the concept of ALARA, but a fundamental problem is that there is no generally accepted definition of the word "reasonable," and the ALARA criteria vary widely from country to country. The ALARA philosophy of radiation protection has been the subject of considerable controversy (Auxier and Dickson, 1983) but is well entrenched in the radiation protection community.

If the number of effects to be expected is a linear function of dose, the same number of effects are to be expected whether 1.0 unit of dose is received by one person or 0.1 units are received by 10 persons. This of course applies only at low levels of exposure, where the dose–response relationship is assumed to be both linear and stochastic. Thus, in addition to the dose received by individuals, the "collective dose" can be estimated and expressed in person-rem (person-Sv in SI units), using the same risk coefficients that are used to calculate the probability of effects in a single individual.

For many types of exposure it becomes possible to estimate the collective dose to the entire population of the world. Radionuclides such as tritium, ^{14}C, ^{129}I, ^{90}Sr, and others are sufficiently long-lived so that they can permeate the global biosphere and the collective dose to the world's population can be estimated. Moreover, for very long-lived radionuclides, it is possible to make estimates of collective dose far into the future. A risk analysis of systems of high-level radioactive waste management projected risks for more than one million years into the future (NAS/NRC, 1983). When this is done, very large numbers of people may be involved, so that even though the risk per individual may be very small, the product of risk per individual times the number of persons results in a finite number of genetic or carcinogenic effects. The EPA requires (EPA, 1985) that high-level nuclear waste repositories should be designed so that there is no more than 0.1 cancer per year over a period of 10^4 years. Assuming that the world's population stabilizes at 10^{10} persons in the twenty-first century, and that there will be 300 generations of persons during the next 10^4 years, the exposed population during that period will total about 3×10^{12} persons. At a rate of 0.1 cancer per year, there would be 10^3 cancers in 10^4 years, and the risk per individual would be 3×10^{-10}, which is an exceedingly small increment of risk.

It certainly seems absurd to carry such low dose calculations so far into the future. The sociopolitical problems that, on the one hand, require such calculations and, on the other hand, are created by them will be discussed in Chapter 16. One possible solution to this dilemma would be to define the lower limit of risk worthy of regulatory concern. Following precedent that has long existed in law, this would require definition of the de minimis dose, a term that is derived from the legal principle that the law should not be concerned with trivialities. This concept is sometimes expressed as "below regulatory concern," or BRC (Party *et al.*, 1989). A number of proposals have been made to define the de minimis dose in terms of the normal variations that exist in exposure to natural sources of radiation (Adler and Weinberg, 1978; Comar, 1979; Eisenbud, 1981b). However the NRC has encountered strong public opposition in setting de minimus or BRC standards.

CHANGES IN CONCEPTS OF STANDARDS SETTING

For about 30 years, standards for internal emitters were based on the dose to the "critical organ," which, for any given radionuclide, is the organ that receives the maximum dose when the radionuclide is ingested or inhaled. Thus, the critical organ for radioiodine would be the thyroid and for ^{90}Sr it would be the skeleton. The maximum permissible concentration of any radionuclide in air or water was based on the criterion that the permissible dose to the critical organ should not be exceeded after 50 years of exposure to the maximum permissible concentration.

A major change took place in 1977 with the release of ICRP Publication 26 (ICRP, 1977), in which it was proposed that the criterion for maximum permissible exposure should be based on quantitative estimates of risk, and that the acceptability of risks from radiation exposure should be based on comparisons with other risks that society considers to be acceptable. This concept was refined in 1991 with the release of ICRP Publication 60 (ICRP, 1991). The NCRP has also proposed a system of standards based on risk (NCRP, 1987e, 1993a), and it is likely that the newest ICRP and NCRP proposals will be adopted eventually by regulatory agencies in the United States and elsewhere.

A basic step in the establishment of standards is the selection of risk coefficients, that is, the number of radiation-induced cancers or genetic effects expected per unit dose, with due allowance for variations in the radiosensitivity of the various organs and tissues of the body. The methods by which this is done were discussed in Chapter 2.

An important tool in dose estimation is the "Reference Man" (ICRP, 1975), which is a detailed compilation of the anatomical and physiological characteristics required for dose calculation. Included in the tabulations are values for the intrapulmonary deposition and clearance of inhaled dusts of varying solubilities, fractional blood uptake of the various chemical elements from the gastrointestinal tract, the fractions of the various elements absorbed by blood that are deposited in various parts of the body, and the rates of elimination. The characteristics of the Reference Man are used to develop the dosimetric models required to relate the dose that the various parts of the body will receive from a given quantity of inhaled or ingested radionuclides. The use of Reference Man permits the dose calculations to be made in a uniform way, but it is important to remember that the required anatomical and physiological parameters vary with age and sex, as well as among individuals of any given age group of either sex. On the basis of experimental data, it has been suggested that, because of individual variability, a few percent of the individuals in a population may receive a dose that is five times the average (Cuddihy *et al.*, 1979).

USE OF ANNUAL LIMITS OF INTAKE

Once the maximum permissible dose has been established, it becomes possible to calculate the allowable limits of annual intake (ALI) of any nuclide or combination of nuclides by either inhalation or ingestion (Vennart, 1981). The calculations are facilitated by the anatomical and physiological constants given for Reference Man. Having calculated the ALI, it then becomes a simple matter to calculate the maximum permissible concentration in air, food, or water, using the standardized inhalation and ingestion rates, which are also given in Reference Man.

The formal mathematics of the procedure for establishing the ALI is complicated by the buildup and elimination of the radionuclides in multiple organs and tissues due to metabolic processes and to the buildup and decay of chains of radionuclides. The reader who wishes to explore the subject more fully is referred to ICRP Publications 26 and 30 (ICRP, 1977, 1979) and to a more comprehensive publication of the NRC (Till and Meyer, 1983). The lists of ALIs published by the regulatory agencies should also be consulted as needed (e.g., Eckerman *et al.,* 1988).

In many cases, the ALI will be difficult to administer directly for releases to the environment. For example, if low-level radioactive wastes are discharged to an estuary, it is impractical to make estimates of exposure of individual members of specific nearby populations to external radiation (e.g., bathers) or to consumers of fish and shellfish. It is more practical to study the exposure pathways and develop models that describe the dose equivalent to hypothetical members of the population that have various assumed recreational and dietary habits. In this way, the dose equivalent per unit emission can be calculated, and the maximum permissible rate of emission to the estuary can be defined as that which will result in the maximum permissible dose equivalent as determined by the exposure pathway model. Compliance with the standard can then be ascertained by monitoring the rate of discharge from the facility in question. Methods of exposure pathway analyses are described in Chapters 4 and 5 and environmental surveillance is treated in Chapter 13.

MONETARY VALUE OF RISK REDUCTION

There have been attempts to assign a lower limit to the need for further dose reduction in terms of the maximum amount of money that should be expended per unit of collective dose averted. When this approach was reviewed by the ICRP (1983), it was noted that there has been a wide range of suggested values, ranging from $10 to $1000 per person-rem ($1000–100,000 per person-Sv). The Health Physics Society (U.S.) recommends $20–$40 per person-rem for health physics decisions (Kathren *et al.,* 1993a). The NRC recommends a value of $1000 per person-rem ($100,000 per person-Sv) averted when considering reductions of effluents from nuclear facilities (NRC, 1993). A more recent review, in the context of decision making for evacuations, found a range of values per person-dose averted of $34,000 to $250,000 per person-Sv ($340–$2500 per person-rem) (Merwin and Darwin, 1991). A value of $100,000 per person-Sv implies an expenditure of $2 million per fatal cancer averted, if the risk figure of the ICRP

(1991) of 5×10^{-2} per Sv (5×10^{-4} per rem) is used. An active debate on this subject continues in the literature (Kathren, 1993a,b; Puskin, 1993).

Some persons find it repugnant to equate death to dollars, but this is done in many ways in the field of public health. The number of ambulances provided for a city, the number of firemen and policemen, or the amount spent for health education are all determined by weighing costs against benefits measured by injuries or deaths averted. Thus, Cohen (1980) has estimated that the cost per fatality averted is about $65,000 for rescue helicopters and $34,000 for improved highway guard rails. It has been estimated that compulsory installation of smoke detectors in bedrooms would save lives at a cost of $40,000 per death averted (Graham and Vaupel, 1981). Society seems prepared to spend far more to prevent a hypothetical death from low-level radiation exposure than from many of the more commonplace hazards that exist. This will be discussed further in Chapter 16.

ORGANIZATIONS INVOLVED IN ESTABLISHING AND IMPLEMENTING RADIATION PROTECTION STANDARDS

Although there is reasonably good agreement among the various scientific groups about many of the quantitative aspects of radiation injury, there is less agreement among the regulating and regulated groups as to how the basic recommendations that issue from the scientific bodies should be administered in practice. In other words, how does one relate the standards based on epidemiological and experimental evidence to the apparatus of public health regulation? Public health administration is more an art than a science. The health administrator is accustomed to starting with a mixture of scientific information of various grades of quality that must be evaluated according to existing concepts of acceptable risks. He or she must then construct a system of regulation that is understandable, that is practical, and that, above all, protects public health.

Public health officials are accustomed to making compromises. Thus, standards for the biological quality of drinking water are often based on an innocuous group of coliform organisms because one is basically interested in certain pathogenic organisms that originate in fecal pollution for which the coliform organisms are a useful indicator. This system of control does not provide absolute safety, but in most cases it provides adequate safety. For many decades this has been a practical system of control that has lent itself to practical systems of enforcement. Other regulatory mechanisms could be designed to provide more safety, but they might not be practical to administer or enforce.

The Nuclear Regulatory Commission

Primary responsibility for regulating the nuclear energy industry in the United States rests with the NRC, which was formed in 1974 and given the responsibilities for health and safety that were originally assigned to the AEC.

The NRC has responsibility for regulating the civilian nuclear power industry, including the design and operation of power reactors, fuel manufacture, spent-fuel processing, and waste management. In addition, the NRC regulates the production and use of radioactive materials ("isotopes") that are produced by reactors and are used in research, industry, and medicine (Minogue, 1978). Under the law, NRC has the authority to transfer to the states its responsibilities regarding the regulation of reactor-produced radionuclides. The radionuclides that occur naturally or are produced in a cyclotron or other particle accelerators are not covered by NRC regulation (Moghissi *et al.,* 1978a).

The NRC and DOE rules and regulations are published in Title 10 of the Code of Federal Regulations, which is divided into many parts, most of which do not involve radiation protection. The parts that are most concerned with radiation protection are listed in Table 3-5.

Environmental Protection Agency

The environmental functions of the federal government were reorganized in 1970 by creation of the EPA, within which a radiation group was established and given responsibility for the development of radiation protection guidelines and environmental radiation standards and regulations. It also administers many of the regulations. This relieved the NRC of the responsibility for the establishment of environmental standards and also took over the functions of the Federal Radiation Council, which was an interagency council established during President Eisenhower's administration. Surveillance functions that were formerly the responsibility of the Public Health Service were also transferred to EPA. Thus, the EPA has the responsibility for establishing federal standards of radiation protection and also maintaining surveillance of air, food, and water. The NRC continues to have responsibility (within its jurisdiction) for issuing the regulations by which the EPA standards will be achieved. If EPA specifies that the dose to members of the general population should not exceed a given level, it is the responsibility

TABLE 3-5

Relevant Regulations Contained in Title 10 of the Code of Federal Regulations[a]

Part	Regulation
20	Standards for protection against radiation
30	Rules of general applicability to domestic licensing of by-product material
31	General domestic licenses for by-product material
32	Specific domestic licenses to manufacture or transfer certain items containing by-product material
33	Specific domestic licenses of broad scope for by-product material
35	Medical use of by-product material
39	Licenses and radiation safety requirements for well logging
40	Domestic licensing of source material
50	Domestic licensing of production and utilization facilities
51	Environmental protection regulations for domestic licensing and related regulatory functions
60	Disposal of high-level radioactive wastes in geologic repositories
61	Licensing requirements for land disposal of radioactive waste
70	Domestic licensing of special nuclear material
71	Packaging and transportation of radioactive material
72	Licensing requirements for the independent storage of spent nuclear fuel and high-level radioactive waste
100	Reactor site criteria
835	Occupational radiation protection (for the DOE)

[a]U.S. Office of the Federal Register.

of the NRC to prepare regulations that will assure compliance with the EPA limit.

FOOD AND DRUG ADMINISTRATION

The Center for Devices and Radiological Health (CDRH), formerly the Bureau of Radiological Health, within the Food and Drug Administration (FDA), has responsibility for developing safety standards for a number of radiation sources not controlled by NRC. These include naturally occurring radioactive materials used in consumer products, X-ray generators, and radionuclides produced by accelerators (in contrast to reactor-produced radionuclides, which are regulated by NRC).

The regulations of the CDRH typically apply directly to the manufacturers of radiological devices. Recommendations applying to the use of devices do not have the force of law, since responsibility for regulation in its area of interest resides in the states. However, CDRH fulfills the important function of assuring uniformity by developing standards that can then be adopted by the states (Neill, 1978).

THE STATES

Under federal law, the NRC is permitted to delegate certain of its health and safety responsibilities to the states, a number of which now control radionuclide use. A basic condition is that the state should have the technical capability to administer its regulations, which must be compatible with those of NRC. Occasionally a state will drop its agreement and regulation then reverts to the NRC. Typically about half of the 50 states have active agreements with the NRC. In some instances, such as the transportation of radioactive materials, local jurisdictions have established regulations without any delegation of authority from the NRC. Dual jurisdiction can thus exist and, although most states have adopted the basic standards proposed by the NRC, differences in interpretation and procedures can result in some confusion. The states are not permitted to issue standards that are less stringent than those of NRC, but the courts have allowed the states to promulgate standards that exceed the NRC requirements.

MUNICIPALITIES

Large cities, such as New York City, may have the authority to regulate radiation sources. In general, they tend to implement authority by means of codes similar to those used by the states.

DEPARTMENT OF TRANSPORTATION

The regulations governing shipment of radioactive materials are the responsibility of the Department of Transportation (DOT) but NRC will enforce these regulations. DOT regulations are published in Title 49 of the Code of Federal Regulations. Equivalent regulations are published for the NRC in Title 10, Part 71 of the Code of Federal Regulations. This is discussed in more detail in Chapter 10.

INTERNATIONAL AGENCIES

A number of international organizations issue recommended safe practices guidelines from time to time. Among these are certain specialized agencies of the United Nations, such as the International Labor Organization, the

World Health Organization, the International Atomic Energy Agency, and the Food and Agriculture Organization. These organizations do not have the authority to require adoption of their recommendations by any of the member countries, although adoption may be made a condition by some agencies for member countries to continue to receive technical assistance or other agency benefits.

CHAPTER 4

Atmospheric Pathways

This chapter will acquaint the nonmeteorologist with the basic principles of the transport, dispersion, and deposition of pollutants that are released to the atmosphere. Atmospheric calculations, computer codes, and comparisons of calculations with measured data will also be discussed. The treatment will be superficial but will provide an understanding of some of the methods used in this highly specialized field. Radiological assessment that uses comprehensive models to determine transport through environmental pathways and to assess dose to human receptors is discussed in Chapter 14.

PROPERTIES OF THE ATMOSPHERE

Until World War II, man-made atmospheric contaminants were injected only into the friction layer, which extends to about 1000 m from the ground and which has properties somewhat different from those of the rest of the atmosphere because of the influences of surface features on atmospheric flow. Transport and mixing within this layer have been studied mainly by meteorologists concerned with dilution of industrial effluents (Randerson, 1984).

The advent of nuclear energy, and more particularly the testing of nuclear weapons, extended the problem of forecasting the fate of atmospheric contaminants considerably above the friction layer to altitudes of

30,000 m or greater. In more recent years, the properties of the upper atmosphere, even to the fringes of outer space, have become important to space scientists. An understanding of the physics and chemistry of the upper atmosphere is also required to predict the physical, chemical, and biological effects of emissions from high-flying aircraft.

Although the atmosphere contains many gases, as shown in Table 4-1, more than 99.9% of its weight is contributed by nitrogen, oxygen, and argon. The relative proportions of these gases remain constant to great heights, but separation due to differences in molecular weight does occur above 60 km. The total mass of the dry atmosphere is estimated to be about 50×10^{17} kg, to which may be added about 1.5×10^{17} kg of water vapor, the most variable constituent of the atmosphere and the one that governs many of its thermodynamic characteristics. Dry air has a density of 1.3 mg cm^{-3} at the surface of the earth, where pressure from the weight of the atmosphere is 101 kPa (760 mm of mercury). At 50 km above sea level, the atmospheric pressure has dropped to 10^{-3} of the sea-level pressure and has dropped by a factor of 10^{-6} at an altitude of 100 km. The thinness of the atmosphere at these altitudes is illustrated by the length of the mean free path between molecules, about 2.5 cm at 100 km and about 25 m at 300 km (Petterssen, 1968). Above an altitude of about 600 km, the molecules are thought to behave as satellites in free elliptical orbits about the earth (Mason, 1982).

TABLE 4-1

Average Composition of the Atmosphere[a]

Gas	Composition by volume (ppm)	Composition by weight (ppm)	Total mass ($\times 10^{22}$ g)
N_2	780,900	755,100	38.648
O_2	209,500	231,500	11.841
A	9,300	12,800	0.655
CO_2	300	460	0.0233
Ne	18	12.5	0.000636
He	5.2	0.72	0.000037
CH_4	1.5	0.9	0.000043
Kr	1	2.9	0.000146
N_2O	0.5	0.8	0.000040
H_2	0.5	0.03	0.000002
O_3[b]	0.4	0.6	0.000031
Xe	0.08	0.36	0.000018

[a]Adopted from Mason (1960). Copyright © 1960, John Wiley and Sons, Inc., reprinted by permission of John Wiley and Sons, Inc.
[b]Variable, increases with height.

The atmosphere contains aerosols that originate from many sources. In addition to the air pollutants introduced as the result of human activities, meteorites, volcanic activity, dust storms, forest fires, and ocean spray contribute great quantities of gases and suspended solids.

A gas, vapor, or aerosol introduced into the atmosphere is diluted by molecular and turbulent diffusion. One can neglect the effect of molecular diffusion, for which the diffusion coefficients are many orders of magnitude smaller than those due to turbulence. The total range of values of the diffusion coefficients that control the rates of atmospheric dilution is enormous, ranging from 0.2 cm^2 s^{-1} for molecular diffusion to 10^{11} cm^2 s^{-1} for the mixing due to large-scale cyclonic storms in the atmosphere (Gifford, 1968). The atmospheric motions that contribute to the mixing processes thus vary from the microscopic to those measured in hundreds of kilometers.

The motions of turbulent diffusion are so complicated that exact mathematical theories are not available to describe the manner in which a contaminant behaves in space and time when it is introduced into the atmosphere. However, methods have evolved that are basically statistical in nature and that make it possible for one to predict the manner in which a contaminant will diffuse in the atmosphere under a given set of meteorological conditions.

The mixing characteristics of the atmosphere are influenced in a major way by its vertical temperature gradient. A typical temperature profile in the temperate zone is illustrated in Fig. 4-1. As height above the ground increases, the temperature normally decreases at a rate, often referred to as the "lapse rate of temperature," of about 6.5°C km^{-1} (3.5°F per 1000 ft). It is seen from Fig. 4-1 that the temperature decreases with height to

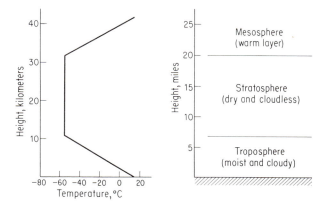

FIGURE 4-1   Idealized profile of the atmospheric temperature gradients. [From Petterssen (1958). "Introduction to Meteorology," 2nd Ed. © 1958 McGraw-Hill.]

about 11 km, which is the approximate beginning of an isothermal region of the atmosphere that, in the example shown, extends to a height of about 32 km. The lower region of the atmosphere, in which the temperature normally decreases with height, is called the troposphere: it contains about 75% of the mass of the atmosphere and almost all its moisture and dust. Above the troposphere, separated by a sharp change in lapse rate at the so-called tropopause, is the isothermal stratosphere. The height of the tropopause varies with latitude and with the season of the year. In contrast to the relatively smooth and quiet motions of the stratosphere, the troposphere is a comparatively unstable, well-mixed region of the atmosphere. Above the stratosphere, beginning at about 30,000 m, is the mesosphere, a region characterized by an increasing temperature gradient. The mesosphere extends upward to the ionosphere, a densely ionized region that begins at a height of about 65,000 m.

During normal daytime conditions, the earth's surface absorbs solar radiation and becomes warmer than the overlying atmosphere. The temperature gradient then becomes superadiabatic; that is, the temperature decreases at a rate exceeding that which would occur if a parcel of air is permitted to expand adiabatically. The dry adiabatic temperature gradient is normally about $-6.5°C$ km^{-1}. The influence of the vertical temperature profile on the stability of the atmosphere can be understood from Fig. 4-2. If a parcel of air having a temperature T_1 at altitude H_1 is raised to altitude H_2, it will cool at the adiabatic rate. Since it is assumed in Fig. 4-2 that

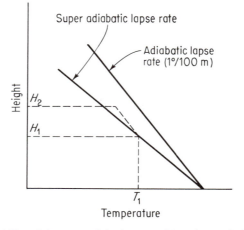

FIGURE 4-2   Instability of the superadiabatic atmosphere. A parcel of air raised in height from H_1 to H_2 cools adiabatically, and its rate of rise is accelerated because it becomes warmer and, therefore, less dense than the ambient atmosphere.

superadiabatic conditions exist, the parcel of air initially at H_1 will be warmer than the ambient atmosphere when it reaches H_2. It will thus be of lower density than the surrounding air and will continue to rise. Similarly, if a parcel of air is lowered in altitude when the atmosphere is in the superadiabatic condition, the parcel will become more dense than the ambient atmosphere and will continue to descend. Under superadiabatic conditions, all vertical motions tend to be accelerated, and the atmosphere is said to be unstable. Figure 4-3 shows that the reverse situation exists when the lapse rate is less than adiabatic.

In fact, as shown in Fig. 4-3, the temperature gradient may increase with height. This results in the highly stable condition called an inversion. Inversions may be caused by the overrunning of warm air over cold (as along a front between air masses), by advection of cool air at low level (as during a sea breeze), or by diurnal cooling of the lower layers of the atmosphere. This last condition usually develops after sunset, when the surface of the earth, having been warmed during the day, begins to cool more rapidly than its overlying atmosphere. As the night proceeds and the surface of the earth becomes cooler relative to the atmosphere, the temperature gradient may in time become positive.

This sequence of events is illustrated in Fig. 4-4 (Holland, 1953), which gives an example of how the vertical temperature gradient can change in a 24-h period. A sharp inversion to a height of about 500 ft (150 m) is evident at 0600 h and has begun to weaken slightly at 0700. By 0800 the

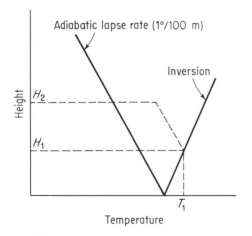

FIGURE 4-3 Inherent stability of the inverted temperature gradient. A parcel of air raised in height from H_1 to H_2 cools adiabatically and sinks to its original position because it becomes more dense than the ambient atmosphere.

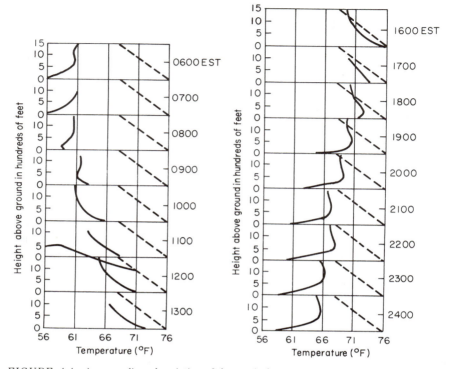

FIGURE 4-4   Average diurnal variation of the vertical temperature structure at Oak Ridge National Laboratory, Oak Ridge, Tennessee, during the period September–October 1950 (solid lines). The dashed lines represent the adiabatic lapse rate. [From Holland (1953).]

morning sun has heated the ground, and the inversion begins to disappear, being replaced by a superadiabatic gradient between 1100 and 1300 h. By late afternoon, as the ground cools, a surfaced-based inversion develops again and persists throughout the night. If the sun rises on a clear day, the superadiabatic condition will again develop by late morning, and the cycle will be repeated. However, should cloud cover limit solar heating of the earth's surface, the inversion may persist through the day.

The change in temperature profiles greatly affects the characteristics of plumes of stack gases, and it can be appreciated from Fig. 4-5 that the concentration of a contaminant at ground level is influenced by the effect of the vertical temperature gradient on atmosphere stability.

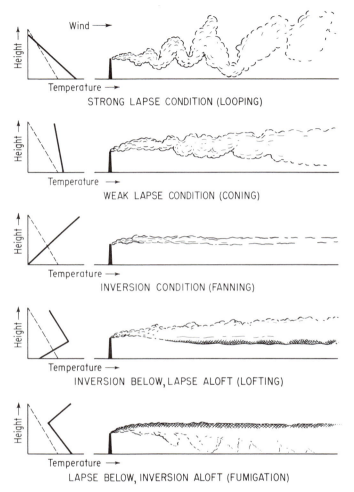

FIGURE 4-5 Schematic representation of stack-gas behavior under various conditions of vertical stability. Actual temperature (solid line) and dry adiabatic lapse rate (dashed line) are shown. [From U.S. Weather Bureau (1955).]

TURBULENT DIFFUSION IN THE FRICTION LAYER FROM A CONTINUOUS-POINT SOURCE

There is no exact mathematical description of turbulent diffusion in the atmosphere, but a number of approximations have been developed during the past 50 years. Most of them have in common the assumption that the mean concentration distribution across the vertical and horizontal axes of the plume cross section follows a Gaussian function, although statistical, similarity, and gradient transport models have also been used in certain instances (Hanna et al., 1982). For the Gaussian approximation, the standard deviations of the mean concentration along the vertical and crosswind axes, σ_y and σ_z, respectively, depend critically on the degree of turbulence. Both these turbulent diffusion-width parameters increase with downwind distance from a point source.

An important application of these equations is to describe the rates of turbulent diffusion from a continuous-point source of emission. Versions of these equations have proliferated over the years with development of computerized methods of handling complex, multifactorial processes.

The uncertainties of the models for prediction of downwind concentrations from a continuous-point source vary with the type of terrain, the period of time over which observations are made, and distance from the source. For flat terrain, the models provide predictions that are valid within a factor of 2 or 3 to distances of 10 km (Crawford, 1978; Little and Miller, 1979; NCRP, 1984a). Carhart et al. (1989) have evaluated eight models designed for longer-range prediction and found that six were able to predict observed concentrations to within a factor of two. In general, models tend to be reasonably accurate for flat areas, but less accurate for complex terrain.

In addition to the assumption of a Gaussian distribution, the turbulent diffusion models all assume that the concentration along the downwind axis of the plume will be directly proportional to the rate of emission of a pollutant and inversely proportional to the average wind velocity. Maximum ground level concentrations are roughly proportional to the inverse square of the effective height of emission (Hanna et al., 1982). The effective height of the emission is the height of the source above the ground plus the rise due to any initial buoyancy of the plume.

One use for turbulent diffusion equations is to estimate the concentration of a pollutant along the centerline of a plume at a point downwind from an elevated point source of emission. If the estimate of the maximum concentrations proves to be well within the regulatory limits, no further calculations may be required.

A frequently used equation was developed during early research on turbulent atmospheric diffusion (Sutton, 1953) to describe the ground-level concentration downwind from a continuously emitting point source:

$$\chi(x,y) = \frac{Q}{\pi\ \sigma_y\ \sigma_z\ \overline{u}} \times exp - \left(\frac{h^2}{2\sigma_z^2} + \frac{y^2}{2\sigma_y^2} \right) \qquad (4\text{-}1)$$

where χ = concentration (Bq or Ci m^{-3}) near ground level at downwind point (x,y); Q = source strength (Bq or Ci s^{-1}); σ_y and σ_z = crosswind and vertical plume standard deviations (m), both functions of x; $\overline{u}$ is the mean wind speed (m s^{-1}) at the stack elevation, h_s (m); h = effective stack height ($h_s + \Delta h$, the plume rise) (m); and x,y are the downwind and crosswind distances (m).

Numerical values for σ_y and σ_z will depend mainly on the degree of atmospheric stability, but are also affected by the roughness and configuration of the terrain, as well as the presence and arrangement of man-made obstacles such as buildings. Therefore, if high accuracy is required, site-specific turbulent diffusion coefficients must be determined (Randerson, 1984). However, the values of σ_y and σ_z can be estimated according to stability classification and distance from the source, by using Figs. 4-6 and

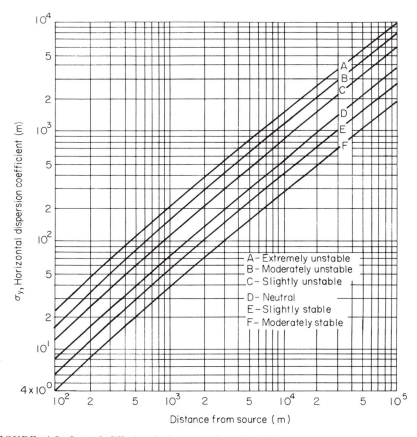

FIGURE 4-6 Lateral diffusion (σ_y) versus downwind distance from source for various turbulence types. [From Gifford (1968).]

4-7 as proposed by Gifford (1968). The turbulence classification, based on the degree of atmospheric stability, can be approximated from Table 4-2.

Many authors have developed curves of concentration normalized to wind speed and emission rate $(\overline{u}\chi/Q)$ for various stack heights. A widely used set of these nomograms can be found in the handbook by Turner (1970). These curves, which are reproduced in Figs. 4-8 to 4-10, illustrate the principal general features of all equations that describe the turbulent diffusion of stack effluents. For a given ambient wind speed, $\overline{u}$, concentration maxima occur at increasing distances with increasing values of h and with increasing degrees of stability. The maximum concentration decreases in inverse proportion to h^2.

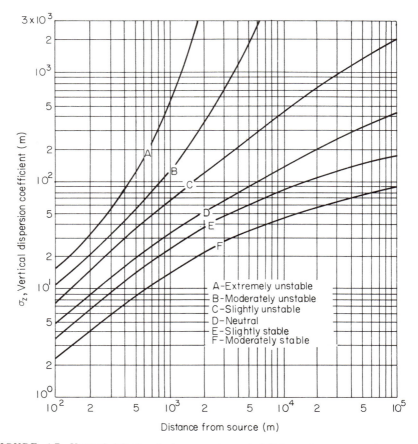

FIGURE 4-7 Vertical diffusion (σ_z) versus downwind distance from source for various turbulence types. [From Gifford (1968).]

TABLE 4-2

Relation of Turbulence Types to Weather Conditions[a,b]

Surface wind speed (m s[-1])	Daytime insolation			Nighttime conditions	
	Strong	Moderate	Slight	Thin overcast or ≥4/8 cloudiness[c]	≤3/8 cloudiness
<2	A	A–B	B	—	—
2	A–B	B	C	E	F
4	B	B–C	C	D	E
6	C	C–D	D	D	D
>6	C	D	D	D	D

[a]From Gifford (1968).
[b]Conditions: A, extremely unstable; B, moderately unstable; C, slighty unstable; D, neutral (applicable to heavy overcast, day or night); E, slightly stable; F, moderately stable.
[c]The degree of cloudiness is defined as that fraction of the sky above the local apparent horizon that is covered by clouds.

The turbulent diffusion equation can be used to predict the maximum ground-level concentration that will occur downwind from a stack. If σ_y and σ_z are assumed to be linearly related, the derivative with respect to x of

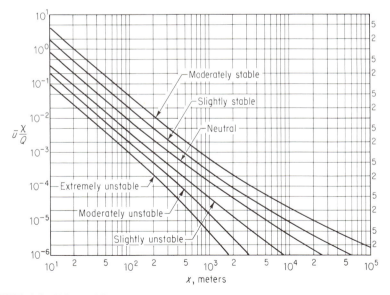

FIGURE 4-8   Values of $\bar{u}\chi/Q$ as a function of downwind distance for a source located at the surface. [From Hilsmeier and Gifford (1962).]

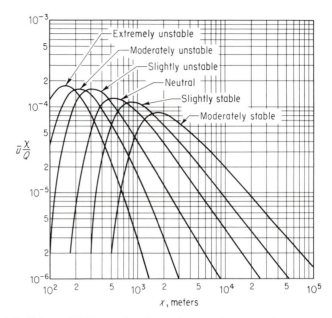

FIGURE 4-9 Values of $\bar{u}\chi/Q$ as a function of downwind distance for a source located at a height of 30 m. [From Hilsmeier and Gifford (1962).]

Eq. (4-1) can be set equal to zero to find the maximum value of the downwind concentration, χ_{max}:

$$\chi_{max} = \frac{Q}{e\pi\bar{u}h^2}\frac{\sigma_z}{\sigma_y} \tag{4-2}$$

where e is the base of the natural logarithms (2.72). The downwind distance from the stack to χ_{max} often is found in the range of 15–20 stack heights, but can occur much closer, as will be discussed.

EFFECTIVE STACK HEIGHT

Pollutants are frequently emitted to the atmosphere at an elevated temperature and with considerable vertical velocity. This is the normal situation for a fossil-fueled power plant as well as for stacks emitting waste gases from many industrial processes. The buoyancy from the combined effects of the temperature and velocity of the exhaust gas may result in a stack height that is considerably higher than the actual height. (Thus, large fossil fuel power plants are designed on the basis of an effective stack height of about $2h$.) The problem of combining the various meteorological parameters

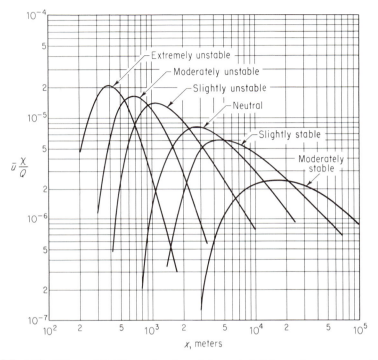

FIGURE 4-10 Values of $\bar{u}\chi/Q$ as a function of downwind distance from a source located at a height of 100 m. [From Hilsmeier and Gifford (1962).]

together with the effluent temperature and velocity is a complicated one because of the way in which the various parameters interact. For example, though the effect of a temperature inversion is to suppress vertical motions and might therefore be expected to reduce the effective stack height, the inversion reduces turbulent mixing to an extent that results in a more persistent temperature differential between the plume and the surrounding atmosphere. Consequently, the plume continues to rise at a slow rate for a considerable time while slowly spreading downwind. The maximum concentration will be many kilometers from the stack. Conversely, during unstable conditions, with greatly enhanced vertical motions, plume "looping" may occur, and the maximum concentrations may be very close to the stack.

Many investigators have attempted to develop theoretical and empirical methods of computing the effective stack height, taking into consideration the amount of momentum and heat in the plume and the ambient atmospheric conditions, including wind speed and vertical temperature gradient

(Briggs, 1984; Electric Power Research Institute, 1982). The equations that have been developed are sufficiently complicated and difficult to apply that their use will ordinarily be limited to the micrometeorological specialist. When turbulent diffusion equations are used without allowing for the effect of plume buoyancy, the estimated concentrations will usually err on the safe side. However, the presence of low pressure in the wake of the stack or nearby downwind buildings can cause a downwash of the plume, actually reducing effective stack height to less than the physical stack height (Hanna *et al.*, 1982).

EFFECT OF BUILDINGS AND TERRAIN ON PLUME DISPERSION

One must be cautious to avoid applying these formulas if the flow patterns are apt to be affected by buildings or local topography. Buildings close to the stack may result in the conditions shown in Fig. 4-11, which is a wind tunnel photograph of the effect of local structures on the dispersion of a plume. Micrometeorology in the vicinity of buildings is often too complex for analytical solutions, and in many instances the question of how high a stack should be to avoid the downwash in the lee of a building is best answered by wind tunnel tests.

FIGURE 4-11 Perturbation of stack plumes by buildings. Note the effect of increased stack height in eliminating the downwash in the lee of the building. (Courtesy of Professor Gordon Strom.)

The general characteristics of the wake downwind of a structure are shown in Fig. 4-12 (Halitsky, 1968), which depicts the flow near a rounded building. Three distinct zones exist: the displacement zone, the wake, and the cavity. The displacement zone is the volume in which the air is deflected around the solid building. Immediately downwind of the building is a region of toroidal circulation known as the cavity or eddy zone in which it is possible for high concentrations to accumulate. Beyond the cavity is the true wake, which is a region of high turbulence within which the contaminant spreads throughout. If the area of the wake cross section is taken as the projected area of the building (A), and if it is assumed that the contaminant is mixed uniformly across the wake section, the maximum concentration at the distance of wake formation is

$$\chi = \frac{Q}{Au}$$

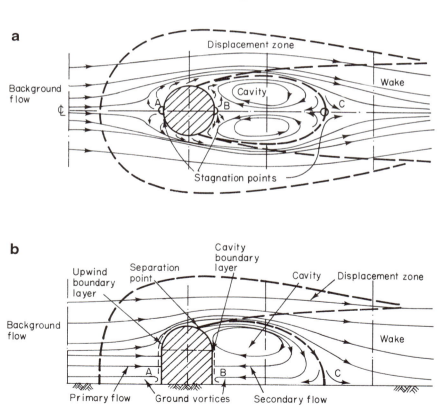

FIGURE 4-12 Flow around a rounded building. (a) Flow in a horizontal plane near the ground; (b) flow in the longitudinal center plane. [From Halitsky (1968).]

This is a highly oversimplified equation and must be modified to match the complexities of real situations. Various approaches to treating flow around buildings, including upwind sources and sources near the building, are discussed by Hanna *et al.* (1982). Hosker (1982) has assembled data on wakes in the lee of arrays of buildings of various shapes in useful handbook form.

Mountains and valleys also tend to distort flow under certain conditions, as shown in Fig. 4-13. There are occasional situations in which inversions in combination with terrain features result in meteorological isolation of an area in the manner shown in Figs. 4-14 and 4-15. Dickerson *et al.* (1984) summarized the research results available up to that time for terrain effects on atmospheric pollutant transport and turbulent diffusion, and studies in this field continue.

COMPUTER MODELS FOR ATMOSPHERIC DISPERSION CALCULATIONS

In practice, computer codes are almost always used to estimate downwind pollution. Many computerized formulations of various atmospheric dispersion models have been developed over the years. Meteorologists at the DOE's Rocky Flats Plant evaluated 37 off-the-shelf dispersion models prior to creating their own site-specific model (Hodgin, 1991).

Codes that are widely used to estimate environmental radioactivity concentrations from point source releases and to demonstrate regulatory compliance with environmental radiation standards are AIRDOS (Moore *et al.,* 1979) and it successor, CAP88 (Parks, 1992). These codes, which are available in versions for personal computers to supercomputers, utilize the straight-line Gaussian turbulent diffusion model. Conservative default values are provided for many meteorological parameters. Users can tailor the codes to site-specific situations by entering site-specific parameter values. Conser-

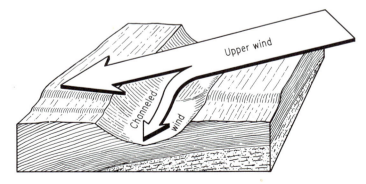

FIGURE 4-13   Wind channeling by valley walls. [From U.S. Weather Bureau (1955).]

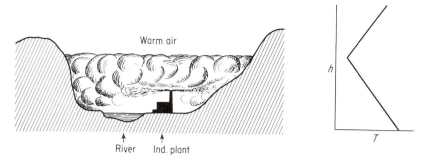

FIGURE 4-14 Fumigation of valley floor caused by an inversion layer that restricts diffusion from a stack. [From U.S. Weather Bureau (1955).]

vative in this context means values of model parameters that will result in larger estimated concentrations than best estimate values. The purpose of using conservative parameters is to avoid underestimating concentrations.

Simpler codes such as COMPLY (EPA, 1989b; Colli, 1990), which is based on recommendations of the NCRP (1989d), are available for facilities that are unlikely to have a problem meeting environmental radiation standards but that must still perform screening calculations to demonstrate compliance. A version of COMPLY (COMPLY-R) is even available for radon emissions. These codes have the advantage of being approved by the EPA.

More Complex Atmospheric Dispersion Models

The relatively simple straight-line Gaussian model embodied in the codes for atmospheric dispersion just discussed is adequate for estimating

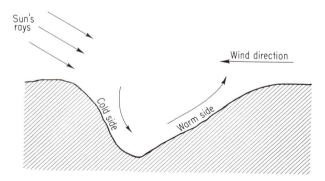

FIGURE 4-15 Atmospheric overturn caused by uneven solar heating of valley walls. [From U.S. Weather Bureau (1955).]

concentrations up to 10 km downwind from a continuous source in relatively flat terrain. There is no possibility for taking into account curvature in the wind direction. However, the wind does not always blow in the same direction, nor is all terrain flat.

The next level of sophistication in atmospheric dispersion models is the puff trajectory model. In this model, releases are modeled as a series of puffs that grow in size as they are carried with the wind, with consequent dilution of the entrained contaminant. Several models for puff growth, including Gaussian and Lagrangian, are used. The concentration of material at the point of interest is modeled as the average concentration of the puffs that pass over the point of interest during a prescribed time interval.

The path of each puff is determined by the wind direction and velocity in the vicinity of the puff. Inherent in this type of model is the need for more detailed meteorological information than for the straight-line Gaussian model. In typical applications, data from tens of meteorological towers are computer-modeled to produce a gridded wind field map that gives, for each grid sector, the wind direction and speed at specified intervals of time. Each puff is moved by the model for the specified time with a speed and direction obtained from the nearest wind field grid sector. The rate of growth of the puff is determined by atmospheric stability.

An early example of the puff trajectory model is MESODIF (MESO-scale Diffusion Model), developed by the National Oceanic and Atmospheric Administration (Start and Wendell, 1974) for Idaho Engineering Laboratory (INEL). MESODIF uses a wind field grid generated from data from 26 meteorological towers. This grid is updated every six minutes and the puffs are moved by the model at six-minute intervals.

Many computer models for atmospheric dispersion calculations with various improvements have been described over the years. The best, of course, are site-specific, which consequently limits their direct use in other locations.

One of the most complex and well-documented site-specific models is the Terrain Responsive Atmospheric Code (TRAC) employed at the DOE's Rocky Flats Plant located near the cities of Boulder and Denver, Colorado (Hodgin, 1985, 1991). Several factors led to the development of this complex model, including the close proximity of buildings to the release points, complex terrain and constantly changing atmospheric conditions along the Colorado front range, the large nearby population densities, the large inventory of hazardous materials at the plant, and continuous pressure from citizens' groups.

Examples of the complex features employed by TRAC, but not utilized by MESODIF, include: (1) treatment of resuspended particles; (2) plume depletion by gravitational settling and wet and dry deposition; (3) accounting for the ingrowth of radioactive decay products; (4) use of multiple atmospheric layers; (5) use of local grid values for mixing depth and atmospheric stability versus hourly climatological values; and (6) explicit treatment of terrain effects.

Comparisons among Models and Environmental Measurements

Field tests of the INEL MESODIF model and comparisons with environmental measurements have given generally satisfactory results (Start *et al.,* 1985). Although the environmental tracer studies indicated some underprediction, opportunities to compare model predictions with several small releases showed that overprediction was the rule, occurring five times out of seven. The ratio of predicted to measured values ranged from 0.5 to 5 (DOE-ID, 1987, 1988, 1989). Winds at INEL are diurnal, blowing generally from the southwest during the day and from the northeast at night, with additional complications caused by drainage winds from northwest-trending mountain valleys. As a result, air contaminants can pass over the same location more than once before dilution renders them inconsequential. Comparisons made between MESODIF and standard straight-line Gaussian models such as AIRDOS or CAP88 invariably show that MESODIF predicts higher maximum concentrations, typically on the order of a factor of five, than the standard models. These results illustrate the limitations of the straight-line Gaussian models in complex meterology.

Several major comparisons have been made among models and between models and environmental measurements of released tracers. (Weber and Garrett, 1985; Lewellen *et al.,* 1985, 1987; Carhart *et al.,* 1989). In an experimental field evaluation of longer-range models at distances from 28 to 600 km, Carhart *et al.* (1989) reported that six of eight models evaluated were able to predict observed concentrations to within a factor of two. The models exhibited a tendency to overpredict the observed values and to underpredict plume spread. Lewellen *et al.* (1985, 1987) also made comparisons among models using the same sets of input data. One general conclusion that can be drawn from the various evaluations of models is that model accuracy is limited to about a factor of two. Another conclusion is that, even though the more complex models include more physically realistic treatments of many of the factors affecting atmospheric dispersion, they offer little if any improvement in ability to predict concentrations over the less complex models. On reviewing the various model evaluations, Ramsdell (1991) concluded that increases in model complexity beyond the Lagrangian puff model were accompanied by increases in time and cost, but not by accuracy.

TURBULENT DIFFUSION OF ACCIDENTAL RELEASES IN THE LOWER ATMOSPHERE

Accidental releases may involve pulsed discharges of radioactive contaminants to the atmosphere. When this happens, the rate of dispersion depends, as outlined earlier, on meteorological conditions and the buoyancy of the emission that, in stable air, is altered by the relative density of the entrained air. Because it is usually not practical to specify the length of time during which the release will take place, one can modify formula (4-1) by substituting Qt, the total number of Bq or Ci released, for Q, Bq or Ci discharged per unit time. This permits calculation of $\overline{\chi t}$ (Bq m^{-3} s or Ci m^{-3} s) in place of χ. This is a convenient form in which to express the calculated consequences of an accident, because $\overline{\chi t}$ is a measure of the total exposure, from which one may proceed directly to a computation of the integrated dose. As for the case of continuous releases discussed earlier, computer codes for computation of concentrations resulting from accidental releases are widely available.

DISPERSION OF AEROSOLS

The turbulent diffusion equations were derived for gaseous effluents but may be applied to plumes in which the contaminants are in the form of aerosols, provided the particle sizes are such that the settling rates are insignificant compared to the scale of vertical motion due to turbulent mixing. In general, it may be assumed that the turbulent diffusion equations will apply to particles small enough to be deposited in the pulmonary tract.

DEPOSITION AND RESUSPENSION

The mechanisms by which particulates will deposit on surfaces vary, depending on whether or not the cloud passage is associated with precipitation. Dry deposition results from gravitational settling and impaction on surfaces exposed to the turbulent atmospheric flow. When precipitation occurs below the rain-forming level, the dust is washed to the surface by falling raindrops (washout). At higher altitudes, dust particles may serve as nuclei for condensing raindrops (rainout), a phenomenon that is responsible for removal of most submicron particles from the atmosphere.

The descent of particles through the atmosphere follows certain well-known physical laws (Drinker and Hatch, 1954; Hinds, 1982) that govern the resistance of the motions of particles moving in viscous media. The terminal velocity of a falling particle will be reached when the "drag" on the particle due to the viscosity of the atmosphere is equal to the force of gravity. The terminal velocity of falling particles in still air is plotted for quartz in Fig. 4-16. Because of their irregular shape, the particles will actually fall at slightly lower velocities than spheres. A useful rule of thumb is that a 10-μm particle having a density of about 2.5 g cm^{-3} falls at a rate of about 0.5 cm s^{-1} (1 foot per minute) at sea level.

Suspended particles may be impacted when turbulent air flows across a solid surface. The efficiency of impaction is defined as the ratio of the number of particles deposited on the surface to those originally contained in the volume of air diverted. When air flows around a cylinder, the efficiency of impaction increases as the velocity increases and the diameter of the cylinder decreases (Chamberlain, 1955). Theories have been developed that explain the manner in which impaction occurs as a function of particle size and velocity when regular shapes such as cylinders and planes are involved, but it is not practical to apply these to the irregular surfaces one frequently finds in practice.

Another method of quantifying the rate of air-to-surface transfer is by use of deposition velocity (v_g), which was originally designed to predict the

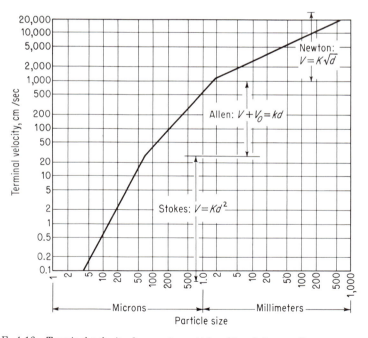

FIGURE 4-16   Terminal velocity for quartz particles (d = 2.6 g cm^{-3}) in air at sea level. [From Drinker and Hatch (1954). "Industrial Dust," 2nd Ed. © 1954 McGraw-Hill.]

transfer of iodine from the atmosphere to the ground (Chamberlain and Chadwick, 1966). The deposition velocity can be useful when settling velocities are inappreciable (i.e., for particle diameters less than about 10 μm), in which case

$$v_g = \frac{\text{amount deposited per square centimeter per second}}{\text{volumetric concentration per cubic centimeter above surface}}$$

where the units of v_g are cm s^{-1}. The deposition velocity incorporates all relevant physical and chemical processes into a single observable parameter.

In the absence of a theoretical basis for estimating v_g, many investigators have undertaken field studies designed to provide measurements of the deposition velocity for a variety of surfaces. These data have been summarized by Sehmel (1979, 1984) for both dusts and gases, and for a variety of meteorological conditions and types of ground cover. Unfortunately, the reported values of v_g are so variable that it is difficult to draw simple conclusions. It has been noted that the variation in v_g is influenced by the type of ground cover, and a "leaf area index" has been used by modelers to correct for this factor. The index varies from 0.5 for grassland to greater than 5 for raw crops. Other factors apparently introduce even greater uncertainty, because the reported values of v_g for dusts range over five orders of magnitude from 10^{-3} to 180 cm s^{-1}. The range of variability reported for the deposition velocities of gases is less (2×10^{-3} to 26 cm s^{-1}). Miller (1984) quotes Heinemann and Vogt (1979) as having determined that v_g for forage grasses is 2 cm s^{-1} for reactive gases such as molecular iodine, 0.1 cm s^{-1} for particles less than 4 μm, and 0.018 cm s^{-1} for unreactive gases. In view of the enormous variability in the reported results, it is not possible to conclude that the published values of deposition velocity are generally useful. However, most studies conclude that v_g for molecular iodine is about 2 cm s^{-1}, and since this is often the most important radioelement released to the atmosphere, the use of v_g can serve a useful purpose.

Washout (scavenging) of pollutants from a plume by precipitation can be important because the concentration in the atmosphere is in this way reduced downwind from the point of washout, where ground deposition is increased. The mechanisms involved in precipitation scavenging have been treated mathematically by many investigators and more than a dozen numerical codes have been published that describe the phenomenon (Slinn, 1984). The mathematical treatments have increased in complexity since the classic work of Chamberlain (1955) and the subject must remain, at least for the present, outside the scope of this text. Useful guidelines for predicting washout as a function of the type and rate of rainfall are not yet available.

After particles have been deposited on the ground, they may become resuspended as the result of wind action. This is another aspect of the general problem of airborne radioactivity that has thus far defied an analytical solution. Sehmel (1984) has listed several dozen factors that influence resuspension, including the specific characteristics of the soil, surfaces, topography, and prevailing meteorology. The resuspension factors (airborne concentration/surface contamination) reported from all field experiments vary by seven orders of magnitude. Sehmel notes that K, the resuspension rate for plutonium (the fraction of a deposit resuspended per second) at the Nevada Test Site, was shown to vary by about four orders of magnitude. However, the spread in the data from Nevada has been shown to be reduced when K is divided by u^3 (the third power of wind speed) (Anspaugh *et al.*, 1975).

It is difficult to see how generalized information about either deposition or resuspension rates can be used with confidence, because of the variety and complexity of the physical processes in any specific case, and the extreme difficulty involved in determining the host of relevant parameters required to model the process. However, in some cases it may be possible to develop useful information for a specific site and for a specific form of contamination.

TROPOSPHERIC AND STRATOSPHERIC BEHAVIOR

The emissions discussed thus far are from near-surface sources that are diffused over distances measured in tens to hundreds of kilometers. The concentrations beyond about a few hundred kilometers are usually of no interest for releases from industrial or research sources of emission. In contrast, when the contamination originates from nuclear explosions or a major violent accident, such as the one at Chernobyl, it is spread throughout the atmosphere on a global scale.

Our knowledge of stratospheric dispersion has been obtained from gas and dust samples obtained by aircraft or balloons penetrating to an altitude of about 35,000 m (Holland, 1959). Studies have been made of the distribution of ozone and water vapor and of a wide variety of radioactive substances, including debris from high-yield nuclear explosions. The transport of naturally occurring radionuclides such as ^{7}Be, ^{32}P, and ^{14}C, all of which are induced by cosmic-ray bombardment of the upper atmosphere, has also been studied. On two occasions, tracers have been incorporated into nuclear weapons exploded at unusually high altitudes. ^{102}Rh was injected into the stratosphere by an explosion 43 km above Johnston Island in the Pacific in August 1958, and the cloud from this nuclear explosion was believed to have risen to about 100 km. In July 1962 an explosion conducted about 400 km above Johnston Island contained a known amount of ^{109}Cd, which has been measured by investigators in many parts of the world (Krey and Krajewski, 1970).

The first significant stratospheric measurements were of the movement of stratospheric water and ozone by Brewer (1949) and Dobson (1956). Their observations were studied by Stewart *et al.* (1957), who developed a model of stratospheric–tropospheric exchange that is consistent with the observed pattern of nuclear weapons fallout that will be discussed in Chapter 9. According to this model (Newell, 1971; UNSCEAR, 1982), air enters the stratosphere in the tropical regions, where it is heated, and rises to an altitude of about 30 km, at which level it begins to move toward the poles. As shown in Fig. 4-17, the tropopause is lower in the polar regions than at the equator, and discontinuities in the tropopause in the temperate regions facilitate transfer from the stratosphere to the troposphere. The westerly jet streams occur at these discontinuities with transport velocities of 100 to 300 km h^{-1} and they are accompanied by vigorous vertical mixing. The rate of transfer from the lower stratosphere is most rapid in the winter and early spring.

The mean residence time of stratospheric aerosols consequently depends on the altitude at which it is introduced, the time of year, and

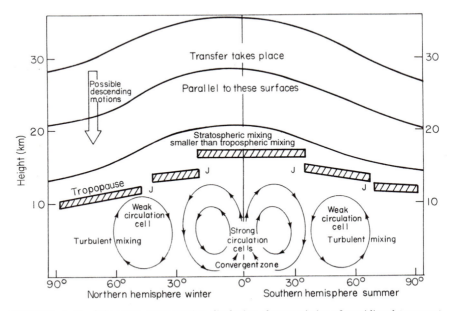

FIGURE 4-17 Schematic cross section displaying characteristics of meridional transport (J indicates tropical jet stream positions). [From UNSCEAR (1964).]

the latitude. Dust injected into the lower polar stratosphere by Russian thermonuclear explosions was found to have a mean residence time of less than 6 months, whereas in tropical latitudes the residence time has been found to be from 2 to 3 years for radionuclides introduced into the middle stratosphere and 5 to 10 years if injected at 100 km (see Chapter 9).

Sedimentation is evidently a significant factor in the thin stratospheric atmosphere since Feely (1960) has shown that ^{14}C injected as carbon dioxide has a longer residence time than ^{90}Sr injected in particulate form (see Chapter 9). It has also been suggested that the long stratospheric residence times of some nuclides may be due to their being reintroduced back into the stratosphere from the troposphere (UNSCEAR, 1977; Reiter, 1974).

Aerosols introduced into the troposphere are distributed by the planetary winds and deposit on the surface of the earth mainly by the scavenging action of rain. A remarkable correlation of ^{90}Sr deposition with rainfall was demonstrated on the Olympic Peninsula by Hardy and Alexander (1962) and will be discussed in Chapter 9. The mean residence time of dust injected into the troposphere is about 30 days on the average, but this can vary from 5 days for dust in the lower rain-bearing region of the troposphere to 40 days in the higher initial altitudes (UNSCEAR, 1977). Rainfall removes aerosols from the troposphere primarily by droplet formation around the particle (rainout) and also by a scrubbing action (washout).

There have been suggestions that ocean spray is effective in scavenging dust near the ocean–atmosphere interface, and that this phenomenon might explain reports that fallout of ^{90}Sr into the oceans is higher than that on land. However, field studies conducted by Freudenthal (1970a,b) concluded that ocean spray is not significant in this respect.

CHAPTER 5

Terrestrial and Aquatic Pathways

Contamination of land and water can occur from deposition of waste material originally introduced into the atmosphere, from discharge directly into surface or subsurface waters, or from wastes placed in or on the ground. Ground contaminants may eventually be mobilized by groundwater or erosion.

The primary reason for being concerned about radioactive contamination of the environment is that it may result in exposure of humans. The two major reasons for focusing concern on humans have to do with human values and the relative radioresistance of many plants and lower animals. Society places great value on individual human lives, which means that risks to individuals must be kept very low. With animal and plant populations, however, value is usually placed on the maintenance of healthy populations rather than on specific individual members of populations. For example, the EPA and other organizations perform ecological evaluations of waste sites that focus on populations and ecosystems rather than individual animals or plants. An exception occurs when threatened or endangered species are involved. In this case, value is placed on individual members of the population.

Many plants and lower animals are orders of magnitude more resistant to the effects of radiation than are humans (Blaylock and Trabalka, 1978;

Auerbach *et al.,* 1971; Templeton *et al.,* 1971). For most mammals, lethality (LD_{50}) can occur with acute doses in the 400–1100 rad (4–11 Gy) range and reproductive effects can be seen in some species at a tenth of these values. Most bird species exhibit higher values of LD_{50} and lower life-forms are much less radiosensitive, with insects typically requiring about 50,000 rad (500 Gy) for lethality.

It has been generally assumed that if we protect our own environment (air, water, land, food) sufficiently to protect ourselves, then natural systems would automatically be protected. This assumption was examined in detail by the IAEA (1992). The known effects of radiation on aquatic and terrestrial species, populations, and ecosystems were reviewed. It was found that although some individual organisms might be affected, measurable effects on populations were unlikely at levels of 1 mGy (100 mrad) per day for the most sensitive terrestrial species and 10 mGy (1 rad) per day for aquatic species. The NCRP (1991a) concluded that aquatic species are protected if their dose rate is limited to 0.4 mGy (40 mrad) per hour, equivalent to 9.6 mGy (~1 rad) per day.

Most national and international authorities recommend that environmental radiation doses to humans in the general population from facilities be limited to a maximum of 1 mSv (100 mrem) per year. Using this value as a starting point, models were used to examine likely doses to plants and animals in natural systems. It was found that doses to natural organisms were unlikely to exceed 1 mGy (100 mrem) per day under the prevailing radiation protection standards. Because the model calculations were conservative, actual doses to natural species are likely to be a factor of 10 or 100 smaller. The current level of protection afforded to humans from routine radioactivity releases appears to be adequate to protect other species as well.

THE FOOD CHAIN FROM SOIL TO HUMANS

Most of the food consumed by human beings is grown on land and, except for elements like carbon and oxygen, which may be obtained from the atmosphere, it is the soil that nourishes the terrestrial ecosystem that supplies human food.

Radionuclides such as ^{40}K, ^{210}Po, and ^{226}Ra that occur naturally in soil are incorporated metabolically into plants and ultimately find their way into food and water. Artificial radionuclides behave in a similar manner, and worldwide contamination of the food chains by radionuclides produced during tests of nuclear weapons in the atmosphere has taken place during the past half century.

In addition to root uptake, deposition on foliar surfaces can occur directly or as a result of resuspension or rain splash (Dreicer *et al.*, 1984), in which case contaminants can be absorbed metabolically or, more likely, transferred directly to animals that consume the contaminated foliage. Foliar deposition is potentially a major source of food-chain contamination by both radioactive and nonradioactive substances (Russell, 1965; Russell and Bruce, 1969; Adriano *et al.*, 1982; Whicker and Schultz, 1982).

SOME PROPERTIES OF SOILS

Soils consist of mineral and organic matter, water, and air arranged in a complicated physicochemical system that provides the mechanical foothold for plants in addition to supplying their nutritive requirements (Department of Agriculture (USDA), 1957; Hillel, 1971; Donahue *et al.*, 1977).

Vertical profiles through soils reveal horizontal layers (horizons) that differ in their physical characteristics and that, in part, determine the kinds and amounts of vegetation that a soil will support. Broadly speaking, three major horizons may be identified. The uppermost, which may be from 30 cm to almost 60 cm in thickness, is the surface soil in which most of the life processes take place. The second horizon is the subsoil, extending to about 1 m below the surface. Still farther below the surface, to a depth of about 1.5 m, is a layer of loose and partly decayed rock, which is the parent material of the soils. These layers are conventionally designated as the A, B, and C horizons and can be differentiated further, as shown in Fig. 5-1. Exceptions to the general model occur when the soil accumulates from wind or water deposition from sources remote from the underlying rock.

The inorganic portion of surface soils may fall into any one of a number of textural classes, depending on the percentages of sand, silt, and clay. Sand consists largely of primary minerals such as quartz and has a particle size ranging from 60 μm to about 2 mm. Silt consists of particles in the range of 2 to 60 μm, and clay particles are smaller than 2 μm in diameter. Figure 5-2 illustrates the textural classes into which soils may be differentiated, depending on the percentages of each of the three principal constituents.

The important physicochemical processes by which soils provide the nourishment for plants are controlled largely within the clay fraction of the soil. An essential characteristic of the platelike particles of secondary aluminum silicates that comprise clay is the abundance of negative surface charges. The resultant ability of the clay particles to attract ions, especially positive ions, to their surfaces is one of the most important properties of soils.

Most of the nutrient ions are not dissolved in the soil water, but are sorbed on the surfaces of the soil particles. A much greater reservoir of

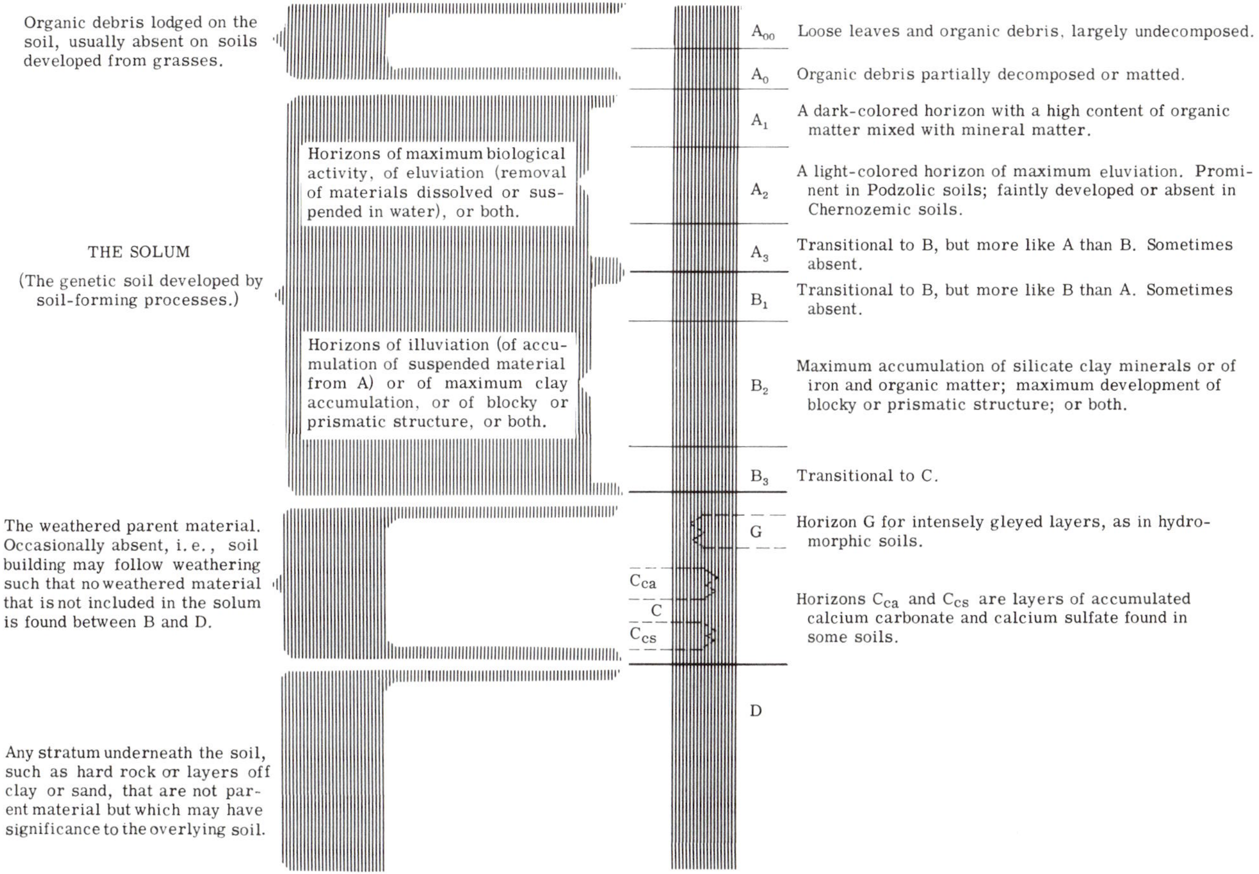

Organic debris lodged on the soil, usually absent on soils developed from grasses.

A_{00}	Loose leaves and organic debris, largely undecomposed.
A_0	Organic debris partially decomposed or matted.
A_1	A dark-colored horizon with a high content of organic matter mixed with mineral matter.
A_2	A light-colored horizon of maximum eluviation. Prominent in Podzolic soils; faintly developed or absent in Chernozemic soils.
A_3	Transitional to B, but more like A than B. Sometimes absent.
B_1	Transitional to B, but more like B than A. Sometimes absent.
B_2	Maximum accumulation of silicate clay minerals or of iron and organic matter; maximum development of blocky or prismatic structure; or both.
B_3	Transitional to C.
G	Horizon G for intensely gleyed layers, as in hydromorphic soils.
C_{ca} and C_{cs}	Horizons C_{ca} and C_{cs} are layers of accumulated calcium carbonate and calcium sulfate found in some soils.

THE SOLUM

(The genetic soil developed by soil-forming processes.)

Horizons of maximum biological activity, of eluviation (removal of materials dissolved or suspended in water), or both.

Horizons of illuviation (of accumulation of suspended material from A) or of maximum clay accumulation, or of blocky or prismatic structure, or both.

The weathered parent material. Occasionally absent, i.e., soil building may follow weathering such that no weathered material that is not included in the solum is found between B and D.

Any stratum underneath the soil, such as hard rock or layers off clay or sand, that are not parent material but which may have significance to the overlying soil.

C_{ca}
C
C_{cs}

D

FIGURE 5-1   Principal horizons in a hypothetical soil profile. Not all horizons are present in any single profile. [From U.S. Department of Agriculture (1957).]

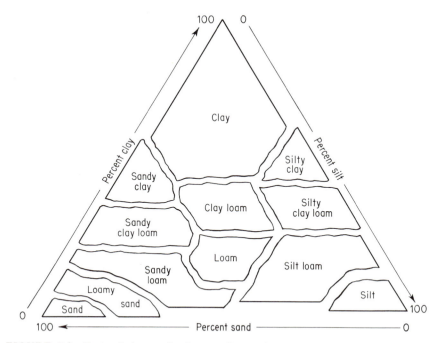

FIGURE 5-2   Textural classes of soils according to the percentage of clay, sand, and silt. [From U.S. Department of Agriculture (1957).]

nutrient elements can be held in this way than in solution. Dissolved nutrients would not remain long in soil but would be leached from it were it not for the extraordinary ability of the clays to bind elements in ionic form. Since it is estimated that 1 m^3 of loam has a surface area of about 2 million m^2, the opportunities for ion exchange are very great.

Cations (positive ions) in water solution are exchanged with cations sorbed on the surface of the clays. Most soils tend to become acidic after a period of time because of the replacement of adsorbed cations by the excess of hydrogen ions in rainwater and, for this reason, such soils must be limed from time to time. This procedure replaces the hydrogen ions with ions of calcium and magnesium, as is illustrated schematically in Fig. 5-3.

The ability of a given soil to exchange cations is quantified by its exchange capacity, customarily expressed as the milliequivalents of cations required to neutralize the negative charge of 100 g of dry soil at pH 7. (An "equivalent" of an ion is the atomic weight divided by the valence.) Montmorillonite clay has such excellent cation-exchange properties that soils rich in this mineral may have cation-exchange capacities of about

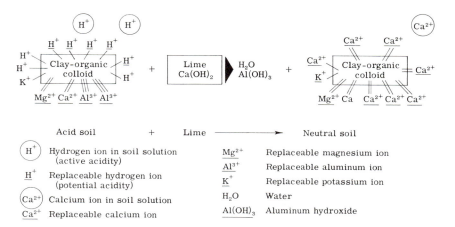

FIGURE 5-3 Cation-exchange reactions when an acid soil is limed. [From U.S. Department of Agriculture (1957).]

100 mEq per 100 g compared to less than 10 mEq per 100 g for predominantly kaolin-type soils. Organic matter derived from the decay of plant material can also furnish a major part of the exchange capacity.

According to Jenne (1968), neither the presence of clays or organic matter nor precipitation as oxides or hydroxides is sufficient to explain the observed binding of transition elements such as Co, Ni, Cu, and Zn, and it is proposed that fixation is due to formation of complexes with hydrous oxides of Fe and Mn that occur as coatings on particles of soils and sediments. Means *et al.* (1978) subsequently found the role of Mn to be predominant for fixation of not only the transition elements, but the actinides as well.

Sandy soils, which are composed of relatively large particles, tend to have a low exchange capacity, which can be increased by adding organic matter.

BEHAVIOR OF RADIONUCLIDES IN SOILS

Uptake of a long-lived radionuclide by plants depends to a considerable degree on whether it remains within the root zone and the extent to which it is chemically available for transport to root endings and translocation to edible portions.

When a radionuclide (in soluble form) comes in contact with soil, it can adsorb to reactive coatings on particles, undergo ion exchange, precipitate as an oxide-hydroxide or sulfide, be complexed with organic compounds, or remain in ionic form (Schulz, 1965). The manner in which the radionuclide partitions among these fractions determines to a large extent the length of time it will remain at the site of deposition and the extent to which it will be available for root uptake.

The manner in which a trace element is partitioned between the soil and interstitial water is usually described by the distribution coefficient K_d, which is expressed as the quantity of the radionuclide sorbed per unit weight of solids divided by the quantity of the radionuclide dissolved per unit volume of water. K_d is usually expressed in units of mL g^{-1}. When radionuclides are transported by water moving through porous or fractured geologic media, the phenomenon of sorption causes the rate of movement of the radionuclide to be reduced relative to the rate of movement of the water. This "retardation factor" is determined by the rates of sorption and desorption on the media surfaces and, at equilibrium (i.e., rate of sorption = rate of desorption), is related directly to K_d. The slowing of radionuclide transport in this way has important implications for radioactive waste management, as will be discussed in Chapter 11. Measured values of K_d for various elements in selected media are given in Table 5-1. Measured values of K_d may vary enormously even for the same element. Some of the factors that affect K_d include whether the study was performed in the laboratory or field, soil or sediment type, temperature, pH, ionic form, physical form, and the presence of other ions.

The ions of some elements are bound so tightly to soil particles that they are nearly immobile. It has been shown that thorium and certain of

TABLE 5-1

Some Reported Values of $K_d{}^a$

Locale	Element	Reported values (mL g^{-1})
Lake Michigan	Cs	510 to $>10^5$
Clinch River	Cs	1360 to 88,000
Freshwater pond	Cs	4600
Different soils	I	0.007 to 52.6
Hudson River	Pu	7×10^5
Lake Michigan	Pu	3.3×10^5
Savannah River	Pu	0.4 to 4.1×10^5
Nevada tuff	Ra	6700
Utah soils	Ra	214 to 467
White Oak Lake, Tennessee	Ru	≤ 1
Laboratory studies	Ru	<1 to 9×10^4
Great Lakes	Sr	30 to 85
White Oak Creek, Tennessee	Sr	100 to 150

^{a}Condensed from NCRP (1984a).

the light rare earth elements sorbed on clays and iron oxides in a massive ore body near the surface of a hill in the state of Minas Gerais, Brazil, are transported from the deposit by groundwater solubilization at a rate of only 10^{-9} per year, despite an annual rainfall that averages 170 cm y^{-1} (Eisenbud et al., 1984; Lei et al., 1986).

When water moves through geologic media, sorption of ions occurs at different rates depending on the physicochemical properties of the ions and the media. Different ionic species sorb at different rates, and in this way form a chromatographic sequence with the most weakly bound species at the leading edge of the pattern. Cations are generally more strongly sorbed than anions (negative ions) because of the preponderance of negatively charged particles on soil surfaces. Smaller multivalent ions usually exhibit larger K_d (are more strongly sorbed) than larger univalent ions, and for unconsolidated media, such as soil, the fraction sorbed is usually inversely proportional to particle size. This is one of the reasons why clays are such effective sorbents (NAS/NRC, 1978).

Large volumes of radioactive waste solutions at Hanford, Washington, have been placed on and under the ground in trenches and cribs. Studies of the migration of the individual fission products toward the water table below the stored wastes have shown that the individual radionuclides form a sorption cone in which the elements are sequentially removed in an order dependent on their relative affinities for the adsorption media. The soils at Hanford are sandy with relatively low cation-exchange capacity, 5–10 mEq per 100 g (Pearce et al., 1960).

Field studies in the former Soviet Union (Spitsyn et al., 1960) showed that ^{90}Sr migrated at a rate of 1.1 to 1.3 cm d^{-1} (less than 5 m y^{-1}) through soils that had moderately high exchange capacities and were permeated with groundwater. Since the average life of a ^{90}Sr atom is about 40 years, the mean distance that would be traversed by an atom of ^{90}Sr before its decay would be less than 200 m under the given conditions. The total amount of ^{90}Sr would diminish to 0.1% of the original quantity in 10 half-lives (280 years), by which time the mean distance traveled would be 1400 m. The capacity of the soils to store some fission products in ionic form is seen to be substantial.

Field and laboratory studies by Spitsyn et al. (1958, 1960), Alexander et al. (1960), and the United Kingdom Agricultural Research Council (1961) all found that the ^{90}Sr fallout from tests of nuclear weapons is held tightly in the upper 10 cm of soil for the first few years after deposition occurs. Alexander (1967) studied the downward movement of ^{90}Sr and ^{137}Cs in fields having a variety of soil types and vegetation characteristics, shortly after the peak fallout of weapons debris occurred in the early 1960s. At all sites studied, the two nuclides were contained mainly in the upper 10 cm of soil.[1] At one site that was examined in more detail than the others, both nuclides were present in appreciable amounts in the leaf litter, but this may have been due to the fact that the sampling was done at a time when weapons testing in the atmosphere had recently occurred. From estimates of the cumulative fallout at the time of sampling, it was concluded that only 50% of the deposit could be accounted for in leaf litter and soil and that the remaining 50% was presumably contained in the vegetative stand. The greater part of the fallout, after 2 or 3 years, is thus contained in the soil cycle: soil–plant–litter–microbial utilization–soil (Ritchie *et al.*, 1970). This is in contrast to findings in sandy loam at another site, 9 years following the maximum fallout pulse in 1963. Only a very small percentage of the ^{90}Sr, ^{137}Cs, and ^{239,240}Pu present in the soil column was found to be associated with leaf litter. The integrated accumulations in the top 30 cm of undisturbed soil were nearly equivalent to the quantities estimated to have been deposited at this latitude, indicating little removal from the site (Hardy, 1974a). Similar findings have been reported for fallout-derived ^{90}Sr, ^{137}Cs, and ^{239,240}Pu in a variety of soils sampled in the late 1970s and early 1980s (Alberts *et al.*, 1980; Linsalata, 1984).

In a study of ^{129}I accumulations in soil near the Savannah River it was estimated that the residence half-time in the top 30 cm of soil during the 25-year period of observation was 30 ± 6 y (Boone *et al.*, 1985).

Alexander's measurements confirmed that ^{137}Cs is more tightly bound by soil than ^{90}Sr. For purposes of forecasting the dose commitment from ^{90}Sr deposited on soil, the UNSCEAR (1969) assumed that ^{90}Sr is leached away from the root zone of cattle fodder at a rate of 2% per year. Whicker and Kirchner (1987) published leaching rates of 6.6×10^{-5} d^{-1} for Sr, 6.6×10^{-6} d^{-1} for Cs, and 6.7×10^{-7} d^{-1} for Pu. The Sr value is consistent with the UNSCEAR estimate. Many estimates of the leaching rate are obtained from laboratory experiments that do not always simulate field conditions properly. Krauskopf (1988) called attention to the reasons why estimates for both solubilities and retardation factors may be in error if based on laboratory simulation.

Plutonium-239/240 present in soil at the Rock Flats Plant near Denver, Colorado, is accompanied by ^{241}Am at an activity level about 18% of that of plutonium. Ibrihim *et al.* (1996) found that the two elements are so tightly held that 80% has remained above 9 cm for 25 years.

Although some generalizations concerning the relative degree of fixation of many radionuclides in soil are possible, their behavior is so dependent

---

[1]Thirty years after the period of heavy testing, the greater mobility of strontium relative to cesium caused the initial ^{137}Cs : ^{90}Sr ratio of 1.6 to increase by more than a factor of ten (Eisenbud, 1991).

on site-specific factors such as rates and amounts of rainfall, drainage, and extent of tillage that more general quantitative forecasts are not practical at this time. The behavior of radionuclides deposited on soils by fallout from nuclear weapons testing is further discussed in Chapter 9.

UPTAKE FROM SOILS

Although almost every naturally occurring element can be identified in soils, only sixteen of them are considered to be necessary for the growth and reproduction of vegetation: carbon, hydrogen, oxygen, nitrogen, phosphorus, sulfur, potassium, calcium, magnesium, iron, manganese, zinc, copper, molybdenum, boron, and chlorine. All are obtained by plants from soil except carbon, hydrogen, and oxygen, which can be supplied by the atmosphere.

It may be assumed that if an ion is present in the soil it will probably be present in plants grown on the soil. Many elements are required for normal metabolism, but some, like iodine, cobalt, uraniun, and radium, are known to be present in plants although they serve no known metabolic function.

The extent to which plants absorb radionuclides from soil depends on the chemical form of the nuclide, its distribution coefficient, the metabolic requirements of the plant, and physicochemical factors in the soil.

There are many compilations of the reported transfer factors from soil to plants (Ng, 1982; Ng *et al.,* 1982). According to Nishita *et al.* (1961), the relative uptake of various radioelements from soils is

$$Sr \gg I > Ba > Cs, Ru > Ce > Y, Pm, Zr, Nb > Pu$$

To provide assistance in estimating plant uptake of various elements in the absence of field data, NCRP (1984a) published the data in Table 5-2, which gives the recommended "default values" that can be used in dose assessment studies in the event that no field data are available. The range of recommended values is huge for some elements and these values must therefore be used with great caution. More detailed lists of values can be found in Ng (1982), Ng *et al.* (1982), and in Till and Meyer (1983).

Radioisotopes of elements that are ordinarily present in soil and that are utilized in plant metabolism are absorbed in a manner independent of the radioactive properties of the element. Thus, ^{45}Ca in soil becomes part of the pool of available calcium in the soil, and the plant will not differentiate significantly between ^{45}Ca and the stable isotopes of calcium. Available calcium refers to that portion of the total soil calcium that exists in exchangeable form and is available for transport to the root system. The roots are sometimes unable to distinguish between chemicals with similar reactive properties called congeners. Thus, plants grown in inorganic solutions con-

TABLE 5-2

Range of Recommended Default Values for Plant
Uptake Factors[a]

Element	Concentration in wet vegetation/dry soil	Max / Min
Co	1×10^{-3}–9.4×10^{-3}	9
Sr	1.7×10^{-2}–1.0	59
Ru	3.8×10^{-3}–6.0×10^{-2}	16
I	2.0×10^{-2}–5.5×10^{-2}	3
Cs	6.4×10^{-4}–7.8×10^{-2}	121
Ra	3.1×10^{-4}–6.2×10^{-2}	200
U	2.9×10^{-4}–2.5×10^{-3}	8.6
Pu	1×10^{-6}–2.5×10^{-4}	250

[a]Adapted from National Council on Radiation Protection and Measurements (1984a).

taining the chemical congeners calcium and strontium are unable to discriminate between the two (FAO, 1960; Whicker and Schultz, 1982).

The elements sorbed in exchangeable form on the surfaces of the soil particles constitute a reservoir that supplies many of the nutritive requirements of the plant. The elements pass to the plant root tips with the soil water in which the dissolved elements are in equilibrium with the sorbed solid phase.

The roots of plants are located at soil depths characteristic of the species: less than 30 cm in the case of spinach, in contrast to alfalfa and asparagus roots, which penetrate to 3 m or more. However, soil type can modify the depth of root structures for a given species (Abbott *et al.*, 1991). The effectiveness of root structures as an absorbing surface is illustrated by winter rye, single plants of which were shown to have a root surface area of over 600 m^2 (Wadleigh, 1957).

FOLIAR DEPOSITION OF RADIONUCLIDES

Radioactive substances can contaminate plants by direct foliar deposition (Russell, 1965, 1966). The radionuclides may then pass directly to grazing animals or humans in the form of superficial contamination or may be absorbed metabolically from the plant surface.

It has also been noted that trace substances present in soil can contaminate plants by the mechanism of rain splash, in addition to root uptake (Dreicer *et al.*, 1984). Particles less than about 100 μm can be deposited on plant surfaces by impacting raindrops up to a height of 40 cm. The importance of soil ingestion by grazing animals has been discussed by Zach and Mayoh (1984), who concluded that the daily intake of radionuclides for cattle predicted by models of the NRC (1977) should be increased by the radionuclides contained in 0.5 kg of ingested soil per day. Resuspension has been shown to be an important consideration in modeling the transport of ^{137}Cs in the milk pathway (Breshears et al., 1992). For certain radionuclides that exhibit high K_d values and are highly discriminated against by plants (i.e., Pu, Th, Am), soil ingestion may be the predominant route of intake. However, these actinide elements are poorly absorbed from the gastrointestinal tract.

An issue that occupied a good deal of interest when nuclear weapons tests were in progress (see Chapter 9) was how to apportion the ^{90}Sr content of cows' milk between that which originated from foliar deposition and that which is taken by the plant from the soil. The importance of this arose from the fact that in order to make forecasts of human exposure from expected deposition of ^{90}Sr, it was necessary to know the extent to which the dose from consumption of milk was due to fresh fallout that would disappear after weapons tests stopped and how much was due to ^{90}Sr that would remain in the soil. A number of investigators (Tajima and Doke, 1956; Russell, 1965) attempted to fit proportionality factors to equations of the type

$$C = p_r F_r + p_d F_d$$

where:

C = 12-month mean ratio of ^{90}Sr to calcium in milk (pCi of ^{90}Sr per g of Ca)
F_r = the annual deposit of ^{90}Sr (mCi km^{-2})
F_d = the cumulative deposit of ^{90}Sr (mCi km^{-2})
p_r = the proportionality factor for the rate-dependent component of C
p_d = the proportionality factor for the deposit-dependent component of C.

To account for the effect of fallout during the previous growing year on the ^{90}Sr content of milk from cows fed stored feed for part of the year, Bartlett and Russell (1966) proposed inclusion of a third term, the "lag factor," thus

$$C = p_r F_r + p_d F_d + p_l F_l$$

in which

F_l = fallout deposition during the last 6 months of the previous year (mCi km^{-2})
p_l = "lag" proportionality factor.

This refinement could be important during periods when the rate of fallout is varying significantly, but Aarkrog (1971b) did not find it necessary to utilize this additional term to relate the observed rates of fallout in Denmark during the period 1962–1968.

During the period of relatively heavy fallout in the early 1960s, direct uptake from soil apparently accounted for as little as 10% of the ^{90}Sr present in milk. The average of the worldwide soil proportionality factor for ^{90}Sr contamination of milk was 0.3 (UNSCEAR, 1964, 1966), and the proportionality factor for the rate-dependent component was 0.8.

The soil pathway is usually unimportant for a radionuclide such as ^{131}I, which has a half-life of only 8 days, and which decays before it is assimilated by the plant.

The significance of surface contamination varies with the growing season, since direct contamination of crops is obviously greater just before or during a harvest, or when active grazing by stock animals (particularly milk-producing animals) is in progress. Conversely, the danger may be lowest in winter months, when there are no standing crops, although it is possible that even during these months direct fallout on the basal structure of grasses in permanent pastures may be stored until the following spring, when contaminants may be absorbed by the growing plants. Retention of this type will be greatest for plants that develop a "mat" of basal parts, old stems, and surface roots (Russell, 1965). In modeling the concentration of ^{131}I in milk, Whicker et al. (1990) estimated that winter levels are about 10% of summer levels. Although no fresh forage is consumed in winter, ^{131}I is ingested together with dust and soil (Zach and Mayoh, 1984).

It has been shown that the major mechanism for contamination of soybeans and wheat grown in soil that is superficially contaminated with plutonium is resuspension due to mechanical harvesting (McLeod et al., 1980; Adriano et al., 1982).

The relative importance of foliar contamination also depends on the structure of the plant and the role of the various parts of the plant in relation to the dietary habits of humans. The inflorescences of wheat have a shape that tends to maximize entrapment of fallout particles. It is possibly for this reason that wheat was found to be a major source of ^{90}Sr from weapons testing fallout in Western diets (AEC, 1960b). It has been reported that cereals generally are subject to relatively higher foliar retention. The influence of dietary practices is illustrated by the fact that white bread has been shown to contain less ^{90}Sr than whole wheat bread, which is made from unsifted flour that contains the brown outer coat as well as the inner white portion of the wheat grain.

Foliar contamination can be removed by radioactive decay, volatilization, leaching by rain, other weathering effects, death and loss of plant parts, and, of course, washing prior to human consumption. Chamberlain (1970) examined the ^{90}Sr data from a number of investigators and found the half-life due to field loss during the growing season to be about 14 days, not considering radioactive decay. This value appears to be a useful "average" value but this half-time is influenced by differences in the physical and chemical properties of the radionuclides as well as by the type of plant (Miller and Hoffman, 1982). The half-time of removal of strontium sprayed on crops varied from 19 days in the summer to 49 days in winter. New growth contained less than 1% of the initial deposit. Krieger and Burmann (1969) found that a fractional loss of 0.05 d^{-1} was satisfactory to describe the process for the first few weeks after fallout, but that field loss was slower thereafter.

Hansen *et al.* (1964), in a 3-year study of the ^{131}I and ^{90}Sr content of milk from dairy herds, found the radionuclide content of the milk from cattle pastured in well-fertilized fields to be 50% lower than that of milk from cattle grazing in badly fertilized fields, because the faster-growing grass diluted the contamination present as foliar deposition.

TRANSPORT OF SOIL PARTICLES BY EROSION

Erosion by rainfall runoff is one mechanism for transport of radionuclides incorporated into surface soil. An understanding of soil erosion is important to soil conservation and agricultural scientists, and formulae for predicting soil loss have evolved over the years under the impetus of the Soil Conservation Service of the U.S. Department of Agriculture (Musgrave, 1947; Wischmeier and Smith, 1978). In addition, a modest amount of more recent research has specifically addressed the mobilization of long-lived nuclides by erosive processes in large watersheds (Foster and Hakonson, 1986; Simpson *et al.*, 1986; Muller and Sprugel, 1978; Linsalata, 1984).

Erosion by wind action causes contaminants that have settled to the surface to become resuspended (Chepil, 1957). This is particularly important for alpha emitters such as plutonium because of the increased likelihood of inhalation of the particles (see Chapter 4).

METABOLIC TRANSPORT THROUGH FOOD CHAINS

From the foregoing it is seen that root uptake and foliar deposition are the two ways by which fallout has contaminated crops that are consumed by humans or that serve as food for stock animals. Much remains to be learned about the mechanisms by which individual radionuclides pass from soil to the root, from the root to the edible portion of the plant, through the body of the stock animal, and into the milk, flesh, internal organs, and eggs.

It is sometimes possible to predict the behavior of a radioelement from knowledge of its chemical congeners. Comar *et al.* (1956) first noted in studies of the transfer of radionuclides and their congeners in food chains that the extent to which discrimination takes place at any step can be described by the observed ratio, OR:

$$OR_{sample-precursor} = \frac{C_e/C_c)_{sample}}{C_e/C_c)_{precursor}}$$

where C_e and C_c are the concentrations of the element and its congener, respectively.

The use of the observed ratio is illustrated by the case of a cow that is fed herbage containing a known ratio of Ca to ^{90}Sr. The concentration of ^{90}Sr that will appear in the milk can be predicted by use of the $OR_{milk-herbage}$, which describes the ability of the cow to discriminate metabolically between Ca and ^{90}Sr in the production of milk. Since this particular OR is known to be about 0.1, we would expect cow's milk to contain 1 Bq ^{90}Sr g^{-1} Ca, if the herbage on which the cow feeds contains 10 Bq ^{90}Sr g^{-1} of Ca.

Of the many artificial radionuclides that have contaminated soils and plants, ^{90}Sr, ^{137}Cs, and ^{131}I have been studied most thoroughly because they usually result in the highest doses to humans from the release of mixed fission products.

Strontium-90

Many discussions of ^{90}Sr in food chains use the ratio of ^{90}Sr to Ca (e.g., AEC, 1972a; NCRP, 1991b). However, the metabolism of strontium is not identical to that of calcium. The calcium content of plants has been found to vary considerably, and contamination reported as activity per gram of food is sometimes less variable than when related to the unit mass of calcium (UNSCEAR, 1969; NCRP, 1991b). Yet, all things considered, the ratio is useful in following ^{90}Sr from one biological level to the next.

However, the ratio may be meaningless when applied to soil under practical conditions because the ^{90}Sr will not normally be homogeneously mixed throughout the soil, and there is no way of expressing the strontium–calcium ratio of the nutrients to which the roots are exposed. Under laboratory conditions where the ^{90}Sr was well mixed with the soil, Frederikson *et al.* (1958) showed that the observed ratio from plant to soil varied only within the narrow range of 0.7 to 0.8 over a wide range in the amount of soil calcium. The discrimination against strontium was apparently unaffected as long as the soil was not oversaturated with calcium ions. This would indicate that only a small degree of differentiation takes place at the soil–root interface, but it is difficult to determine in practice exactly how the ^{90}Sr is distributed in the soil, and the use of the OR may not always be practical.

Roberts and Menzel (1961) found that a portion of the ^{90}Sr may become unavailable to plants as a result of reactions in the soil. Studies over a 3-year period showed this fraction to be variable from about 5 to 50%.

Measurements of $OR_{human\ bone-diet}$ have been reported from seven countries (UNSCEAR, 1969) and have been shown to average about 0.15, with little variation from country to country. In the United States, $OR_{human\ bone-diet}$ is 0.18 and ranges from 0.15 (Chicago) to 0.22 (San Francisco). There appears to be no systematic difference between dairy and nondairy foods.

The observed ratio in passing from plants to milk by way of the cow has been shown (Comar and Wasserman, 1960) to be about 0.13. Thus, the overall observed ratio ($OR_{human\ bone-plant}$) would be about 0.15 if the plants are consumed directly and about 0.020 if the calcium is consumed from milk. In short, the lactating animal has been shown to be a strontium decontaminator with an efficiency of about 85%. Lough et al. (1960) showed that the observed ratio in passing from diet to human milk is also about 0.1.

The overall effect of metabolic differentiation between strontium and calcium in passing from soil to human bone can be summarized for milk and vegetable diets as follows, starting with soil containing one unit of activity per gram of calcium:[2]

$$
\begin{array}{llll}
1\ \text{SU} & 1\ \text{SU}^3 & 0.13\ \text{SU} & 0.020\ \text{SU} \\
\text{in} \rightarrow & \text{in} \rightarrow & \text{in} \rightarrow & \text{in} \\
\text{soil} & \text{plant} & \text{milk} & \text{human bone}
\end{array}
$$

$$
\begin{array}{lll}
1\ \text{SU} & 1\ \text{SU} & 0.15\ \text{SU} \\
\text{in} \rightarrow & \text{in} \rightarrow & \text{in} \\
\text{soil} & \text{plant} & \text{human bone}
\end{array}
$$

The net ^{90}Sr ratio in human bone will thus vary, depending on whether the dietary calcium is derived primarily from dairy foods or from other sources. The U.S. population derives almost all its calcium from milk or milk products, but this is not true of all other countries. In Chile, the population receives only 8% from dairy products, compared to 77% in the United States and 87% in Finland. The cereals, legumes, and vegetables are the dominant sources of calcium in the Far Eastern countries (FAO, 1960).

In some parts of the world, the calcium naturally present in food is supplemented with mineral calcium, a practice that would tend to lower

[2]One pCi (37 mBq) or ^{90}Sr g^{-1} Ca has been referred to in the literature as a strontium unit or SU.

[3]This holds only when the strontium and calcium are uniformly mixed throughout the root zone of the plant.

the ratio ^{90}Sr : Ca. Calicum carbonate has been added to the maize used in the preparation of tortillas in Mexico and to the flour used for making bread in the United Kingdom (FAO, 1960).

Several methods have been proposed by which the plant uptake of ^{90}Sr from soils can be reduced (Menzel, 1960). These include the application of lime, gypsum, fertilizer, and organic matter. According to Menzel, these techniques are only moderately effective and could not be expected to reduce the plant uptake of ^{90}Sr by more than 50% in productive soils. This is a modest diminution in uptake considering the rather large quantities of required soil amendments. For example, at the levels of exchangeable calcium ordinarily found in productive soils, it would require several tons of lime per acre to affect a measurable change in strontium uptake. However, unproductive soils that have low cation-exchange capacity and low exchangeable calcium have relatively large uptakes of ^{90}Sr that can be reduced appreciably by the addition of calcium in available form. In the United Kingdom, the ^{90}Sr uptake in herbage was shown to increase manyfold as the amount of exchangeable calcium diminished from 3 to 4 g kg^{-1} of soil to less than 2 g kg^{-1} (U.K. Agricultural Research Council, 1961).

Cesium-137

It has been well established that cesium is so tightly bound by the clay minerals of the soil that root uptake is slight, and foliar absorption is, therefore, the main portal of entry of ^{137}Cs to the food chains during periods of active fallout. As late as 18 years after the cessation of tests in 1963, ^{137}Cs was still confined to the upper 24 cm of silty clay and exhibited a half-value depth (that depth at which the concentration is 50% of the surface contamination) of about 6 cm in undisturbed soil (Linsalata, 1984). In a study conducted on tilled and untilled soils of five midwestern watersheds, 10 years after cessation of atmospheric weapons tests, it was found that ^{137}Cs remained largely in the upper 5 cm of untilled soil, but was evenly mixed in the upper 20 cm of tilled soil (Ritchie and McHenry, 1973).

Cesium is a congener of potassium, but the Cs : K ratio is not as constant in biological systems as is the ratio Sr : Ca. In general, however, Cs concentrations and the OR tend to increase with trophic level (Pendleton *et al.*, 1965). The uptake of cesium from soil has been shown to be inversely proportional to the potassium content of soils in which there is a potassium deficiency (Nishita *et al.*, 1961; Menzel, 1964), and Broseus (1970) showed that this inverse dependence on the potassium content of soil explains the high cesium content of milk from cows grazing in certain parts of the island of Jamaica. A similar observation was made in the Tampa, Florida, milk shed, where the high concentration of ^{137}Cs in milk was traced to the practice of using pangola grass for cattle fodder. This grass has a lower potassium content than other feeds (Porter *et al.*, 1967). Davis (1963) provided a comprehensive review of the ecological relationships between Cs and K.

Although cow's milk is the largest single contributor of ^{137}Cs to the United States adult diet, other foods containing grain products, meat, fruit, and vegetables contribute about two-thirds of the dietary cesium intake (Gustafson, 1969). Measurements made by Gustafson on representative United States diets from 1961 through 1968 are given in Table 5-3. A recent study in Germany following the Chernobyl accident also showed that milk, milk products, and meat contributed the largest share of ^{137}Cs to the diet (Voigt and Paretzke, 1992).

Wilson *et al.* (1969) studied the transport of ^{137}Cs from the atmosphere to milk and concluded that root uptake was so slight that it could be neglected in any model designed to forecast the dose to humans from ^{137}Cs in fallout. This negligible root uptake was reconfirmed by Breshears *et al.* (1992). For reasons not known, the crude fiber content of the forage was found to influence uptake by the cow, with transfer coefficients varying from 0.0025 for alfalfa and corn silage to 0.01 for mixed grain.

Wilson and associates developed a simple model for predicting the ^{137}Cs content of milk from cattle fed on stored feed. It was assumed that for the first 6 months in a given year the cows are fed stored feed that had been exposed to the previous year's fallout and consume feed contaminated by the current year's fallout during the second 6 months. The mean concentra-

TABLE 5-3

Sources of ^{137}Cs in the U.S. Adult Diet[a]

Year	Milk (%)	Grain products (%)	Meat (%)	Fruit (%)	Vegetables (%)
1961	31	17	12	20	15
1962	38	17	13	15	14
1963	39	21	22	8	6
1964	34	26	21	8	5
1965	28	23	26	9	4
1966	25	30	23	11	5
1967	24	28	17	8	7
1968	31	19	19	10	9

[a]From Gustafson (1969).

tion of milk in picocuries per liter was then calculated for each of the half-years, using the equation

$$C_m = BC_a$$

where:

C_m = the average concentration in cow's milk (pCi L^{-1})
C_a = the average concentration in air (pCi m^{-3})
B = a coefficient obtained by multiplying the dry weight intake, fallout contamination factor and transfer coefficient shown in Table 5-4.

This technique was shown to be useful for forecasting the ^{137}Cs content of milk from seven milk sheds across the nation from 1962 to 1967, although for reasons that were not understood by the investigators, the model did not work well for the milk sheds that supply Tampa, Florida, and Seattle, Washington. The excellent correlation between the observed ^{137}Cs concentration in air during the growing season (May–July) and the mean quarterly ^{137}Cs content of milk is shown in Fig. 5-4, from which the Tampa and Seattle data are excluded. The ^{137}Cs levels in Tampa were anomalously high, a fact that has been attributed to the pangola grass prevalent in the area, as noted earlier. A recent and more sophisticated food chain model that includes the milk pathway has been developed by Whicker and associates (Whicker and Kirchner, 1987; Whicker et al., 1990).

When ^{137}Cs is ingested by humans, about 80% is deposited in muscle and about 8% in bone (UNSCEAR, 1969; Spiers, 1968). The half-life in adults depends on body weight, sex, and dietary habits and has been shown to vary between 19 ± 8 days for infants and 105 ± 25 days for men. Women have a more rapid Cs turnover than men (NCRP, 1977a). A concentration of 1 pCi (37 mBq) per gram of potassium in the human body produces an annual absorbed dose of 18 μrad (0.18 μGy) (NCRP, 1987a).

One must be careful that any generalization about pathways to humans is not negated by some special dietary consideration. In the case of ^{137}Cs, it has been shown that Laplanders and other residents of the far north are

TABLE 5-4

Summary of Information for a Model of ^{137}Cs in Milk for a Dry Lot Herd[a]

	Dry weight intake (kg d^{-1})	Fallout contamination factor (m^3 kg^{-1})	Feed to milk transfer coefficient (d L^{-1})
Hay	14	9100	0.0025
Grain	7	1470	0.010
Silage	5	6740	0.0025

[a]From Wilson et al. (1969).

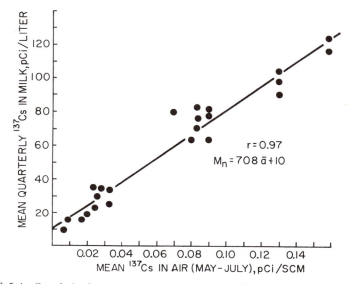

FIGURE 5-4   Correlation between surface air activities of ^{137}Cs and quarterly averaged ^{137}Cs in milk from seven milk sheds across the nation from 1962 to 1967. [From Wilson *et al.* (1969).]

subject to relatively high ^{137}Cs intake owing to their dependence on reindeer, which feed on lichens, which in turn have a tendency to concentrate a variety of trace substances present in the atmosphere (Miettinen, 1969). About 25% of the ^{137}Cs contained in lichens is absorbed by reindeer (Holleman *et al.*, 1971). As a result of this phenomenon, the Lapps contained about 50 times the ^{137}Cs body burden of Finns in the southern part of the country. This phenomenon has been said to be characteristic of all arctic and subarctic regions in the Northern Hemisphere (Rahola and Miettinen, 1973).

Because of their high potassium content, wood ashes have remarkably high concentrations of ^{137}Cs relative to concentrations observed in most samples of environmental media. Farber and Hodgdon (1991) measured ^{137}Cs in 47 samples of wood ash from 14 states in the United States and reported that the concentrations ranged from 79 to 21,100 pCi kg^{-1} (3 to 780 Bq kg^{-1}). The largest concentrations were found in ash from the New England states. Wood ash is used as a fertilizer and may contribute to ^{137}Cs concentrations in agricultural soils.

Radioiodine

Iodine is an essential nutrient that is required for the functioning of the thyroid hormones. It is generally believed that sea spray serves as the source of iodine in the earth's atmosphere and soils, but it has also been proposed that iodine may be released by the action of atmospheric ozone in reducing the iodide in rocks and soils (Whitehead, 1984).

Because of its 8-day half-life, ^{131}I is not a significant environmental contaminant insofar as uptake from the soils is concerned. The decay rate is relatively rapid in relation to the growing time of crops, and significant contamination by means of root uptake would be improbable. On the other hand, radioiodine deposited on the surfaces of plants can be ingested directly by cattle and pass in this way to milk or other dairy products. Since the time is usually short from the collection of milk to its consumption, the possibility of iodine contamination of fresh milk must be considered. If the milk is processed into powdered form or aged cheese, radioiodine contamination will be less of a problem because longer storage time will permit decay of the isotope. However, cottage cheese reaches the consumer almost as quickly as fresh milk and may be contaminated with short-lived nuclides.

Mathematical models for the transfer of radioiodine to food are handicapped by the fact that since radioiodine has a short half-life, contamination must of necessity take place over a relatively short period of time. Thus, transfer from air to forage is apt to be variable because of meteorological factors that tend to average out in the case of longer-lived nuclides such as ^{137}Cs or ^{90}Sr. Field observations of the various parameters needed for mathematical modeling have been found to be variable. Chamberlain found the values of deposition velocity (v_g) to range from 0.1 to 0.4 cm s^{-1} following the Windscale accident (Chamberlain, 1960; Chamberlain and Chadwick, 1966). Hawley et al. (1964) found the deposition velocity to be 0.6 cm s^{-1} on the average for radioiodine released experimentally at the National Reactor Test Station in Idaho, but the individual observations varied from 0.17 to 1.1 cm s^{-1}.

Soldat (1963), reporting on studies at Hanford Laboratories, found that over a 2-year growing season the ratio Bq kg^{-1} grass : Bq m^{-3} air was about 4200, and that the ratio Bq L^{-1} milk : Bq kg^{-1} grass was about 0.15. Various estimates have been made of the extent to which the dose to the human thyroid is increased by the grass–cow–milk pathway compared to inhalation. A widely used multiple of 700 has been justified by Burnett (1970) and is intermediate among various proposed factors.

Chamberlain (1970) concluded from an analysis of data from various investigators that the ^{131}I foliar deposition data conformed to the concept that vegetation behaves as a filter, with deposition dependent on the density of the foliage.

Once the radioiodine deposits on foliage, it is removed by weathering and death of plant parts at a rate of about 5% per day. Given the 8-day radioactive half-life of ^{131}I, this leaves an effective half-life of removal from grass ranging from 3.5 days in the Idaho experiments to about 6 days at Windscale. An effective half-life of about 5 days is usually used for purposes of risk estimation. Some data suggest that long-term weathering is better described by two-component exponential decay with half-times of 5 days and 48 days (Till and Meyer, 1983). Soldat (1965) found that radioiodine in milk reached a peak 3 days after an accidental release. The long-term behavior of radioiodine in soil has been studied by Boone et al. (1985) using [129]I from worldwide fallout. They reported that the half-time of radioiodine in the top 30 cm of soil during a 25-year period was 30 ± 6 years.

According to Garner (1960), a cow consumes about 20 kg d^{-1} of grass on the average, and an average vegetative stand per square meter of pasture is about 0.125 kg. It follows that a cow grazes an area of about 160 m^2 d^{-1}. It can be assumed that about 5–11% of the ingested radioiodine will be secreted in cow's milk (Lengemann, 1966, Whicker and Kirchner, 1987).

The fraction of ingested iodine that reaches the human thyroid following ingestion is usually taken as 0.3 and the effective half-life for elimination from the thyroid as 7.6 days (ICRP, 1960). However, thyroidal radioiodine uptake can be blocked almost completely by 50–200 mg of stable iodine, as the iodide or iodate (Blum and Eisenbud, 1967; NCRP, 1977b). Substantial dose reduction will result from administration of KI as long as 2 h after exposure. The side effects of administrations are minor and infrequent, with incidence estimated to lie between 10^{-7} and 10^{-6} per therapeutic dose (NCRP, 1977b). Stable iodine was administered to the population around the Chernobyl reactor following the accident in an attempt to reduce the thyroid dose (Mettler et al., 1992). Knowledge of the behavior of the radioiodine in the environment was advanced significantly by studies conducted during the period of weapons testing, as is discussed in Chapter 9.

Uranium, Thorium, and the Transuranic Actinide Elements

An understanding of the food-chain behavior of uranium, thorium, and the artificially produced transuranic actinide elements is important because of their long half-lives, the fact that they are alpha emitters, and their persistence in the environment. Although some of these elements, such as plutonium, were unknown until World War II, they were subsequently produced in great quantity. The general reduction of tensions between the world's superpowers has led to greatly reduced production of plutonium for military purposes, but plutonium is still produced in reactors, deliberately in the few extant breeder reactors and to a much lesser extent in conventional light-water reactors. As a rule of thumb, 1 g of [239]Pu per megawatt-day (MWD) is produced in light-water reactors. Because of their radiotoxicity, the physical and chemical properties of the transuranic elements have been under intensive investigation for many years. and much has been learned about their behavior in the environment (Hanson, 1980; Watters *et al.*, 1980; Litator *et al.*, 1994).

The usefulness of the available information on transfer factors from soil to plant and from plant to animals is limited by the great variability in the reported data. For example, it has been noted by Pimpl and Schuttelkopf (1981) that the concentration ratios (CRs) for plutonium, measured by different groups, vary from 10^{-9} to 10^{-3} for the same plant! The range is from 10^{-6} to 10^{-1} for americium and 10^{-4} to 10^{-1} for curium. Thus, a fundamental weakness in the risk assessment models now in use to predict the effects of the actinide elements is the inadequacy of reliable information on transfer factors, particularly under field conditions.

The range of reported values can be narrowed considerably by selecting carefully designed experiments in which surface contamination is not a factor. Thus, the concentration ratios (activity per gram of dry plant divided by activity per gram of dry soil) for actinide element analogues in 43 samples of various vegetables grown in lateritic soils near Morro do Ferro, Brazil, have been reported by Linsalata *et al.* (1987). Unweighted mean CRs were found to decrease in the order ^{228}Ra (6×10^{-2}) $\geq$ ^{226}Ra (3×10^{-2}) > La (2×10^{-3}) = Nd (2×10^{-3}) $\geq$ Ce (8×10^{-4}) > Th (1×10^{-4}). Vegetable uptake of these elements from soil can be simplified as Ra^{2+} > rare earth elements^{3+} > Th^{4+}. Similar concentration ratios have been published, based on field data by others (Bondietti *et al.*, 1979; Schreckhise and Cline, 1980a; Romney *et al.*, 1982; Pimpl and Schuttelkopf, 1981; Trabalka and Garten, 1983).

Because the transuranic elements do not exist in nature, their environmental behavior has also been investigated by the use of naturally occurring chemical analogues. Thorium has been shown to be a suitable analogue for plutonium in the quadrivalent state, and the light rare earth elements lanthanum and neodymium can be used as analogues for americium and curium. It has also been suggested that, under certain reducing conditions, thorium may be a suitable analogue for neptunium (Krauskopf, 1986). It was noted earlier in this chapter that, based on analogue studies, the mobilization rates of Th and La at Morro de Ferro by groundwater is about 10^{-9} y^{-1}, which means that under certain conditions even the longest-lived transuranic nuclides would decay in place if the analogue concept is valid.

Uranium   Considering the fact that uranium mining and processing have flourished for many years, remarkably little is known about the ecological transport mechanisms that govern the movement of this element in the food chains. Data on soil–plant or plant–animal relationships are relatively scarce, although it has been known since the 1940s that the practice of using phosphate fertilizers results in the presence of uranium in food in concentrations up to 8 ng g^{-1} (Reid *et al.*, 1977). Garten (1978) and Linsalata (1994) reviewed the transport of uranium in food chains and reported soil-to-plant CRs that range from 3×10^{-5} to 9×10^{-2}. Till and Meyer (1983) distinguished between CRs that include or exclude foliar contamination. When foliar contamination is excluded, the CRs range from 1.6×10^{-4} to 5×10^{-3}. When foliar contamination is included, the CRs are larger, ranging from 1×10^{-3} to 1.7×10^{-2}.

Thorium   Thorium behaves quite differently in natural systems than in some laboratory experiments or in disturbed systems such as surface mine environments and mill tailings (Till and Meyer, 1983; Ibrahim and Whicker, 1992a,b). The mobilization rate of thorium in solution from a highly weathered, near-surface ore body in Brazil has been reported to be on the order of 10^{-9} y^{-1}, as discussed in Chapters 6 and 11. Low values of the CR (10^{-3} to 10^{-4}) are seen in natural systems and in plants grown in soil in the laboratory. CR values for plants growing near the thorium deposit in Brazil were on the order of 10^{-4}. Limited data indicate that CR values for Th are lower when foliar deposition is excluded, as is the case for uranium. However, a high CR value (0.9) was reported for ^{230}Th in beans grown in spiked nutrient solution in the laboratory. Thorium in a nutrient solution is obviously much more available to plants than is thorium in normal soils. For plants growing on exposed uranium tailings, CR values ranged from 0.1 to 0.69. Exceptionally high values (1.9–2.9) were reported for ^{230}Th in vegetation growing on a uranium tailings impoundment edge at a conventional acid leach uranium production operation. Soil acidity, the presence of thorium in the sulfate form, and the saturation condition near the tailings impoundment are thought to explain the high plant uptake (Ibrahim and Whicker, 1992a).

Radium   The behavior of radium in the environment has been the subject of reviews by Williams (1982), McDowell-Boyer *et al.* (1980), and Sheppard (1980). Interest has been stimulated by the potential hazards from ^{226}Ra in mill tailing piles (see Chapter 7).

Campos *et al.* (1986) studied the transport of ^{226}Ra from the Morro do Ferro in Brazil and reported that the annual mobilization rate by groundwater solubilization is on the order of 10^{-7}. One of the principal reasons for the slow rate of mobilization was shown to be the tenacity with which radium is sorbed on clays and organic materials. Prantl *et al.* (1980) found the uptake of ^{226}Ra in 11 types of root and leafy vegetables grown on soil contaminated with uranium tailings to be 1.1×10^{-3}. McDowell-Boyer *et al.* (1980) reviewed 12 published reports of the uptake of ^{226}Ra by vegetables, grains, and feed hay and found a wide range of concentration ratios. The

CR (dry weights) for grains ranged from 6×10^{-2} to 2×10^{-5}. Till and Meyer (1983) reported values ranging as high as 0.44 for leafy vegetables (also dry weights).

Plutonium   Although the hazard potential of airborne plutonium is of concern, plutonium does not move readily in soil or the food chain. As mentioned earlier, 80% of the plutonium in soil near Rocky Flats was found to reside in the top 9 cm 25 years after contaminating events (Ibrihim *et al.*, 1996). A small fraction of the plutonium was found deeper along decayed root channels (Litaor, 1994). The CRs for plutonium in foodstuffs are low, ranging from 1.5×10^{-6} to 1.4×10^{-3} if foliar deposition is excluded and from 1×10^{-4} to 1.4×10^{-2} if deposition is included. Plutonium in the form of microspheres is even less available, with CRs ranging from 3.9×10^{-9} to 2.9×10^{-8}. The transfer factors from fodder to milk (2.7×10^{-9}) and fodder to beef (1.0×10^{-6}) for plutonium are the lowest of the many elements tabulated by Till and Meyer (1983).

Neptunium   Neptunium is a transuranic actinide element produced in reactors and bombs by neutron reactions with uranium (NCRP, 1986b). Its short-lived nuclides, which have half-lives measured in days, are prominent in the early fallout that results from nuclear weapon explosions, but ^{237}Np, with a half-life of 2.1×10^{6} y, does not become important until after about 10^{5} y, when almost all other nuclides have decayed.

The nuclides of neptunium received very little attention until the release of ICRP Publication 30 (ICRP, 1979–1988), which recommended that the assumed gastrointestinal tract-to-blood absorption factor for ingested Np be increased to 10^{-2} from 10^{-4}, which had been the value in use for many years. Since the half-life of ^{237}Np is one of the longest in high-level nuclear wastes, the increase in the assumed transfer factor affected the risk assessments for waste repositories in the period beyond about 10^{5} years (NAS/NRC, 1983; Thompson, 1982; Cohen, 1982).

There is relatively little information about the behavior of neptunium in the environment in the pentavalent state, but there is evidence that this element is more rapidly transferable from soils than in the quadrivalent state. Since it has been reported that neptunium is reduced to its quadrivalent state under the anaerobic conditions that would be expected to exist in waste repositories (Bondietti and Francis, 1979), the true potential significance of ^{237}Np for long-range risk assessment is uncertain.

The relative ease of transfer of neptunium from soil is illustrated by the report that the concentration ratio (concentration in dry weight of plant divided by concentration in dry weight of soil) is about 0.4 compared to 10^{-4} for Pu and 2×10^{-3} for Am and Cm (Schreckhise and Cline, 1980b).

However, this ease of transfer has not been demonstrated in animal uptake studies (Thompson, 1982; NCRP, 1986b).

TRANSPORT IN GROUNDWATER SYSTEMS

Water at various depths underlies most landmasses. This groundwater is a very important source of water for domestic, agricultural, and industrial use. In rural areas and in much of the arid western United States, groundwater provides the only practical supply of water for many uses. Worldwide there is more than 30 times as much groundwater as fresh water in lakes and rivers. In the United States, groundwater provides 20% of all drinking water, 40% of irrigation water, and 80% of rural and livestock water (Cain and Boothroyd, 1983; Anderson and Woessner, 1992). With rare exception, groundwater does not occur in underground lakes or pools but rather as saturated geological formations called aquifers. The depth of aquifers ranges from the surface to hundreds of meters. The water may be fresh or salt and may be hundreds or thousands of years old. Aquifers may be trapped between impermeable layers, in which case they are called confined aquifers.

Groundwater is part of the overall hydrological cycle. Aquifers are recharged by infiltration of surface water. After passing through the aquifer, water is discharged from springs, which may be on the surface or under lakes or rivers. Water may be removed from aquifers by wells that flow under their own pressure (artesian wells) or that require pumping (Cain and Boothroyd, 1983).

The behavior of water in aquifers is affected by three important factors. Porosity, which is defined as the percentage of pore or void space, determines how much water can be contained per unit volume of aquifer. Many aquifer materials have porosities of 30 to 40%. Another important property is permeability, the degree to which the pores are connected to one another. Permeability determines the rate at which water will flow through the aquifer under a given pressure. The hydraulic gradient or slope of the aquifer affects the pressure, which, together with the permeability, determines the rate of flow of the water in the aquifer (Fetter, 1980).

The flow of water through permeable substances was first studied systematically by the French engineer Henry Darcy (Fetter, 1980). The rate of flow of water through a bed of permeable material in one dimension can be described by the equation known as Darcy's law:

$$Q = -KA\left(\frac{dh}{dl}\right)$$

where Q is the rate of flow, A is the area of flow, and K is the hydraulic conductivity, which is related to the permeability. In older literature, K is sometimes called the coefficient of permeability. The term dh/dl is the hydraulic gradient, where dh is the change in the hydraulic head (h) corresponding to a small change in length (l). The negative sign is used because dh/dl is negative in the direction of flow. Groundwater flow rates are usually less than 0.3 m d^{-1} but may range as high as 120 m d^{-1} under unusual circumstances. In some cases, groundwater is essentially still, moving as little as 1 m y^{-1}. Modeling the flow of groundwater and its contaminants in actual aquifers involves three-dimensional flow in nonhomogeneous media and is a very complex task. Anderson and Woessner (1992) have prepared a useful book on this subject.

Groundwater is usually more difficult to contaminate than is surface water, but unlike surface water it tends to retain some forms of dissolved contamination rather than flush them toward the oceans (Cain and Boothroyd, 1983). The partition of contaminants between groundwater and the surrounding geological formation can be described by K_d, as described earlier in this chapter for soils. Many contaminants are removed as water passes through the material of the aquifer. As a result, much groundwater is so clean that it requires no treatment prior to use for human consumption.

Groundwater has become contaminated with low concentrations of radioactivity at several sites in the United States where nuclear energy for both civilian and military purposes was developed in the early years. Groundwater contamination has resulted most often from leakage of liquid waste tanks and from percolation ponds and injection wells used to dispose of very low level liquid radioactive waste.

An example of extensive, but very low level, contamination of groundwater has occurred at the Idaho National Engineering Laboratory (INEL) located west of Idaho Falls. The aquifer that underlies the INEL originates in Yellowstone National Park and Grand Teton Mountain areas, flows generally in a southwesterly direction, and emerges in springs in the vicinity of Twin Falls, Idaho. The INEL site has been used to develop nuclear reactors and for nuclear fuel reprocessing. Both activities resulted in large volumes of slightly contaminated water that required disposal. Percolation ponds and direct injection wells were used for this purpose for approximately 40 years. Figure 5-5 gives the yearly activity of tritium disposed of via injection wells and percolation ponds at the INEL (Mann and Cecil, 1990). Other radionuclides, such as ^{90}Sr, ^{137}Cs, ^{60}Co, and isotopes of Pu and Am, were disposed of in much smaller amounts (Bartholomay et al., 1995). Of these isotopes, tritium and ^{90}Sr have developed distinct plumes of contamination. Cesium-137, ^{60}Co, and isotopes of Pu and Am are detectable in monitoring wells near the point of injection but have not developed extensive plumes.

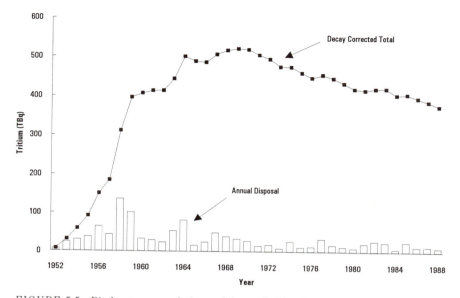

FIGURE 5-5   Discharge, accumulation, and decay of tritium in the Snake River Plain Aquifer as a result of waste management operations at the Idaho National Engineering Laboratory, west of Idaho Falls, Idaho.

Figures 5-6a and 5-6b illustrate the extent of the tritium and ^{90}Sr plumes. The actual area of contamination is no doubt somewhat larger; the indicated plume size is defined by the lower limit of detection of the analytical methods used in routine monitoring. Drinking water supplies on the INEL site are maintained below EPA standards for the general public and contamination in water off site, if present at all, is order of magnitude below the standards.

TRANSPORT IN SURFACE WATER SYSTEMS

Surface waters are coupled to subsurface aquifers, soils, and the atmosphere. Trace contaminants that somehow find their way into deep underground aquifers may in time reach surface waters and become incorporated into the biosphere. Atmospheric pollutants eventually deposit on soils or surface waters and the main mechanisms for removal of contaminants from soil involve transport by water by a sequence of processes, including surface runoff and leaching into soil water that eventually seeps to streams.

Because warm water is discharged by the condenser cooling systems of some power stations, the question is sometimes raised of possible synergistic

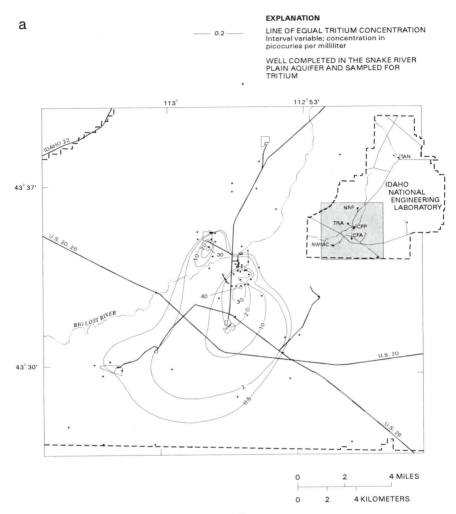

FIGURE 5-6   Extent of the (a) tritium and (b) ^{90}Sr groundwater plumes at the Idaho National Engineering Laboratory, Idaho Falls, Idaho. [From Bartholomay *et al.* (1995).]

ecological effects of temperature and ionizing radiation. No such synergism has been demonstrated over a wide range of temperatures or dose rates, except that in some cases organisms grown in warm water have been shown to absorb radionuclides as much as 50% faster owing to increased growth rates (Harvey, 1970). Ophel and Judd (1966) administered ^{131}I and ^{90}Sr

b

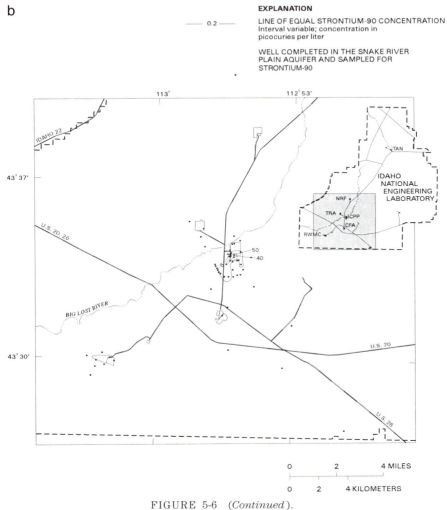

FIGURE 5-6   (*Continued*).

to goldfish (*Carassius auratus*) and found no impairment of ability to withstand near-lethal temperatures at doses of 10,000 rad (100 Gy) to bone and 100,000 rad (1000 Gy) to the thyroid. Angelovic *et al.* (1969) reported that the estuarine fish *Fundulus heteroclitus* has lowered tolerance to heat and salinity when irradiated, but the administered doses were very high, in the range of 2000 to 6000 rad (20 to 60 Gy).

MIXING WITHIN AQUATIC SYSTEMS

It was seen in Chapter 4 that the concentration of an atmospheric pollutant downwind of a source of known strength can be calculated conveniently if a few readily obtainable meteorological data are available. The same diffusion equations can be used under many topographical conditions and, if the source strength is known, one need only measure wind direction and velocity and estimate the degree of atmospheric stability to approximate the downwind concentration within reasonable limits of uncertainty.

Although much progress has been made (NCRP, 1984a; Peterson, 1983; Jirka *et al.,* 1983; Till and Meyer, 1983), a generalized approach to the dispersion of pollutants introduced into a body of water is not possible given the state of our knowledge. The rate of mixing is dependent on depth of water, type of bottom, shoreline configuration, tidal factors, wind, temperature, and depth at which the pollutant is introduced, among other factors. Each stream, river, bay, lake, sea, and ocean has its own mixing characteristics that vary from place to place and from time to time. Hydrologists have developed useful dispersion equations that are valuable for the specific situations for which they are intended, but site-specific parameters are usually required that limit the general applicability of the equations without prior field studies.

The mixing process in the aquatic environment is also complicated by the fact that the fate of a pollutant may be dependent on other physical and biological processes (Figs. 5-7 and 5-8). If a pollutant is a suspended solid, it can settle to the bottom, be filtered by organisms, or become attached to plant surfaces. Pollutants in solution can sorb on suspended organic and inorganic solids or can be assimilated by the plants and animals. The suspended solids, dead biota, and excreta settle to the bottom and become part of the organic-rich substrate that supports the benthic community of organisms. The sediments more often act as a sink (temporary or permanent) for pollution, but they may also become a source, as when they are resuspended during periods of increased turbulence or are dredged and deposited elsewhere. The sediments may also serve as a secondary source of pollution when desorption occurs. Lentsch *et al.* (1972) showed that the role of estuarine sediments as a source for Mn and ^{54}Mn is related to salinity and that the K_d for ^{54}Mn may oscillate with the tidal cycle under estuarine conditions. Figure 5-9 illustrates how the K_d for ^{137}Cs can be influenced by the presence of cations that compete for sorption sites. It is seen that the K_d for ^{137}Cs in the tidal portion of the Hudson River estuary varies inversely with Cl$^-$ over a range of two orders of magnitude (Jinks and Wrenn, 1976; Linsalata *et al.,* 1985). The chloride ion itself plays no role in competing with ^{137}Cs for sorption sites on particulate matter, but is simply a measure

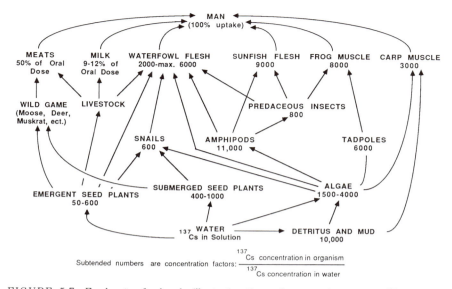

FIGURE 5-7 Freshwater food web, illustrating the pathways to humans for ^{137}Cs in the aquatic environment. [From Pendleton and Hanson (1958).]

of salinity and the associated presence of cations that are normally present in seawater and compete with Cs for sorption sites. Laboratory studies have shown that K^+ is the primary cation in competition with ^{137}Cs for sorption sites (Hairr, 1974).

Prediction of the dispersion of pollutant species that have large K_d values and thus favor the particulate phase is more difficult than for those that remain in solution. Elements that tend to remain in solution include Sr, Cr, and Sb. Elements that are easily sorbed on sediments and suspended matter include Cs, Mn, Fe, Co, and the actinide elements.

If sufficient information can be obtained about physical characteristics of a body of water, it is possible to estimate with some degree of certainty its capacity to receive radioactive nuclides without exceeding the permissible limits of human exposure. However, in the absence of site-specific information, the uncertainties that exist are likely for some time to result in a highly conservative approach (i.e., one that errs on the safe side) to the discharge of radionuclides into the aquatic environment. Proper planning of liquid-waste management practices often requires that dispersion and ecological studies be conducted at each site where a radioactive waste outfall is to be located.

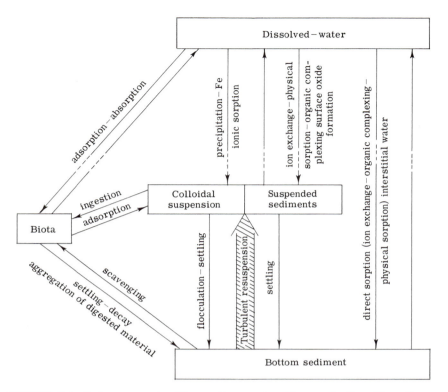

FIGURE 5-8   Basic components of the aquatic ecosystem. The complexities of the pathways among the biota are illustrated in Fig. 5-7. [From Hairr (1974).]

MIXING CHARACTERISTICS OF RECEIVING WATERS

The Oceans

The oceans (Sverdup *et al.*, 1963; Kinne, 1970) cover an area of 3.6×10^8 km^2 and, with an average depth of 3800 m, they contain a total volume of 1.37×10^9 km^3 (Revelle and Schaefer, 1957). Bordering the oceans are the continental shelves, which skirt most of the coastlines to a depth of about 150 m and in some places extend seaward for more than 150 km.

The near-surface waters of the oceans, to a depth that varies geographically from 10 to 200 m, is a region in which rapid mixing occurs as a result of wind action. Because of this mechanical mixing, the vertical gradients of temperature, salinity, and density are nearly uniform.

About 75% of the ocean volume is cold, deep water at a temperature of 1–4°C and a salinity of 3.47%, but between the surface water and the deep water is an intermediate zone characterized by decreasing temperature

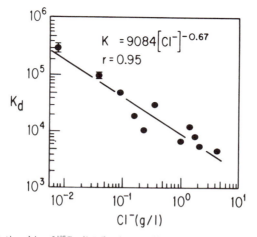

FIGURE 5-9   Relationship of ^{137}Cs distribution coefficients (K_d) and chloride concentrations in continuous water samples at Indian Point, New York, in 1971. [From Jinks and Wrenn (1976).]

and increasing salinity and density with depth. The fact that the salinity is increasing with depth reduces vertical motions because lighter water overlies dense water. The intermediate zone thus tends to restrict exchange between the surface waters and the deep waters (Revelle *et al.,* 1956; Pritchard *et al.,* 1971). The zone where the density changes are greatest, sometimes called the *thermocline* (when due to temperature) or *pycnocline* (when due to density), may be as much as 1000 m in depth, but is usually less.

The characteristic currents of the ocean surface are due primarily to wind action and tend to be related to the surface wind patterns. The movement of surface water has been shown to be as much as 144 km d^{-1} in the Florida current and 66 km d^{-1} in the Kuroshio current of the western Pacific (NAS/NRC, 1957b). Radioactivity from Bikini Atoll was found to drift westward at a rate of about 14 km d^{-1} after tests of nuclear weapons in 1954 (Miyake and Saruhashi, 1960). The major surface currents of the oceans are shown in Fig. 5-10.

A number of studies on the manner in which radioactive substances diffuse vertically and horizontally in the mixed layer have been made in connection with the United States weapons testing program in the Marshall Islands (see Chapter 9) and in the Irish Sea, where extensive studies of the fate of radioactive wastes discharged from Sellafield (formerly known as Windscale) have been conducted by the British (see Chapter 11).

Studies of mixing of radioactive fallout in the Pacific indicated a persistent holdup of radioactivity near the surface for as long as 5 years after

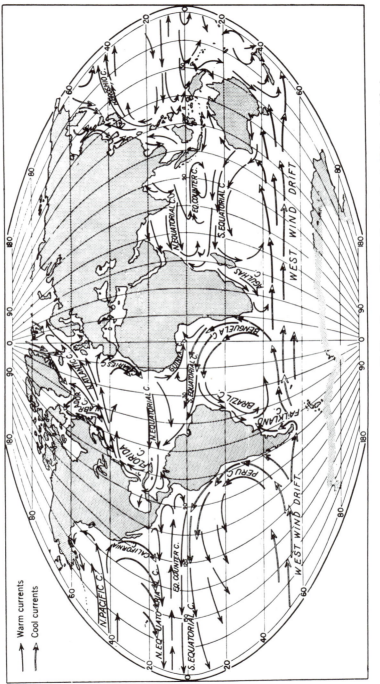

FIGURE 5-10 Principal ocean currents of the world. [From Petterssen (1958). "Introduction to Meteorology," 2nd Ed. © 1958 McGraw-Hill.]

cessation of atmospheric tests of nuclear weapons (Volchok and Kleinman, 1971). Strontium-90 and ^{137}Cs have been the nuclides most thoroughly studied, and it has been found that the ratio of the two nuclides has been constant with respect to time, depth, and sampling location. The complexities of the subject are well illustrated in the report by Bowen *et al.* (1980) of the vertical and horizontal distribution of fallout radionuclides in the Pacific.

After a series of tests in the Marshall Islands in 1954, extensive surveys were made of the spread of radioactivity in the northern Pacific. The general course of the contamination was initially in a westerly direction to the region of the Asiatic mainland, where the contamination turned north into the Kuroshio current. The data from these surveys have been summarized by Harley (1956) and by Miyake and Saruhashi (1960) and are illustrated in Fig. 5-11. Seventeen years after the 1954 tests at Bikini Atoll, the annual flux of actinide elements from the lagoon was estimated to be 3 Ci (110 GBq) of ^{241}Am and 6 Ci (220 GBq) of ^{239,240}Pu (Nevessi and Schell, 1975).

Fission products introduced into surface waters near Bikini in the Marshall Islands (Folsom and Vine, 1957) were found to have moved 225 km in 40 days and to have diffused during this time only to a depth of 30 to 60 m. The horizontal area was found to be about 40,000 km^2. The dilution was such that had 1000 Ci (37 TBq) been introduced, the average concentration at the end of 40 days would have been $1.5 \times 10^{-10} \mu\text{Ci}\,\text{mL}^{-1}$ (5.6 Bq m^{-3}).

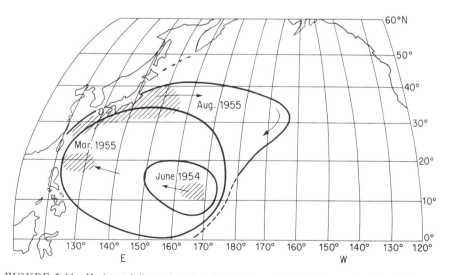

FIGURE 5-11   Horizontal dispersion of nuclear weapons debris in the western Pacific Ocean after tests by the United States in the Marshall Islands in 1954. [From Miyake and Saruhashi (1960).]

The mechanisms and rates of dispersion of water from the deep ocean bottom can be inferred from measurements of the vertical distributions of ^{226}Ra and ^{230}Th, both of which are produced from decay of ^{238}U in the bottom sediments. From various data, Koczy (1960) developed the model of vertical diffusion shown in Fig. 5-12. Dissolved substances released from the ocean floor diffuse slowly through a friction layer 20 to 50 m in depth, within which the rates of mixing are controlled by molecular diffusion. Mixing is most rapid (3 to 30 cm^2 s^{-1}) just above the friction layer and decreases rapidly with height above the ocean floor to a level about 1000 m below the surface, where a secondary minimum ($\sim10^{-2}$ cm^2 s^{-1}) is thought to exist. Diffusion rates then increase as one approaches the mixed layer, where the diffusion coefficients range from 50 to 500 cm^2 s^{-1}.

Koczy (1960) estimated the vertical velocity in the Atlantic Ocean to be between 0.5 and 2 m y^{-1} at depths between 750 to 1750 m. If these values apply at greater depths, a radioactive solution placed at a depth of 3000 m would not appear in the surface water for more than 1000 y.

The vertical motions of the oceans have also been studied by Pritchard *et al.* (1971) using vertical profiles of ^{14}C concentrations. A vertical velocity of about 6.6 m y^{-1} was found to be typical in the northeast Pacific Ocean to a depth of about 1000 m.

Pritchard (1960) had earlier calculated the rate at which a radioactive substance would diffuse to the surface if released from the ocean depths. He used a simplified model in which a 5-m layer of contamination was

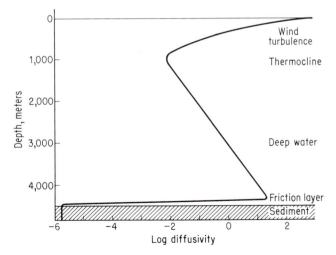

FIGURE 5-12   Vertical diffusion from the ocean depths according to Koczy (1960).

assumed to rise by 5 m each year and its position occupied by a new contaminated layer. No exchange between the 5-m layers was assumed, but horizontal diffusion was assumed to occur, so that as the radioactivity rises, each layer has a greater horizontal spread and, therefore, a lower concentration than the layer beneath.

The horizontal velocity for the deep ocean was not known but was assumed by Pritchard to lie between 2×10^{-3} and 2×10^{-4} m s^{-1}, so that horizontal spread of the material over most of the area of the North Atlantic would occur between 40 and 400 years, which is somewhat less than the 600 years required to rise from a depth of 4000 m to a depth of 1000 m. Based on Pritchard's model, at the end of 600 years the radioactivity released during a 1-year period would be contained in a layer 5 m thick having a horizontal area of 3×10^{13} m^2. From this it was calculated that 5×10^{11} Ci (1.85×10^{22} Bq) of ^{90}Sr could be placed on the bottom each year at a depth of 4000 m without exceeding a concentration of 10^{-9} μCi mL^{-1} (37 Bq m^{-3}) at the base of the 1000-m layer. An alternative estimate, based on the assumption that rapid mixing occurs in the bottom 1000 m, reduces Pritchard's estimate to 9×10^8 Ci (3.3×10^{19} Bq), which is in close agreement with the figure suggested by Koczy (1960). These estimates were based on very few data. More recent information may become available as a result of an international research program that was conducted to study the feasibility of using the deep ocean sediments as a repository for high-level radioactive wastes (Robinson and Marietta, 1985; see Chapter 11). An effort to computer model ocean behavior is showing promise (Kerr, 1993).

Rivers, Estuaries, and Coastal Waters

Knowledge of the behavior of trace substances in rivers, estuaries, and coastal waters is important because these aquatic systems are major receptors of effluents from industrial plants and municipalities (Fisher *et al.*, 1979). The term estuary is usually applied to the tidal reaches of a river, and is a semi-enclosed coastal body of water that has a free connection to the open sea, and within which the marine water is measurably diluted with fresh water from land drainage (Pritchard, 1967).

The estuarine waters, in which tidal action brings about a mixing of salt and fresh water, are of special importance because of their high biological productivity. In addition to providing habitat for large populations of shellfish, estuaries also serve as the nursery grounds for many species of fish that later move to offshore waters where they are harvested (Reid and Wood, 1976).

Each estuary has its own physical characteristics that must be studied in detail on an individual basis. One kind of study that can be made is the investigation of diffusion and convection as was done in the Delaware River basin (Parker *et al.,* 1961). A scale model of the Delaware basin was constructed at the United States Army Waterways Experiment Station at Vicksburg, Mississippi (Fig. 5-13). The model was 1200 m long and about 200 m wide. Figure 5-14 illustrates the type of information obtained by dye studies over a period of 58 tidal cycles (about 1 month). In this particular study, which involved instantaneous injection of a given dose, the concentration at the end of 58 tidal cycles remained at approximately 1% of the maximum concentration during the initial tidal cycle. The conditions of the experiment were conservative: there was no radioactive decay, sedimentation, or biological uptake. Thus, only the diminution in concentration due to mixing was measured.

The most extensive set of river mixing measurements in the United States is probably on the Columbia River, where the large plutonium-producing reactors at Hanford used river water for cooling and discharged traces of induced radioactivity, notably ^{32}P and ^{65}Zn, in the effluents. Studies of these two isotopes have been made in the water, sediments, and biota of the river. An example of the manner in which dilution of the Hanford

FIGURE 5-13   Scale model of the Delaware River, one of several models of rivers, estuaries, and bays by which flow characteristics are studied by the U.S. Army Corps of Engineers at their laboratory in Vicksburg, Mississippi. (Courtesy of U.S. Army Corps of Engineers.)

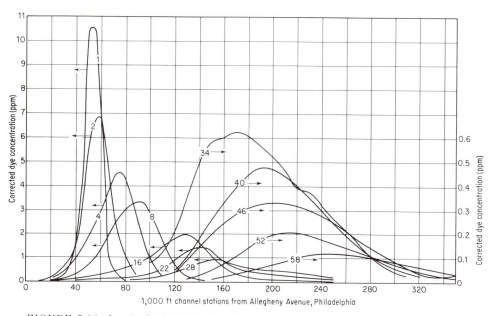

FIGURE 5-14 Longitudinal distribution of contaminants after a designated number of tidal cycles in the Delaware River near Philadelphia. [From Parker *et al.* (1961); reproduced from *Health Physics,* Vol. 6, by permission of the Health Physics Society.]

effluents takes place is shown in Fig. 5-15 (Foster, 1959). Additional data on the dosimetric implications of the Hanford releases are given in Chapter 14.

Many models have been published to describe the dispersion of pollutants introduced into aquatic systems. The models take into consideration the location of the source, the form and rate of input of the pollutant, removal and transformation mechanisms, and transport processes in the water and sediments (Peterson, 1983; NCRP, 1984a). The aquatic transport and diffusion equations provide estimates of the radionuclide concentrations within a water body, the rate of deposition and accumulation of radionuclides on the shoreline and bottom, and movement of the radionuclides through biotic pathways to humans (Jirka *et al.*, 1983). The methods by which the models and computer codes are developed are discussed elsewhere (Little, 1983). The factors with which the model must deal are numerous, as can be seen from Fig. 5-16, which identifies some of the variables that must be included in the development of any mathematical model or computer code. It has been noted elsewhere (NCRP, 1984a) that despite all the measurements that have been made in the most thoroughly studied rivers of the

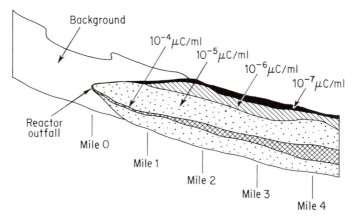

FIGURE 5-15   Horizontal mixing of radioactive reactor effluents in the Columbia River. [From Foster (1959).]

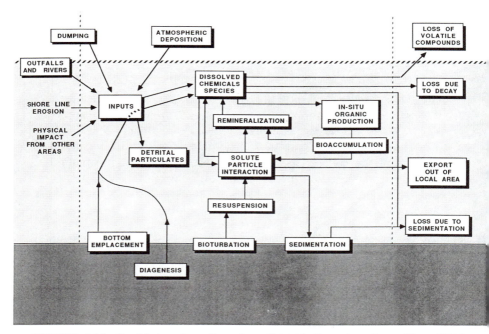

FIGURE 5-16   Possible transformations of a pollutant in the water column. [From NOAA (1979).]

United States during the past 30 years, there are insufficient data to validate the diffusion models that have been published.

BIOLOGICAL UPTAKE OF RADIONUCLIDES

In an aquatic system, the supply of basic compounds such as carbon dioxide and of elements such as oxygen, calcium, hydrogen, and nitrogen is either contained in solution or held in reserve in the bottom sediments (Reid and Wood, 1976). These nutrients are absorbed and metabolized through the utilization of solar energy by two main types of food-producing organisms: rooted or large floating plants and minute floating plants called phytoplankton. Since food production by the higher plant forms is limited to relatively shallow water, it is the phytoplankton that have the main responsibility for converting the mineral resources of the aquatic environment into food for the higher organisms.

The phytoplankton serve as food for small zooplankton, which in turn serve as the basic nourishment of several higher trophic levels. The phytoplankton also serve as food for certain filter-feeding fish and bottom-dwelling animals. Sedimentation of the excrement of aquatic animals and the action of organisms that decompose dead plants and animals eventually complete the cycle by returning the nutrient elements to abiotic forms in which they again become available to the phytoplankton (Lowman *et al.,* 1971).

Some elements, including copper, manganese, zinc, and iron, are present in varying amounts in water and sediments that may or may not be essential to life processes. The aquatic organisms sequester these trace elements to a varying degree, and this phenomenon must be taken into consideration when any decision is made as to the rate at which any given radionuclide can be released to any body of water.

Onishi *et al.* (1981) thoroughly reviewed the transport of radionuclides in aquatic systems. The ratio of the concentration of an element in the organism to the concentration in the water is known as the concentration factor (CF). This ratio should be measured under equilibrium conditions, since an organism that has just entered a contaminated environment would obviously provide an inappropriate sample and would yield a spuriously low CF.

The CF can also be greatly influenced by the presence of chemical congeners in the water. One would expect the CF for ^{90}Sr to be greatly affected by the concentration of Ca in the water. Because of the relatively high concentration of K in seawater, the CF for ^{137}Cs in freshwater biota is orders of magnitude higher than in marine or estuarine organisms.

Considering all of the foregoing, it is not surprising that for each element of interest there is a wide range of CFs reported in the literature. The variability could undoubtedly be reduced if reported CFs could be limited to those in which the organism was in equilibrium with its environment and if other sources of variability could be taken into consideration. In most cases, the reports in the literature do not provide sufficient information to permit such adjustments (NCRP, 1984a). If the reported data could be normalized to either wet weight or dry weight, the variability would be reduced by a factor of 5 (Blaylock, 1982). The NRC has issued recommended CFs for use in dose assessment (NRC, 1977), but for most elements it is difficult to justify single recommended values in view of the great variability in the CFs reported in the literature.

The significance of the presence of radionuclides in marine and freshwater foods depends on the part of the organism in which the radionuclide is located. A radionuclide is more important in risk assessment when it concentrates in an organ that is consumed by humans than if it deposits in the portion that is not eaten. Thus, although it is known that clams, oysters, and scallops concentrate ^{90}Sr, as do certain crabs, this element is stored in the shell, which is not ordinarily consumed. On the other hand, ^{65}Zn and ^{60}Co are known to concentrate in the edible tissues of seafood. A CF obtained from analysis of whole fish can be an order of magnitude higher than a CF based on analysis of the fish muscle (NCRP, 1984a).

In British studies of the fate of radioactive effluents from Windscale (now called Sellafield) (Dunster, 1958; Preston and Jeffries, 1969), it was found that the factor that limited the quantity of waste discharged into nearby coastal waters was the accumulation of radioactivity in seaweed harvested in Cumberland about 20 km from the point of discharge. The local inhabitants who regularly consumed substantial quantities of seaweed were the limiting factor in determining the maximum amount of radioactivity that could be discharged into this particular environment.

At the Idaho National Engineering Laboratory west of Idaho Falls, ponds have been used for decades to receive low-level liquid wastes. Waterfowl resting or feeding on these ponds incorporate radioactivity and transport it to uncontrolled areas as they migrate. Halford *et al.* (1982) estimated that about 700 waterfowl visited the Test Reactor Area annually and that they stayed for an average of 6 days. The mean dose commitment to a hunter who shot and consumed a bird shortly after it left the pond was estimated to be 12 mrem (120 μSv) (Halford *et al.*, 1981). The maximum dose was estimated to be 54 mrem (540 μSv). Studies of radioactivity in ducks have continued (Markham *et al.*, 1988), and more recently the doses were found to be lower by more than a factor of 10 because of shorter waterfowl stay times and changes in waste management practices (Morris, 1993). Mourning doves obtain water at radioactive waste management ponds and can serve as a potential pathway of radioactivity from the environment to humans (Markham and Halford, 1982).

Another example in which local factors influence the significance of river water contamination was found at Hanford, where Columbia River water irrigates thousands of square miles of land downstream from the Hanford reservation, resulting in ^{65}Zn contamination of farm produce (Davis *et al.*, 1958). The concentration in beef and cow's milk was particularly noteworthy.

ROLE OF THE SUSPENDED SOLIDS AND SEDIMENTS

The sediments play a predominant role in aquatic radioecology by serving either as a sink or as a temporary repository for radioactive substances, which can then pass by way of the bottom-feeding biota or by resuspension or dissolution to the higher trophic levels (Duursma and Gross, 1971).

Figure 5-17 (Linsalata *et al.*, 1986) illustrates the high degree of concentration of ^{137}Cs that takes place in Hudson River sediments ($K_d = 10^4$–10^5) relative to that in fish, rooted plants, or dissolved in water. These data were collected in the northern (freshwater) reaches of the Hudson River estuary beyond the range of influence of a nuclear power reactor located at Indian Point. Figure 5-17 shows: (a) the annual concentrations of this nuclide in filtered river water, surficial (0–5 cm) bottom sediments, aquatic plants, and indigenous fish; and (b) the rate of fallout deposition and reactor discharge of ^{137}Cs. In general, the long-term trend of ^{137}Cs concentrations in fish (and to a lesser extent in plants) tends to follow the concentrations in water rather closely (simple correlation coefficient, r, = 0.50). However, between 5 and 20% of the ^{137}Cs content of fish has, for various times, locations, and species, been attributed to a sedimentary source (Wrenn *et al.*, 1972; Jinks, 1975).

Radionuclides introduced into a body of water reach the bottom sediment primarily by sorption on suspended solids that later deposit on the bottom. The deposited remains of biota that have absorbed or adsorbed pollutants may also be an important source.

In studies of the Clinch River below Oak Ridge, Tennessee, the amounts of radioactivity contained by the suspended solids were found to be variable (Parker *et al.*, 1966), which is not surprising considering that the load and composition of particulate matter vary from place to place in a river. The tendency of the sediments to remove pollutants will depend on their capacity for sorption and ion exchange, as well as

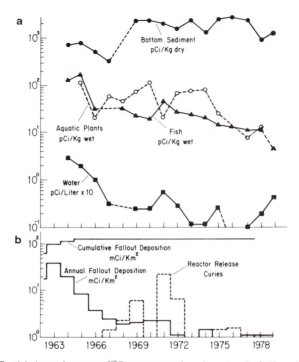

FIGURE 5-17 (a) Annual average ^{137}Cs concentrations in water (<0.45 μm), surface sediment (0–5 cm), rooted vegetation, and various indigenous fish species collected from the freshwater reaches of the Hudson River estuary. The apparent increase in ^{137}Cs concentration in bottom sediments (1967–1969) was probably a result of changes in the sampling locations. (b) Rate of fallout deposition and reactor discharge of ^{137}Cs. [From Linsalata et al. (1986).]

the salinity of the overlying water. It was estimated that of the total quantity of radioactive materials released into the river during a 20-year period, the sediments contained 21% of the ^{137}Cs, 9% of the ^{60}Co, 0.4% of the ^{106}Ru, about 25% of the rare earth radionuclides, and about 0.2% of the ^{90}Sr (Pickering et al., 1966). The depth distribution of ^{137}Cs in cores several feet deep were found to be well correlated with the annual releases to the river. This was found not to be the case for ^{60}Co. In other river systems, the sediments might be less quiescent owing to stirring or scouring during periods of high runoff.

Lentsch et al. (1972) studied the behavior of ^{54}Mn released into the Hudson River from reactors located at Indian Point. The manganese content of several species of rooted aquatic plants was found to be proportional to the dissolved manganese concentration in water, but this in turn was shown to be highly variable owing to the periodic intrusion of salt water, which was found to release manganese bound in the river sediments. In this investigation, ^{54}Mn was found to behave similarly to stable manganese present in the river system (Lentsch *et al.*, 1972).

CHAPTER 6

Natural Radioactivity

Natural radioactivity originates from extraterrestrial sources as well as from radioactive elements in the earth's crust. About 340 nuclides have been found in nature, of which about 70 are radioactive and are found mainly among the heavy elements. All elements having an atomic number greater than 80 possess radioactive isotopes, and all isotopes of elements heavier than number 83 are radioactive.

The radioactivity of the earth includes three major categories. Primordial radionuclides have half-lives sufficiently long that they have survived since their creation. Secondary radionuclides are derived from radioactive decay of the primordials. Cosmogenic radionuclides are continuously produced by bombardment of stable nuclides by cosmic rays, primarily in the atmosphere. A much larger number of radioactive isotopes than now exist were produced when the matter of which the universe is formed first came into being several billion years ago, but most of them have decayed out of existence. The primordial radionuclides that now exist are those that have half-lives at least comparable to the age of the universe. Radioisotopes with half-lives of less than about 10^8 years have become undetectable in the 30 or so half-lives since their creation, whereas radionuclides with half-lives greater than 10^{10} years have decayed very little up to the present time.

In most places on earth the natural radioactivity varies only within narrow limits, but in some localities there are wide deviations from normal levels because of abnormally high soil concentrations of radioactive minerals.

NATURALLY OCCURRING RADIOACTIVE SUBSTANCES

The naturally occurring radionuclides can be divided into those that occur singly (Tables 6-1a and 6-1b) and those that are components of three chains of radioactive elements. The uranium series (Table 6-2) originates with ^{238}U, the thorium series (Table 6-3) orignates with ^{232}Th, and the actinium series (Table 6-4) originates with ^{235}U. Each table provides the isotope, half-life, and principal radiations. Tables 6-1a and 6-1b also provide typical concentrations. A fourth family, the neptunium series, which originated in the parent element ^{241}Pu, is known to have existed at one time, but this nuclide has a half-life of only 14 years and existed only briefly after its formation. Other members of that series also have relatively short half-lives. The only surviving member of the neptunium family is the nearly stable nuclide ^{209}Bi, which has a half-life estimated to be about 2×10^{18} years.

The three remaining chains of radioactive elements and the long-lived pimordial nuclide ^{40}K account for much of the external background radiation dose from radioactivity to which humans are exposed. Of the 22 identified cosmogenic nuclides (Table 6-1a), only four, ^{14}C, ^{3}H, ^{22}Na, and ^{7}Be, are of any consequence whatsoever from the perspective of dose to humans. Only 2 of the 17 nonseries, primordial nuclides, ^{40}K and ^{87}Rb, are of most interest (Table 6-1b). In nature, ^{235}U and a few other nuclides of uranium and thorium undergo fission spontaneously or as a result of interactions with neutrons that originate from cosmic rays or other natural sources. The half-life of ^{235}U due to spontaneous fission is between 10^{15} and 10^{16} years, which means that decay by this process proceeds at a rate less than 10^{-7} of that due to α emission. Others of the heavy nuclides undergo spontaneous fission with half-lives that range from 10^{14} to 10^{20} years (Rankama, 1954).

Many transuranic elements, such as plutonium, neptunium, and americium, which now exist because they have been produced artificially (Seaborg, 1958), must have existed in nature at one time, but their half-lives are so short that they disappeared long ago. However, some of the transuranic elements are produced in minute amounts by naturally occurring neutrons that result from cosmic radiation and spontaneous fission of uranium isotopes. Plutonium-239 has been detected in pitchblende in a ratio to ^{238}U of 10^{-11} to 10^{-13}, and ^{237}Np has been identified in uranium minerals in a ratio to ^{238}U of 1.8×10^{-12}.

URANIUM

The uranium normally found in nature consists of three isotopes having mass numbers 234, 235, and 238. In the earth's crust, ^{238}U is present in

TABLE 6-1a

Radionuclides Induced in the Earth's Atmosphere by Cosmic Rays[a]

Radionuclide	Half-life	Major radiations	Target nuclides	Typical concentrations (Bq kg^{-1})[b]		
				Air (troposphere)	Rainwater	Ocean water
^{10}Be	1,600,000 y	β	N, O			2×10^{-8}
^{26}Al	720,000 y	β^+	Ar			2×10^{-10}
^{36}Cl	300,000 y	β	Ar			1×10^{-5}
^{80}Kr	213,000 y	K X ray	Kr			
^{14}C	5730 y	β	N, O			5×10^{-3}
^{32}Si	~650 y	β	Ar			4×10^{-7}
^{39}Ar	269 y	β	Ar			6×10^{-8}
^{3}H	12.33 y	β	N, O	1.2×10^{-3}		7×10^{-4}
^{22}Na	2.60 y	β^+	Ar	1×10^{-6}		
^{35}S	87.4 d	β	Ar	1.3×10^{-4}	2.8×10^{-4}	
^{7}Be	53.3 d	γ	N, O	0.01	$7.7–107 \times 10^{-3}$	
^{37}Ar	35.0 d	K X ray	Ar	3.5×10^{-5}	0.66	
^{33}P	25.3 d	β	Ar	1.3×10^{-3}		
^{32}P	14.28 d	β	Ar	2.3×10^{-4}		
^{38}Mg	21.0 h	β	Ar			
^{24}Na	15.0 h	β	Ar		$3.0–5.9 \times 10^{-3}$	
^{38}S	2.83 h	β	Ar		$6.6–21.8 \times 10^{-2}$	
^{31}Si	2.62 h	β	Ar			
^{18}F	109.8 m	β^+	Ar			
^{39}Cl	56.2 m	β	Ar		$1.7–8.3 \times 10^{-1}$	
^{38}Cl	37.29 m	β	Ar		$1.5–25 \times 10^{-1}$	
^{34m}Cl	31.99 m	β^+	Ar			

[a]Adapted from NCRP (1987a); Perkins and Nielson (1965). Reproduced from *Health Physics* Vol. 11, by permission of the Health Physics Society.
[b]One Bq = 27 pCi.

TABLE 6-1b

Nonseries Primordial Radionuclides[a]

Radionuclide	Half-life (y)	Major radiations	Typical crustal concentration (Bq kg^{-1})[b]
^{40}K	1.26×10^9	β, γ	630
^{50}V	6×10^{15}	γ	2×10^{-5}
^{87}Rb	4.8×10^{10}	β	70
^{113}Cd	$>1.3 \times 10^{15}$	Not reported	$<2 \times 10^{-6}$
^{115}In	6×10^{14}	β	2×10^{-5}
^{123}Te	1.2×10^{13}	X rays	2×10^{-7}
^{138}La	1.12×10^{11}	β, γ	2×10^{-2}
^{142}Ce	$>5 \times 10^{16}$	Not reported	$<1 \times 10^{-5}$
^{144}Nd	2.4×10^{15}	α	3×10^{-4}
^{147}Sm	1.05×10^{11}	α	0.7
^{152}Gd	1.1×10^{14}	α	7×10^{-6}
^{174}Hf	2.0×10^{15}	α	2×10^{-7}
^{176}Lu	2.2×10^{10}	e^-, γ	0.04
^{187}Re	4.3×10^{10}	β	1×10^{-3}
^{190}Pt	6.9×10^{11}	α	7×10^{-8}
^{192}Pt	1×10^{15}	α	3×10^{-6}
^{209}Bi	$>2 \times 10^{18}$	α	$<4 \times 10^{-9}$

[a] Adapted from NCRP (1987a).
[b] One Bq = 27 pCi.

the amount of 99.28% and is usually in radioactive equilibrium or near-equilibrium with ^{234}U, which is present in the amount of 0.0058%. Uranium-235, the parent isotope of the actinium series, is present in the amount of 0.71%.

Uranium is found in all rocks and soils. Some example concentrations are listed in Table 6-5, which shows that the acid igneous rocks contain concentrations on the order of 3 ppm, about 100 times greater than that in the ultrabasic igneous rocks. The phosphate rocks of Florida and southeastern Idaho and neighboring areas contain as much as 120 ppm and have been used as a commercial source of uranium (Clegg and Foley, 1958; NCRP, 1993b). The high uranium content of phosphate rocks is reflected in correspondingly high uranium concentrations in commercial phosphate fertilizers. Typical concentrations for the more prevalent rock classes, the crustal average, and soil concentrations are listed in Table 6-6, which shows that uranium concentrations range from 0.5 to 4.7 ppm in the common rock types. The overall effect of soil development results in the average soil concentration of uranium being less than the average rock concentration.

TABLE 6-2

Uranium Series[a]

Nuclide	Historical name	Half-life	Major radiations
^{238}U	Uranium I	4.47×10^9 y	α, <1% γ
^{234}Th	Uranium X$_1$	24.1 d	β, γ
^{234m}Pa	Uranium X$_2$	1.17 m	β, <1% γ
^{234}Pa	Uranium Z	21.8 y	β, γ
^{234}U	Uranium II	244,500 y	α, <1% γ
^{230}Th	Ionium	7.7×10^4 y	α, <1% γ
^{226}Ra	Radium	1600 y	α, γ
^{222}Rn	Emanation radon	3.8 d	α, <1% γ
^{218}Po	Radium A	3.05 m	α, <1% γ
^{214}Pb (99.98%)	Radium B	26.8 m	β, γ
^{218}At (0.02%)	Astatine	2 s	α, γ
^{214}Bi	Radium C	19.9 m	β, γ
^{214}Po (99.98%)	Radium C'	164 μs	α, <1% γ
^{210}Tl (0.02%)	Radium C″	1.3 m	β, γ
^{210}Pb	Radium D	22.3 y	β, γ
^{210}Bi	Radium E	5.01 d	β
^{210}Po (~100%)	Radium F	138.4 d	α, <1% γ
^{206}Tl (0.00013%)	Radium E″	4.20 m	β, <1% γ
^{206}Pb	Radium G	Stable	None

[a]Adapted from Bureau of Radiological Health (1970).

The uranium content of air in New York State was reported to range from 35 to 470 aCi m^{-3} (the prefix a- refers to atto-, or 10^{-18}) (1.3 to 17 μBq m^{-3}) and to be correlated with the concentration of suspended particulates (McEachern et al., 1971). Airborne soil and possily coal fly ash were thought to be the most likely sources. Subsequent measurements of uranium in air are somewhat lower. A value of 0.3 μBq m^{-3} (8 aCi m^{-3}) was reported near Chicago (Goldchert et al., 1985) and a value of 0.6 μBq m^{-3} (16aCi m^{-3}) was reported for New York City air (Fisenne et al., 1987). A reduction in coal burning associated with changes in fuel use patterns and emission control likely account for the lower values of uranium in air. The NCRP (1987a) reviewed the limited data on uranium in air and adopted a representative value for the United States of 0.7 μBq m^{-3} (20 aCi m^{-3}).

Uranium occurs in traces in many commercial products. Pre-World War II samples of steel analyzed by Welford and Sutton (1957) contained uranium in the range of 0.01 to 0.2 ppm. Surprisingly, photographic emulsions and other photographic materials contain from 0.2 to 1 ppm of uranium (Smith

TABLE 6-3

Thorium Series[a]

Nuclide	Historical name	Half-life	Major radiations
^{232}Th	Thorium	1.4×10^{10} y	α, <1% γ
^{228}Ra	Mesothorium I	5.75 y	β, <1% γ
^{228}Ac	Mesothorium II	6.13 h	β, γ
^{228}Th	Radiothorium	1.91 h	α, γ
^{224}Ra	Thorium X	3.66 d	α, γ
^{220}Rn	Emanation thoron	55.6 s	α, <1% γ
^{216}Po	Thorium A	0.15 s	α, <1% γ
^{212}Pb	Thorium B	10.64 h	β, γ
^{212}Bi	Thorium C	60.55 m	α, γ
^{212}Po (64%)	Thorium C′	0.305 μs	α
^{208}Tl (36%)	Thorium C″	3.07 m	β, γ
^{208}Pb	Thorium D	Stable	None

[a]Adapted from Bureau of Radiological Health (1970).

TABLE 6-4

Actinium Series[a]

Nuclide	Historical name	Half-life	Major radiations
^{235}U	Actinouranium	7.038×10^8 y	α, γ
^{231}Th	Uranium Y	25.5 h	β, γ
^{231}Pa	Protoactinium	2.276×10^4 y	α, γ
^{227}Ac	Actinium	21.77 y	β, <1% γ
^{227}Th (98.62%)	Radioactinium	18.72 y	α, γ
^{223}Fr (1.38%)	Actinium K	21.8 m	β, γ
^{223}Ra	Actinium X	11.43 d	α, γ
^{219}Rn	Emanation actinon	3.96 s	α, γ
^{215}Po	Actinium A	1.78 ms	α, <1% γ
^{211}Pb (~100%)	Actinium B	36.1 m	β, γ
^{215}At (0.00023%)	Astatine	~0.1 ms	α, <1% γ
^{211}Bi	Actinium C	2.14 m	α, γ
^{211}Po(0.273%)	Actinium C′	0.516 s	α, γ
^{207}Tl (99.73%)	Actinium C″	4.77 m	β, <1% γ
^{207}Pb	Actinium D	Stable	None

[a]Adapted from Bureau of Radiological Health (1970).

TABLE 6-5

Average Uranium Concentration in Various Rocks[a]

Rock type	Uranium concentration		
	ppm	pCi g^{-1}	Bq kg^{-1}
Acid igneous	3.0	0.99	37
Intermediate igneous	1.5	0.50	18
Basic igneous	0.6	0.20	7.3
Ultrabasic igneous	0.03	0.010	0.37
Meteorites	0.003	0.0010	0.037
Phosphate rock (Florida)	120	40	1500
Phosphate rock (N. Africa)	20–30	6.6–10	240–370
Bituminous shale (Tennessee)	50–80	17–26	610–980
Normal granite	4	1.3	49
Limestones	1.3	0.43	16
Other sedimentary rocks	1.2	0.40	15

[a]From Lowder and Solon (1956).

TABLE 6-6

Ranges[a,b] and Averages of the Concentrations of ^{40}K, ^{232}Th, and ^{238}U in Typical Rocks and Soils[c]

Material	Potassium-40		Thorium-232		Uranium-238	
	% total K	Bq kg^{-1}	ppm	Bq kg^{-1}	ppm	Bq kg^{-1}
Igneous rocks						
Basalt (crustal ave.)	0.8	300	3–4	10–15	0.5–1	7–10
Mafic	0.3–1.1	70–400	1.6, 2.7^d	7, 10^d	0.5, 0.9^d	7, 10^d
Salic	4.5	1100–1500	16, 20^d	60, 80^d	3.9, 4.7^d	50, 60^d
Granite (crustal ave)	>4	>1000	17	70	3	40
Sedimentary rocks						
Shale sandstones	2.7	800	12	50	3.7	40
Clean quartz	<1	<300	<2	<8	<1	<10
Dirty quartz	2?	400?	3–6?	10–25?	2–3?	40?
Arkose	2–3	600–900	2?	<8	1–2?	10–25?
Beach sands	<1	<300	6	25	3	40
Carbonate rocks	0.3	70	2	8	2	25
All rock (range)^a	0.3–4.5	70–1500	1.6–20	7–80	0.5–4.7	7–60
Continental crust (ave.)	2.8	850	10.7	44	2.8	36
Soil (ave.)	1.5	400	9	37	1.8	22

[a]Examples of materials outside the ranges can be found, but quantities are relatively small.
[b]One Bq kg^{-1} = 0.027 pCi g^{-1}.
[c]Adapted from NCRP (1987a).
[d]Mean and median, respectively.

and Dzuiba, 1949). Uranium in commercial products is discussed in more detail in Chapter 7.

Because uranium occurs in soils and fertilizers, the element is present in food and human tissues. The daily intake of uranium and other alpha-emitting radionuclides in a sample New York City diet is given in Table 6-7 (Fisenne et al., 1987). These data are consistent with earlier reports (Fisenne and Keller, 1970; Holtzman, 1980; NCRP, 1975c). On average, the annual intake of uranium from all dietary sources is about 13 Bq (320 pCi) (NCRP, 1987a).

The intake of uranium from tap water can be a small or large fraction of the total intake depending on concentrations in local water supplies. Hess et al. (1985) reported population-averaged domestic water concentrations in the United States that range typically from 0.4 Bq m^{-3} (0.01 pCi L^{-1}) in many eastern states, to 13 Bq m^{-3} (0.35 pCi L^{-1}) in midwestern

TABLE 6-7

Radionuclides in a New York City Diet[a]

Type of food	Annual intake (kg y^{-1})	Daily intake (mBq d^{-1})[b]					
		^{234}U	^{238}U	U total	^{230}Th	^{232}Th	^{226}Ra
Fresh vegetables	48	3.0	3.1	6.3	2.6	2.3	7.2
Canned vegetables	22	0.27	0.25	0.52	0.08	0.07	1.3
Root vegetables	10	0.34	0.21	0.55	0.03	0.16	0.41
Potatoes	38	0.11	0.09	0.20	0.02	0.01	0.68
Dry beans	3	0.25	0.23	0.49	0.26	0.22	0.46
Fresh fruit	59	0.32	0.32	0.64	0.02	0.02	7.5
Canned fruit	11	0.08	0.03	0.11	0.01	0	0.20
Fruit juice	28	0.04	0.04	0.09	0.05	0.03	2.1
Bakery products	44	3.5	2.7	6.4	1.2	0.34	9.7
Flour	34	0.43	0.44	0.89	0.46	0.24	8.2
Whole-grain products	11	0.75	0.51	1.3	0.19	0.08	3.1
Macaroni	3	0.03	0.03	0.06	0.02	0.01	0.57
Rice	3	0.02	0.02	0.05	0.01	0	0.06
Meat	79	0.40	0.50	0.90	0.64	0.43	0.43
Poultry	20	0.04	0.04	0.09	0.03	0	1.1
Eggs	15	0.04	0.08	0.12	0.04	0	3.5
Fresh fish	8	0.38	0.29	0.68	0.03	0.03	0.65
Shellfish	1	6.0	5.2	11.4	0.08	0.08	0.16
Dairy products	200	0.54	0.40	0.96	0.20	0.14	3.0
Rounded total		16	14	32	6	4	50

[a]From Fisenne et al. (1987).
[b]One mBq = 0.027 pCi.

states, and to 130 Bq m^{-3} (3.5 pCi L^{-1}) in the western states. Assuming an annual water intake of 0.7 m^3, Hess's values correspond to annual intakes of 0.3, 9, and 90 Bq (7, 240, and 2400 pCi), respectively. Above-normal uranium concentrations in water have been reported in France, Finland, and the former Soviet Union (UNSCEAR, 1982).

In water, ^{238}U and ^{234}U can be found out of equilibrium owing to geochemical processes. As a decay product, ^{234}U resides in the immediate vicinity of the parents, ^{238}U, ^{234}Th, and ^{234m}Pa. The energy released from the decays is thought to weaken the crystal structure in the immediate vicinity of the ^{234}U, thus increasing its mobility relative to the primordial ^{238}U.

Newer studies have shown that the skeleton contains less uranium than previously thought (NCRP, 1987a; Fisenne and Welford, 1986). In the United States, the typical concentration in wet skeleton is about 8 mBq kg^{-1} (0.22 pCi kg^{-1}), which may be compared with previous estimates that ranged from 12 to 200 mBq kg^{-1} (0.32 to 5.7 pCi kg^{-1}). Lung, kidney, and bone receive the highest annual doses from uranium, estimated to be 11, 9.2, and 6.4 μSv (1.1, 0.92, and 0.64 mrem), respectively, for U.S. and Canadian residents (NCRP, 1987a).

Radium-226

Radium-226 and its decay products are responsible for a major fraction of the dose received by humans from the naturally occurring internal emitters. [For an exhaustive review of the subject see IAEA (1990).] Referring to Table 6-2, it is seen that ^{226}Ra is an α emitter that decays, with a half-life of 1622 years, to ^{222}Rn, with a half-life of 3.8 days. The decay of ^{222}Rn is followed by the successive disintegration of a number of short-lived α- and β-emitting progeny. After six decay steps, in which isotopes are produced that range in half-lives from 1.6×10^{-4} seconds to 26.8 minutes, ^{210}Pb is produced, which has a half-life of 22 years. This nuclide decays through ^{210}Bi to produce ^{210}Po (half-life 138 days), which decays by α emission to stable ^{206}Pb.

Radium, being an α emitter, does not add directly to the γ activity of the environment but does so indirectly through its γ-emitting decay products.

Radium-226 Content of Rocks and Soils

Radium-226 is present in all rocks and soils in variable amounts. Igneous rocks tend to contain somewhat higher concentrations than sandstones and limestones. Rankama and Sahama (1950) give a mean concentration of 0.42 pCi g^{-1} (16 Bq kg^{-1}) in limestone and 1.3 pCi g^{-} (48 Bq kg^{-1}) in igneous rock, as listed in Table 6-8. Radium-226 is generally in approximate equilibrium with ^{238}U and these early radium values are consistent with the range of 7–60 Bq kg^{-1} for ^{238}U compiled by the NCRP (1987a) (see Table 6-6).

Radium-226 in Water

The radium content of public water supplies has been reviewed comprehensively by Hess *et al.* (1985), who described the geological and geochemical factors that influence the concentration of the two principal radium isotopes, ^{226}Ra and ^{228}Ra, which are progeny of uranium and thorium, respectively. There is more ^{232}Th than ^{238}U in nature on an activity basis, but there are geochemical factors that cause local concentrations of uranium, which often result in greater amounts of ^{226}Ra relative to ^{228}Ra. Thus, it is has been generally assumed that the ratio ^{226}Ra : ^{228}Ra is greater than unity, although until recently most of the reported measurements have been of ^{226}Ra, with fewer measurements of ^{228}Ra. As discussed next, however, newer data from a stratified random survey suggest that average concentrations of ^{226}Ra and ^{228}Ra in water may be more nearly equal.

The radium content of surface waters is low, 0.1 to 0.5 pCi L^{-1} (4 to 19 Bq m^{-3}), compared to most groundwaters (Hess *et al.*, 1985). Dissolved radium sorbs quickly to solids and does not migrate far from its place of release to groundwater. It has been suggested that radium transport in groundwater is even less than that of ^{222}Rn, which has a half-life of only 3.8 days (King *et al.*, 1982; Krishnaswami *et al.*, 1982).

Interest in the radium content of water supplies increased following publication in 1976 of the National Primary Interim Drinking Water Standards (EPA, 1991a), which included a standard for total radium of 5 pCi L^{-1}

TABLE 6-8

Average Radium, Uranium, Thorium, and Potassium Contents in Various Rocks[a]

Type of rock	^{226}Ra		^{238}U		^{232}Th		^{40}K	
	pCi g^{-1}	Bq kg^{-1}	pCi g^{-1}	Bq kg^{-1}	pCi g^{-1}	Bq kg^{-1}	pCi g^{-1}	Bq kg^{-1}
Igneous	1.3	48	1.3	48	1.3	48	22	810
Sedimentary								
Sandstones	0.71	26	0.4	15	0.65	24	8.8	330
Shales	1.08	40	0.4	15	1.1	41	22	810
Limestones	0.42	16	0.4	15	0.14	5	2.2	81

[a]Adapted from UNSCEAR (1958).

(185 Bq m^{-3}). Surveys of water supplies from many states (Cothern and Lappenbusch, 1985) show that the radium limit is exceeded in many communities that obtain water from wells, including communities of about 600,000 persons in Illinois, Iowa, Missouri, and Wisconsin (Lucas, 1982). About 75% of the supplies that exceeded 5 pCi L^{-1} (185 Bq m^{-3}) were located in two areas of the United States: (1) the piedmont and coastal plain areas of the Middle Atlantic states and (2) the north central states of Minnesota, Iowa, Illinois, Missouri, and Wisconsin. The findings in the Atlantic coastal plain and piedmont areas are summarized in Table 6-9. There is wide variability, with ^{226}Ra concentrations being generally higher than the ^{228}Ra concentrations. The concentration of ^{226}Ra is in some cases as high as 25 pCi L^{-1} (930 Bq m^{-3}), with ^{228}Ra concentrations up to about 17 pCi L^{-1} (630 Bq m^{-3}). It can be estimated that a population consuming 2 liters daily of water that contains 5 pCi L^{-1} (185 Bq m^{-3}) of ^{226}Ra would receive an average effective dose equivalent of approximately 5 mrem y^{-1} (50 μSv y^{-1}). The dose would be increased by the presence of ^{228}Ra.

Subsequently the EPA (1991a) conducted a random survey of radioactivity in 1000 drinking water supply systems that obtain the water supply from groundwater. The random sample was stratified by system size. The results for both ^{226}Ra and ^{228}Ra are given in Table 6-10. The data also allowed estimation of the numbers of persons affected. For ^{226}Ra, 3.4 million are likely exposed to >5 pCi L^{-1} (185 Bq m^{-3}) and 890,000 to >20 pCi L^{-1} (740 Bq m^{-3}). The corresponding numbers are 1.3 million and 164,000 for ^{228}Ra. The fact that the average values for ^{228}Ra are a factor of two larger than for ^{226}Ra is probably an artifact of the larger minimum reporting level for ^{228}Ra.

The EPA regulations, which limit the total radium content of potable water to 5 pCi L^{-1} (185 Bqm $^{-3}$), require that some well water supplies be treated (Hanslick and Mansfield, 1990). A study of the effectiveness of water treatment methods for the removal of radium was undertaken by Brinck et al. (1978) in areas of Iowa and Illinois where the EPA limit was exceeded. Four different water treatment methods were studied. Reverse osmosis and sodium ion-exchange processes were generally about 92% effective. The removal efficiency of the lime–soda ash softening process varied from 75 to 95%. Systems designed to remove iron only were found to have removal efficiencies from 11 to 53%. A detailed analysis of the cost and effectiveness of various methods of removing radioactivity from drinking water is given by Reid et al. (1985) and by the EPA (1991a).

States in the United States have been trying to implement and enforce the 5 pCi L^{-1} (185 Bq m^{-3}) standard since the implementation of the regulation in 1976 with mixed results. A large percentage of the water systems with radium problems have chronically exceeded the limit and continue to

TABLE 6-9

Summary of ^{226}Ra and ^{228}Ra Distributions in Groundwater by Aquifer Type for the Atlantic Coastal Plain and Piedmont Provinces[a]

Aquifer type	Number of samples	^{228}Ra Geometric mean pCi L⁻¹	Bq m⁻³	Maximum[b] pCi L⁻¹	Bq m⁻³	^{226}Ra Geometric mean pCi L⁻¹	Bq m⁻³	Maximum[b] pCi L⁻¹	Bq m⁻³
Igneous rocks (acidic)	42	1.39	51.4	22.6	836	1.80	66.6	15.9	588
Metamorphic rocks	75	0.33	12.2	3.9	144	0.37	13.7	7.4	274
Sand	143	1.05	38.9	17.6	651	1.36	50.3	25.9	958
Arkose	92	2.16	79.9	13.5	500	2.19	81.0	23.0	851
Quartzose	50	0.27	10.0	17.6	651	0.55	20.4	25.9	958
Limestone	16	0.06	2.2	0.2	7.4	0.12	4.4	0.3	11.1

[a]Adapted from Hess et al. (1985). Reproduced from Health Physics Vol. 48, by permission of the Health Physics Society.
[b]In all cases the reported minimum value was zero.

TABLE 6-10

Summary of the [226]Ra and [228]Ra Data for Drinking Water Systems and Estimates
of the Number of Water Supply Systems in the United States That Exceed
Specific Limits[a]

Quantity	[226]Ra		[228]Ra	
	pCi L^{-1}	Bq m^{-3}	pCi L^{-1}	Bq m^{-3}
Minimum reporting level (MRL)	0.18	7	1	37
Average of all samples[b]	0.4	15	0.7	26
Average of all samples > MRL	0.87	32	2.0	74
Number of water systems in excess of:	Number of systems			
1 pCi L^{-1} (37 Bq m^{-3})	5600		7500	
5 pCi L^{-1} (185 Bq m^{-3})	600		500	
20 pCi L^{-1} (740 Bq m^{-3})	70		40	

[a]Adapted from EPA (1991a).
[b]Concentrations less than the minimum reporting level (MRL) were counted as one-half of
the MRL.

do so (EPA, 1991a). EPA recognized this difficulty, as well as the disproportionate cost of removing radium, and published a proposed rule increasing
the radium standard to 20 pCi L^{-1} (740 Bq m^{-3}). The new standard applies
separately to [226]Ra and [228]Ra rather than to the combined total (EPA, 1991a).
The liberalized radium standard has not been adopted as of this writing,
primarily because the proposed rule also contained a controversial limit on
[222]Ra of 300 pCi L^{-1} (11 kBq m^{-3}), which would require many water suppliers
to install radon treatment equipment.

Radium-226 in Ocean Water

Little variation is found in the [226]Ra content of Atlantic Ocean surface
water outside Antarctica (Broecker et al., 1976). Eighty samples analyzed
by Broecker and coworkers averaged 0.03 pCi L^{-} (1.1 Bq m^{-3}). Deep ocean
water contains somewhat greater amounts of [226]Ra, which was explained by
the inflow of O_2-deficient bottom water from the polar regions. The distribution of radium in the oceans and seas was reviewed by Okubo (1990).

Radium-226 in Food

Radium is chemically similar to calcium and is absorbed from the soil
by plants and passed up the food chain to humans. Because the radium in

food originates from soil and the radium content of soil is known to be variable, there is considerable variability in the radium content of foods. In addition, it is reasonable to expect that chemical factors such as the amount of exchangeable calcium in the soil will determine the rate at which radium will be absorbed by plants.

One of the earliest attempts to estimate the radium content of food was undertaken by Mayneord and associates (1958, 1960), who made α-radiation measurements of ashed samples of foods and differentiated the α activities of the thorium and uranium series by counting the double α pulses from the decay of ^{220}Rn and ^{212}Po, whose disintegrations are separated by only the 0.158-second half-life of ^{212}Po. These early measurements served to approximate the total ^{226}Ra and ^{228}Ra content of foods and were highlighted by the fact that Brazil nuts were found to be much more radioactive than other foods. This was later investigated by Penna Franca *et al.* (1968), who showed that the anomaly is due to the tendency of the Brazil nut tree (*Bertholletia excelsa*) to concentrate barium, a chemical congener of radium. Penna Franca found the radium content of Brazil nuts to range between 273 and 7100 pCi kg^{-1} (10-260 Bq kg^{-1}), with only 3 out of 15 samples assaying less than 1000 pCi kg^{-1} (37 Bq kg^{-1}). The radioactivity was about equally divided between ^{226}Ra and ^{228}Ra and was not related to the radium or barium content of the soil in which the tree is grown. The radium concentration of Brazil nuts is on the order of 1000 times greater than in the foods that comprise the average diet in the United States. However, Gabay and Sax (1969) have shown that most of the radium from ingested Brazil nuts is not retained.

Fisenne and Keller (1970) estimated the ^{226}Ra intake of inhabitants of New York City and San Francisco to be 1.7 and 0.8 pCi d^{-1} (0.07 and 0.03 Bq d^{-1}), respectively. However, this twofold difference is not reflected by differences in the ^{226}Ra content of human bone from the two cities, which were found (Fisenne *et al.,* 1981) to be 0.013 pCi g^{-1} (0.5 mBq g^{-1}) of ^{226}Ra in bone ash for both cities. Studies of this kind involve highly sophisticated and sometimes uncertain food and bone sampling techniques that may involve errors that can obscure differences of a factor of two. The ^{226}Ra content of the New York City diet is shown in Table 6-11. The curious, relatively large contribution from eggs seen in 1966 and 1968 dropped somewhat in the 1978 sample but still has not been explained.

Radium-226 Content of Human Tissues

A small fraction of ingested radium is transferred across the small intestine and most is deposited in bone, which contains about 70–95% of the total body radium (ICRP, 1973a). A number of investigators in various

TABLE 6-11
Radium-226 in New York City Diets[a]

Diet category	Food intake (kg y⁻¹)	1966 Conc.[b] (mBq kg⁻¹)	1966 Intake[c] (Bq y⁻¹)	1968 Conc. (mBq kg⁻¹)	1968 Intake (Bq y⁻¹)	1978 Conc. (mBq kg⁻¹)	1978 Intake (Bq y⁻¹)	Average ²²⁶Ra intake (Bq y⁻¹)	Average ²²⁶Ra intake (pCi y⁻¹)
Fresh vegetables	48	19	0.89	59	2.8	55	2.6	2.1	57
Canned vegetables	22	24	0.53	25	0.55	22	0.47	0.52	14.0
Root vegetables	10	52	0.52	44	0.44	15.0	0.15	0.37	10.0
Potatoes	38	104	3.9	63	2.4	6.5	0.25	2.2	59
Dry beans	3	41	0.12	36	0.11	56	0.17	0.13	3.6
Fresh fruit	59	15.9	0.94	7.4	0.44	46	2.7	1.37	37
Canned fruit	11	6.3	0.07	5.9	0.07	6.6	0.07	0.07	1.9
Fruit juice	28	15.5	0.44	33	0.93	27	0.77	0.71	19.2
Bakery products	44	104	4.6	63	2.8	81	3.5	3.6	98
Flour	34	70	2.4	85	2.9	88	3.0	2.8	75
Whole-grain products	11	81	0.90	100	1.10	103	1.13	1.04	28
Macaroni	3	78	0.23	52	0.16	69	0.21	0.20	5.4
Rice	3	28	0.08	122	0.37	7.3	0.02	0.16	4.3
Meat	79	0.37	0.03	0.74	0.06	2.0	0.16	0.08	2.2
Poultry	20	28	0.56	4.1	0.08	20	0.40	0.35	9.4
Eggs	15	226	3.4	518	7.8	85	1.28	4.1	112
Fresh fish	8	25	0.20	41	0.33	30	0.24	0.25	6.9
Shellfish	1	30	0.03	33	0.03	58	0.06	0.04	1.1
Dairy products	200	9.3	1.85	9.3	1.85	5.5	1.10	1.60	43
Yearly intake	637		22		25		18	22	587

[a]Data for 1966 and 1968 from Fisenne and Keller (1970); data for 1978 from Fisenne *et al.* (1987).
[b]One mBq = 0.027 pCi.
[c]One Bq = 27 pCi.

parts of the world have provided measurements of the ^{226}Ra content of bone. These measurements are summarized in Fig. 6-1, in which the concentrations range from about 0.3 to 3.7 Bq kg^{-1} of calcium (8–100 fCi g^{-1} Ca) (the prefix f- refers to femto-, or 10^{-15}). These concentrations result in contents of about 0.3 to 3.7 Bq (8 to 100 pCi) in an adult skeleton containing 1.0 kg of Ca (NCRP, 1984b). The population-weighted skeletal content is 0.85 Bq (23 pCi). Table 6-12 is a summary of observations of radium concentrations in bone in the United States. There is an association between elevated ^{226}Ra concentration in bone and the elevated ^{226}Ra drinking water concentrations found in the midwestern United States.

Estimates of the dose delivered by ^{226}Ra and its decay products require knowledge of the fraction of ^{222}Rn retained by the tissue in which the radium is deposited, since most of the dose is due to α emissions from the decay of ^{222}Rn and its decay products. A ^{222}Rn retention factor of one-third is conventionally used (UNSCEAR, 1982; NCRP, 1987a). The dose estimate is also complicated by the fact that the α energy is deposited at loci determined by the pattern of deposition within the tissue. Assuming an average skeletal content of 0.85 Bq (23 pCi), NCRP (1987a) calculated the average annual dose equivalent to cortical and trabecular bone of 170 μSv (17 mrem). The bone lining cells receive 90 μSv (9 mrem), the red marrow receives 15 μSv (1.5 mrem), and soft tissues receive 3 μSv (0.3 mrem).

THORIUM-232

Typical concentrations of ^{232}Th for the more prevalent rock classes, the crustal average, and soil average are listed in Table 6-6. Thorium-232 con-

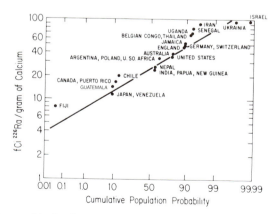

FIGURE 6-1   Geographic distribution of measured ^{226}Ra concentration in human bone. [From National Council on Radiation Protection and Measurements (NCRP, 1984b).]

TABLE 6-12

Radium-226 Concentrations and Contents in Human Bone from the
United States and Puerto Rico[a]

Region	Number of cases	Concentrations[b]			Skeletal content	
		Bq kg^{-1} ash	Bq kg^{-1} calcium	Bq kg^{-1} bone	Bq	pCi
Boston	77	0.52	1.34	0.16	1.34	36
Houston	23	0.85	2.2	0.26	2.2	59
Midwest	128					
Low-Ra water		0.56	1.44	0.17	1.44	39
High-Ra water[c]		1.4	3.6	0.42	3.6	98
Midwest[c]	32	1.0	2.6	0.30	2.6	70
New England	18	0.59	1.52	0.18	1.52	41
New York	143	0.37	0.95	0.11	0.95	26
Puerto Rico	42	0.18	0.46	0.05	0.46	12.5
Puerto Rico	27	0.22	0.57	0.07	0.57	15.3
San Francisco	71	0.41	1.06	0.12	1.06	29
Wisconsin	75	0.44	1.14	0.13	1.14	31

[a]Adapted from NCRP (1987a).
[b]One Bq kg^{-1} = 0.027 pCi g^{-1}.
[c]These results include samples from persons living in areas with levels of ^{226}Ra in drinking water greater than 20 Bq m^{-3} (0.5 pCi L^{-1}).

centrations range from 1.6 to 20 ppm in the common rock types with a crustal average of 10.7 ppm. As can also be seen in Table 6-6, the crustal and soil average concentrations of ^{232}Th are about four times those of uranium. However, the specific activity of ^{232}Th is 0.11 pCi g^{-1} compared to 0.33 pCi g^{-1} for ^{238}U, so the radioactivity due to the two nuclides is approximately equal. Later in this chapter we will consider areas of the world in which the ^{232}Th content of rocks and soils is very much greater than normal.

The characteristics of the thorium series are different from those of the uranium series in a number of respects (see Tables 6-2 and 6-3).

1. ^{228}Ra has a shorter half-life than ^{226}Ra (5.8 years compared to 1620 years).

2. ^{228}Ra is a β emitter that decays to α-emitting ^{228}Th, which has a half-life of 1.9 years. ^{228}Th, in turn, decays through a series of α emitters, including the noble gas ^{220}Rn (thoron), which has a half-life of only 54 seconds compared to 3.8 days for ^{222}Rn. Because of its short half-life, ^{220}Rn has less opportunity than ^{222}Rn to diffuse from the matrix in which it is found.

3. The solubility of ^{228}Ra is comparable with that of ^{226}Ra, but the dose rate to an organism from assimilated ^{228}Ra, a β emitter, is time dependent

because the dose depends on ingrowth of α-emitting ^{228}Th and its short-lived descendants.

4. In the ^{228}Ra chain, there is no long-lived "stopping" nuclide comparable to ^{210}Pb ($T_{1/2}$ = 22 years). The longest-lived nuclide beyond ^{228}Th is ^{212}Pb, with a half-life of 10.6 hours. The dosimetry and radiochemistry of the thorium series tend to be complicated by these characteristics (Fresco *et al.*, 1952).

Because of its relative insolubility and low specific activity. ^{232}Th is present in biological materials only in insignificant amounts. Linsalata (1994) reported ^{232}Th concentrations (fresh weight) in the edible portions of washed vegetables grown in silty clay and peaty soils of New York State to range between 0.001 pCi kg^{-1} (0.037 mBq kg^{-1}) for carrots and 0.093 pCi kg^{-1} (3.4 mBq) for a variety of squash. The mean value for 25 vegetable samples, including potatoes, corn kernels, carrots, beans, and squash, was 0.018 $\pm$ 0.022 pCi kg^{-1} (0.7 $\pm$ 0.8 mBq kg^{-1}). Wrenn and coworkers measured the ^{232}Th, ^{230}Th, and ^{228}Th in the tissues of residents of New York and Colorado (Wrenn *et al.*, 1981; Ibrihim *et al.*, 1983). Thorium was found to be present in the highest concentrations in pulmonary lymph nodes and lungs, indicating that the principal source of human exposure is inhalation of suspended soil particles. Because thorium is removed from bone very slowly, the concentrations of both ^{230}Th (which is found in the ^{238}U decay series) and ^{232}Th were found to increase with age. Average concentrations of ^{232}Th, ^{228}Ra, and ^{228}Th in tissues have been assembled by NCRP (1987a) and are given in Table 6-13.

RADIUM-228 (MESOTHORIUM)

Although ^{228}Ra (referred to as mesothorium in some earlier writings) frequently occurs in soil and water in approximately a 1 : 1 ratio to ^{226}Ra, there is surprisingly little information about its occurrence in foods or in human tissues. Systematic ^{228}Ra measurements in food and water have not been made on a scale comparable to ^{226}Ra. Early data suggested that under normal circumstances the ^{228}Ra content of food, water, and human tissues is from one-half to one-fourth of the ^{226}Ra content (UNSCEAR, 1966). More recent data suggest that ^{228}Ra content of food is only slightly less than the ^{226}Ra content. The NCRP (1987a) estimates that the daily intake of ^{228}Ra is about 40 mBq d^{-1} (1.1 pCi d^{-1}), which may be compared to their estimate of 50 mBq d^{-1} (1.4 pCi d^{-1}) for ^{226}Ra. Where elevated levels of ^{226}Ra have been noted in drinking water, ^{228}Ra levels are often comparable with ^{226}Ra. Elevated levels have been found in Illinois and the Atlantic coastal plain and piedmont areas (Gilkeson *et al.*, 1984; Hess *et al.*, 1985). The data in

TABLE 6-13

Average Concentrations of Selected ^{232}Th
Series Radionuclides in Tissues[a]

Organ or tissue	Concentration (mBq kg^{-1}, wet)[b]		
	^{232}Th	^{228}Ra	^{228}Th
Gonads	0.15	4	0.5
Breast	0.15	4	0.5
Lungs	20	4	15
Cortical bone	12	50	100
Trabecular bone	4	50	28
Red bone marrow	0.15	4	0.5
Thyroid	0.15	4	0.5
Kidneys	3	4	10
Liver	2	4	5
Other tissues	0.15	4	0.5

[a]From NCRP (1987a).
[b]One mBq = 0.027 pCi.

Tables 6-9 and 6-10 also suggest that ^{228}Ra and ^{226}Ra concentrations in water may be more comparable than previously believed.

RADON-222 AND RADON-220 (THORON)

When ^{226}Ra decays by α emission, it transmutes to its decay product ^{22}Rn, with a half-life of 3.8 days. Similarly, ^{224}Ra, which is a descendant of the ^{232}Th chain, decays by α emission to 54-second ^{220}Rn, historically known as thoron. Radon-219, historically called actinon, is a member of the ^{235}U chain and decays most rapidly, having a half-life of about 3.92 seconds. All radon isotopes are noble gases, occurring as nonpolar, monatomic molecules, and are inert for practical purposes. The 3.8-day ^{222}Rn isotope has a greater opportunity than the shorter-lived radon isotopes to escape to the atmosphere. The mechanisms by which ^{222}Rn diffuses from soil into the atmosphere have been discussed by Tanner (1964, 1980). Tanner (1992) has also assembled a useful bibliography on ^{222}Rn in the outdoor environment and its mobility in the ground.

When the parent radium decays in rock or soil, the resulting radon atom recoils. Some fraction of the recoiling atoms come to rest in geologic fluids, most likely water in the capillary spaces. A fraction of the radon in soil water enters soil gas, primarily by diffusion, and then becomes more mobile. Radon reaches the atmosphere when soil gas at the surface exchanges with atmospheric gas. A less important mechanism is diffusion from soil gas to atmospheric gas. The concentration of ^{222}Rn in typical soil ranges from 4 to 40 kBq m^{-3} (10^2–10^3 pCi L^{-1}), several orders of magnitude higher than ^{222}Rn concentrations found in the outdoor atmsophere.

The average outdoor ^{222}Rn concentration in Washington, D.C., has been shown by Lockhart (1964) to be more than 100 times greater than the average concentration in ice-covered Little America, Antarctica, and more than 12 times the values observed at Kodiak, Alaska. The same investigator reported wide variability from day to day. For example, the mean daily concentrations varied more than a 100-fold in Washington, D.C., during 1957 (Lockhart, 1958). Gesell (1983) reviewed the reported data from various parts of the United States and estimated the annual average concentration to range from 0.016 pCi L^{-1} (0.6 Bq m^{-3}) in Kodiak, Alaska, to 0.75 pCi L^{-1} (28 Bq m^{-3}) in Grand Junction, Colorado. The average annual concentrations varied from 0.22 to 0.30 pCi L^{-1} (8.1–11.1 Bq m^{-3}) at four more normal localities.

Other reports from several countries indicate that the average concentrations of ^{222}Rn in outdoor air may be taken normally to be in the range of 0.1–0.5 pCi L^{-1} (4–19 Bq m^{-3}). A number of investigators have observed periodicity in hour-to-hour observations of the ^{222}Rn and ^{220}Rn content of outdoor air (UNSCEAR, 1982). Maximum concentrations are observed in the early hours and the lowest values are found in the late afternoon, when the concentrations are about one-third that of the morning maxima (Gold et al., 1964; Gesell, 1983). Over the course of a year, ^{222}Rn levels tend to peak in the fall or winter months and have minima in the spring. This variation is consistent with the pattern of atmospheric turbulence, which tends to be greater in spring (Gesell, 1983).

It is likely that the variations at any given locality are dependent on meterological factors that influence both the rate of emanation of the gases from the earth and the rate of dilution in the atmosphere (Wilkening, 1982). Thus, the rate of emanation from soil may increase during periods of diminishing atmospheric pressure, although the evidence is not consistent on this point.

It is also likely that the history of an air mass for several days prior to observation influences its ^{222}Rn and ^{220}Rn content (Barreira, 1961). Passage of the air over oceans, and possibly precipitation, would tend to reduce the concentration of these gases, whereas periods of temperature inversion should cause the concentrations to increase by limiting the volume of the atmosphere within which dilution can take place. The gases can be expected to be present in greater amounts over large masses of igneous rock than over large bodies of water or over sedimentary formations.

Radon-222 concentrations decrease with sufficient elevation above the terrain (Bakulin, 1970), but the data near the surface are not consistent. Pearson (1967) observed that ^{222}Rn levels decreased by a factor of two from 1 cm to 1 m above ground at a U.S. location. Doi and Kobayashi (1994) found no change in ^{222}Rn concentration over the range from 4 cm to 1 m at a Japanese location.

Atmospheric ^{22}Rn levels are also likely to be influenced by soil moisture. In a Canadian study, ^{222}Rn detectors were installed outdoors in 78 communities across the country for the summers of 1990 and 1991 (Grasty, 1994). The summer of 1990 was unusually dry and the summer of 1991 was wet. Large regional variations were seen in 1990, with the higher concentrations occurring in the arid regions. Average outdoor concentrations in Manitoba and Saskatchewan were about 60 Bq m^{-3} (1.6 pCi L^{-1}), much higher than previously reported average outdoor ^{222}Rn levels. In 1991 the average ^{222}Rn concentrations were much lower, about 12 Bq m^{-3} (0.3 pCi L^{-1}), which is consistent with reported continental values worldwide. Most of the sites with high ^{222}Rn concentrations were on glacial lake clays. Grasty suggested that the clays developed cracks during the hot dry summer of 1990, increasing transport of ^{222}Rn from soil gas to the atmosphere.

Because the decay products of ^{222}Rn and ^{220}Rn are electrically charged when formed, they tend to attach themselves to dusts that are normally present in the atmosphere. If the radioactive gases coexist with the dust in the same air mass for a sufficiently long time, the parents and their various decay products will achieve radioactive equilibrium. The growth of the ^{222}Rn decay products approaches an equilibrium in about 2 hours and beyond that time further growth is slowed by the presence of 22-year ^{210}Pb. Blifford and associates (1952) investigated the relationships between the concentrations of ^{222}Rn and its various decay products in the normal atmosphere and found, as would be expected, that the atmosphere is markedly depleted in ^{210}Pb relative to the precursors of this isotope. This is because the inert dust of the atmosphere, the ^{222}Rn, and the ^{222}Rn decay products coexist long enough under normal circumstances for equilibria to be reached between ^{222}Rn and the more short-lived decay products. Since the radionuclide with the longest half-life prior to ^{210}Pb is 26.8-minute ^{214}Pb, equilibrium is reached in about 2 hours. The ^{210}Pb, which has a 22-year half-life, would take about 100 years to reach equilibrium. Various mechanisms exist for removing dust from the atmosphere, and the ratio of ^{210}Pb to its shorter-lived ancestors was shown by Blifford to be indicative of the length of time the dust resides in the atmosphere. He concluded by this method of analysis that the mean life of the atmospheric dust to which the ^{222}Rn decay products are attached is 15 days. In later work, Moore et al. (1972) estimated that the mean aerosol residence time for particles in the troposphere was 7 days and that the increase in residence time with altitude within the troposphere was less than a factor of three.

Wilkening (1964) found the concentration of ^{222}Rn decay products in the atmosphere to be depleted during passage of a thunderstorm. He attributed this to the action of electric fields, which changed from a normal value to about 1.8 V cm^{-1} to -340 V cm^{-1}. Deposition of the decay products during rainstorms may temporarily increase the γ background.

The thorium series below ^{220}Rn has no long-lived member. The equilibrium between ^{220}Rn and its decay products will be achieved at a rate governed by the time required for the buildup of ^{212}Pb (half-life 10.6 hours).

The natural radioactivity of atmospheric dust, owing primarily to the attached decay products of ^{222}Rn, can be demonstrated readily. When air is drawn through filter media, the ^{222}Rn decay products attached to the filtered atmospheric dust cause both the α and β activity of the filter media to rise. Curve A of Fig. 6-2 illustrates the manner in which the increase in α radioactivity occurs in the case of normal outdoor air containing 10 Bq m^{-3} (.27 pCi L^{-1}) of ^{222}Rn in equilibrium with its decay products. The rise in α activity continues for about 2 hours, at the end of which time the accumulated decay products decay at a rate compensated by the newly

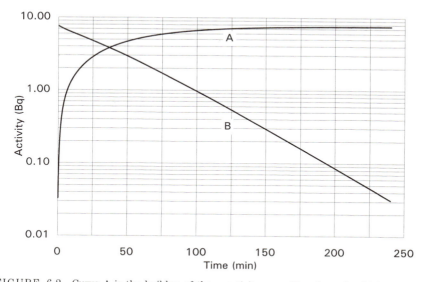

FIGURE 6-2   Curve A is the buildup of the α activity on a filter through which is drawn 0.01 m^3 per minute of air containing 10 Bq m^{-3} of radon in equilibrium with its decay products up to but not included ^{210}Pb. Curve B is the decay of the accumulated α activity after the flow of air is stopped.

deposited decay products. The radioactivity of the filter media will not increase beyond this equilibrium unless the rate of air flow or the concentration of ^{222}Rn is increased. When air flow ceases, the α radioactivity of the filter will diminish with an effective half-life of about 40 minutes, as shown in curve B of Fig. 6-2.

Thus, the adsorbed ^{222}Rn decay products endow the ordinary dusts of the atmosphere with radioactivity. Wilkening (1952) found that the ^{222}Rn decay products tend to distribute themselves on atmospheric dust in a manner that depends on the particle size of the dust, and that the bulk of the activity is contained on particles having diameters less than 0.035 μm. Anderson and associates (1954; Anderson and Turner, 1956) found a close correlation between the concentration of ^{222}Rn decay products in the atmosphere and the concentration of suspended solids. This suggests that the dust concentration of the atmosphere, like the ^{222}Rn concentration, is an indication of the history of an air mass: air that has been recently over the oceans would be depleted in both ^{222}Rn and dust.

When air that contains ^{222}Rn or ^{220}Rn in partial or total equilibrium with their decay products is inhaled, the inert gases are largely exhaled immediately. However, a fraction of the dust particles will be deposited in the lung, with the place of deposition and the manner of clearance from the lung dependent on the factors discussed in Chapter 2. With each breath, additional inert dust will be deposited until radioactive equilibrium is reached, at which point the amount of activity deposited in unit time equals the amount eliminated from the lung by the combination of physiological clearance and radioactive decay. In the case of ^{222}Rn in equilibrium with its decay products, the total energy dissipation in the lungs from the decay products is about 500 times greater than that derived from ^{222}Rn itself. On the basis of the model developed by Harley (1984), NCRP estimated the average annual dose to the basal cells of the bronchial epithelium of the lung of U.S. and Canadian residents from radon decay products to be 24 mSv (2400 mrem). This calculation assumed an average radon concentration of 30 Bq m^{-3} (0.8 pCi L^{-1}) and an equilibrium fraction of 0.5 (NCRP, 1987a). The annual effective dose equivalent received under these conditions was estimated to be 2 mSv (200 mrem).

Blanchard and Holaday (1960) showed that, as in the case of ^{222}Rn, the dose delivered to the lung by ^{220}Rn is about one-thousandth of the dose from the decay products with which it is in equilibrium. However, the dose to the lung from ^{220}Rn and its decay products does not add significantly to the dose received from the ^{222}Rn series.

Radon-222 is partially absorbed when inhaled, and the decay products deliver a dose to the soft tissue of the whole body that has been estimated by the NCRP (1987a) to be 7 μSv (0.7 mrem) per year from 30 Bq m^{-3} (0.8 pCi L^{-1}).

Radon in Groundwater

Radon-222 dissolved in potable water is another source of human exposure, mainly because the ^{222}Rn is released from solution at the tap and enters the home atmosphere (Gesell and Prichard, 1975; Prichard and Gesell, 1983; Cross et al., 1985; Nazaroff et al., 1987, 1988; Watson and Mitsch, 1987; Kahn and Rosson, 1993).

Elevated ^{222}Rn levels have been found only in groundwater and are related to the occurrence of elevated levels of the Ra^{226} precursor in rock. Crystalline rock formation, especially granite, can be a source of elevated levels, whereas sedimentary rock usually yields lower levels. National surveys by the EPA show low ^{222}Rn concentrations in the larger public groundwater supplies, which are often drawn from sedimentary sources. On the other hand, studies of individual and small community supplies have found levels in the 10^3–10^4 kBq m^{-3} (10^4–10^5 pCi L^{-1}) range, particularly in the Appalachian piedmont province (Kahn and Rosson, 1993). Some of the highest concentrations of ^{222}Rn in drinking water have been observed in Maine and New Hampshire, where concentrations as high as 10^5 pCi L^{-1} (3700 kBq m^{-3}) have been reported (Smith et al., 1961; Hess et al., 1980, 1981). Well waters in the central United States ranged from 117 to 287 pCi L^{-1} (4.3 kBq m^{-3}), with a mean of 197 pCi L^{-1} (7.3 kBq m^{-3}). As part of the same stratified-random survey of 1000 groundwater systems discussed earlier for radium, the EPA also obtained ^{222}Rn analyses. Concentrations in water ranged from below the minimum recording level of 100 to nearly 26,000 pCi L^{-1} (3.7–960 Bq m^{-3}). About 11% of the systems surveyed had concentrations above 1000 pCi L^{-1} (37 kBq m^{-3}) and 1% above 10,000 pCi L^{-1} (370 kBq m^{-3}). EPA estimates that 17 million people in the United States are served by systems that provide water containing more than 300 pCi L^{-1} (11.1 Bq m^{-3}) of ^{222}Rn.

Cross et al. (1985) suggested that a standard of no less than 10,000 pCi L^{-1} (370 Bq m^{-3}) could be supported for ^{222}Rn in drinking water, based on comparability with established standards. The EPA (1991a) proposed a limit of 300 pCi L^{-1} (11.1 kBq m^{-3}) for ^{222}Rn in drinking water but the limit was widely opposed. The Safe Drinking Water Act Amendments passed in August 1996 give the EPA three years to develop a new standard. The subject of natural radioactivity in drinking water is reviewed comprehensively by Lappenbusch and Cothren (1985).

INDOOR RADON

In confined spaces, especially those bounded by radon-emitting materials, ^{222}Rn concentrations can be orders of magnitude higher than outdoors. Examples include underground mines, especially uranium mines, caves, and structures. One of the surprising developments in recent years has been the finding that in many homes the concentration of ^{222}Rn (and its decay products) is so high as to involve potential risks far greater than those from many other pollution hazards that have attracted attention. Excellent reviews of the indoor ^{222}Rn knowledge can be found in Nazaroff and Nero (1988) and in Nero *et al.* (1990).

Scientific interest in indoor ^{222}Rn developed worldwide in the mid-1970s, when convenient instrumentation became available. This happened to coincide with the emphasis placed on the need to conserve energy by weatherproofing homes, thereby reducing the rate of infiltration of outside air. It was widely speculated that the high ^{222}Rn concentrations found were the result of the energy conservation programs, since the ^{222}Rn concentrations would be expected to be inversely proportional to the rate of ventilation, which, in U.S. homes, ranges generally from 0.5 to 1.5 air changes per hour. However, subsequent studies have shown that the ventilation rate does not have a major influence on the ^{222}Rn concentration in homes because the rate varies only within a factor of about three. Other factors, including the ^{226}Ra concentration in soil, soil permeability, and pressure differences between indoors and outdoors, are believed to be more important variables (Lourelro *et al.*, 1990).

Widespread public and governmental interest in indoor ^{222}Rn was stimulated in part by the discovery of the now-famous Watras house in Pennsylvania. Mr. Watras was a construction engineer in the nearly completed Limerick nuclear power plant in Pottstown, Pennsylvania. In December of 1984, before the plant went critical, radiation monitors were installed throughout. Mr. Watras tripped every monitor in all areas of the plant that he entered. Plant health physicists determined that the activity on Mr. Watras and his clothing was of natural origin. After ruling out other possibilities, the Watras home was tested for ^{222}Rn. The measurements ranged to 2700 pCi L^{-1} (100 kBq m^{-3}), an extraordinarily high value. The Watras family moved to a motel on the advice of the Pennsylvania Department of Energy Resources (Cole, 1993). The event caused a great deal of publicity and is widely viewed as triggering the scientific and political activity that led to the Indoor Radon Abatement Act. This act provides for studies of indoor radon, research on its control, and risk communication, but is not regulatory in nature because indoor radon is recognized as being of natural origin.

The indoor ^{222}Rn problem exists mainly in residential dwellings because the ^{222}Rn originates primarily from the soil, which has its greatest effect on one- or two-story buildings. The building materials themselves are a minor source of ^{222}Rn compared to soil, except where the materials contain relatively high concentrations of radium and have sufficient permeability and porosity to allow the ^{222}Rn to escape. This is true, for example, where gypsum board or other building materials have been manufactured as a by-product of phosphate fertilizer production (Paredes et al., 1987; Lettner and Steinhäusler, 1988). On the other hand, the slag by-product from elemental phosphorus production, though containing elevated [226]Ra, is a glassy material with low permeability and has not been identified as a source of indoor [222]Rn (EPA, 1990b).

Radon-222 availability to structures from adjacent rock and soil is related mainly to the concentration of [222]Rn in the spaces in rock fractures and soil pores and to the permeability of the ground to gases. Permeability and diffusion are reduced as the sizes of those spaces are reduced and as the proportion of the spaces filled by liquids is increased. Rock types that usually have above-average concentrations of [222]Rn in the pore and fracture spaces include granites, some gneisses, phosphatic rocks, marine shales, and some recrystallized limestones and dolomites. Ground with coarse grain size (such as gravels and coarse sands), particularly if well drained, is highly permeable and apt to make more [222]Rn available than would be expected on the basis of its radium content. Muds and clays tend to be of low permeability, especially if wet, and make less [222]Rn available. Buildings located on hillsides and ridges are more apt to be located on soils that are coarser and better drained than those in adjacent valleys. Other things being equal, [222]Rn availability should be greater on hillsides and ridges (Tanner, 1986, 1989).

Radon-222 can enter the indoor atmosphere in a number of ways, including diffusion from materials of construction or diffusion from soil through breaches in the foundation. However, there is evidence that diffusion of [222]Rn from soil is a minor source compared to the movement of soil gases directly through the foundation as a result of slight pressure differentials that can result from barometric changes, temperature differentials, or wind velocity. Steinhäusler (1975) showed that meteorological factors in particular can influence the indoor concentrations of [222]Rn and its decay products. Water supplies ordinarily make a small contribution to the indoor [222]Rn concentration, but can be the predominant source in areas where the [222]Rn content of groundwater is unusually high. Studies in Maine showed [222]Rn in water to be a significant contributor (Hess et al., 1981). In a later study of 28 houses near Conifer, Colorado, the water supply was also shown to contribute significantly to levels of indoor Rn-222 in many of the houses (Lawrence et al., 1992). The approximate contributions of various sources to the indoor [222]Rn concentrations are given in Table 6-14 (Nero et al., 1986, 1988).

In a study of the indoor [222]Rn concentrations in an area of Maine in which the groundwater contained up to about 50,000 pCi L^{-1} (1.85 MBq m^{-3}) of [222]Rn, it was found that the concentrations of [222]Rn in water and air in single-family dwellings were weakly correlated ($r = 0.50$; $N = 70$), with a

TABLE 6-14

Approximate Contributions of Various Sources to Indoor ^{222}Rn Concentrations[a]

Source	Single-family homes		Apartments (high rise)	
	pCi L^{-1}	Bq m^{-3}	pCi L^{-1}	Bq m^{-3}
Soil (based on flux measurements)	1.5	55	>0	>0
Water (public supplies)[b]	0.01	0.4	0.01	0.4
Building materials	0.05	2	0.1	3.7
Outdoor air	0.25	10	0.25	10
Observed indoor concentrations	1.5	55	0.3?	12?

[a]Adapted from Nero *et al.* (1986, 1988). Excerpted with permission from Nero, A. V., Jr., Schwehr, M. B., Nazaroff, W. W., and Revzan, K. L. (1986). Distribution of airborne radon-222 concentrations in U.S. houses. *Science* **234,** 992–997. Copyright 1986 American Association for the Advancement of Science.

[b]Applies to residences served by public supplies. Contributions can be much larger in some cases.

regression slope of 1.3 pCi L^{-1} of air (48 Bq m^{-3}) per 10,000 pCi L^{-1} (370 kBq m^{-3}) of water (Hess *et al.*, 1983). Based on other measurements (Hess *et al.*, 1981), it was estimated that the contribution of groundwater to indoor air could be expressed as a ratio of ^{222}Rn in air to ^{222}Rn in water of 0.8×10^{-4}, standardized for one air change per hour. From these data, the contribution of groundwater to indoor air could be approximated by the ratio of ^{222}Rn in air to ^{222}Rn in water of 10^{-4}, a value first estimated by Gesell and Prichard (1975). Subsequently, Nazaroff *et al.* (1988) applied a statistical approach to the data available at that time and concluded that the ratio could be best described as having a geometric mean of 0.65×10^{-4} with a geometric standard deviation of 2.88.

The first effort to estimate the U.S. national distribution of indoor ^{222}Rn was made by Nero and colleagues (Nero, 1985; Nero *et al.*, 1986), who used various approaches to aggregating data from 38 reports comprising 1377 houses.[1] After eliminating studies in areas with prior expectation of elevated ^{222}Rn, they concluded that the distribution of ^{222}Rn in U.S. housing stock could be characterized by an arithmetic mean of 1.5 pCi L^{-1} (55 Bq m^{-3}) and a long tail, or by a log-normal distribution with a geometric mean of 0.9 pCi L^{-1} (33 Bq m^{-3}) and a geometric standard deviation of 2.8. About

[1]There are numerous methods for measuring radon indoors over periods of time ranging from instantaneous, to averages of a few days, up to a calendar quarter or even a year (George, 1996). Screening methods for average indoor radon in common use today include alpha particle track detectors, charcoal cartridges that adsorb radon and are subsequently analyzed in a laboratory , and electret ionization chambers. Charcoal cartridges are capable only of measuring average radon concentrations over a period of a few days. The other two methods can measure average radon concentration over many months.

1–3% of the houses exceed the NCRP (1984b) action level of 8 pCi L^{-1} (300 Bq m^{-3}) and about 7% exceed the EPA action level of 4 pCi L^{-1} (150 Bq m^{-3}) (EPA, 1991b). A simple aggregate of the 552 houses in 19 studies conducted in areas without expectation of elevated ^{222}Rn, and for which individual data were available, is shown in Fig. 6-3. This grouping of the data conforms to a log-normal distribution with an arithmetic mean of 1.66 pCi L^{-1} (60 Bq m^{-3}) and a geometric mean of 0.96 pCi L^{-1} (36 Bq m^{-3}). Nero recognized that his review had potential biases because the studies available to him comprised mostly middle-class, single-family dwellings.

Shortly after Nero's work became available, two other data sets on indoor ^{222}Rn were published (Cohen, 1986; Alter and Oswald, 1987). Cohen made measurements in the homes of 453 physics professors in 42 states and the District of Columbia. He obtained an arithmetic average of 1.47 pCi L^{-1} (54 Bq m^{-3}) and a geometric mean of 1.03 pCi L^{-1} (38 Bq m^{-3}). The principal bias of the Cohen study was that the residences were also primarily middle-class, single-family units. The results were quite similar to those of Nero. Since the Alter and Oswald study was a compilation of data from a ^{222}Rn testing service, the residences tested were selected by the owners or local investigators rather than randomly. This study has a recognized bias built in because the measurements were largely made in response to knowledge or concern about indoor ^{222}Rn in specific areas. As might be expected, the average concentration found was considerably higher than the value of approximately 1.5 pCi L^{-1} (55 Bq m^{-3}) found by Nero and by Cohen. The Alter and Oswald data had a mean of 4.3 pCi L^{-1} (158 Bq m^{-3}).

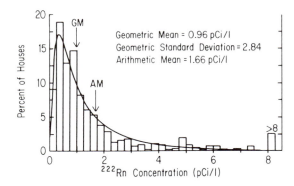

FIGURE 6-3 Probability distribution from aggregation of 552 individual data in 19 sets. The smooth curve is the log-normal functional form corresponding to the indicated parameters, calculated directly from the data. [From Nero (1985).]

Subsequent to these efforts, a number of national and regional surveys of indoor ^{222}Rn were made in the United States and other countries (Cohen, 1989, 1991; Langroo et al., 1991; White et al., 1992; Faisca et al., 1992; Marcinowski and White, 1993; Field et al., 1993; Sabbarese et al., 1993; Marcinowski et al., 1994).

In the United States, the most current, representative survey of indoor ^{222}Rn is the National Residential Radon Survey (EPA, 1992b; Marcinowski et al., 1994). The primary objectives of this study were to compile a national frequency distribution of ^{222}Rn levels in residences and to develop frequency distributions for each of the 10 EPA regions in order to determine which regions of the country may have indoor ^{222}Rn problems. Secondary objectives included breakdowns by residence type and correlations with construction types. The study covered single-family detached homes, multiunit structures (e.g., apartments), and mobile homes. It did not cover group quarters such as dormitories and other types of institutional housing. A stratified, random sampling plant was used to ensure broad geographic coverage and to eliminate sampling bias.

Each residence surveyed was supplied with a ^{222}Rn detector for each lived-in level. Residences with only one level were supplied with two detectors. The detectors were of the alpha track type and were left in place for one year in order to obtain an annual average. Data were obtained from more than 15,000 detectors placed in 5694 residences. The ^{222}Rn level for each residence was estimated by simply averaging the results from all of the detectors in the residence. Weighing for time spent on the various levels was not done because EPA could not get accurate data on occupancy times.

The arithmetic average national ^{222}Rn concentration was found to be 1.25 pCi L^{-1} (46 Bq m^{-3}) and the median was 0.67 pCi L^{-1} (25 Bq m^{-3}). Figure 6-4 provides the average and median ^{222}Rn concentrations for each of the 10 EPA regions. It is clear that areas of the Midwest and Intermountain West have the highest indoor ^{222}Rn levels, averaging about twice the national average, whereas the Northwest has the lowest levels. Figure 6-5 gives the frequency distribution for the entire country. Approximately 6% of the residences surveyed had ^{222}Rn concentrations over 4 pCi L^{-1} (150 Bq m^{-3}). Applied nationally, this figure implies that 5.8 million residences may have such ^{222}Rn levels. The results from Nero's review, Cohen's 1986 study, and those from the carefully designed National Residential Radon Survey are similar, suggesting that the distribution for indoor ^{222}Rn levels in the United States can be considered to be well characterized.

World indoor ^{222}Rn levels do not necessarily follow the pattern seen in the United States. Levels are often higher in Scandinavian countries such as Denmark (National Institute of Radiation Hygiene, 1987), where the average of the summer and winter ^{222}Rn levels is 93 Bq m^{-3} (2.5 pCi L^{-1}).

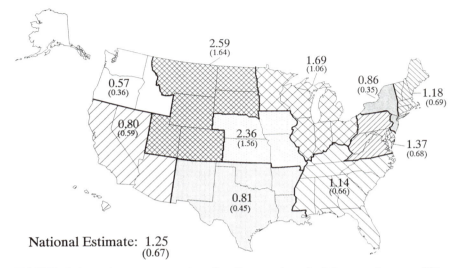

FIGURE 6-4 Arithmetic mean and median (in parentheses) of the annual average ^{222}Rn concentrations over all living areas by EPA region (in pCi L^{-1}; 1 pCi L^{-1} equals 37 Bq m^{-3}). [From EPA (1992b).]

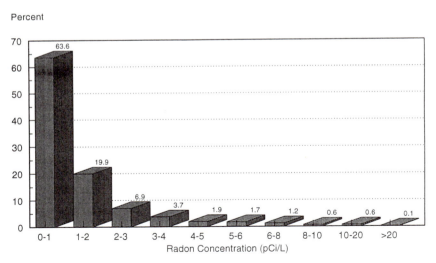

FIGURE 6-5 National distribution of annual average ^{222}Rn levels in the U.S. housing stock averaged over all living areas (in pCi L^{-1}; 1 pCi L^{-1} equals 37 Bq m^{-3}). [From EPA (1992b).]

A very low result was seen in Australia, where a nationwide ^{222}Rn survey of homes was recently conducted using solid-state track detectors (Langroo *et al.*, 1991). In this survey, dosimeters were placed in approximately 3400 randomly distributed homes (representing about 1 in 1400 occupied dwellings) for 12 months. The average Rn concentration in Australian homes measured over a year was only 11 Bq m^{-3} (0.3 pCi L^{-1}), much lower than that found in the United States and within the range of continental outdoor ^{222}Rn background levels.

Radon-222 levels in commercial and industrial structures have been generally believed to be much lower than levels found in residences. This was based on limited measurements in a few buildings (Cohen *et al.*, 1984; Turk *et al.*, 1986) and on reasoning that multistory buildings with forced ventilation would be less likely to reach high ^{222}Rn levels (Nero, 1988). However, indoor ^{222}Rn in commercial and industrial structures may warrant reconsideration. Elevated ^{222}Rn has been found in underground workplaces in Germany (Schmitz and Fritsche, 1992). Scott (1992) reported on a systematic study of indoor ^{222}Rn in which 5700 alpha track ^{222}Rn detectors were distributed in 3100 of the 11,000 occupied buildings at all 74 DOE locations in the United States. The screening measurement results identified 86 buildings at seven sites that may exceed the EPA action level of 4 pCi L^{-1} (150 Bq m^{-3}) for residences. This amounts to 2.7% of the structures, only a factor of two lower than the percentage of U.S. residences estimated to exceed this level. Although this screening study does indicate lower indoor ^{222}Rn levels for commercial and industrial structures, 2.7% of such structures should not be considered an insignificant number. The public health implications of indoor radon were discussed in Chapter 2.

Natural underground caves have limited ventilation and are bounded by rock and soil capable of emanating radon into the air space. Radon is also carried into caves by water. Radon-222 levels are typically much higher in caves than outdoors (Wilkening and Watkins, 1976). Many caves are operated as tourist attractions and the public receives small ^{222}Rn exposures from this source. Larger exposures are received by guides and others who work in the caves. In a study of caves operated by the U.S. National Park Service, Yarborough (1980) identified numerous locations in several caves with radon decay product levels greater than 1 WL,[2] which corresponds to

---

[2]One Working Level Month (WLM) is approximately equal to annual exposure to an average of 4 pCi L^{-1} (150 Bq m^{-3}) of ^{222}Rn if the decay products are in 50% equilibrium with the ^{222}Rn. One WLM exposure would occur from being exposed to 1 Working Level (WL) for a period of one working month, that is, 170 hours. One WL is defined as that concentration of radon decay products that has a potential alpha energy release of 1.3×10^5 MeV L^{-1} (2×10^{-5} J m^{-3}) of air. The working level was so named because it was the original radon decay product standard for uranium miners.

about 50 pCi L^{-1} (1.9 kBq m^{-3}) assuming 50% equilibrium. Average annual occupational exposures ranged from 0.14 to 2.1 WLM. The maximum annual occupational exposure noted was 3.8 WLM at Mammoth Cave, Kentucky. In comparison, the limit for uranium miners in the United States is 4 WLM per year. In a Minnesota cave, radon levels were observed to vary by a factor of 20 with highs of 20 kBq m^{-3} (540 pCi L^{-1}) in winter and 25 kBq m^{-3} (675 pCi L^{-1}) in summer (Lively and Krafthefer, 1995). Control of ^{222}Rn levels in caves is difficult because artificially increasing ventilation alters cave microclimates and may destroy cave formations or diminish their attractiveness. Measurements of ^{222}Rn levels in caves are also useful for determining ventilation patterns (Cunningham and LaRock, 1991).

Methods of Reducing Indoor ^{222}Rn Concentrations

Several methods are available to reduce the concentration of ^{222}Rn in indoor air (NCRP, 1984b, 1989a; Moeller and Fujimoto, 1984; Scott, 1988; Jonassen and McLaughlin, 1988; EPA, 1989a). It was originally believed that the reduced ventilation rates within homes as a result of weather-proofing were a major factor in increasing the concentration of ^{222}Rn. As noted earlier, this has been shown not to be the case: the overriding factor is the rate at which ^{222}Rn enters the home with soil gases, except in places where homes are supplied by water that has a high ^{222}Rn content. As a consequence, remedial efforts are focused mainly on the sources of ^{222}Rn and only secondarily on increasing ventilation. Radon-222 in water can be reduced by aeration or granular activated carbon (Shapiro and Sorg, 1988). Methods for reducing indoor ^{222}Rn are discussed in the following.

"Sub-slab suction" is used to draw ^{222}Rn from under and around homes, reducing the rate of entry through cracks and other openings. The suction has the effect of lowering the pressure of the soil gas relative to the pressure inside the house, thus preventing ^{222}Rn from seeping into the house. Sub-slab suction is accomplished by installing piping that penetrates the slab and applying negative pressure to the piping either with a fan or by convection. Sumps can be vented in this fashion as well.

Another important method for reducing ^{222}Rn is to seal openings in the foundation. These include floor cracks, joints between the floor and wall, wall cracks and openings, floor drains, and sumps. Usually sealing and sub-slab suction are used together. If these methods do not achieve the desired ^{222}Rn reduction, it may be necessary to increase ventilation. This approach has energy conservation consequences, so some form of heat exchange is usually warranted.

A potentially simpler approach is to remove the ^{222}Rn decay products from indoor air with filtration (Curling *et al.*, 1990a,b) or by use of fans supplemented with ion generators (Moeller *et al.,* 1988), but these alternative methods have not gained widespread acceptance to date. Although these devices clearly remove a portion of the ^{222}Rn decay products from the indoor atmosphere, the actual dose reduction achieved is not fully understood. A complication is that the devices may increase the fraction of the ^{222}Rn decay product atoms that are unattached to atmospheric aerosol particles. Unattached ^{222}Rn decay products deliver a larger dose to lung than the same concentration of ^{222}Rn decay products attached to aerosol.

The EPA has estimated that 5.8 million U.S. homes have concentrations that exceed their recommended action level of 4 pCi L^{-1} (150 Bq m^{-3}). Nero (1985) estimated that as many as one million U.S. homes have concentrations that exceed the 2 WLM y^{-1} ($\sim$8 pCi L^{-1}; 300 Bq m^{-3}) limit recommended by NCRP. The EPA urges that every home in the nation be tested for ^{222}Rn, and that concentrations be reduced if they exceed the action level. This policy, if fully implemented, would cost homeowners an estimated $8 to $20 billion. EPA also recommends that residents consider reducing levels if they exceed 2 pCi L^{-1} (75 Bq m^{-3}), which would add to the cost.

Beneath the surface of an apparently straightforward policy lies some disagreement that has not been obvious to the public (Cole, 1993). Epidemiological studies, taken as a whole, do not demonstrate a measurable link between residential ^{222}Rn and cancer. In encouraging people to protect themselves, the EPA has conducted an aggressive radon awareness campaign that does not highlight scientific uncertainties. Despite the campaign, public enthusiasm for ^{222}Rn testing has been lackluster. Only about 5% of the homes in the country had been tested by 1992, despite urging in the 1986 Citizen's Guide to Radon (EPA, 1986a)[3] and the vigorous public information campaign. Most states have accepted support from the EPA for radon programs and some have appropriated state funds for studies and for radon contractor certification programs, but most state funding has declined recently. This illustrates a central dilemma about the responsibility of government when a health or environmental risk is uncertain and where there is no corporate or governmental entity responsible for causing the risky condition.

Indoor ^{222}Rn raises unprecedented questions of public policy, although the EPA and NCRP recommendations on ^{222}Rn do not have the force of law. Should there be regulations for indoor ^{222}Rn or should the effort be left on a voluntary basis? Should people be required to reduce the ^{222}Rn levels to accepted standards in buildings they own? Should there be different policies in owner-occupied buildings than in rented buildings? Will prospective buyers of real estate require that ^{222}Rn levels be certified? Will the ^{222}Rn levels

---

[3]The EPA Citizen's Guide to Indoor Radon has since been updated, but the basic recommendation of 4 pCi L^{-1} (148 Bq m^{-3}) remains unchanged (EPA, 1992a).

within a building be an important factor in determining its value? These questions are beginning to be answered. In most cases, radon reduction is voluntary, but the state of Florida (1994) has promulgated a standard of 0.02 WL (~4 pCi L^{-1}; 150 Bq m^{-3}). Radon tests are increasingly required by buyers and transfer companies during real estate transactions, just like tests for termite infestation.

Radon-222 in Natural Gas

A radioactive gas that had properties similar to those of the "radium emanation" was first separated from petroleum in 1904 (Burton, 1904). Thus, it is not surprising that ^{222}Rn is also present in natural gas. It is found at the wellhead in concentrations that average about 40 pCi L^{-1} (1.5 kBq m^{-3}), but samples from some fields contain more than 1000 pCi L^{-1} (37 kBq m^{-3}) (Johnson et al., 1973; Gesell, 1975). Natural gas at the wells contains from 55 to 98% methane, and a much smaller percentage of other heavy hydrocarbons (ethane, propane, butane), as well as carbon dioxide, nitrogen, helium, and water vapor. The gas is blended and processed to produce liquefied petroleum gas (LPG), a product that consists mainly of propane with lesser amounts of ethane. The boiling point of radon is close to that of propane and ethane, which has the effect of increasing the ^{222}Rn concentration in LPG while reducing the concentration in the methane-rich gas pumped into the pipelines. The long-lived decay products of ^{222}Rn (^{210}Pb and ^{210}Po) tend to accumulate on the interior surfaces of the LPG plant machinery and constitute a potential source of exposure of maintenance personnel (Gesell et al., 1975; Summerlin and Prichard, 1985).

Since ^{222}Rn has a half-life of only 3.8 days, exposure of consumers is influenced by storage time in the case of LPG as well as the pipeline transit time. In 1987 it was estimated that ^{222}Rn from natural gas results in an average annual effective dose of about 4 μSv (0.4 mrem) to the 125 million persons living in homes in which natural gas is used for cooking, and 18 μSv (1.8 mrem) for those in the 16 million homes using nonvented gas space heaters (NCRP, 1987d). The same report lists a dose of 1.3 Sv (0.4 mrem) for users of LPG.

LEAD-210 AND POLONIUM-210

Lead-210 (sometimes called by its historic designation, radium D) is a 22-year β emitter separated from its antecedent ^{222}Rn by six short-lived α and β emitters (see Table 6-2). The longest-lived radionuclide between ^{222}Rn and ^{210}Pb is ^{214}Pb, which has a half-life of only 26.8 minutes. The ^{210}Pb decays to 138-day ^{210}Po via the intermediate ^{210}Bi, which has a 5-day half-life. Thus, following the decay of 3.8-day ^{222}Rn in the atmosphere, ^{210}Pb is produced rapidly, but its long half-life allows very little to decay in the atmosphere before it precipitates to the earth's surface, mainly in rain or snow.

The ^{210}Pb content of the atmosphere has been found to vary from 0.2 to 1.5 mBq m^{-3} (5 to 40 $\times$ 10^{-3} pCi m^{-3}), with the lowest values at island stations such as San Juan, Puerto Rico, and Honolulu, Hawaii, and the higher values in the interior of the United States (NCRP, 1987a). Since the mean residence time of dust suspended in the troposphere is about 15 days, there is little time for ^{210}Po to be formed in suspended dust, and the concentration of ^{210}Po near ground level is smaller than that of ^{210}Pb. For purposes of estimating dose in the United States, NCRP (1987a) adopted nominal ground level concentrations for ^{210}Pb and ^{210}Po of 0.7 and 0.07 mBq m^{-3} (20 $\times$ 10^{-3} and 2 $\times$ 10^{-3} pCi m^{-3}), respectively. On the basis of limited measurements, Fisenne (1993) estimated that ^{210}Pb concentrations indoors are about one-fourth of outdoor concentrations.

Radioactive disequilibria are found in the upper portions of rocks and soil profiles from which ^{222}Rn diffuses. Atmospheric transport and deposition of ^{210}Pb cause it to be distributed in a more uniform pattern than the ^{226}Ra from which it is derived. The ratio of ^{210}Pb to ^{238}U in surface soil is about 2 (Fisenne et al., 1978). It would also be expected that broad-leaved plants on which ^{210}Pb can be deposited would be enriched in this radionuclide. This in fact is observed, for example, in the case of tobacco leaves.

The ^{210}Pb : ^{210}Po ratio within a matrix will depend on the initial ratio, the length of time ^{210}Pb exists, and whether the polonium is selectively removed from its site of production by chemical or biological mechanisms. Since ^{210}Po has a half-life of only 138 days, appreciable ingrowth in vegetation can take place during a single growing season, and additional buildup can occur during food storage, with equilibrium being reached in about 1 year. When ^{210}Pb is absorbed into the body, ingrowth of ^{210}Po can occur because ^{210}Pb is deposited in the skeleton, from which it is removed slowly, with a half-life of about 10^4 days.

Jaworowski (1967) reported that rainwater contains from 1 to 10 pCi L^{-1} of ^{210}Pb (37 to 370 Bq m^{-3}) with a mean of about 2 pCi L^{-1} (74 Bq m^{-3}). In an area having 1 m of rainfall per annum, this would indicate a wet deposition of about 2 mCi km^{-2} (74 MBq km^{-2}) per year.

The concentration of ^{210}Pb in a "standard" diet was found by Magno et al. (1970) to be not significantly different between locations in the United States and averaged 0.80 pCi kg^{-1} (0.03 Bq kg^{-1}). His values were in good agreement with those reported by Morse and Welford (1971), which ranged from 0.70 to 1.0 pCi kg^{-1} (0.026 to 0.037 Bq kg^{-1}) of ^{210}Pb in eight U.S. cities. UNSCEAR (1982) estimated that the average daily intake of ^{210}Pb from dietary sources is about 3 pCi (0.11 Bq) under normal circumstances.

Holtzman (1980) found lower values for U.S. residents, a mean of 1.4 pCi (0.052 Bq), with a relatively narrow range of 1.3–1.6 pCi d^{-1} (0.048–0.06 Bq d^{-1}). NCRP (1987a) estimated that the mean dietary intake of ^{210}Pb is about 0.05 Bq d^{-1} (1.4 pCi d^{-1}) and that the ^{210}Po contents of the standard diet are slightly larger than ^{210}Pb with an average ratio of 1.3. Food and water ingestion is a more important contributor to blood levels of ^{210}Pb than inhalation, even among smokers. Clemente et al. (1984) studied the relationship of ^{210}Pb in teeth and the ambient level of radon. A statistically significant relationship was found, with a lifetime exposure of 1 WLM resulting in ^{210}Pb addition of 1 mBq g^{-1} (0.03 pCi g^{-1}). In the United States, ^{210}Pb and its decay products are estimated to contribute annually about 0.14 mSv (14 mrem) to soft tissue and bone marrow and 0.7 mSv (70 mrem) to bone surface (NCRP, 1987a). Table 6-15 gives the contributions of various sources to the body content of ^{210}Pb.

There are two notable groups in which the dose from ^{210}Po is apt to be exceptionally high: cigarette smokers and residents of northern lands who subsist on reindeer that consume lichens.

Marsden and Collins (1963) originally noted the presence of α activity in tobacco, following which Radford and Hunt (1964) measured the concentrations of ^{210}Po in cigarettes, and Little et al. (1965) demonstrated that ^{210}Po was present in the lungs of cigarette smokers. Subsequent investigators have studied this phenomenon (Cohen et al., 1980) and concluded that the dose to the basal cells of the bronchial epithelium in cigarette smokers is increased by 2–12 mrem y^{-1} (20–120 μSv y^{-1}). Autoradiographs of the

TABLE 6-15

Contributions of Various Sources to the Body Content of ^{210}Pb[a]

Mode of entry	Intake (mBq d^{-1})[b]		Fractional absorption to blood	Amount reaching blood (mBq d^{-1})[b]	
	Nonsmokers	Smokers		Nonsmokers	Smokers
Gut (food and water)	50	50	0.18	9	9
^{222}Rn in body	0.07	0.07	1.0	0.07	0.07
^{226}Ra in body	0.3	0.3	1.0	0.3	0.3
^{210}Pb inhaled	15	47	0.16	2.6	6.1
^{222}Rn decay products inhaled	0.3	0.3	0.16	0.05	0.05
Total	66	98		12	16

[a]Adapted from NCRP (1987a).
[b]One mBq equals 0.027 pCi.

bronchial epithelium from one smoker showed the ^{210}Po to be highly local-ized. If the ^{210}Po remained fixed in position, the dose to the tissue surround-ing the "hot spot" would be much higher than the average dose to the basal calls of the bronchial epithelium. On the basis of α-track measurements of autoradiographs of the lung tissue of 13 smokers, Rajewsky and Stahlhofen (1966) had previously estimated that the basal cells of the subsegmental bronchi of cigarette smokers may receive as much as 86 mrem y^{-1} (0.86 mSv y^{-1}) and the basal cells of the terminal bronchi may receive as much as 150 mrem y^{-1} (1.5 mSv y^{-1}). These values are higher than those reported by Cohen et al. (1980).

Compared to nonsmokers, about twice as much ^{210}Pb and ^{210}Po is found in the ribs of smokers (Holtzman and Ilcewicz, 1966). Ashed ribs from smokers contained 0.28 pCi g^{-1} (10.4 Bq kg^{-1}) of ^{210}Pb and 0.25 pCi g^{-1} (9.3 Bq kg^{-1}) of ^{210}Po. The lungs contained 5.9 pCi kg^{-1} (0.22 Bq kg^{-1}) of ^{210}Po (wet weight). From these data, Holtzman estimated that the dose to the total skeleton is elevated by about 30% in cigarette smokers and the dose to the cells of the bone surface is increased by about 8%.

Polonium-210 is believed to enter tobacco by ingrowth in ^{210}Pb deposited on tobacco leaves from the atmosphere (Martell, 1974). Direct uptake of ^{210}Po from soil is probably not significant. The transfer coefficient to vegeta-bles grown in ^{210}Po-contaminated alkaline soil has been found to range from about 6×10^{-4} to 10^{-6} (Watters and Hansen, 1970). Hill (1966) found a close correlation between the ^{210}Po and ^{137}Cs contents of human tissues from Canadian subjects, thus strengthening the suggestion that dietary habits that tend to favor consumption of broad-leaved vegetables or other foods subject to surface deposition may influence the ^{210}Po content of the tissues. Cesium-137 is known to be absorbed by humans mainly from surface deposition on plants (see Chapters 5 and 9). Others have noted that pipe tobacco contains less ^{210}Po than do cigarettes, apparently because the to-bacco used for smoking pipes is not aged as long as cigarette tobacco (Harley and Cohen, 1980).

The dose from ^{210}Po is increased in Laplanders and some Eskimos who eat reindeer and caribou that feed on lichens that absorb trace elements in the atmosphere (Beasley and Palmer, 1966). Kauranen and Miettinen (1969) and Persson (1972) found the ^{210}Po content of Lapps living in north-ern Finland to be about 12 times higher than that of residents of southern Finland, where more normal dietary regimes exist. These investigators found the liver dose in Laplanders to be 170 mrem y^{-1} (1.7 mSv y^{-1}) compared to 15 mrem y^{-1} (0.15 mSv y^{-1}) for residents of southern Finland who do not regularly eat reindeer meat. Unlike other naturally occuring α emitters, ^{210}Po deposits in soft tissues and not bone.

Holtzman (1964) measured the ^{210}Pb and ^{210}Po contents of Illinois potable waters known to be high in radium and found them to be low relative to ^{226}Ra. This seems to indicate a loss of ^{210}Pb owing to chemical precipitation, biological activity, or other factors.

Polonium-210 is emitted to the atmosphere during the calcining (or "nodulizing") of phosphate rock as part of the process of production of elemental phosphorus (Guimond and Windham, 1985; Pennders *et al.*, 1992). The relatively volatile polonium is released during these thermal operations, whereas the other natural radionuclides tend to remain in the solid form. Levels of ^{210}Po in soils and small mammals near a phosphorus plant were found to be three to four times the control levels but elevated ^{210}Pb was not statistically significant in vegetation (Arthur and Markham, 1984). Atmospheric emissions of ^{210}Po from calciners and nodulizing kilns at elemental phosphorus plants in the United States are limited to 2 Ci y^{-1} (74 GBq y^{-1}). Exceptions are provided if the plant has installed specified emission control equipment and still does not meet the limit (EPA, 1994). In any case, plants may not emit more than 4.5 Ci y^{-1} (167 GBq y^{-1}).

POTASSIUM-40

Of the three naturally occurring potassium isotopes, only ^{40}K is unstable, having a half-life of 1.3×10^9 years. It decays by β emission to ^{40}Ca, followed by K capture to an excited state of ^{40}Ar, and γ-ray emission to the ^{40}Ar ground state. Potassium-40 occurs to an extent of 0.0118% in natural potassium, thereby imparting a specific activity of approximately 800 pCi g^{-1} potassium (30 kBq kg^{-1}). Representative values of the potassium content of rocks, as summarized in Table 6-6, indicate a wide range of values, from 0.3 to 4.5% for various rock types. Certain basalts and sands are low in potassium, whereas granites and other basalts are high. Seawater contains ^{40}K in a concentration of about 300 pCi L^{-1} (11 kBq m^{-3}).

The potassium content of soils of arable lands is strongly influenced by the use of fertilizers. It is estimated that about 3000 Ci (110 TBq) of ^{40}K is added annually to the soils of the United States in the form of fertilizer (Guimond, 1978).

A person who weighs 70 kg contains about 140 g of potassium, most of which is located in muscle. From the specific activity of potassium, it follows that the ^{40}K content of the human body is on the order of 0.1 μCi (4 kBq). This isotope delivers a dose of 18 mrem y^{-1} (0.18 mSv y^{-1}) to the gonads and other soft tissues and 14 mrem y^{-1} (0.14 mSv y^{-1}) to bone (NCRP, 1987a). Because of its relative abundance and its energetic β emission (1.3 MeV), ^{40}K is easily the predominant radioactive component in normal foods and human tissues. It is important to recognize that the potassium content of the body is under strict homeostatic control and is not influenced by variations in environmental levels. For this reason, the dose from ^{40}K within the body is constant.

RUBIDIUM-87

Of the two rubidium isotopes found in nature, ^{85}Rb and ^{87}Rb, only the latter is radioactive, with a half-life of 4.8×10^{10} years. Rubidium-87 is a pure β emitter, and it is present in elemental rubidium in the amount of 27.8%, which endows this element with a specific activity of 0.02 μCi g^{-1} (0.74 kBq g^{-1}). Pertsov (1964) quotes Vinogradov in listing the rubidium content of all but highly humic soils as about 0.01%. The ^{87}Rb content of ocean water has been reported to be 2.8 pCi L^{-1} (104 Bq m^{-3}), with marine fish and invertebrates ranging from 0.008 to 0.08 pCi g^{-1} (0.3 to 3 Bq kg^{-1}) wet weight (Mauchline and Templeton, 1964). It is estimated (NCRP, 1987a) that the whole-body dose from ^{87}Rb is 0.3 mrem y^{-1} (3 μSv y^{-1}) on average.

NATURAL RADIOACTIVITY IN PHOSPHATE FERTILIZERS

Phosphate fertilizers are used in huge amounts around the world and are essential for food production. The natural resource from which agricultural phosphorus is obtained is phosphate rock, found in sedimentary formations, usually interbedded with marine shales or limestones. The United States is a major phosphate user, with extensive mines located in Florida, North Carolina, Tennessee, Idaho, Montana, Utah, and Wyoming. In recent years, more than 90% of the phosphate rock mined in the United States has come from Florida. The location of the Florida phosphate deposits is shown in Fig. 6-6. They average 4 m in thickness and lie under sand and clay overburdens that vary from about 1 to 10 m in thickness (Roessler et al., 1980). As the surface mining operations proceed, the overburdens, sand tailings, and other waste products are returned to the land.

It has been known since early in this century that the phosphate rocks contain relatively high concentrations of uranium. Depending on the economics of the uranium industry, uranium has periodically been extracted as a sidestream of phosphoric acid production (NCRP, 1993b). The concentration of uranium in U.S. phosphate ores ranges from 8 to 400 ppm. The phosphate rocks from the important Florida deposits average 41 ppm with ^{226}Ra in secular equilibrium. Phosphate rock is mined in huge quantities: it is reported that in 1974, about 26 million tons of ore were sold for fertilizer production in the United States, and that the ore contained about 1000 Ci

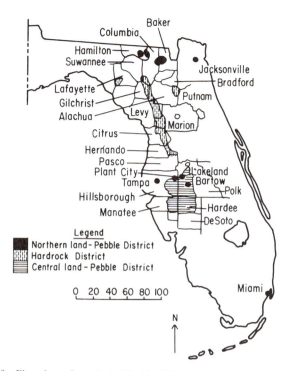

FIGURE 6-6   Phosphate deposits in Florida. [From Guimond and Windham (1980).]

(37 TBq) of ^{226}Ra in equilibrium with ^{238}U (Guimond, 1978). By 1975, 2 billion tons of ore had been removed from 50,000 hectares (Stowasser, 1977).

Several types of phosphate fertilizers are produced, and these are usually blended for application in the field. The radium and uranium tend to separate in the process of producing phosphoric acid, an important step in fertilizer manufacture (Roessler *et al.,* 1979; Guimond, 1978; NCRP 1993b). The radium passes to the gypsum produced as a by-product. A smaller fraction of the radioactivity is rejected in the process and ends up in the mounds of stored waste products.

Elevated concentrations of ^{222}Rn are found within structures located on land reclaimed from phosphate mines (Guimond and Windham, 1980; Roessler *et al.,* 1983). A survey of indoor ^{222}Rn and ^{222}Rn decay product exposures in buildings on or near the Florida phosphate deposits was conducted by Guimond and Windham, who found that 71% of homes built on reclaimed land contained more than 0.01 WL and 23% contained more than 0.05 WL. High levels were also found in structures built over mineral deposits that had not been mined, but relatively few measurements were made. Although these [222]Rn concentrations are above the average concentrations found in U.S. homes, it is noteworthy that many homes in areas of the United States contained higher concentrations than are found in the phosphate regions.

The high concentrations of [226]Ra also result in increased γ-radiation exposure in homes built on reclaimed land. Johnson and Bailey (1983) found excess γ dose to average 17 mrem y^{-1} (0.17 mSv y^{-1}), with values ranging as high as 166 mrem y^{-1} (1.66 mSv y^{-1}).

The radioactivity associated with the land reclaimed prior to the 1940s tends to be higher than that in land reclaimed more recently, because methods have been developed for separating the phosphate-rich clay fraction of the deposit. A major fraction of the radium and uranium associated with the phosphate minerals was formerly returned to reclaimed land (Roessler et al., 1980). In 1978, the Florida Department of Health and Rehabilitation Services ruled that some sort of rehabilitation would have to be applied to homes in which the [222]Rn decay product concentrations exceeded 0.029 WL. This is about three times the remedial action level suggested by the Surgeon General for application to homes constructed over uranium mill tailings in Grand Junction, Colorado (see Chapters 7 and 15). It has been estimated that of the approximately 4000 buildings on reclaimed land in 1978, 6–10% would require some kind of corrective action. Roessler et al. (1980) found that most of these structures were located on land reclaimed prior to the 1940s. The radium and radon contents of water supplies from the phosphate rock region of North Carolina were investigated by Watson and Mitsch (1987), who found the levels to be well within acceptable limits.

The contribution of uranium and radium to agricultural lands owing to the application of phosphate fertilizers does not significantly affect the dose received from the general population (Kirchmann et al., 1980; Ryan, 1981). However, phosphorus, in the form of mineral phosphate rock, is sometimes added to cattle feed and this practice can result in increased levels of uranium and radium in cows' milk (Reid et al., 1977). However, continued application of phosphate fertilizers to soil over a period of many years could eventually double the radium and uranium content of the soil, which would result in a corresponding doubling of the dose to bone from this source. Spalding and Sackett (1972) found that the uranium content of North American rivers is higher than in the past, which they attribute to increased runoff of phosphate fertilizers.

NATURAL RADIOACTIVITY IN BUILDING MATERIALS

The Federal Republic of Germany conducted a survey of 30,000 dwellings and found that, on average, the external radiation exposure was 33% higher within the dwellings than outdoors. Thus, although the building materials absorb the radiation that originates outside the building, exposure within the building is more than compensated by the presence of radionuclides in the materials of construction (Kolb and Schmier, 1978). Data on radioactivity in selected building materials are given in Table 6-16. Since most people spend 90% of their time indoors, the dose they receive from external natural radiation is increased somewhat. According to Kolb and Schmier, the absorbed dose to the population of the Federal Republic of Germany ranges from 40 to 80 mrad y^{-1} (0.4–0.8 mGy y^{-1}) for continuous indoor exposure compared to 30 to 60 mrad y^{-1} (0.3–0.6 mGy y^{-1}) for continuous outdoor exposure. The natural radioactivity of building materials in Taiwan was studied by Chang *et al.* (1974).

Wood-frame residences, on the other hand, contain very little radioactivity and provide some shielding and isolation from natural sources in the underlying soil. The NCRP (1987a) concluded that for the United States and Canada, indoor dose rates to γ radiation can be assumed to equal outdoor rates in the vicinity, on average. For individual structures, however, indoor dose rates can vary considerably from those outdoors.

Two important by-products from the processing of phosphate rock are gypsum and calcium silicate slag, both of which are used in the construction industry. Use of these products in buildings may increase the γ-radiation dose (EPA, 1990b; Laiche and Scott, 1991). Building materials can also be a source of indoor ^{222}Rn, as was discussed earlier in this chapter, but are usually much less important than infiltration from soil.

NATURAL RADIOACTIVITY IN FOSSIL FUELS

Coal contains radionuclides of the uranium and thorium series, as well as ^{40}K. The uranium and thorium concentrations in mined coal have been assembled from various sources by Beck *et al.* (1980) and are given in Table 6-17, which includes data on most of the coal beds used in the United States. The mean value for all coals sampled is 1.7 μg g^{-1} for uranium and 4.5 μg g^{-1} for natural thorium, which is similar to the average concentrations found in soils and rocks, also shown in Table 6-17. Subsequently the EPA estimated slightly lower average concentrations of 1.3 ppm for uranium and 3.2 ppm for thorium based on 5000 samples from all the major coal-producing areas (EPA, 1984a). Most investigators have found that the various radionuclides of the uranium and thorium series are in secular equilibrium. Beck *et al.* (1980) noted that there have been exceptions, but did not consider the evidence for disequilibria to be conclusive.

The quantity of radionuclides discharged to the atmosphere per ton of coal consumed depends on the concentration of radionuclides in the coal,

TABLE 6-16

Estimates of Concentrations of Uranium, Thorium, and Potassium in Building Materials[a]

Material	Uranium			Thorium			Potassium		
	ppm	mBq g⁻¹	pCi g⁻¹	ppm	mBq g⁻¹	pCi g⁻¹	%	mBq g⁻¹	pCi g⁻¹
Granite	4.7	63	1.7	2	8	0.22	4.0	1184	32.0
Sandstone	0.45	6	0.2	1.7	7	0.19	1.4	414	11.2
Cement	3.4	46	1.2	5.1	21	0.57	0.8	237	6.4
Limestone concrete	2.3	31	0.8	2.1	8.5	0.23	0.3	89	2.4
Sandstone concrete	0.8	11	0.3	2.1	8.5	0.23	1.3	385	10.4
Dry wallboard	1.0	14	0.4	3	12	0.32	0.3	89	2.4
By-product gypsum	13.7	186	5.0	16.1	66	1.78	0.02	5.9	0.2
Natural gypsum[b]	1.1	15	0.4	1.8	7.4	0.2	0.5	148	4
Wood[b]	—	—	—	—	—	—	11.3	3330	90
Clay brick[c]	8.2	111	3	10.8	44	1.2	2.3	666	18

[a]From NCRP (1987a), except where noted.
[b]From Chang et al. (1974).
[c]From Hamilton (1971).

TABLE 6-17

Uranium and Thorium Concentrations in Coal as Mined (Fraction of Dry Weight)[a]

Region, type	No. of samples	U (μg g^{-1}) Range	Geo. mean	Mean	Th (μg g^{-1}) Range	Geo. mean	Mean
Pennsylvania, anthracite	53	0.3–25.2	1.2	1.5	2.8–14.4	4.7	5.4
Appalachian[b]	331	<0.2–10.5	1.0	1.4	2.2–47.8	2.8	4.9
Midwest[b]	143	0.2–43	1.4	3.3	<3–79	1.6	5.2
Northern Great Plains[b,c]	93	<0.2–2.9	0.7	0.9	<2–8.0	2.4	2.7
Gulf Coast, lignite	34	0.5–16.7	2.4	3.2	<3–28.4	3.0	8.3
Rocky Mountains[b,c]	134	<0.2–23.8	0.8	1.6	<3–34.8	2.0	3.6
Alaska[c]	18	0.4–5.2	1.0	1.2	<3–18	3.1	4.4
Illinois basin	56	0.31–4.6	1.3	1.5	0.71–5.1	1.9	2.1
Appalachian	14	0.4–2.9	1.3	1.5	1.8–9.0	4.0	4.5
Western	22	0.3–2.5	1.0	1.2	0.62–57	1.8	2.3
Western	19	0.11–3.5	0.85	0.9			
All samples	910		1.04	1.74		2.40	4.47
		Typical range	Avg.		Typical range	Avg.	
Soil		0.9–4.0	1.8		2–12	6	
Rocks		0.5–5	2.7		1.6–20	9.6	

[a]From Beck et al. (1980).
[b]Bituminous.
[c]Subbituminous.

177

the method of combustion, and the efficiency of fly ash recovery. About 90% of the mass of coal is consumed during combustion and the radioactive nuclides tend to concentrate in the nonvolatile fraction or "ash." The ash then undergoes partitioning, according to whether it separates within the furnace and stack or passes with the hot gases to the "fly ash." Modern furnaces in the United States now burn pulverized coal and realease 60–85% of the coal ash content into fly ash (Fay and Parker, 1977). In former times, it was common practice to allow the fly ash to escape to the environment, but this is no longer permitted in the United States and other countries where electrostatic precipitators, bag houses, and scrubbers are employed to reduce the amount of fly ash released with the stack gases. Modern coal-fired plants achieve 99.5% ash retention (NCRP, 1987d). Because enrichment of some radionuclides occurs in the stack, the concentrations of ^{210}Pb and ^{210}Po in the fly ash are 5–10 times greater than in the original coal (Beck et al., 1980).

The normal levels of uranium and thorium in the environs are sufficiently high that changes due to the emissions from coal-fired power stations are barely detectable: some investigators (Bedrosian et al., 1970) found no changes in the vicinity of an old, relatively poorly controlled plant, and Beck et al. (1980) could find only slight changes in a carefully designed survey of several plants. It has been found, however, that the ^{226}Ra content of snow downwind of coal-fired power plants is higher than the concentration in rainwater (Jaworowski et al., 1975). The concentrations of naturally occurring radionuclides in *airborne* particulates collected in the vicinity are elevated above background and increase the dose to the lung. The dose equivalent to lung for a maximally exposed individual from emissions of radionuclides from a typical modern 1000-MW electric power plant and a 1972 reference plant are given in Table 6-18. It is seen that there has been a major reduction in the dose equivalent received from a modern plant compared to the 1972 reference plant, which operated without stringent fly ash control. The NCRP (1987d) recognizes the large uncertainties in dose estimates from coal emissions and has estimated that the average effective dose equivalent to a member of the U.S. population from identified pathways lies in the range of 0.3 to 3 μSv (0.03 to 0.3 mrem) per year. Because the NCRP estimate includes doses from the plume, resuspended material, direct radiation, and the food chain, it is not directly comparable to the data in Table 6-18.

The first report on the radioactive emissions from coal-burning power plants (Eisenbud and Petrow, 1964) concluded that when the data were normalized for their radiotoxicity relative to the emissions from the first of the commercially operated pressurized water reactors, which had just begun operation, the dose from the fossil fuel emissions was greater than that

TABLE 6-18

Estimated Dose Equivalents to Lung of Maximum Exposed Individuals
from Model Coal-Fired Power Plant Releases[a]

Nuclide	Dose equivalent			
	Modern plant		1972 reference plant	
	mrem y^{-1}	μSv y^{-1}	mrem y^{-1}	μSv y^{-1}
^{238}U–^{234}U	0.009	0.09	0.46	4.6
^{230}Th	0.015	0.15	1.55	15.5
^{226}Ra	0.003	0.03	0.23	2.3
^{210}Pb	0.005	0.05	0.22	2.2
^{210}Po	0.075	0.75	3.1	31
^{232}Th	0.009	0.09	0.93	9.3
^{228}Th	0.018	0.18	1.85	18.5
^{228}Ra	0.004	0.04	0.38	3.8
Total	0.14	1.4	8.7	87

[a]From Beck et al. (1980).

from the nuclear reactors. However, the stringent requirements of the federal Clean Air Act have since resulted in substantial reductions in the lung dose from the fly ash emissions, as shown in Table 6-18. Radioactivity from coal combustion continues to be a topic of interest and numerous studies, many from countries other than the United States, have been published (e.g., Corbett, 1983; Nakaoka et al., 1985; Alvarez and Garzon, 1989; Nowina-Konopka, 1993).

Municipal incinerator ash also contains natural radioactivity. Radionuclide concentrations of the ^{238}U and ^{232}Th decay series, as well as ^{40}K, were determined by Kitto (1992) in fly and bottom ashes from two municipal incinerators. Cesium-137 was also detected. Secular equilibrium was observed for the series radionuclides, although ^{222}Rn losses during combustion were evident. The fly ash samples showed increasing enrichment of ^{40}K and ^{210}Pb as particle diameters decreased. Radon-222 emanation from the bottom ashes was not detected but the fly ashes had an emanating power near 7%. Overall concentrations indicated that municipal incinerator ashes contain the equivalent radioactivity of surface soil and substantially less than ashes from coal combustion. Kitto concluded that an incinerator's contribution to atmospheric radioactivity is minimal.

Radioactivity has long been known to exist in petroleum gas (Burton, 1904) and the possible occupational impacts of contaminated gas-processing equipment has been noted (Gesell et al., 1975; Summerlin and Prichard,

1985). However, natural radioactivity in oilfield equipment and its disposal has only more recently received attention in the literature as an environmental radioactivity problem worthy of regulation (Rogers, 1991; Wennerberg, 1991, 1992). Louisiana, Mississippi, and Texas have promulgated regulations for natural radioactivity related to the oil and gas industry, and other states have regulations in draft form (Bernhardt *et al.,* 1996).

Radium-226 and, to a lesser degree, ^{228}Ra, and their decay products have recently been discovered in the scale that forms in oilfield piping and equipment. The activity is a potential internal hazard to workers from the inhalation and ingestion of the dust produced during descaling or pipe-cleaning operations. A higher-than-normal background γ dose rate is seen where the pipe cleaning was done routinely (Wilson and Scott, 1992).

"Tank battery sites" have been used for the intitial processing of crude oil to separate out water and sediment (Hebert and Scott, 1993). One or more wellheads are connected to a tank battery site consisting of storage and separation tanks, and a holding pit for increased separation of oil, water, and sediment. The sediment remaining in the pit is composed of a tarlike sludge. For reclamation, soil and sludge from the pits are removed and spread on land to allow biodegradation of the hydrocarbons. Natural radionuclides accumulate in these sites, predominantly ^{226}Ra and its decay products. Radium-228 is seen to a lesser extent. The γ dose rates measured by Hebert and Scott at six reclaimed tank battery sites ranged from 0.25 to 1.9 μGy h^{-1} (25 to 185 μrad h^{-1}), well above a typical background dose rate of 0.05 to 0.1 μGy h^{-1}. The radium concentration measured over the top 50 cm ranged from 0.09 to 7.8 Bq g^{-1} (2.5 to 210 pCi g^{-1}). Typical normal soil has a radium content of about 0.04 Bq g^{-1} (1 pCi g^{-1}). One approach to managing the wastes arising from natural radioactivity in the petroleum industry is hydraulic fracturing, suggested by Woods *et al.* (1993). With this approach, waste drill cuttings, scales, sludges, and platings containing elevated amounts of natural radioactivity are ground to form slurries suitable for injection to a well. Injection takes the form of hydraulic fracturing into a zone in the reservoir that is depleted in petroleum, thus returning the radioactivity to the reservoir from which it came.

INDUCED RADIONUCLIDES

A number of radionuclides that exist on the surface of the earth and in the atmosphere have been produced by the interaction of cosmic rays with atmospheric nuclei. The most important of these are tritium (^{3}H), ^{14}C, and ^{7}Be, and of lesser importance are ^{10}Be, ^{22}Na, ^{32}P, ^{33}P, ^{35}S, and ^{39}Cl. The properties of thse isotopes and the extent to which they have been reported in various media are listed in Table 6-1.

Carbon-14 is formed by ^{14}N capture of neutrons produced in the upper atmosphere by cosmic-ray interactions. The incident cosmic-ray neutron flux is approximately 1 neutron per second per square centimeter of the earth's surface, and essentially all these neutrons disappear by ^{14}N capture (Anderson, 1953). The incident neutron flux integrated over the surface of the earth yields the natural rate of production of ^{14}C atoms, which has been estimated (UNSCEAR, 1977) to be 38 kCi y^{-1} (1400 TBq y^{-1}), and is believed to have been unchanged for at least 15,000 years prior to 1954, when nuclear weapons testing began to alter the normal ^{14}C inventory to a noticeable extent (NCRP, 1985b).

Carbon-14 of natural origin is present in the carbon of all biota at the historically constant amount per gram of carbon of 6 pCi g^{-1} (0.22 Bq g^{-1}). After death of an organism, the ^{14}C equilibrium is no longer maintained, and the ratio of ^{14}C to ^{12}C diminishes at a rate of 50% every 5600 years, which makes it possible to use the ^{14}C content of organic materials for the purpose of measuring age (Libby, 1952).

Because the ^{14}C originally present in coal and oil has decayed almost completely, the introduction into the atmosphere of carbon from combustion of these fuels tends to reduce the specific ^{14}C activity of atmospheric CO_2. Suess (1958) first noted that the $^{14}C : {}^{12}C$ ratio of tree rings had diminished during the past century, which he attributed to the dilution of atmospheric ^{14}C by combustion of ^{14}C-free fossil fuels. For this reason, the concentration of ^{14}C in atmospheric carbon tends to be lower in urban and industrial areas.

Clayton et al. (1955) and Lodge et al. (1960) have used the ^{14}C content of particulate atmospheric carbon to estimate the fractions of the dust that originate from garbage incineration and combustion of fossil fuels. Incineration of food residues, textiles, paper, and other organic constituents of garbage produces smoke in which the ^{14}C is in contemporary equilibrium, because comparatively little refuse is so old that a significant fraction of the ^{14}C has decayed. Because there has been total decay of the ^{14}C present in the organic matter from which fossil fuels have formed, the relative contributions of the two sources of particulate carbon could be estimated. This technique may become less useful in time because the increasing fraction of plastics contained in municipal refuse is introducing a new source of ^{14}C-free carbon.

The total carbon content of the body is approximately 18%, or 12.6 kg for a 70-kg man, and the ^{14}C body burden from natural sources is thus on the order of 0.1 μCi (3.7 kBq), but the dose is small owing to the low energy of the ^{14}C β particles (0.01 MeV). It is estimated that the dose from ^{14}C is

3 mrem y^{-1} (30 μSv y^{-1}) to the skeletal tissues of the body and 1 mrem y^{-1} (10 μSv y^{-1}) to the soft tissues (NCRP, 1987a).

Tritium a radioactive isotope of hydrogen, is formed from several interactions of cosmic rays with gases of the upper atmosphere (Suess, 1958). Tritium exists in the atmosphere principally in the form of water vapor and precipitates in rain and snow. Like ^{14}C, it is produced in thermonuclear explosions, and this has increased the atmospheric content of tritium in a manner that will be discussed in a subsequent chapter. The natural production rate of ^{3}H is estimated (NCRP, 1979) to be about 0.19 atom cm^{-2} s^{-1}, corresponding to a steady-state global inventory of about 26 MCi (9.6 $\times$ 10^5 TBq).

The natural concentration of tritium in lakes, rivers, and potable waters was reported to have been 5–25 pCi L^{-1} (185–925 Bq m^{-3}) prior to the advent of weapons testing (UNSCEAR, 1982). The annual absorbed dose from tritium of natural origin is estimated to be about 1 μrem y^{-1} (0.01 μSv y^{-1}) uniformly distributed in all tissues (NCRP, 1987a).

The other nuclides formed from cosmic-ray interactions with the atmosphere may be potentially useful as tracers for studying atmospheric transport mechanisms, but relatively few observations have been reported.

NATURAL SOURCES OF EXTERNAL IONIZING RADIATION

The dose received from external sources of ionizing radiation originates from cosmic rays and from γ-emitting radionuclides in the earth's crust. UNSCEAR (1988) estimated the external annual effective dose equivalent from all naturally occurring radiation in "normal" parts of the world to be 0.36 mSv y^{-1} (36 mrem y^{-1}) from cosmic sources and 0.41 mSv y^{-1} (41 mrem y^{-1}) from terrestrial radiation. For the United States, the estimates are 0.27 mSv y^{-1} (27 mrem y^{-1}) for cosmic sources and 0.28 mSv y^{-1} (28 mrem y^{-1}) from terrestrial radiation (NCRP, 1987a).

Solon et al. (1958) and later Beck (1966) and Beck and de Planque (1968) made extensive measurements of the natural γ-radiation background in a number of cities throughout the United States. The data of Solon et al. were about 30% higher than those reported by Beck, probably because of the greater effect of fallout during the period when their measurements were made. The techniques used by Beck and associates permitted differentiation between fallout and natural radiation. This was not possible at the time Solon et al. made their measurements because γ spectrometry was not yet practical for use in the field.

The mean dose rate in the 124 locations measured by Solon and coworkers (1958) was 81 ± 20 mrad y⁻¹ (0.81 ± 0.2 mGy y⁻¹), compared to the Beck (1966) mean of 61 ± 23 mrad y⁻¹ (0.61 ± 0.23 mGy y⁻¹) at 210 locations. Solon *et al.* showed that their data were well correlated with barometric pressure, indicating the effect of cosmic sources of radiation. The data gathered by Beck at the principal cities where measurements were made are summarized in Fig. 6-7.

TERRESTRIAL SOURCES OF EXTERNAL RADIATION

The terrestrial sources of γ radiation are ^{40}K and nuclides of the ^{238}U and ^{232}Th series. If the concentrations of these three nuclides in soil are known, the dose can be estimated, using methods developed originally by Hultqvist (1956) and further developed by Beck (1972). The absorbed dose in air, 1 m above soil having unit concentrations of the three nuclides, is given in Table 6-19. If these data are adjusted for the typical concentrations of the three nuclides in soil (NCRP, 1975c), it will be found that ^{232}Th and ^{40}K each contribute 10–12 mrad y⁻¹ (0.1–0.12 mGy y⁻¹). The absorbed dose above various kinds of rocks is given in Table 6-20. According to Kohman (1959), the similarity of dose rates from the various isotopes listed in Table 6-20 is a coincidence arising from the fact that these isotopes happen to be present in rocks in amounts that are approximately inversely proportional to their specific activities.

The external γ radiation from radionuclides in the earth's crust is thus influenced by the kind of rock over which the measurements are made. The actual doses to people cover a somewhat narrower range owing to the fact

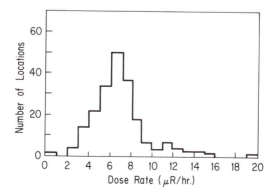

FIGURE 6-7 Frequency distribution of the γ dose rate from natural emitters at 210 different locations in the United States. [From Beck (1966); reproduced from *Health Physics*, **12**, by permission of the Health Physics Society.]

TABLE 6-19

Absorbed Dose Rate 1 m above Soil Containing
1 pCi g^{-1} (37 mBq g^{-1}) of ^{40}K, ^{238}U, and ^{232}Th,
Including Decay Products[a]

Nuclide	Soil concentration		Absorbed dose rate in air	
	pCi g^{-1}	mBq g^{-1}	mrad y^{-1}	μGy y^{-1}
^{40}K	1	37	1.4	14
^{238}U + decay products	1	37	13.9	139
^{232}Th + decay products	1	37	21.6	216

[a]Adapted from Beck (1972).

that most people live on soil rather than rock. The soils tend to be less variable in their radioactive content because the igneous rocks, which are high in radioactive content, weather more slowly and, therefore, contribute fewer radionuclides to soils than the softer sedimentary rocks.

Beck *et al.* (1966) also reported on the variation with time of the natural γ dose rates at two communities in Westchester County, New York, during 1963–1965. The individual observations at one location that averaged 6.9 ± 2.1 μrad h^{-1} (0.069 ± 0.021 μGy h^{-1}) ranged from 6.0 to 8.2 μrad h^{-1} (0.06 to 0.082 μGy h^{-1}). At the second location, which averaged 7.1 ± 2.1 μrad h^{-1} (0.071 ± 0.021 μGy h^{-1}), the range was similar, from 6.0 to 8.3 μrad h^{-1} (0.06 to 0.083 μGy h^{-1}). These investigators attributed the

TABLE 6-20

Gamma Radiation Dose Rates from Typical Values of Radium, Uranium,
Thorium, and Potassium in Rocks[a]

Type of rock	Dose equivalent rate							
	mrem y^{-1}				μSv y^{-1}			
	^{226}Ra	^{238}U	^{232}Th	^{40}K	^{226}Ra	^{238}U	^{232}Th	^{40}K
Igneous	24	26	37	35	240	260	370	350
Sedimentary								
Sandstones	13	7.7	18	15	130	77	180	150
Shales	20	7.7	31	36	200	77	310	360
Limestones	7.7	8.4	4	4	77	84	40	40

[a]From UNSCEAR (1958).

principal source of variation to the effect of soil moisture, which can account for 30% by weight during wet periods. Snow cover will also reduce the natural γ dose rate. Whereas the soil moisture might be expected to attenuate the radiation from thorium and potassium, the radiation from the uranium series might be expected to increase because soil water would inhibit the diffusion of ^{222}Rn. However, an examination of the data published in the report of Beck *et al.* fails to show an inverse correlation of this type.

In addition to the direct ground-level measurements and estimates of external dose that can be made from knowledge of the radionuclide content of soils, dose estimates can be made from aircraft by using sensitive γ detectors designed to provide estimates of ground-level dose (see Chapter 13). Many such surveys have been made by government agencies, either to explore for uranium or to provide information about the levels of radiation in the vicinity of nuclear facilities. These data have been analyzed by Oakley (1972), who made estimates of the population dose distribution in the United States. The data are grouped by three geographical regions: (1) the Atlantic and Gulf coastal plain, for which the mean absorbed dose rate is 23 mrad y^{-1} (0.23 mGy y^{-1}); (2) a portion of the eastern slope of the Rocky Mountains, where the absorbed dose averages 90 mrad y^{-1} (0.9 mGy y^{-1}); and (3) the remainder of the United States, where the average absorbed dose is 46 mrad y^{-1} (0.46 mGy y^{-1}).

Cosmic Radiation

The primary radiations that originate in outer space and impinge isotropically on the top of the earth's atmosphere consist of 87% protons, 11% α particles, about 1% nuclei of atomic number Z between 4 and 26, and about 1% electrons of very high energy. An outstanding characteristic of the cosmic radiations is that they are highly penetrating, with a mean energy of about 10^{10} eV and maximum energies of as much as 10^{20} eV. The primary radiations predominate in the stratosphere above an altitude of about 25 km (NCRP, 1987a).

It is now known that these radiations originate outside the solar system and that only a small fraction is normally of solar origin. However, the solar component becomes very significant outside the atmosphere following flares associated with sunspot activity, which follows an 11-year cycle.

The interactions of the primary particles with atmospheric nuclei produce electrons, γ rays, neutrons, and mesons. At sea level the mesons account for about 80% of the cosmic radiation flux and electrons account for about 20%. It has been estimated that 0.05% of primary protons penetrate to sea level (Myrloi and Wilson, 1951). With the development of high-altitude aircraft and manned space probes, the dose from primary cosmic radiations attracted interest (Curtis *et al.,* 1966; O'Brien and McLaughlin, 1972), which continues to the present (NCRP, 1989c, 1995; Reitz *et al.,* 1993). Sophisticated radiation measurements are now an important component of the scientific programs using satellite probes into space (Tobias and Todd, 1974; Akopova *et al.,* 1987).

The dose from cosmic radiation is markedly affected by altitude. The annual cosmic-ray dose equivalent is about 29 mrem (0.29 mSv y^{-1}) at sea level. For the first few kilometers above the earth's surface, the cosmic-ray dose rate doubles for each 2000-m increase in the altitude (Fig. 6-8). However, for the first 1000 m, the *total* dose rate actually decreases with altitude above the surface, because attenuation of the γ rays from terrestrial sources occurs more rapidly than the increase in cosmic radiation (Schaefer, 1971). Residents of Denver (altitude 1600 m) receive nearly twice the dose at sea level and in Leadville, Colorado (altitude 3200 m), the residents receive about 125 mrem y^{-1} (1.25 mSv y^{-1}) from cosmic rays, which is more than four times the annual dose at sea level. The increase in dose rate at higher altitudes is shown in Fig. 6-9, where it is seen that at polar latitudes, rates in excess of 1.5 mrem h^{-1} (15 μSv h^{-1}) are received at altitudes of 50,000–80,000 feet, the upper limit of high-performance aircraft such as the SST. On rare occasions, once or twice during the 11-year cycle, a giant solar event may deliver dose equivalents in the range 1–10 rem h^{-1} (10–100

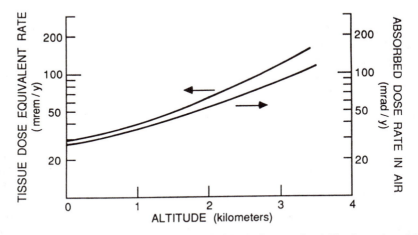

FIGURE 6-8 Variation of cosmic-ray dose with altitude above sea level. The charged particle absorbed dose rate in air or tissue is shown in the lower curve and the total dose equivalent rate (charged particles plus neutrons) is shown in the upper curve at 5 cm depth in a 30-cm-thick slab of tissue. These dose rates may be converted to mSv y^{-1} or mGy y^{-1} by dividing by 100 [From NCRP (1987a).]

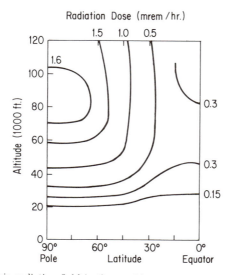

FIGURE 6-9   Cosmic radiation field in the earth's atmosphere, from sea level to 120,000 ft. [Reprinted from Schaefer (1971). Radiation exposure in air travel. *Science* **173,** 780–783. Copyright 1971 American Association for the Advancement of Science.]

mSv h^{-1}), with a peak as high as 5 rem (500 mSv) during the first hour (Upton *et al.,* 1966). During a well-documented solar flare in February 1956, dose rates well in excess of 100 mrem h^{-1} (1 mSv h^{-1}) existed briefly at altitudes as low as 35,000 ft (Schaefer, 1971).

Because of the effect of altitude, the passengers and crew of high-flying aircraft are subject to additional dose from cosmic rays. A transcontinental flight has been estimated to result in a dose of about 2.5 mrem (0.025 mSv), or 5 mrem (0.05 mSv) per round trip (NCRP, 1987a). The cabin attendants and aircraft crew have been estimated to receive an average incremental dose of about 160 mrem y^{-1} above that received at sea level (NAS/NRC, 1980). A more recent study determined that air crews who worked an exceptionally heavy schedule (1100 hours per year) could receive annual doses of between 0.3 and 9 mSv (30 and 900 mrem) depending on the routes flown (O'Brien *et al.,* 1992). The U.S. Federal Aviation Administration has explicitly recognized flight crew members as occupationally exposed to radiation (Barish, 1995).

Unusual solar activity is capable of injecting huge quantities of high-energy protons into interplanetary space and can increase the dose to hazardous levels that might require evasive action by aircraft flying above about 13 km. An international commission concerned with high-flying aircraft, as well as the Federal Aviation Administration (FAA) in the United

States, has recommended continuous monitoring for solar flares so that high-flying aircraft can take evasive action if necessary (Wilson, 1981).

Above the earth's atmosphere, the dose consists of two main components. One is the dose from highly energetic cosmic radiation trapped in the earth's magnetic field as illustrated in Fig. 6-10. A second portion is received beyond the earth's magnetic field and is due to the background cosmic radiation, on which may be superimposed very sharp peaks of radiation due to solar flares.

On entering the earth's magnetic field, some of the primary particles are deflected toward the polar regions, resulting in a somewhat lower radiation flux at the equator. This phenomenon becomes more accentuated with altitudes above a few kilometers, as shown in Fig. 6-11. The difference in the dose rate due to geomagnetic latitude varies from 14% at sea level to 33% at 4360 m (Pertsov, 1964).

The cosmic-ray doses to passengers on supersonic and subsonic aircraft for representative flights are given in Table 6-21. Although the dose received per hour is greater when flying in an SST, the dose per trip is less than that for subsonic flight because of the reduced flying time. The dose equivalent received during supersonic flight is less certain than the dose because the quality factor for cosmic rays at high altitudes is not well known. The quality factor probably lies in the range of 1.4 to 3.3 (NCRP, 1995).

The geomagnetically trapped radiations consist mainly of protons and electrons produced by backscatter of the primary cosmic-ray beam on the earth's atmosphere and protons of solar origin. Because of differences in the mass-to-charge ratios of protons and electrons, the trajectories for the two particles diverge, giving rise to an inner radiation belt consisting mainly of protons and an outer belt consisting mainly of electrons, as illustrated in Fig. 6-10. The proton region begins about 1000 km, above the geomagnetic equator and ends at an altitude of about 3000 km.

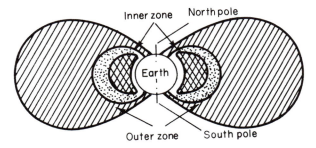

FIGURE 6-10   The geomagnetically trapped corpuscular radiations.

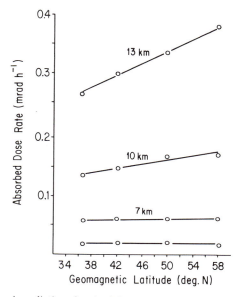

FIGURE 6-11   Cosmic radiation absorbed dose index rate at four different altitudes as a function of latitude. The lowest line corresponds to sea level. [From UNSCEAR (1977).]

TABLE 6-21

Calculated Cosmic Ray Doses to a Person Flying in Subsonic and Supersonic Aircraft under Normal Solar Conditions[a]

Route	Subsonic flight at 11 km			Supersonic flight at 19 km		
	Flight duration (h)	Dose per round trip		Flight duration (h)	Dose per round trip	
		mrad	μGy		mrad	μGy
Los Angeles–Paris	11.1	4.8	48	3.8	3.7	37
Chicago–Paris	8.3	3.6	36	2.8	2.6	26
New York–Paris	7.4	3.1	31	2.6	2.4	24
New York–London	7.0	2.9	29	2.4	2.2	22
Los Angeles–New York	5.2	1.9	19	1.9	1.3	13
Sydney–Acapulco	17.4	4.4	44	6.2	2.1	21

[a]From UNSCEAR (1977), adapted from Wallace (1973).

When astronauts travel into outer space they are exposed to the intense radiation of the two belts of trapped electrons, the primary cosmic radiation particles and the radiation from solar flares (UNSCEAR, 1982; NCRP, 1989c). The largest portion of the dose to the Apollo astronauts was received when they were passing through the earth's radiation belts, and the dose can be influenced by the trajectory through the belts. The Apollo missions lasted from about 6 to 12 days and resulted in average crew doses of 1.6 to 11.4 mGy (160–1140 mrad). The Skylab missions lasted from 28 to 90 days and resulted in doses of 16 to 77 mGy (1.6–7.7 rad). During the first 25 shuttle flights (NCRP, 1989c), the indiviudal crew doses ranged from about 0.06 to 6 mGy (6–600 mrads). The doses would be much higher in the event of a solar flare. It has been estimated that the dose from a flare that occurred in July 1959 could have been between 40 and 360 rad (0.4 to 3.6 Gy). Space radiation is undoubtedly an important constraint on long-term space travel.

Dose calculations from the solar flare particles are very difficult owing to the wide range of energies and the complex shielding geometries presented by space capsules. Conversion of cosmic-ray dose estimates from absorbed dose (Gy or rad) to dose equivalent (Sv or rem) is made difficult by the range of energies and linear energy transfer (LET). The LET of the incident protons varies from about 0.21 to 58 keV μm^{-1}. For high-Z nuclides such as iron, the LET can be as high as 3500 keV μm^{-1}. The absorbed dose for the centrally traversed cells varies from 0.07 rad (0.7 mGy) for relatively low-LET protons to 1200 rad (12 Gy) for high-LET iron ions. Estimates of the relative biological effectiveness (RBE) for the carcinogenic effect of the heavy ions range from 5 for helium to 27 for iron and argon (Fry *et al.*, 1985).

TECHNOLOGICAL DEVELOPMENTS THAT INCREASE HUMAN EXPOSURE

Aeorspace activities are only one way in which the dose from nature is increased by technological developments. The dose can also be increased by use of building materials that have high levels of natural radioactivity and, as discussed earlier, by living in houses in which ^{222}Rn and its decay products accumulate. It has also been mentioned that the dose received from the ^{222}Rn decay products, especially ^{210}Po, can be increased by smoking cigarettes. Other ways in which the dose from natural sources can be increased include burning natural gas (which may contain ^{222}Rn), mining phosphate minerals (which are often associated with uranium), and injudiciously disposing of uranium mill tailings. The exposures from mill tailings will be discussed in Chapter 7. Other sources of technological enhanced natural radioactivity (Gesell and Prichard, 1975) will be reviewed in Chapter 10.

AREAS HAVING UNUSUALLY HIGH NATURAL RADIOACTIVITY

Except for brief mention of the occurrences of high levels of ^{226}Ra in well water, the discussion of natural radioactivity has thus far been limited to levels to which people are normally exposed. Yet there are places in the world where the levels of natural radiation exposure are abnormally high.

THE MINERAL SPRINGS

It has long been known that many mineral springs contain relatively high concentrations of radium and ^{222}Rn, and in many places in the world the radioactivity of springs has been exploited for their alleged curative powers.[4] Spas in South America, Europe, Japan, and elsewhere have commercialized the high radioactive content of local waters, and in some places research laboratories are operated in which the physiological basis for the alleged curative effects is studied. Visitors are encouraged not only to drink and bathe in the radioactive waters but also to sit in "emanatoria," where they can breathe ^{222}Rn emanating from surrounding rock, as shown in Fig. 6-12 (Pohl-Rüling and Scheminzky, 1954; Pohl-Rüling, 1993).

Published values of ^{226}Ra in mineral waters (UNSCEAR, 1958) ranged to more than 100 pCi L^{-1} (3.7 kBq m^{-3}), which is several hundred times greater than the values normally reported for public water supplies. However, to illustrate the fact that the values reported for spring waters are not typical of the drinking water of the region, the ^{226}Ra concentration in tap water of Bad Gastein, Austria, is reported to be 0.62 pCi L^{-1} (23 Bq m^{-3}) compared to 100 pCi L^{-1} (3.7 kBq m^{-3}) for some of the local springs (Muth et al., 1957).

About 5 million liters of water that contain high concentrations of ^{222}Rn are discharged daily from hot springs in Bad Gastein. The water is conveyed to hotels and bathhouses, where 58 Ci y^{-1} (2.1 TBq y^{-1}) are released to the atmosphere. Uzinov et al. (1981) described the manner in which ^{222}Rn emanations from this water result in exposure to the city residents, tourists

---

[4]Mineral waters were perceived as being beneficial to health in Roman times, many centuries before any knowledge of radioactivity existed. When the phenomenon of radioactivity was discovered, high levels were found to exist in the mineral waters. This resulted in the following fallacy: the waters are good for health, the waters are radioactive, hence radioactivity is good for health.

FIGURE 6-12 Gasteiner Heilstollen Radon spa at Bad Gastein, Austria. Clockwise from upper left: the main resort building; train used to transport guests into the mountain; guests entering the radon area; guests "bathing" in the

who visit the spas, and employees of the many bathhouses. They estimate that exposure of some of the attendants is as high as 40 WLM y^{-1}. A cytogenetic survey of persons subject to a gradient of ^{222}Rn exposure, including residents of Bad Gastein and several groups of employees of the spas, demonstrated dose-dependent increases in the frequency of chromosomal aberrations in the white blood cells (Pohl-Rüling and Fischer, 1979).

One of the most popular mineral springs localities in the United States is at Saratoga, New York, where reports of the medicinal value of the water go back to the early eighteenth century. It is known that many of the springs contain ^{226}Ra in amounts that exceed the EPA limit of 5 pCi L^{-1} (185 Bq m^{-3}) by a factor of 100. A survey has been made of the persons using one spring, Hathorn No. 1 (Aulenbach and Davis, 1976), which attracts many visitors and contains an average of about 200 pCi L^{-1} (7.4 kBq m^{-3}). Twenty-seven long-time users of the water from this spring who were selected for study stated that they consumed the water for 5 to 65 years, in amounts that ranged from about 0.2 to 3 L d^{-1}. Based on the information obtained in this way, it was calculated that the ^{226}Ra body burdens of the 27 individuals questioned ranged from slightly above normal to a high of 0.11 μCi (4.1 kBq), which is slightly above the limit of 0.1 μCi (3.7 kBq) for industrial workers. More comprehensive surveys are needed to evaluate the public health significance of mineral springs such as these. There are few studies of the extent to which the radioactivity contained in mineral spring waters has been absorbed by human beings.

Monazite Sands and Other Radioactive Mineral Deposits

Major anomalies in the concentrations of radioactive minerals in soil have been reported in three countries, Brazil, India, and China.

In Brazil, the radioactive deposits are of two distinct types: the monazite sand deposits along certain beaches in the states of Espirito Santos and Rio de Janeiro, and the regions of alkaline intrusives in the state of Minas Gerais (Roser *et al.*, 1964; Brazilian Academy of Sciences, 1977).

Monazite is a highly insoluble rare earth mineral that occurs in beach sand together with the mineral ilmenite, which gives the sands a characteristic black color. The black sands in Brazil are advertised for their radioactivity and are much sought by some tourists for their perceived benefits to health. The external radiation levels on these black sands range up to 5 mrad h^{-1} (50 μGy h^{-1}), and people come from long distances to relax on the sands and to enjoy the many hotels that have been constructed to care for their needs (Fig. 6-13). The most active of these Brazilian vacation towns is Guarapari, which has a stable population of about 7000 people and an annual influx of 10,000 vacationers. Some of the major streets of Guarapari have

FIGURE 6-13   Beach of the black sands at Guarapari, state of Espirito Santos, Brazil. The dark areas contain monazite sands over which the exposure rate is as high as 5 mrad h^{-1}. Tourists travel to these beaches because of local beliefs about the beneficial effects of radioactivity.

radiation levels as high as 0.13 mrad h^{-1} (1.3 μGy h^{-1}), which is more than 10 times the normal background. Similar radiation levels are found inside some of the buildings in Guarapari, many of which, in parts of the village that are not built on monazite sand, are nevertheless elevated in radioactivity because the beach sands were incorporated into the building materials. Roser and Cullen (1964) undertook extensive external radiation measurements throughout the black sands region and concluded that almost all of the approximately 60,000 inhabitants of the region were exposed to abnormally high radiation levels, but that only a small number (about 6600) were exposed to more than 0.5 rem y^{-1} (5 mSv y^{-1}). The population is too small to permit meaningful epidemiological investigations to be conducted, but cytogenetic studies have shown a higher than normal frequency of abnormalities in blood chromosomes (Barcinski *et al.,* 1975).

The principal radionuclides in monazite are from the ^{232}Th series, but there is also some uranium present and, therefore, some opportunity for ^{226}Ra uptake. However, very little food is grown in the monazite areas of Brazil, and the diets of the local inhabitants are derived principally from outside sources. The exposures in the monazite areas are due primarily to external radiation, and the internal dose is not believed to be significant (Eisenbud *et al.,* 1964; Penna France *et al.,* 1970).

In the state of Kerala, on the southwest coast of India, the monazite deposits are more extensive than those in Brazil, and about 100,000 persons inhabit the area. The dose from external radiation is, on average, similar to the doses reported in Brazil, 500–600 mrad y^{-1} (5–6 mGy y^{-1}), but individual doses up to 3260 mrad y^{-1} (32.6 mGy y^{-1}) have been reported (Sunta *et al.*, 1982), The epidemiological studies that might be possible with a population of this size have not yet been undertaken (Eisenbud, 1992), but some reports have been published. A higher than normal incidence of mongolism (Down's syndrome) has been reported in the high-background area (Kochupillai *et al.*, 1976), but this has been refuted (Sundaram, 1977). Down's syndrome has not been previously identified as radiogenic. In contrast to Brazil, there have been no reports of cytogenetic studies in humans. Mistry *et al.* (1979) reported that uptake of ^{228}Ra by food grown in the monazite area is greater than the uptake reported from Brazil.

Another area of elevated natural radiation has been found on Ullal Beach near Managalore, India (Radhakrishna *et al.*, 1993). Dose rates in this area range to about 2 μGy h^{-1} (200 μrad h^{-1}) and are related to monazite. Characterization of this area has begun and both thorium and uranium series radionuclides contribute to the dose (Narayana *et al.*, 1995).

A distinctly different source of exposure to natural radioactivity exists near the city of Araxa in the Brazilian state of Minas Gerais. The soil is generally poor in this area except in patches where it contains apatite, a phosphate mineral associated with a number of radioactive minerals that contain both uranium and thorium. The food grown in this area contains relatively high amounts of ^{228}Ra and ^{226}Ra, but the number of exposed persons is small. Penna Franca *et al.* (1970) surveyed the dietary habits of the indigenous population of this region and undertook radiochemical analysis of the foods. Of the 1670 persons who live in this area, nearly 200 ingest radium in amounts that are 10–100 times greater than normal. The ratio ^{228}Ra : ^{226}Ra is about 6 to 1 in the diet.

A unique anomaly located near Poços de Caldas, also in the state of Minas Gerais, is the Morro do Ferro, a hill that rises about 250 m above the surrounding plateau (Fig. 6-14). Near the summit of the hill is a near-surface ore body that contains about 30,000 metric tons of thorium and an estimated 100,000 metric tons of rare earth elements. The ambient γ-radiation levels near the summit of the hill range from 1 to 2 mrad h^{-1} (0.01 to 0.02 mGy h^{-1}) over an area of about 30,000 m^2 (Eisenbud *et al.*, 1984). The flora from this hill have absorbed so much ^{228}Ra that they can readily be autoradiographed, as shown in Fig. 6-15.

Studies have been undertaken of the exposures of rats living underground on the Morro do Ferro. Of particular interest is the dose to these rodents due to inhalation of ^{220}Rn, which was found by Drew and Eisenbud

FIGURE 6-14   The Morro do Ferro (Mountain of Iron) in Minas Gerais, Brazil.

(1966) to be present in the rat burrows in concentrations up to 100 pCi cm^{-3} (3.7 MBq m^{-3}). The dose to basal cells of the rat bronchial epithelium was estimated to be in the range of 3000 to 30,000 rem y^{-1} (30 to 300 Sv y^{-1}). Of 14 rats trapped and sacrificed for pathological study, none was observed to show any radiation effects. This is of little significance, since the Morro do Ferro is a relatively small area and if rats were affected by this exposure, they could be replenished rapidly from the surrounding normal areas. Using thermoluminescent dosimeters implanted into trapped rats that were released and later recaptured, the external radiation dose to the rats was estimated to be between 1.3 and 6.7 rad y^{-1} (13 and 67 mGy y^{-1}).

Other studies being undertaken at the Morro do Ferro will be discussed in Chapter 11.

THE CHINESE MONAZITE AREA

About 73,000 persons live in an area of monazitic soils in Guangdong Provice, China, where the external radiation dose rate is about 330 mrad y^{-1} (3.3 mGy y^{-1}). The radiation levels are between three and four times normal, and Wei and associates have undertaken a comprehensive long-term study of the area, beginning in 1972 (Wei, 1980). The ^{226}Ra body burdens in the

FIGURE 6-15 Autoradiograph of species of *Adiantum* from the Morro do Ferro in the state of Minas Gerais, Brazil. (Courtesy of Dr. Eduardo Penna Franca.)

high-background area were reported to be 281 pCi (10.4 Bq), about three times higher than in the control area. As noted in Chapter 2, the Chinese investigators could not find any effect of the increased radiation on the health of the inhabitants (Wei *et al.*, 1990).

THE NATURAL REACTOR AT OKLO

Much to the surprise of the scientific world, it was discovered in 1972 that the site of an open-pit uranium mine called Oklo in the West African republic of Gabon was the fossil remains of a 2-billion-year-old natural reactor. Although contemporary uranium contains 0.72% of the fissionable isotope ^{235}U, this nuclide has a half-life of only 700 million years compared to the

TABLE 6-22

Estimated Annual Effective Dose Equivalent Rates from Natural Sources in Normal Regions[a]

Source	Effective dose equivalent rate (mrem y^{-1})			Effective dose equivalent rate (μSv y^{-1})		
	External	Internal	Total	External	Internal	Total
Cosmic including neutrons	36		36	360		360
Cosmogenic nuclides		1.5	1.5		15	15
Primordial nuclides						
^{40}K	15	18	33	150	180	330
^{87}Rb		0.6	0.6		6	6
^{238}U series						
^{238}U → ^{234}U		0.5	0.5		5	5
^{230}Th		0.7	0.7		7	7
^{226}Ra	10	0.7	10.7	100	7	107
^{222}Rn → ^{214}Pb		110	110		1100	1100
^{210}Pb → ^{210}Po		12	12		120	120
^{232}Th series						
^{232}Th		0.3	0.3		3	3
^{228}Ra → ^{224}Ra	16	1.3	16.3	160	13	173
^{220}Rn → ^{208}Pb		16	16		160	160
Total (rounded)	80	160	240	800	1600	2400

[a]Adapted from UNSCEAR (1988).

TABLE 6-23

Annual Estimated Average Effective Dose Equivalent Received by a Member of the Population of the United States from Natural Radiation[a]

Source	Average annual effective dose equivalent	
	μSv	mrem
Inhaled (radon and decay products)	2000	200
Other internally deposited radionuclides (^{40}K, ^{210}Po)	390	39
Terrestrial radiation	280	28
Cosmic radiation	270	27
Cosmogenic radioactivity (^{14}C)	10	1
Total (rounded)	3000	300

[a]Adapted from NCRP (1987a).

TABLE 6-24

Annual Estimated Average Dose Equivalent Received by Selected Tissues of a Member of the Population of the United States from Natural Radiation[a]

Source	Annual dose equivalent (mSv)				Annual dose equivalent (mrem)			
	Bronchial epithelium	Other soft tissues	Bone surfaces	Bone marrow	Bronchial epithelium	Other soft tissues	Bone surfaces	Bone marrow
Cosmic	0.27	0.27	0.27	0.27	27	27	27	27
Cosmogenic	0.01	0.01	0.01	0.03	1	1	1	3
Terrestrial	0.28	0.28	0.28	0.28	28	28	28	28
Inhaled	24	—	—	—	2400	—	—	—
In the body	0.35	0.35	1.1	0.5	35	35	110	50
Total (rounded)	25	0.9	1.7	1.1	2500	90	170	110

[a]Adapted from NCRP (1987a).

4.5-billion-year half-life of ^{238}U. The percentage of ^{235}U present in the ore was thus very much greater 2 billion years ago, and the conditions for initiating fission reactions apparently existed. Criticality was sustained for on the order of 10,000 years, during which time an estimated 15,000 MW-years of energy was released by the consumption of 6000 kg of ^{235}U. Studies of the migration of the fission products and transuranic elements produced during this period have shown that, with few exceptions, migration was minimal. The implications of this finding for radioactive waste management are discussed in Chapter 11 (Cowan, 1976; IAEA, 1975).

SUMMARY OF HUMAN EXPOSURES TO NATURAL IONIZING RADIATION

The annual effective dose equivalent received by persons living in most parts of the world, where the natural radioactivity is within normal limits, is estimated in Table 6-22. In this compilation, ^{222}Rn and its short-lived decay products contribute about 40% of the total effective dose equivalent, a term that, as explained in Chapter 2, takes into consideration the dose equivalent to the tissue, the volume of tissue irradiated by the nuclides, and its relative radiosensitivity of the tissue. Radon-222 and its short-lived decay products irradiate the basal cells of the bronchial epithelium, which has a mass of only a few grams.

Somewhat different values of the annual effective dose equivalent have been estimated for residents of the United States (Table 6-23) (NCRP, 1987a). The major difference between the two estimates is the average effective dose equivalent due to ^{222}Rn. This is a difficult quantity to estimate because world average ^{222}Rn levels are not well known and there are several models used to convert ^{222}Rn exposure to lung dose. Estimates of dose equivalent (as distinguished from *effective* dose equivalent) for several tissues are given in Table 6-24. In Europe, the estimated average annual effective dose equivalent varies from a low of 2 mSv y^{-1} for the United Kingdom to a high of over 7 mSv y^{-1} for Finland (Green *et al.*, 1992). Most of the country-by-country variation is attributed to differences in indoor radon levels.

CHAPTER 7

Production and Reprocessing of Nuclear Fuels

The production of nuclear energy is based mainly on the fission of ^{235}U, which is present in natural uranium to the extent of 0.7%. Uranium-238, which is the most abundant isotope of uranium (99.28%), is not readily fissionable but does transmute by neutron capture to ^{239}Pu, which is a fissile isotope. Thus, the source materials of atomic energy may be either fissile, like ^{235}U, or fertile, like ^{238}U. The production and reprocessing of nuclear fuels and their relationships to the other components of nuclear energy are illustrated as a diagram in Fig. 1-2 of the Introduction.

Thorium is a possible source of nuclear energy for the future. It occurs in nature almost entirely as the isotope ^{232}Th, which is a fertile nuclide that can be transmuted to the fissile ^{233}U. An important disadvantage of ^{233}U is that it is a gamma emitter, which makes it more difficult to handle.

URANIUM

When interest in uranium first developed in World War II, commercially exploitable deposits were thought to be comparatively rare and to occur only in a few districts rich in the mineral pitchblende. The sources then known to exist were in rich deposits located in the former Belgian Congo, the Great Bear Lake region of Canada, and the former Czechoslovakia. It was also known that much lower grade deposits existed on the Colorado plateau, and these deposits had for some years been used as a source of radium. However, the full extent of the uranium resources of the Southwest United States was not then appreciated. Following an intensive exploratory program conducted by the U.S. government, the domestic uranium industry grew rapidly during the post-World War II period. The United States also encouraged exploration and mining in other Western countries. However, the large civilian market for uranium then forecast did not materialize because of the slackening of requirements for nuclear power that began in the United States in the mid-1970s.

The present-day U.S. reserves of uranium are in the form of sandstone ores that have a U_3O_8 content of only 0.1%, compared to 20 to 60% in the ores imported during and immediately following World War II. However, the amount of uranium available from the much larger quantities of low-grade ores is far greater than what was obtained from the limited supplies of high-grade material.[1]

In 1992 the U.S. uranium reserve was estimated to be 114,000 metric tons (MT), most of which is located in the Wyoming basin and the Colorado plateau (Underhill and Muller-Kahle, 1993). As a practical matter, the availability of mineral reserves depends on price. DOE (1991a) estimated the 1990 U.S. reserves at 118,000 MT at $67 per kilogram and 420,000 MT at $111 per kilogram. The "reasonably assured" world supply of uranium at $80 per kilogram was estimated to be 1.6×10^6 MT in 1987 (Uranium Institute, 1989).

In 1992 the world requirement for reactor-grade uranium was 56,800 MT, somewhat greater than the worldwide production of 35,525 MT. This was a major change from 1988, when demand was for 51,000 MT and production was 60,800 MT (Underhill and Muller-Kahle, 1993). The 1992 demand is expected to continue through 2010. The supply is being affected by the sudden availability of uranium from the former Soviet Union, and demand is being influenced by the cessation of nuclear weapons production in the United States.

The U.S. uranium industry has indeed been passing through a declining period. From peak employment of 21,000 person-years in 1979, there has been a steady decrease to 682 person-years in 1992 (Nuclear Energy Institute, 1995). By 1993, all U.S. mills were decommissioned or in standby status (Nuclear Energy Institute, 1995). The output of these mills in the past was transported to uranium refineries where the concentrates were

---

[1]Potentially significant reserves are also present in phosphate deposits. The radiological health implications of phosphate rock mining are important whether or not uranium is recovered and are discussed in Chapter 6.

converted into uranium compounds having the required high degree of chemical purity. From there the uranium went to isotopic enrichment plants where the ^{235}U content was increased. Currently enrichment is taking place in huge gaseous-diffusion plants located in Ohio and the Tennessee Valley, but newly developed laser separation processes may eventually replace gaseous diffusion.

If isotopic enrichment is not required, the uranium may be shipped as the oxide or metal to a number of privately or government-owned installations where natural uranium reactor fuel elements are fabricated. The enriched uranium from the gaseous-diffusion plants may also be shipped to fuel-fabrication facilities. In the past, a substantial fraction of the metallic natural uranium was shipped to plutonium-production facilities located at Hanford and Savannah River.

URANIUM MINING

Uranium mining has taken place in both underground workings and open pits. The size of the mines varies from relatively small workings that employ one or two men who use hand tools to thoroughly mechanized enterprises that employ more than 100 men. Uranium mining by *in situ* underground leaching (solution mining) has also been demonstrated to be feasible and may become more important when uranium production is resumed (Hunkin, 1980). Solution mining involves solubilizing uranium by pumping water, to which oxidizing and complexing agents have been added, into an ore body and then pumping the solution to the surface for processing (Brown and Smith, 1980). The method reduces the risks to the miners and has minimal visible environmental impact. However, there is concern that groundwater contamination may occur.

The concentration of radon and its decay products in the air of underground uranium mines can be hazardous to the workers unless proper ventilation is employed. As was discussed in Chapter 2, it has been known since early in this century that the high incidence of lung cancer among miners in Bohemia and Saxony was due to radon exposure. Despite that experience, adequate protective measures against radon were not adopted when uranium mining began in the United States. Because of a technicality in the Atomic Energy Act, the AEC did not assume responsibility for the health of the uranium miners when the agency was formed in 1947. The Act specified that the AEC preempted responsbility for occupational health and safety from the states, but only after the ore is removed from its place in nature. AEC unwisely interpreted the law literally, which left control to the mining states, where enforcement of the rules of safe practice was inadequate (Eisenbud, 1990). Because of this decision, little was done to evaluate or improve conditions, and many preventable radon-induced cases of lung cancer have since developed among the miners. The federal government did not begin enforcement until the late 1960s (Holaday, 1969), when the first radon standards specifically for miners were promulgated.

Radon and its decay products are the only significant gaseous radioactive contaminants discharged from the mines to the general environment. The required ventilation rates for the mines vary from 1000 cfm (cubic feet per minute) to over 200,000 cfm, and the discharged air contains radon in concentrations that range from 0.5 to 20 μCi per 1000 ft^3 (0.65–26 kBq m^{-3}), which is no greater than the normal radon flux from about 1 km^2 of earth's surface (Holaday, 1959). The risks to the general public from the radioactive emissions from uranium mines are insignificant (Blanchard *et al.*, 1982).

URANIUM MILLS

The uranium ore is shipped from the mines to mills, where a uranium concentrate, U$_3$O$_8$ (yellowcake), is produced. Milling begins with a grinding process that reduces the ore to the consistency of fine sand, following which the uranium is separated by acid or alkaline leaching and concentrated by either ion exchange or solvent extraction. Although more than 95% of the uranium is removed from the ore by this process, almost all of the radioactive decay products in the uranium series remain with the tailings contained in slurries that are discharged into holding areas. Most of the U.S. mills are located in the arid regions of the southwest, where the impounded slurries dry rapidly and over time become sizable mounds of residue (Clegg and Foley, 1958; Goldsmith, 1976; NRC, 1980a,b). It has been estimated that by 1983, approximately 1.75×10^8 MT of uranium mill tailings had accumulated at 52 active and inactive sites, all but one of which (in Canonsburg, Pennsylvania) were located in the western states (DOE, 1985a).

Uranium milling in the United States reached a peak between 1960 and 1962, when there were 25 mills in operation, but this number has been greatly reduced by lessened demand for uranium. By the end of 1990, there were only 2 operating mills in the United States, although 14 mills were said to be operable (DOE, 1991b). All of the U.S. mills are now either decommissioned or in standby status (Nuclear Energy Institute, 1995).

The piles of mill tailings are potential sources of environmental problems because of (1) emanation of radon, (2) dispersion by wind and water, and (3) the use of the tailings in building construction. Public Law 92-314 gave the DOE the responsibility for stabilization of the tailings at 24 existing inactive uranium processing sites, according to the standards established by the EPA (1983). The tailings piles and other contaminated lands associated with these 24 sites range in size from 21 to 612 acres, for a total of 3894 acres (DOE, 1995b). The actions taken under this law are discussed in Chapter 15.

Radon Emanation

The ^{226}Ra concentrations of mill tailings piles have been found to vary from 50 to 1000 pCi g^{-1} (1.85–37 kBq kg^{-1}) of dry tailings (EPA, 1982). Radon production within the tailings piles will continue for hundreds of thousands of years because they contain almost all of the ^{230}Th originally present in the ore, and this long-lived nuclide ($T_{1/2} \approx 80,000$ y) is the parent of ^{226}Ra, which in turn decays to ^{222}Rn. Most models for the production of radon in tailings piles assume that radon enters the interstitial gases within a dry tailings pile by alpha recoil. The radon then diffuses to the surface of the pile, where it enters the atmosphere at highly variable emission rates (Tanner, 1980; NAS/NRC, 1986). The DOE (1981) gathered data for 24 inactive mill tailings sites, and the measured emission rates showed great variability not only between sites but also in serial measurements at individual sites. There was poor correlation between the predicted and measured emission rates, and no correlation was found between the radon flux and the radium content of the tailings or cover. Such correlations must surely exist and will no doubt be found when account is taken of parameters such as soil moisture, wind velocity, barometric pressure, and other factors that would be expected to influence the emission rate and radon concentration.

It has been reported that the emanation rate is sensitive to variations in atmospheric pressure and that a change of only 1% in barometric pressure can result in emission rate variations of 50 to 100% (Clements and Wilkening, 1974). The relationship between atmospheric pressure and diffusion rate has been explained by some investigators as being due to the more rapid release of radon during periods of low atmospheric pressure and dilution of interstitial soil radon by air pumped into the ground during periods of high pressure (Kraner et al., 1964; Clements and Wilkening, 1974).

A generic environmental impact statement on uranium milling prepared by the NRC assumed the flux from dry tailings (NRC, 1980a) to be 1.0 pCi m^{-2} s^{-1} of ^{222}Rn for dry tailings containing 1 pCi g^{-1} of ^{226}Ra (1.0 Bq m^{-2} s^{-1} of ^{222}Rn for dry tailings containing 1000 Bq kg^{-1} of ^{226}Ra). Schiager (1986) noted that this emission rate per unit activity in the tailings is comparable with the rates of radon emission from normal soils. The increased radon flux from the tailings piles is due to the higher ^{226}Ra content of the tailings, which may contain several hundred pCi g^{-1} ($\sim$10 kBq kg^{-1}) compared to 1–2 pCi g^{-1} (0.04–0.08 kBq kg^{-1}) for normal soils. Since the area covered by the tailings piles is small, usually on the order of 10 to 100 acres, the piles themselves do not make a significant contribution to atmospheric radon in the general environment. The influence is localized and can be detected only within about 1 mile (Healy, 1981; Shearer and Sill, 1969; Schiager, 1974).

It is important to note that the emission rate applies only to radon, the parent of the radon series, since the decay products are not gaseous and do not diffuse from the ground. Decay product ingrowth occurs after release to the atmosphere, when the radon is being diluted by turbulence. Thus, for some distance downwind, the fraction of the decay products associated with the radon will be much lower than at equilibrium.

Based on the various studies performed, the radon emissions themselves are a minor source of risk to individual persons. However, the small risk applies to a large number of persons and persists for a long time. The NRC (1980a) estimated that the radon generated by the tailings piles, if all mills were to be in full operation until the year 2000, would result in about six premature deaths (from cancer) per year during the period of 1979–2000 in the United States, Mexico, and Canada. If we assume that the populations of these countries will total 400 million persons during this period, the calculated risk that any one individual will die prematurely because of cancer produced by this source of radon exposure in a 70-year lifetime would be about 1.5×10^{-8}. The calculated risk to individuals located within about 3 km of the piles would be much higher, about 4×10^{-4} per lifetime, but only a few such individuals would be at risk. The hypothetical risks from the emissions of radon are thus minimal under even the worst conditions.

Dispersion by Wind and Water

Materials from the tailings piles can be dispersed into the environment by either impoundment failure or erosion by flowing water or wind action. Impoundment failure can be prevented during the foreseeable future by applications of available engineering methods, and if cover is provided for the piles, erosion by wind or water can be avoided in the semiarid regions in which the tailings piles are located (Webb and Voorhees, 1984). However, considering the long half-life of ^{230}Th, the parent of ^{226}Ra, there can be no assurance that these protective features will be effective for the hundreds of thousands of years during which radium production will be sustained. One cannot rule out the possibility that future changes in climate or population density may result in increased risks.

Erosion by wind and water can be reduced by contouring the pile, providing cover, stabilizing the surface with rock or other material, and constructing dikes to divert flood waters. Erosion of new tailings sites can be minimized by burying the tailings in shallow pits or by locating them away from sites subject to flooding. Radon inhibition can be accomplished by covering the pile with compacted earth or less permeable materials such as asphalt, clay, or cement. The reduction in radon emission that results from a given thickness of soil cover will depend on the moisture content. The thickness of the required soil cover is reduced for soils having higher moisture contents.

Groundwater contamination can be reduced by use of underlying plastic or clay barriers. These are of course feasible only for new tailings sites. Should this be done in areas of high precipitation and low evaporation, it would be necessary to seal the surface of the pile. Otherwise, the underlying liner may act as a bowl that can cause contaminants to accumulate and move toward the surface.

EPA has established an emission limit for ^{222}Rn of 20 pCi m^{-2} s^{-1} (0.74 Bq m^{-2} s^{-1}) and requires that the disposal method be designed to provide "reasonable assurance" that radon emissions will not exceed that limit, averaged over the disposal area for 1000 years. This leaves open the possibility that the standards would be exceeded beyond 1000 years. The EPA estimated that the cost of complying with its standards will be \$310–\$540 million by the year 2000 (EPA, 1983). This seems to be a large expenditure, considering the minimal risks inovlved.

In past years, liquid tailings were allowed to seep into nearby streams. It had been shown (Gahr, 1959) that the ^{226}Ra content of water in the Colorado River below Grand Junction was 30 pCi L^{-1} (1100 Bq m^{-3}) compared to 0.3 pCi L^{-1} (11 Bq m^{-3}) upstream. The San Miguel River below Uravan, Colorado, where a mill was located, contained 86 pCi L^{-1} (3200 Bq m^{-3}) compared to 4.9 pCi L^{-1} (180 Bq m^{-3}) in water immediately upstream of the mill.

The Animas River in southwestern Colorado serves as a public water supply for the cities of Aztec and Farmington, and the water is also used for irrigation. In 1955, the radium concentration below Durango, where a mill was located, was found to be 3.3 pCi L^{-1} (120 Bq m^{-3}) compared to 0.2 pCi L^{-1} (7 Bq m^{-3}) upstream. That considerable concentration was taking place in the stream biota was shown by the fact that vegetation below Durango contained 660 pCi g^{-1} (24,000 Bq kg^{-1}) compared to 6 pCi g^{-1} (220 Bq kg^{-1}) above Durango. Stream fauna below Durango contained 360 pCi g^{-1} (1300 Bq kg^{-1}) compared to 6 pCi g^{-1} (220 Bq kg^{-1}) above the mill (Tsivoglou et al., 1960a,b).

It was found that using the then-existing ICRP recommendations as a guide, consumers of untreated river water received about three times the maximum permissible daily intake of radium (Tsivoglou et al., 1960a,b). This included food grown on land irrigated with river water, as well as the water consumed. About 61% was due to the radium content of food resulting from the contaminated irrigation water.

Studies conducted by Shearer and Lee (1964) concluded that the radium was leachable from both the tailings piles and the river sediments. Thus, radium was entering the rivers not only in untreated liquid wastes but in surface runoff during rainfall as well. Steps were taken by the mill operators to correct the problem, and by 1963 the radium content of the Animas River sediments had been reduced to three times background, compared to several hundred times background several years earlier. The Animas River experience illustrates the potential danger, if precautions are not taken, of contamination of potable water by radium and other radionuclides and toxic chemicals present in tailings piles.

It has been estimated that there are approximately 1.75×10^8 MT of tailings located at sites in the United States. Assuming an average of 280 pCi g^{-1} of ^{225}Ra in dry tailings (NRC, 1980a), the ^{226}Ra content of the tailings piles is about 50,000 Ci (about 2 PBq; P, peta, a multiple of 10^{15}) being produced by ^{230}Th decay at a rate of about 20 Ci y^{-1} (0.74 TBq y^{-1}). The tailings also include substantial quantities of the nuclides of the ^{226}Ra decay series, which are in secular equilibrium except to the extent that some radon is lost by diffusion from the piles. Lead-210, ^{210}Po, and ^{210}Bi are assumed to be present in a concentration of 250 pCi g^{-1} (9.3 kBq kg^{-1}) compared to 280 pCi g^{-1} (10.4 kBq kg^{-1}) for ^{230}Th and ^{226}Ra.

Breslin and Glauberman (1970) measured the airborne dust downwind from unstablized tailings piles and demonstrated clear relationships between the distance from the tailings piles and the concentrations of uranium and ^{210}Pb. For the three tailings piles sampled, the air concentrations were well below permissible levels in two cases, but approached the upper limits recommended in 10 CFR 20 at a distance of about 1000 ft in the third case.

Uranium mill tailings piles are being stabilized and otherwise remediated in several ways, depending on site-specific conditions. These remedial actions are discussed in Chapter 15.

Use of Mill Tailings for Construction

Many residents of Grand Junction, Colorado, were exposed to both gamma radiation and radon because mill tailings were used to construct homes and public buildings (DOE, 1980; Committee on Armed Services, 1981). That practice was not discontinued until 1966. Under Public Law 92-314, enacted in 1978, Congress provided financial assistance to the state of Colorado to limit the radiation exposures that existed because of the use of tailings for construction purposes. Construction of buildings in many other western towns near uranium mills also included tailings, although the numbers were much smaller than in Grand Junction. The remediation of these buildings and associated properties is discussed in Chapter 15.

Refining

The mill concentrates in this country are sent to any of several locations in which the uranium is converted to either the metal or some intermediate uranium compounds, such as orange oxide (U_3O_8) or green salt (UF_4). The principal steps in converting the concentrates to a form that is of acceptable chemical purity are shown in Table 7-1. These processes involve potential exposure of the employees to alpha-emitting dusts. In past years, when high-grade ores were processed, significant levels of radon and gamma radiation were reported (Eisenbud and Quigley, 1956).

The refining operations involve the mechanical processing of dry powders of uranium compounds, which can result in the discharge of uranium dust to the environment. Present-day plants are equipped with filtration equipment that effectively removes the uranium dust, and the monetary

TABLE 7-1

Principal Steps in the Refining and Conversion of Uranium[a]

Feed
Miscellaneous uranium concentrates
(approximately 75% by weight U_3O_8 or equivalent)
↓
Production of orange oxide (UO_3)
1. Digestion of the uranium concentrates in nitric acid
2. Solvent extraction to remove impurities and reextraction into water
3. Boil-down of the uranyl nitrate solution from (2) to a molten uranyl nitrate hexahydrate
4. Denitration of the molten salt by calcination to produce orange oxide powder
↓
Conversion to green salt (UF_4)
1. Reduction of the orange oxide to brown oxide (UO_2) by contacting with hydrogen
2. Conversion of the brown oxide to green salt by contacting with anhydrous hydrofluoric acid
↓
Reduction to metal
1. Reduction of the green salt to massive uranium metal (derbies) by a thermite-type reaction using magnesium as the reducing agent
2. Vacuum casting of several uranium "derbies" from (1) to produce a uranium ingot
↓
Product
High-purity uranium metal in ingot form

[a]Adapted from U.S. Atomic Energy Commission (AEC, 1957b).

value of uranium is such as to preclude the possibility of its being discharged to the atmosphere in large quantities for sustained periods of time. However, the hastily constructed plants during World War II had insufficient control over dusts contained in exhaust air, and relatively large amounts of uranium were discharged to the outside atmosphere. Nevertheless, uranium is so abundant in the environment that the element was undetectable above the natural background at moderate distances from the plants (Klevin *et al.*, 1956).

The kinds of wastes produced by the refineries depend on the type of feed material that is processed. During World War II, and for a few years thereafter, when the high-grade ores contained as much as 100 mCi ^{226}Ra per ton of ore (4 MBq kg^{-1}), some of the sludges contained as much as 1 Ci ^{226}Ra per ton (40 MBq kg^{-1}). However, the uranium industry has been operating with ores of much lower grade for several decades, and the uranium is separated at the mills, thus sparing the refineries the problem of disposing of waste products that contain large amounts of radium. The limiting factor in the discharge of wastes from uranium refineries is apt to be the chemical wastes rather than their radioactive constituents.

During the World War II era, there were 31 plants, laboratories, and storage sites involved with production of uranium from ore. Operation of those plants was discontinued by the mid-1950s, and they were cleaned up according to the then-existing understanding of what constituted adequate decontamination. As time went on, more conservative criteria were developed and many of the sites were found to be above the limits considered suitable for unrestricted access. In 1974, the AEC, predecessor of the DOE, initiated a program of decontamination known as the Formerly Utilized Sites Remedial Action Program (FUSRAP). Currently 46 sites are estimated to have 2.3 million cubic yards (1.76 m^{-3}) of soil and building rubble requiring removal or other management. In 1995 it was estimated that the cost of the cleanup program would be $2.5 billion and that the project would be completed in the year 2016.(DOE, 1995f). Remedial actions under the FUSRAP program are discussed in Chapter 15.

The story of the waste sludges from processing the rich ores during and shortly after World War II provides insight into the difficulties of anticipating problems before they arise. The highest-quality ore came from the Belgian Congo (now Zaire). Under the procurement contract between the U.S. government and the Belgian mine operator, the latter retained title to the waste sludges, which at the time were intended to be shipped back to Europe for further processing. There was reason to believe the sludges would serve as a valuable source of radium, which was then selling for $20,000 per gram. What was not foreseen was that by-product radionuclides from the atomic energy program would provide inexpensive substitutes for the radium used in medicine, research, and industry. The Belgian owner had the option of abandoning the stored sludges, by the exercise of which thousands of curies (1000 Ci = 37 TBq) of radium in sludges were stored in silos and steel drums. A half-century later this became part of the legacy of the past inherited by the DOE from its predecessor agencies, the Manhattan District and AEC.

From 1952 until 1989, the U.S. government operated a centralized uranium materials production plant in Fernald, Ohio. This plant, known as the Feed Materials Production Center (FMPC), received concentrates from the mills and produced the uranium compounds required for the nuclear weapons program. At various times during the past decade, six additional plants have been privately operated to supply uranium for civilian nuclear power plants. The liquid and airborne releases from the private facilities have been under regulation, and doses to members of the public have been minimal. The NCRP (1987b) estimated that the doses received from these operations by the maximally exposed individuals were less than about 0.03 mSv y^{-1} (3.0 mrem y^{-1}). Residents in the vicinity of Fernald brought a class action suit against the contractor based on concerns over health risks. The claim was denied in court, but the DOE awarded $78 million as compensation for the anxiety caused by the existence of the plant. Results from a dose reconstruction at Fernald indicate higher doses than those estimated for the civilian facilities (see Chapter 14).

ISOTOPIC ENRICHMENT

The green salt (UF$_4$) produced by the refineries is converted to uranium hexafluoride (UF$_6$), which in the past has been shipped for isotopic enrichment to the large gaseous-diffusion plants located in Portsmouth, Ohio; Paducah, Kentucky; and Oak Ridge, Tennessee. The UF$_6$ is pumped through cascades of porous barriers. Each stage of diffusion results in a slight enrichment in ^{235}U, the number of stages being determined by the degree of enrichment required.

The gaseous-diffusion plants are now considered to be obsolete, and centrifugal systems are being used in several countries. Electromagnetic separation from plasmas and chemical separation of isotopes have also been proposed, as well as the use of lasers (Tait, 1983). Some specialists believe that enrichment by lasers will be the future method of choice in the United States. Bombardment of UF$_6$ or U vapor with multiple laser photons can result in up to 50% enrichment in a single pass (Knief, 1981). Laser systems are being used in some countries (Ansoborlo et al., 1994).

Even more so than in the refineries, the economic value of the enriched uranium precludes the likelihood of widespread environmental contamination from these plants. In addition, as the uranium progresses through the diffusion plant, it becomes of increasing importance as a source of material for weapons and must be subject to strict accountability.

The enriched uranium from the diffusion plants is destined either for assembly into weapons or into fuel elements for reactors. Depleted uranium is used for shielding, armor, projectiles, and counterweights.

Until 1992, uranium enrichment plants in the United States were owned by the federal government. At that time, a quasi-governmental corporation was formed, the U.S. Enrichment Corporation (USEC), with a mission to improve the enrichment enterprise and to eventually sell it to the private sector. The company manages the gaseous-diffusion plants in Paducah, Kentucky, and Portsmouth, Ohio. USEC has reportedly cut costs and improved capacity and eventually is expected to be on the market for total privatization (Nuclear Energy Insight, 1995b).

FUEL-ELEMENT MANUFACTURE

Fuel-element manufacture is currently performed at a number of government and private facilities. Again, the relatively high cost of the uranium and the requirements for strict accountability make it unlikely that significant environmental contamination can occur from these plants, but the possibility of accidents cannot be discounted entirely. Uranium chips are pyrophoric, and if fires or explosions occur, more than normal amounts of activity may be released.

REPROCESSING SPENT REACTOR FUEL

When a reactor core has reached the end of its useful life, only a small percentage of the ^{235}U will have been consumed in fission and an additional small fraction of the ^{238}U will have been transmuted to ^{239}Pu and other transuranic elements. If reprocessing of commercial reactor fuel was viable in the United States, spent fuel would first be removed from the reactor and stored in underwater pools to allow the intense, short-lived radioactivity to decay. Then it would be transported (Fig. 7-1) to a fuel-reprocessing plant in which the spent fuel would be chemically treated to (1) recover the remaining ^{235}U and the transuranic elements and (2) convert the fission products into a form suitable for long-term storage (see Chapter 11). The presence of large amounts of fission products and transuranic actinide elements in the irradiated fuel greatly complicates processing procedures and makes it necessary to adopt elaborate measures both to protect the operating personnel and to avoid environmental contamination.

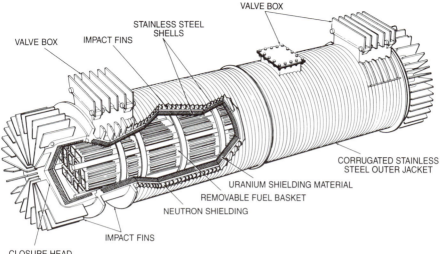

VALVE BOX

STAINLESS STEEL
SHELLS

VALVE BOX   IMPACT FINS

CORRUGATED STAINLESS
STEEL OUTER JACKET

URANIUM SHIELDING MATERIAL

REMOVABLE FUEL BASKET

NEUTRON SHIELDING

IMPACT FINS

CLOSURE HEAD

FIGURE 7-1   General Electric IF 300 spent fuel cask, designed for transportation of all types of light-water-moderated fuels. During normal operation the cask is filled with water, which provides heat transmission to the walls by natural circulation. The fins are intended for impact protection. The outer surface of the cask is cooled by air blowers (not shown). (Courtesy of General Electric Company.)

History of Commercial Fuel Reprocessing in the United States

From World War II until 1966, all fuel reprocessing in the United States was performed at four government-owned centers of atomic energy development and production. Two of the plants, at Hanford and Oak Ridge, were built during World War II; the other two were built in the 1950s at Savannah River, South Carolina, and at the Idaho National Engineering Laboratory (formerly the National Reactor Testing Station) at Idaho Falls, Idaho. These plants were constructed to meet military needs and were not intended for the processing of civilian fuel. In recent decades the principal fuel-reprocessing centers for the nuclear weapons program have been at Savannah River and Hanford, but these facilities ceased operation with the end of the cold war. The plant in Idaho reprocessed fuel from the U.S. Navy's nuclear propulsion program and from some research and demonstration reactors.

There was an apparent need for commercial fuel reprocessing, but the initial attempts by private industry to meet the needs of the projected nuclear power industry were failures for a combination of technical, economic, and political reasons (Colby, 1976). The first and only privately owned nuclear fuel-reprocessing plant in the United States, Nuclear Fuel Services Inc. (NFS), went into operation in 1966 in West Valley, New York, with a daily processing capacity of 1 MT of low-enriched uranium oxide fuel. Other privately owned plants were constructed by General Electric Company in Morris County, Illinois, and the Allied Chemical Corporation at Barnwell, South Carolina (Unger *et al.*, 1971; AEC, 1970a), but these plants never operated. The General Electric plant was not placed in operation because it proved to be technically and economically deficient, and plans for the Allied Chemical plant in South Carolina were abandoned because of serious licensing difficulties. The plant in West Valley developed licensing and economic problems that resulted in the 1972 shutdown of the plant and the subsequent abandonment of the venture by its owners.

Any incentive to correct the deficiencies of the three plants for the purpose of starting or restarting fuel reprocessing was eliminated in 1977 when President Carter announced a U.S. policy of prohibiting fuel reprocessing as a means of preventing the diversion of plutonium and the possible uncontrolled proliferation of nuclear weapons (Gilinsky, 1978). The question of whether this was a justifiable policy subsequently became moot because lessened requirements for nuclear power eliminated the economic incentive to recycle the fuel. In 1970, the estimated need for nuclear power by countries with market economies in the year 2000 was 2250 gigawatts electric (GWe). By 1990, the estimate had been reduced to about 400 GWe. Instead of fuel reprocessing, the spent fuel in some countries is being stored in geological repositories. In the United States, a repository site is under study, but not yet selected, and the spent fuel is being stored in pools (Semenov and Oi, 1993). Despite the fact that federal law required that a central repository be established by 1998, there is no prospect that this can be done by the end of the century. Thus, lack of storage space for spent fuel may necessitate building a reprocessing plant in the United States, and in 1994 a consortium of several electric utilities and engineering companies received preliminary permission to construct a reprocessing plant in Louisiana that would be capable of supplying 15% of the enriched uranium fuel needed in this country (Nuclear Energy Institute, 1994).

In contrast, the French Atomic Energy Commission included fuel reprocessing as an integral part of its nuclear power program. One reprocessing center at LaHague is dedicated to spent fuel from light-water reactors and has two plants, one built 25 years ago and a new one that completed its first full year of operation in 1991. The LaHague center reprocessed 660 MT of fuel in 1991, and an upgrade of the older plant will bring the capacity of the center to about 1000 MT per year. The other reprocessing center is at Maracoule, which serves French defense programs and which also reprocessed gas-cooled reactor fuel (COGEMA, 1992).

The fuel-reprocessing methods vary, depending on the materials from which the fuel is fabricated, but all fuel-reprocessing plants have used some form of the Purex process, a solvent extraction system using tributyl phosphate (TBP) diluted with kerosene (Etherington, 1958; Stoller and Richards, 1961). The fuel elements are sheared into small lengths, after which the cladding is removed chemically. Zirconium cladding may be removed by dissolution in ammonium fluoride solutions in a process that produces about 4600 gal of liquid waste per ton of fuel (19 m^3 per metric ton). The fuel is then dissolved in nitric acid, following which a series of TBP extraction and stripping steps result in separation of the original solution into the required fractions, transuranic elements, uranium, plutonium, and fission products. The waste products are in the latter fraction and are highly radioactive. After being concentrated by evaporation, the wastes may be stored in underground tanks until the decay heat has subsided, following which the wastes can be prepared for ultimate disposal as will be discussed in Chapter 11.

By far the greatest amount of fuel processing in the United States took place at the major plutonium-production facilities at Hanford and Savannah River. About 10,000 tons of irradiated uranium have been processed at Hanford since 1944 to produce about 120 tons of plutonium (Gephart and Lundgren, 1995; Lentsch *et al.*, 1995). Chapter 11 will discuss progress that has been made toward arranging for permanent disposal of the huge quantities of highly radioactive waste produced.

The large fission-product and transuranic actinide inventories, together with the very nature of the fuel-reprocessing methods, present opportunities for major environmental contamination unless strict procedures are followed to avoid release of radioactive substances to the vicinity of the plant. The potential hazards of these operations were not understood when the first chemical reprocessing plants were built during World War II and it was necessary to rely on the remoteness of the sites to provide protection. Storage tanks were known to have leaked, some wastes were stored in near-surface, unprotected "cribs," and huge quantities of ^{131}I were released directly to the atmosphere. Because of secrecy, and the urgent need to maintain production in the plants of the nuclear weapons complex in the United States, many problems have developed for which remedial measures will be necessary (see Chapter 15). The quantities of ^{131}I were so large that concern developed in the early 1990s that health effects might have developed among residents in Washington State. This led to the Hanford Dose Reconstruction Project described in Chapter 14.

Sources of Radioactive Emissions

As in other components of the nuclear industry, the opportunities for environmental contamination from fuel reprocessing can be divided into those that occur in normal plant operation and those that might result from accidents so severe as to overwhelm the defenses against uncontrolled releases. The nature of the process is such that chemical explosions are conceivable, and since fissionable materials are processed, it is possible for critical masses to be assembled accidentally. However, the potential hazards have been offset by conservative processes designed to minimize the probability of a serious accident. Only a handful of criticality accidents have actually occurred at sites such as Los Alamos, Hanford, Oak Ridge, and Idaho and none has been reported in the United States since 1977 (Lushbaugh, 1990). It is nevertheless assumed that an accident can happen in fuel reprocessing facilities, and compensating safeguards are provided to minimize releases to the environment. Although none are currently operating in the United States, the safety of commercial fuel-reprocessing plants was regulated by the NRC under a system similar to that used to regulate reactor design and operation.

Accidental criticality is prevented by limiting the amount of fissionable material being processed. The geometrical design of process vessels and storage containers is carefully controlled and is a basic method by which the opportunity for assembling a critical mass is prevented. In some cases, neutron-absorbing components have been placed in tanks in the form of rings or parallel plates (Unger *et al.,* 1971).

Dispersal of airborne material in the event of an accident is controlled by constructing the process building as a series of shells having gradations of negative atmospheric pressure. The physical design of the processing cells is such that they can withstand the blast effects from the maximum credible explosion.

The reprocessing process results in enormous quantities of fission products, the management of which will be discussed in Chapter 11. In this chapter, we will limit our discussion to the low-level wastes that may be discharged to the atmosphere or nearby surface waters.

The primary sources of gaseous wastes are the fuel-element chopping and dissolution processes (Logsdon and Chissler, 1970). Because of the relatively long storage time before the fuel is processed, most of the ^{131}I (half-life 8 days) decays. This is determined by the age of the fuel, and for the foreseeable future ^{131}I will not be a problem if fuel processing is resumed. As mentioned earlier in this chapter, the fuel from the Hanford reactors was so quickly processed during wartime that large quantities of ^{131}I remained in the fuel and was discharged in the untreated gaseous wastes.

A potential long-range problem exists because of ^{129}I, a nuclide with a half-life of 1.7×10^7 years. It is produced in fission with a yield of 1% and is thus present in reprocessing wastes in relatively large quantities. Because of its long half-life, it will accumulate in the environment, become part of the iodine pool, and deliver a thyroid dose to the general population that could increase in proportion to the rate of nuclear power production (NCRP, 1983). The radioiodines can be removed chemically with caustic scrubbers or other means, such as by reactions with mercury or silver.

Other radioactive gaseous releases from a nuclear fuel-reprocessing plant include ^{85}Kr and ^{3}H. The gaseous release of tritium has been small compared to the releases in liquid form. The stack releases of ^{85}Kr are substantial but have not been a source of significant exposure in the vicinity of the processing plants. The doses to individuals from ^{85}Kr and ^{3}H emissions from reprocessing plants would be expected to be small, but they could accumulate in the general environment in a manner that would expose very large numbers of people. The collective dose to be expected is of course sensitive to the assumptions made regarding the amount of nuclear power generation. If spent-fuel reprocessing is resumed in the future, it will be necessary to develop policies concerning the need to separate these nuclides from waste streams. The various available options are discussed by NCRP (1975b, 1979, 1983) and OECD (1980). The long-lived gaseous wastes are discussed further in Chapter 11.

THORIUM

Thorium (Cuthbert, 1958; Albert, 1966) is estimated to be three times more abundant in the earth's crust than uranium and may ultimately become an important source of nuclear energy as techniques are developed for converting the ^{232}Th to ^{233}U in breeder reactors. However, this seems unlikely for the foreseeable future. The most important known occurrences of thorium minerals are in the monazite sands of Brazil and India. Although thorium has been used for many years in the manufacture of gas mantles and welding electrodes and at the present time has a limited application in the atomic energy industry, the production capacity for thorium is so small that there is little potential for general environmental contamination with this material or its decay products (Meyer *et al.,* 1979). There is, however, some thorium contamination resulting from activities in the 1930s and 1940s that is being managed under DOE's Formerly Utilized Sites Remedial Action Program (DOE, 1995f).

AVERAGE DOSES FROM THE PRODUCTION AND REPROCESSING OF NUCLEAR FUELS

Estimates have been made of the doses received by the U.S. public from the nuclear fuel industry (NCRP, 1987b). The average annual dose received by the general public is on average only a fraction of a microsievert (μSv), in part because of the relatively few facilities involved. The maximum dose is received by residents in the vicinity of mines and mills, notably from the emission of radon from tailings piles. Given the drastic reduction of mining and milling activities and the ongoing stabilization of tailings piles in the United States, these doses are likely to be even smaller today.

CHAPTER 8

Power Reactors

The first nuclear reactor was operated briefly by Enrico Fermi and his associates at the University of Chicago on December 2, 1942, less than 4 years after the discovery of nuclear fission. Under wartime pressure, reactor technology continued to develop rapidly, and only 1 year later a 3.8 thermal megawatt (MWt) research reactor began operation at Oak Ridge. It remained in service for more than 20 years and served as the chief source of radioisotopes for much of the Western world (Hewlett and Anderson, 1962; Tabor, 1963). Even more remarkable, the first of several reactors designed for plutonium production began operation in the state of Washington in 1944 at an initial power level of 250 MWt. These and additional units, at considerably higher power levels, also remained in service for more than 20 years. At present, about 1000 land-based reactors have been built and operated in various parts of the world. These include reactors built for research as well as production of power, radioisotopes, and plutonium. More than 200 vessels of the American and Russian navies have been powered by nuclear reactors.

From 1945 until 1954 the reactor program in the United States was dominated entirely by the government and was closely identified with the military applications of nuclear energy. All reactors constructed in the United States prior to 1953 were located on government sites, but in that year a research reactor was placed in operation at North Carolina State University, the first reactor to go into operation outside an AEC facility.

Statutory changes that occurred in 1954 with passage of a revised Atomic Energy Act made it possible, for the first time, to disseminate information about reactor technology to private industry as well as to the world at large. This change coincided with the start of President Eisenhower's Atoms for Peace program, which expressed the policy of the United States to make available the civilian benefits of atomic energy on a global basis. Great impetus to the development of a civilian reactor industry came from the First United Nations International Conference on the Peaceful Uses of Atomic Energy, which took place in Geneva in the fall of 1955 (United Nations, 1956).

The 1955 conference was followed by entry of many private companies into the business of designing and building reactors. It was generally recognized that it would take another decade for nuclear power to become economically viable, but meanwhile markets were developing for research reactors as well as for experimental "demonstration power reactors" that were financed in part by the U.S. government in cooperation with electrical utilities and manufacturers. In parallel with the civilian program, a joint AEC–U.S. Navy program to develop nuclear propulsion was also proceeding rapidly, and the first nuclear power submarine, the *Nautilus*, was launched in 1954. The use of nuclear power for naval propulsion was highly successful and accelerated the development of civilian nuclear power in many ways.

The use of electricity in the United States had been increasing at an average rate of about 6% per year for several decades prior to the 1970s, necessitating that the installed generation capacity be doubled every 10 to 12 years. The economics of nuclear power generation tended to favor construction of large plants, and as the plants increased in size [from 200 electrical megawatts (MWe) in 1950 to 1000 MWe in 1970], nuclear power became more desirable from an economic point of view. In addition, national concern with increasing levels of air pollution in many communities during the 1960s and early 1970s also favored installation of nuclear units, as did the increasing costs of fossil fuels and political uncertainties regarding the long-range availability of these fuels, particularly those that are imported.

In 1963 the New Jersey Central Power Company announced that it was purchasing a General Electric Company 620-MWe reactor. This was the first power reactor to be purchased in the United States purely on economic grounds and the decision heralded the birth of a new major industry. There followed a 15-year period of growth for the nuclear industry until the mid-1970s, when installation of additional nuclear capacity came to a sudden stop owing to a number of interacting factors, of which one was the 1979 accident at Three Mile Island (TMI) Unit II, operated by the General Public Utilities Corporation near Harrisburg, Pennsylvania (see Chapter 12). The accident resulted in more than $1 billion in damage to the power plant and increased the already distrustful attitude of the public toward nuclear power. Moreover, the accident occurred during a period of worldwide economic recession that had already resulted in a reduction in the rate of growth of electrical demand, in both the United States and other countries. By the late 1970s, annual growth in electricity demand had dropped from its peak of 7% in 1974 to 2%, which meant that surplus capacity existed on the drawing boards and in plants under construction. To complicate things further, interest rates in the United States rose precipitously, which affected the economics of nuclear power plants because they are much more capital-intensive than power plants that use fossil fuels. The more stringent regulatory requirements imposed after the TMI accident also increased the cost of nuclear power. Finally, the TMI accident, the reduced demand for electricity, the higher costs of nuclear power plant construction, and the emerging difficulties of radioactive waste disposal were all seized upon by antinuclear organizations to further their opposition to nuclear power. This resulted in additional delays in licensing, which further increased the cost of the nuclear plants. The antinuclear pressure received more impetus from the disastrous accident in Chernobyl in April 1986.

As of the end of 1996, no new nuclear plant has been ordered in the United States since the accident at TMI, and orders for about 100 plants have been canceled, some of which were nearing completion. Whether or not there is a viable future for nuclear power in the United States remains to be seen and will depend in a large measure on whether public confidence in nuclear energy can be restored. Increased awareness of the political and economic problems associated with petroleum production, and the increased attention being given to the environmental problems associated with use of fossil fuels, seems to have resulted in a more favorable attitude toward nuclear power as an option for electricity production that should be preserved. The prospects for global warming due to the accumulation of carbon dioxide in the atmosphere has been particularly important in this regard.

Despite its setbacks, nuclear power production is a major industry at the present time. By 1994 there were nearly 500 nuclear power plants either in operation or under construction in the world. A total 5500 civilian power reactor operating years had been accumulated. The United States was generating 22% of its electricity from nuclear fuel overall, but nuclear power accounted for about 50% of production in New England and parts of the Middle West. France generated 73% of its electricity from nuclear fuel in 1993, and 12 other countries (Lithuania, Belgium, Slovak Republic, Finland, South Korea, Sweden, Spain, Slovenia, Switzerland, Germany, Hungary, and Bulgaria) generated more than 30% from nuclear fuel.

The health physicist, physician, engineer, or health officer concerned with the effects of reactors on the environment should possess a general understanding of those aspects of design that may affect the kind and quantity of radioactive effluents discharged to the environment under normal and abnormal conditions. It is also important to appreciate the basic methods used to prevent uncontrolled release of radioactivity to the environment and to understand the monitoring activities required to assure compliance with the applicable limitations on discharges of radioactive wastes. Finally, in the event of an accidental release to the environment, the environmental specialist should be prepared to advise the measures that can be taken to minimize the consequences of the release. The discussion that follows is intended to provide the general reader with some of the basic aspects of reactor design and operation that affect safety. The reader who wishes to pursue the subject more comprehensively is referred to several excellent texts on the subject (Marshall, 1983; Weinberg and Wigner, 1958; Thompson and Beckerley, 1964; LaMarsh, 1966, 1975, 1983; Glasstone, 1994; Okrent, 1981; Knief, 1992).

SOME PHYSICAL ASPECTS OF REACTOR DESIGN AND OPERATION

Contemporary reactors, with only a few exceptions, use either natural uranium or uranium in which the amount of isotope 235 has been enriched. The amount of ^{235}U enrichment may vary from 0.7% (natural) to more than 90%, but the fuel in most civilian power reactors is enriched to about 3%. Plutonium can also be used as a fuel for reactors, as will be discussed later in this chapter.

Because of a number of difficulties in the use of uranium metal, UO_2 is used in most present-day reactors. The metal tends to swell after intense neutron bombardment and metallic uranium is also very active chemically and can react exothermally with both air and water. Uranium dioxide is a ceramic that melts at about 2200°C, but has the disadvantage of being a rather poor heat conductor so that the heat transfer requirements necessitate that the rods must have rather small diameters, typically ½ to ¾ inches (13 to 19 mm) in diameter.

The fuel may be fabricated as rods, pins, plates, or tubes and is protected by a cladding whose function is to prevent the escape of fission products and protect the fuel from the eroding effect of the coolant. The cladding may be zirconium, stainless steel, or other special alloys. In most power reactors, the fuel is in the form of sintered UO_2 pellets less than 0.5 inch (13 mm) in diameter and about 1 inch (25 mm) long. The pellets are aligned within tubes of zircalloy or stainless steel about 4 m in length, and the tubes

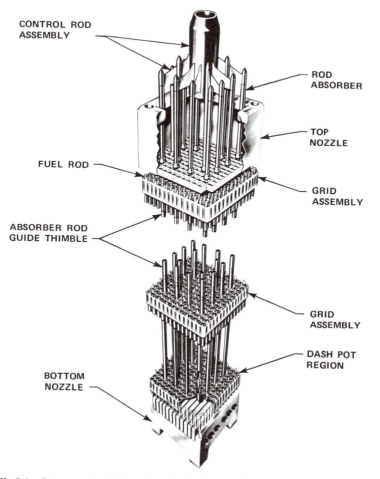

FIGURE 8-1   Cutaway of a fuel and control rod assembly. (Courtesy of Westinghouse Electric Corporation.)

are arranged in bundles as shown in Fig. 8-1, within which, at selected positions, control rod guide channels replace the fuel pins.[1]

Fission results from the capture of a neutron by the nucleus of a fissionable atomic species. Because more than one neutron is released in the process, a multiplication of neutrons may be achieved, which allows addi-

---

[1]The pellets are mounted within the tubes in such a way that helium-filled spaces exist. This improves the fuel-to-cladding heat transfer. The spaces serve as a plenum within which volatile fission products that diffuse from the fuel accumulate.

tional atoms of uranium to be split, which in turn will yield additional neutrons to continue the multiplicative fission process.

Some of the neutrons produced by fission will escape from the reactor system and be lost. This can be minimized by surrounding the reactor core with a reflector that tends to scatter escaping neutrons back into the system. Other neutrons may be captured by the nuclei of nonfissionable materials in the reactor system: this can be minimized by selection of materials that have a low capture cross section[2] for neutrons and by careful control over high cross-section impurities in the materials from which the system is constructed. The nuclear reaction becomes self-sustaining when, for every atom that fissions, one fission-producing neutron remains after allowing for escape by leakage or loss by nonfission capture. At this point the reactor is said to be in a "critical" condition.

The neutrons produced in fission are relatively energetic and, in most commercial power reactors, the probability that they will produce fission is increased by reducing their energies to less than 0.1 eV. This is accomplished by distributing the fuel in discrete components between which is placed a moderator of low atomic number. The neutrons are reduced in energy as a result of elastic collision with moderator atoms. Most reactors are designed for thermal neutrons and are called "thermal reactors." A fast reactor has no moderator and depends for its operation on the production of fission by fast neutrons (i.e., greater than 100 keV). A lmited number of gas-cooled power reactors that use graphite as a moderator have been built, principally in the United Kingdom.

Almost all power reactors use water to reduce ("moderate") the neutron energies to the thermal region. Water-cooled power reactors are of two main types. Boiling-water reactors (BWR) produce steam that is used to drive the reactor turbines directly. Pressurized-water reactors (PWR) are maintained at pressures sufficiently high to prevent boiling in the reactor itself. Hot water from a PWR is circulated through a heat exchanger called a steam generator, where water in the secondary circuit is boiled for steam.

The state of criticality of a reactor is expressed by k, the neutron multiplication factor, which is the ratio of the number of neutrons produced by fission in any one generation to the number of neutrons produced in the preceding generation. When criticality exists, k equals unity, and $dn/dt = 0$, where n is the neutron flux, usually expressed as neutrons per square centimeter per second. The heat produced in a reactor, and hence its power level, is directly proportional to the neutron flux.

---

[2]The cross section is a quantitative expression of the probability of occurrence of a given reaction between a nucleus and an incident particle. The unit of cross section is the barn, equal to 10^{-24} cm^2 per nucleus. The cross section of a nucleus for fission or capture depends on the energy of the incident neutrons.

When k is less than unity, the reactor is subcritical, and a chain reaction cannot be sustained. When k is greater than unity, the reactor is supercritical, $dn/dt > 0$, and the power tends to increase with time. Unless k is reduced to ≤ 1, the heat produced would increase to a level that would destroy the reactor. When $k < 1$, the power level will diminish and cannot be sustained at any given level without improving the neutron economy. This is usually done by adjusting the position of neutron-absorbing control rods.

A reactor is designed so that it is possible to maintain $k = 1$ for various neutron densities corresponding to the desired power levels. To increase the power level, k is made slightly greater than unity for a brief period of time. Another way of describing neutron kinetics at any given time is the reactivity, ρ, defined as

$$\rho = (k - 1)/k \qquad (8\text{-}1)$$

With this formulation, a value of $\rho = 0$ corresponds to the steady-state condition.

The control rods of a thermal reactor contain neutron-absorbing materials such as cadmium, indium, boron, or hafnium. Insertion of the rods into the core reduces the number of neutrons available for fission, thus reducing the value of ρ, which causes the power level to diminish. Conversely, the power level will increase when the rods are withdrawn from the core.

The reactivity of a reactor is also affected by its temperature and the radiation history of the core. The effect of core temperature on reactivity is an important characteristic of a reactor and determines its ability to self-regulate should the power increase inadvertently. The overall effect of temperature on reactivity is the result of a number of factors. For one, an increase in temperature results in increases in nonfission capture cross sections. Another effect is that the coolant density is decreased by a rise in temperature, and this affects reactivity by reducing the moderating ability of the coolant and increasing the number of neutrons that escape from the reactor core or are removed from the system by nonfission capture. Another way in which temperature may reduce reactivity is by causing vapor bubbles to form in the liquid coolant-moderator. A negative temperature coefficient of reactivity is a basic requirement for power reactor safety and serves to self-regulate the reactor. With a positive coefficient, dangerous instabilities would result from the fact that any power excursion that caused an increase in core temperture would cause a further increase in power and temperature because of increased reactivity and so on, until the destruction of the reactor.

Reactivity is lost as the core ages owing to fuel burnup and because some of the accumulating fission fragments or their decay products have a high cross section for thermal neutrons and therefore increase the fraction of neutrons lost by capture. Xenon-135 and ^{149}Sm are particularly important in this regard. The concentration of ^{135}Xe increases for several hours after reactor shutdown and may reduce reactivity to such an extent as to prevent restart of the reactor for a day or more.

Thus, because of the effects of temperature and fission-product poisons, a hot, aged core is less reactive than a cold, fresh core, which requires that there should be compensating loss in reactivity at start-up. In most power reactors, this is accomplished by the use of control rods. Another technique is the use of ^{10}B, which has a high capture cross section for thermal neutrons and which can be used to reduce reactivity when added in small amounts to the reactor core or coolant. The boron concentration in the cooling water can be controlled by the use of an ion exchanger. This technique is often used in PWRs in which the boron concentration is gradually lowered as the uranium in the fuel is used up. In the case of BWRs, one cannot use dissolved boron because it leaves deposits on the fuel as the water boils. Instead, the poisons are fixed in solid form and are gradually depleted by neutron capture at just the right rate to correct for the loss of reactivity due to fuel burnup.

When the control rods are removed, the value of k increases from $k < 1$ to $k > 1$ and the power level begins to rise. At the desired flux density (power level) the rod positions are adjusted until $k = 1$, and the steady-state critical condition is achieved.

The rate at which the power will rise or fall depends on the amount by which k is greater or less than unity. If n neutrons are present per unit volume at the beginning of each generation and L is the generation time (or average neutron lifetime), one can roughly approximate the change in the number of neutrons as

$$dn/dt = n(k - 1)L \qquad (8\text{-}2)$$

Integrating, we obtain $n = n_0 e^{t(k-1)/L}$, where n_0 is the initial neutron flux (and thus a measure of the initial power level) and n is the flux at any time t.

If we call the ratio $L/(k-1)$ the reactor period T, so that $n = n_0 e^{t/T}$, then T is equal to the time it takes to increase the power by a factor of e, sometimes called the e-folding time.

Under most conditions, L is constant, being characteristic of the core. The period is dependent solely on $k-1$, and reactor power will rise or fall in an exponential manner, depending on whether $k-1$ is positive or negative.

These relationships show that the reactor period is sensitive to changes in the generation time L. The average time that elapses between the production of a fission neutron and its ultimate capture is about 10^{-3} s in a natural uranium reactor. Assuming a value of $k = 1.005$ and a value of $L = 10^{-3}$ s, the power level in each second would increase by a factor of e^5, which is about 150. This would cause the reactor to be uncontrollable.

Fortunately, production of a small fraction of the fission neutrons is delayed. The size of this fraction depends on which fissionable material is involved and the energy of the fission-causing neutrons, and varies from 0.0024 for ^{233}U to about 0.007 for ^{235}U. The delay fraction actually controls the value of L, which is about 0.1 s for ^{235}U. The delay time itself varies—some neutrons are delayed by many tens of seconds, others by only a few milliseconds. The delayed neutrons have a controlling effect on the rate of power increase for cases where $1 < k < 1 + \beta$, where $\beta =$ the fraction of fission neutrons delayed. In this range of values for k, criticality depends on the delayed neutrons, and they greatly slow the rate of power increase to a doubling time of many seconds, depending on the exact value of k. However, should k exceed $1 + \beta$, then the reactor will become critical without the delayed neutrons. In such a case the reactor is said to be *prompt critical* and the rate of increase is limited only by the time required for the fission neutrons to thermalize (10^{-3} to 10^{-4} s). A basic objective of reactor design is that a prompt critical condition must not be possible.

The reactor designer must select materials that have the desired nuclear, thermal, and structural properties. Corrosion must be minimized, because corrosion products become radioactive in passing through the reactor and complicate its operation. The particularly difficult problem of "hot particles" has emerged in the last decade, in which small fragments of alloys containing high proportions of cobalt circulate with the coolant and become activated in the core, producing discrete, intensely radioactive particles (NCRP, 1989e). Small particles of fissioned fuel are also found.

One of the objectives of design is to contain the fission products within the fuel elements by maintaining the integrity of the cladding under normal operating conditions. However, should cladding failure occur, there must be provision for operating safely, even though some fission products may escape from the fuel into the coolant.

The reactor must be equipped with sufficient instrumentation so that the operator can know when the system is not functioning properly and to permit him or her to take whatever remedial measures are required. When prompt response is essential, the information obtained by instruments is fed into electrical systems designed to take corrective action automatically. The challenge to the designer is to provide a plant that will produce power efficiently and yet at all times be under complete control.

The choice of materials for the core, moderator, and coolant must be such as to minimize the possibility of exothermic reactions. Many of the materials that are potentially useful in reactor construction react exothermally with each other under certain conditions. Examples include metallic uranium, sodium, and zirconium, all of which react exothermally with water.

Some of the energy that becomes available during a serious reactor accident may be due to chemical reactions between these materials.

One of the most fundamental estimates that must be made in the course of design is the temperature of the hottest channel in the core. Systematic peaking in both the radial and axial directions does occur and can be predicted by calculation and experiment. The product of the radial and axial power-peaking factors can be used to determine the amount of power that will be produced in the hottest part of the core, assuming that the peaking factors in both directions coincide. In addition to this type of peaking, which is determined by design factors, it is necessary to consider variations from the average power density that will result from small manufacturing deviations in fuel-rod diameter, the amount of enrichment, fuel density, the extent to which fuel is out of round, and other similar factors. The most conservative approach is to assume that the least favorable of all such deviations coincide spatially. The product of all such factors gives a so-called "hot channel factor" that becomes a fundamental limitation in thermal design. The amount of power that can be produced by the reactor is limited by the temperature of the hot channel and the requirement that failure of the cladding be avoided.

In a pressurized-water reactor, boiling does not normally occur but may be present owing to insufficient pressure or temperature rise. A moderate amount of boiling is not deleterious and may, in fact, increase the amount of heat transfer from the fuel. However, if the boiling, instead of being nucleate, produces a vapor film along the surface of the fuel element, heat transfer can be greatly reduced, and the temperature of the fuel and its cladding may rise dangerously. The reactor is designed so that the temperature in the hot channel does not approach the temperature at which departure from nucleate boiling (DNB) occurs.

The severe thermal and radiation environment within a reactor can affect reactor materials by producing changes in physical properties such as thermal conductivity, resistivity, hardness, and elasticity. In addition, dimensional changes in some materials may occur, and in graphite these changes may be associated with the storage of relatively large amounts of energy, which under certain conditions can be released as heat. This phenomenon was responsible for an accident to an air-cooled, graphite-moderated reactor at Windscale in the United Kingdom in 1957 (see Chapter 12).

The effects on materials of intense radiation must be considered and can be studied by subjecting test specimens to conditions simulating those expected in the reactor. The reactor pressure vessel, fuel, cladding, welds, and various pieces of hardware that may be included in fuel-element design are particularly important. For example, irradiation may affect the rate at which fission gases such as xenon and krypton are released from the fuel into the helium-filled gap that in light-water reactors separates the fuel and the cladding. This could affect the rate of heat transfer to the cladding. The effect of irradiation on the integrity of the cladding must also be studied. If the physical properties are changed so that the cladding erodes or otherwise becomes unable to withstand the rigors of the thermal and radioactive environment within the core, fission products may be released. These and similar questions can best be answered by actual irradiation of test specimens. It is of utmost importance that the structural integrity of the pressure vessel not be compromised (Pugh, 1992).

During every phase of reactor design the engineer and physicist must examine each decision for its possible effect on the safety of the system. Decisions regarding materials, dimensions, equipment, operating temperatures, and other aspects of design must be carefully reviewed.

TYPES OF REACTORS

Reactors can be constructed to serve as sources of either radiation or heat. Included among those used as radiation sources are: (1) "production" reactors in which products of neutron irradiation such as plutonium are produced; (2) research reactors such as those located on university campuses and other research centers; and (3) industrial-type test reactors that are used to study the effects of radioactivity on materials of construction and equipment components. The reactors used as sources of heat are used primarily as sources of power for electric generators but may, in principle, be used to generate steam for space heating or industrial processing. The discussion here will be limited to power reactors.

LIGHT-WATER REACTORS

With few exceptions, the power reactors constructed by U.S. power companies have been of the so-called light-water type in which the fuel is enriched to about 3% and water is used as both moderator and coolant. Prior to about 1963, when the light-water reactors (LWR) emerged as economically viable and operationally practical, a number of other systems were demonstrated, some in the laboratory and some as power generators. These included designs in which the fuel and coolant were mixed homogeneously, reactors that were moderated and cooled by organic fluids, and other types, all of which gave way to the light-water reactors, which were adopted for use by the U.S. Navy. The LWRs include two types, the pressurized-water reactor and the boiling-water reactor.

In the pressurized-water reactor (Fig. 8-2a), the reactor core is enclosed in a 6- to 8-inch steel pressure vessel through which water is circulated and absorbs the heat produced. This cooling loop is maintained under pressure in excess of 2000 psi and is called the primary system. The water does not boil because of the high pressure, but passes through the heat exchanger to produce steam in the secondary loop, which is maintained at a lower pressure to permit boiling. The steam thus produced is pumped through a steam dryer and then to a turbine generator from which the steam tailings are condensed and returned to the boiler (steam generator). Thus, in the PWR the steam used to drive the turbine generator does not pass through the reactor but receives its heat via the steam generator.

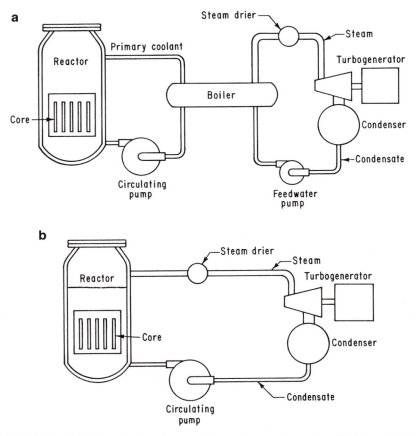

FIGURE 8-2   (a) Schematic flow diagram of pressurized-water reactor. (b) Schematic flow diagram of boiling-water reactor.

Radioactivity is retained in the primary system except to the extent that there is leakage to the secondary system.

In the boiling-water design (Fig. 8-2b), the water is allowed to boil within the reactor core, and the steam passes through a dryer directly to the turbine generator from which the steam tailings are condensed and returned to the reactor.

HIGH-TEMPERATURE GAS-COOLED REACTORS

In the past, the British (Kaplan, 1971; Goodjohn and Fortescue, 1971) have placed considerable emphasis on the development of high-temperature gas-cooled reactors (HTGR) and have had a number of such reactors in operation since the mid-1950s. This system has received relatively little attention in the United States, but one HTGR was operated between 1979 and 1989 by the Public Service Company of Colorado at Fort St. Vrain. It was shut down because of poor operating performance (NAS/NRC, 1992).

FAST REACTORS

Reactors can also be designed to operate without a moderator, using fast neutrons to initiate fission. Among the advantages is that since no moderator is required, it is possible to design cores that are small in comparison with thermal reactors. In addition, the capture cross sections of most substances for fast neutrons are relatively low, permitting a wider choice in the selection of construction materials. The fission products also have low capture cross sections for fast neutrons, and therefore greater burnup of fissionable material is possible because certain fission products do not act as poisons in fast reactors. Finally, because the nonfission capture cross sections of fast neutrons by the fertile material are relatively high, the fast reactor is inherently favorable for breeding plutonium in the manner to be described in the next section.

One of the disadvantages of the fast reactor is the fact that the selection of a coolant is restricted to those having no moderating effect. Liquid sodium, which is frequently used for cooling fast reactors, has obvious disadvantages because of its chemical reactivity with air and water, but this is offset by excellent heat transfer properties and the fact that high pressures are not required when sodium is used as a coolant. The small size of the liquid-cooled fast reactor reduces the surface areas available for heat transfer and, thus, further limits the choice of coolants to those having high conductivity and specific heat.

Because of the favorable neutron economy referred to earlier, it is possible to design fast reactors with relatively small amounts of excess reactivity, but unintended changes in the configuration might increase the reactivity and cause core damage. Should sufficient core melting occur, it is possible that the molten fuel could arrange itself in such a way as to produce a critical mass. Although this can probably be avoided by proper design, this factor has led to a great deal of conservatism in the use of fast reactors for civilian power.

BREEDER REACTORS

The breeder reactor is one in which some of the neutrons that are not used for fission are absorbed in "fertile" nuclei such as ^{238}U or ^{232}Th, which transmute to ^{239}Pu and ^{233}U, which are in turn fissionable. This results in a fuel economy in which more fissionable material is produced than is consumed.

Light-water reactors are inherently inefficient and convert only 1 to 2% of the potentially available energy of the uranium mined into heat. In contrast, a breeder reactor can economically release up to about 75% of the energy contained in uranium, thereby achieving fuel efficiencies about 40 times greater than those of light-water reactors (Seaborg and Bloom, 1970).

Prior to the worldwide reduction in the requirements for electricity that began in the mid-1970s, it was commonly believed that limitations in the supply of uranium required that the breeder reactor be developed. Most of the countries that use nuclear power, including the United States, the former Soviet Union, Japan, the United Kingdom, and, most notably, France, were active in breeder reactor development. However, these programs have been drastically curtailed as of this writing. Following two decades of research and development, the United States decided to construct an advanced 380-MWe breeder at Clinch River, Tennessee, but considerable public opposition and the changing economic picture caused the program to be dropped.

LOW-LEVEL DISCHARGES FROM LIGHT-WATER REACTORS

The radionuclides that accumulate in reactors are primarily those produced by fission within the reactor core and, secondarily, the activation products formed when traces of corrosion products and other impurities contained in the coolant-moderator undergo neutron bombardment in passing through the core. In this way, radionuclides of elements such as chromium, cobalt, manganese, and iron are produced. The fission product inventory is very much larger than the inventory of corrosion products, but the nature of reactor operation is such that the corrosion products may be present in relatively greater quantities in aqueous wastes.

Uranium oxide has the desirable characteristic that fission products are trapped efficiently within its crystal structure, thereby minimizing their escape to the coolant. The helium-filled gap between the fuel and the cladding (provided to improve heat transfer) contains volatile fission products (iodines and noble gases) that migrate from the fuel.

The inventory of each fission product can be calculated during any period of reactor operation for any given power level history, and their decay can be calculated during periods of shutdown or reduced power. If a reactor operates sufficiently long so that equilibrium with the short-lived nuclides has taken place, the distribution of the radionuclides as a function of time after reactor shutdown is given in Fig. 8-3. The total amount of radioactivity as a function of time after shutdown for various radiation histories is given in Fig. 8-4, in which it is seen that a 1-MW reactor will accumulate about 21 MCi (780 PBq) of radioactive products after 500 days at full power. The principal radionuclides contained in a reactor core that has been shut down for 1 day subsequent to 2 years of continuous operation are given in Table 8-1.

Light-water reactors produce both gaseous and liquid radioactive wastes during normal operation, some of which are discharged directly to the environment. The wastes are produced either by the fission process or by activation of elements in the primary coolant. The fission products may originate from traces of uranium present on the surfaces of fuel elements or other reactor components, but the major source is by leakage or diffusion through the fuel cladding. Radionuclides can accumulate in gaseous form or as dissolved or suspended solids. Chemical-processing systems are provided that remove some of the radionuclides and concentrate them into a form for ultimate disposal (Fig. 8-5) (Blomeke and Harrington, 1968; Cottrell, 1974). However, a small fraction may be discharged directly to the environment.

Contaminated coolant is accumulated in storage tanks and then passed through a waste-gas stripper. Steam that is generated in the stripper scrubs incoming water free of gas as it passes downward through sections of porcelain saddles. The stripped gases, after passing through a condenser, are then routed to the waste-gas system. The effluent from the waste-gas stripper is passed to an evaporator from which the vapors are condensed and passed to a demineralizer having cation and mixed-bed resins. The sludge from the evaporator bottom is passed to storage tanks from which it goes to drumming stations. The effluent from the demineralizer passes to storage tanks, where it can be sampled. If it is of satisfactory quality, it

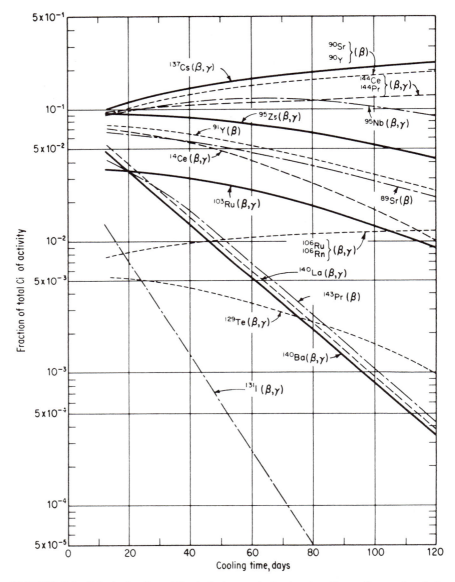

FIGURE 8-3 Principal radionuclides in fission products at various times after reactor shut-down. It is assumed that the reactor has operated for an extended period and that an approxi-mate equilibrium has been attained prior to shutdown. [From Glasstone (1955).]

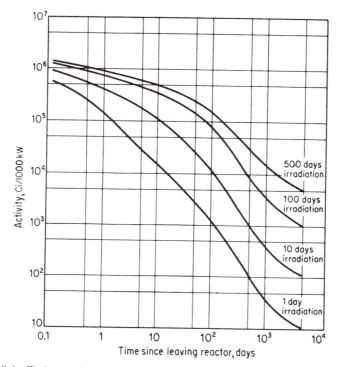

FIGURE 8-4 Fission-product inventory in a reactor core after various periods of irradiation and shutdown. [From Parker and Healy (1956).]

can be discharged via the condenser discharge canal (Fig. 8-6) or it may enter the purification system for reuse in the primary system. In the event the quality is not satisfactory, the water may be returned for further processing.

The induced activities that occur in the primary coolant depend on the materials of construction of the core, pressure vessel, pumps, piping, and other components in contact with the water. Induced activation of impurities ordinarily present in water is minimized by removal of the contaminants by water treatment before the coolant is introduced into the reactor.

Stainless steel is an important material in reactor construction and results in formation of a number of nuclides that find their way into the coolant. These include ^{60}Co, ^{59}Fe, ^{51}Cr, ^{54}Mn, and ^{55}Fe, all of which have moderately long half-lives and ultimately present a waste-disposal problem in addition to being an operational problem because of the gamma radiation they emit.

In-pile tests of uranium oxide fuel have shown that the various radioac-

TABLE 8-1

Inventory of Selected Radionuclides 1 Day
after 2 Years of Reactor Operation[a]

Selected radionuclides	Half-life	Activity in fuel (kCi MWt^{-1})[b]
^{3}H	12.3 y	0.0043
^{85}Kr	10.7 y	0.25
^{89}Sr	51 d	24
^{90}Sr	28.9 y	1.8
^{90}Y	64 h	1.8
^{91}Y	58.8 d	32
^{99}Mo	66.6 h	40
^{131}I	8.06 d	28
^{133}Xe	5.3 d	54
^{134}Cs	2.06 y	0.61
^{132}Te	78 h	34
^{133}I	20.8 h	22
^{136}Cs	13 d	0.74
^{137}Cs	30.2 y	2.4
^{140}Ba	13 d	46
^{140}La	40.2 h	49
^{144}Ce	284.4 d	35

[a] Reprinted with permission from "Engineering for the Resolution of the Energy–Environment Dilemma. Committee on Power Plant Siting." Copyright 1972 by the National Academy of Sciences. Courtesy of the National Academy Press, Washington, D. C.
[b] One kCi = 37 TBq

tive elements diffuse from the fuel at different rates, depending on their relative mobility. Table 8-2 summarizes the escape-rate coefficients measured in pressurized-water reactor fuel. Elements such as cesium, iodine, xenon, krypton, rubidium, and bromine have escape-rate coefficients that are a factor of 10^3 greater than those of strontium and barium.

LIQUID WASTES

The fission and activation products that diffuse from the fuel into the circulating water can be removed continuously by means of a purification system that is designed to have sufficient capacity to allow continued reactor operation in the event of minor failures in 1% of the fuel rods.

Tritium is produced in light-water reactors in quantities that are relatively copious compared to other radionuclides. Fortunately, tritium, which emits beta particles of very low energy, ordinarily enters the environment in the form of water. It does not concentrate significantly in biological systems and has a relatively rapid turnover rate. The maximum contaminant

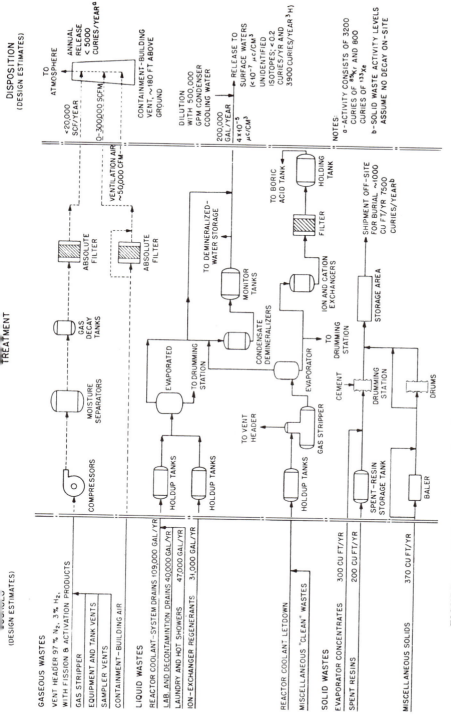

FIGURE 8-5 Waste management flow diagram for a large pressurized-water reactor. [From Blomeke and Harrington (1968).]

FIGURE 8-6   Condenser discharge for a pressurized-water reactor. Low-level liquid wastes can be released into this tunnel, through which condenser coolant is flowing at a rate of 300,000 gal min^{-1}. The pipe at the left edge of the canal is a sampling manifold. Sampling intake pipes can also be seen. (Courtesy of Consolidated Edison Company.)

level (MCL) in drinking water is relatively high, and large amounts of tritium can be discharged to the environment without exceeding the permissible levels of human exposure (Moghissi and Carter, 1973; NCRP, 1979).

The tritium originates in two ways. Albenesius (1959) first demonstrated that tritium is produced in fission at a rate of about 1 atom per 10,000 fissions. Tritium is also produced by spallation following neutron irradiation

TABLE 8-2

Escape-Rate Coefficients[a]

Elements	Escape-rate coefficient (s^{-1})
Cs, I, Xe, Kr, Rb, Br	1.3×10^{-8}
Sr, Ba	1.0×10^{-11}
Zr, Ce, and rare earths	1.6×10^{-12}
Te	1.0×10^{-9}
Mo	2.0×10^{-9}

[a]From U.S. Atomic Energy Commission (1960a).

of boron, a major source of tritium in reactors that use ^{10}B as a burnable poison.

Differences in the diffusion rates of the various radionuclides from intact fuel cladding cause the coolant radioactivity in most reactors to be relatively depleted in less labile fission products such as ^{90}Sr or ^{89}Sr. In the event of fuel cladding failure, direct exposure of the fuel to the coolant can increase the amounts of radiostrontium in the coolant but the radioactivity will always be dominated by the more soluble fission products such as Cs and I.

The exact composition of the liquid wastes from light-water reactors will vary from reactor to reactor depending on the materials of construction and the condition of the fuel. Table 8-3 lists the relative radionuclide composition of the primary coolant of a pressurized-water reactor. The concentrations listed are relative to ^{137}Cs, which is the most labile of the nonvolatile long-lived fission products. The effect of the relatively low diffusivity of the strontium nuclides is shown by their low concentration relative to ^{137}Cs.

The light-water coolant can be purified continuously by drawing off a fraction of the coolant for treatment. Following filtration to remove suspended radionuclides, the water then passes through cation- and anion-exchange resin beds. If necessary, the coolant can be passed through a gas stripper in which the water is percolated over plates across which a countercurrent stream of steam is passed, which serves to remove dissolved gases such as air, fission gases, and hydrogen (Coplan and Baron, 1978; Moghissi et al., 1978a).

Apart from the coolant purification process, there are many other ways in which radioactive liquid wastes are produced. Leaks of coolant from valves, flanges, and pumps result in contamination of sump water. Components that are removed for repair must sometimes be decontaminated, and this will result in contaminated water, as will washing casks, sluicing resin beds, laundering contaminated clothes, and washing contaminated labora-

TABLE 8-3

Composition of Primary Coolant Relative to
the Concentration of ^{137}Cs in a Typical
Pressurized-Water Reactor

Nuclide	Relative concentration[a]
^{3}H	5.3×10^{2}
^{51}Cr	$<1.7 \times 10^{-2}$
^{54}Mn	2.8×10^{-1}
^{55}Fe	8.1×10^{-2}
^{59}Fe	$<2.4 \times 10^{-3}$
^{58}Co	2.1×10^{-2}
^{60}Co	1.1×10^{-1}
^{65}Zn	$<3.1 \times 10^{-3}$
^{90}Sr	6.6×10^{-4}
^{91}Y	2.6×10^{-3}
^{95}Zr-Nb	$<2.8 \times 10^{-2}$
^{103}Ru	$<3.1 \times 10^{-2}$
^{106}Ru	$<8.6 \times 10^{-2}$
^{131}I	$<5.2 \times 10^{-3}$
^{134}Cs	4.8×10^{-1}
^{137}Cs	1.0×10^{0}
^{144}Ce	$<3.8 \times 10^{-2}$

[a] ^{137}Cs $= 1.0$.

tory ware. Provision must be made for containment of these waste liquids and their treatment and ultimate disposal. The system for treating such wastes at a PWR is shown in Fig. 8-5, in which low-level contaminated water is accumulated for periodic processing.

GASEOUS WASTES

Different methods of managing gaseous wastes are used depending on whether the source is a PWR or BWR. In a PWR, the waste fission-product gases that are stripped from liquid wastes can be passed through a condenser that removes the condensible portion, passing the remainder to holdup tanks and then to catalytic recombiners that remove the radiolytic hydrogen. The residual gases can then be pumped to holdup tanks or can be stored on activated charcoal to permit decay, following which the gases can be passed to the exhaust stack through a high-efficiency filter. After 3 months of holdup the radioiodine has decayed, and the remaining radioactivity is due mainly to ^{85}Kr, which can be released to the atmosphere or, in special circumstances, condensed cryogenically or removed for off-site disposal.

Other sources of PWR gaseous wastes are leaks from the primary system. Small quantities of gases from the primary coolant can leak directly to containment and be vented to the atmosphere when the containment building is purged to permit access by personnel. Small leaks can also develop in the heat exchanger, causing quantities of primary coolant to pass to the secondary loop. When this happens, radioactive gases can pass to the atmosphere without treatment via the boiler blowdown.

In the BWR, the fission and radioactive noble gases boil off with the steam, pass through the turbine, and then enter the condenser from which they are removed by the air ejector. Leakage of air into the condenser occurs because it is normally operated under vacuum. Small quantities of air that leak into the condenser mix with radiolytic hydrogen and oxygen as well as the gaseous fission products. When the first BWRs began operation in the late 1960s, the air ejector vented to the environs through a stack that allowed a holdup time of only a few minutes, too short to permit decay of the short-lived noble gases. The amount of radioactivity in the gaseous discharges from BWRs has been greatly reduced by catalytic recombination of the radiolytic hydrogen and oxygen, which account for about 80% of the air ejector exhaust. Recombination of the oxygen and hydrogen thus results in a fivefold reduction in the gas volume, which slows the passage time through the exhaust system and provides time for additional noble gas decay.

Two short-lived radionuclides, ^{89}Kr ($T_{1/2} = 3.2$ min) and ^{137}Xe ($T_{1/2} = 3.8$ min), decay to ^{89}Sr and ^{137}Cs, which are trapped on high-efficiency filters. The gases in modern BWRs are then passed through refrigerated activated charcoal beds that remove the remaining radioiodines and noble gases, thereby allowing additional time for radioactive decay. Another option is to compress the gases for storage in tanks to allow sufficient time for almost complete decay of all but ^{85}Kr, which has a 10.8-year half-life. The gaseous discharges from a typical BWR for various levels of waste-gas treatment are given in Table 8-4 (Collins et al., 1978).

Stigall et al. (1971) found that the inherent barriers to ^{131}I transport through a BWR system function with a high degree of efficiency and that it is not a significant nuclide in the gaseous releases from this type of reactor.

RADIATION EXPOSURE OF THE PUBLIC FROM REACTOR EMISSIONS

The radioactive emissions from operating power reactors in the United States are routinely reported by the NRC (e.g., Tichler et al., 1989). The emissions have resulted in insignificant doses to the general population, and in some cases are less than the dose from natural radioactivity discharged by power plants fueled with coal (see Chapter 6).

TABLE 8-4

Typical BWR Annual Releases for 30-min Holdup, Treatment by
Charcoal Delay, and Treatment by Cryogenic Distillation[a]

Radionuclide	Half-life	Ci y^{-1} per 3400-MWt reactor[b]		
		Base case (30-min holdup)	Charcoal delay	With cryogenic distillation
^{83m}Kr	1.9 h	44,000	c	12
^{85m}Kr	4.5 h	84,000	80	22
^{85}Kr	10.8 y	290	290	280
^{87}Kr	76 min	240,000	c	72
^{88}Kr	2.8 h	280,000	5	76
^{89}Kr	3.2 min	2,800	c	60
^{131m}Xe	12 d	220	18	
^{133m}Xe	2.2 d	4,300	c	c
^{133}Xe	5.3 d	120,000	460	13
^{135m}Xe	15 min	11,000	c	3
^{135}Xe	9 h	330,000	c	34
^{137}Xe	3.8 min	9,700	c	39
^{138}Xe	14 min	390,000	c	90
^{131}I	8.0 d	5.0	c	0.0026
^{133}I	21 h	2.1	c	0.011

[a]From Collins et al. (1978).
[b]One Ci = 37 GBq
[c]Less than 1 Ci y^{-1} for noble gases, less than 10^{-4} Ci y^{-1} for radioiodine.

The doses received by members of the public from routine power reactor emissions are a fraction of the ICRP and NCRP recommendations. The NRC design guidelines for emissions from power reactors (Appendix I of 10 CFR 50) were the first formal application of the ALARA principle by a regulatory agency and were based on what was practicable. Plants were to be designed so that the doses to individuals beyond the site perimeter were less than 5 mrem y^{-1} (50 μBq y^{-1}) from liquid discharges, less than 10 mrem y^{-1} due to gamma radiation from gaseous emissions, and less than 20 mrem y^{-1} due to beta radiation in gaseous emissions (10 CFR 50). As discussed in Chapter 3, however, doses due to air emissions from nuclear power plants are now limited by the EPA to 10 mrem y^{-1} from all pathways (40 CFR 61).

The NRC issues annual reports that summarize the dose received by people who live between 2 and 80 km from nuclear power reactors. In 1988, when 110 nuclear power plants were operating at 70 sites in the United States, the mean collective effective dose commitment from all pathways ranged from a low of 0.0011 person-rem (1.1 × 10^{-5} person-Sv) to a high of 16 person-rem (0.16 person-Sv). All told, 150 million persons lived within the 2- to 80-km annuli and their collective dose commitment was 75 person-rem (0.75 person-Sv) from the emissions during the year (Baker, 1992).

No estimates are given for people who live less than 2 km from the reactor because there are very few such individuals, and since the annual limit of 10 mrem (0.1 mSv) is readily met, their collective dose would be insignificant.

Reports that summarize emissions from the operation and maintenance of reactors on U.S. naval vessels are also issued annually (e.g., Mangino *et al.*, 1995).

There has been a gradual reduction in both liquid and gaseous emissions from the power reactors owing to consistent improvement in fuel quality of all LWRs. Additional reductions in BWR emissions have resulted from installation of stack gas treatment systems, such as those described earlier, that allow longer decay times before emission to the atmosphere. Despite the fact that large-scale testing of nuclear weapons had been terminated in 1963 (see Chapter 9), the man-made radioactivity in the Hudson River estuary was dominated through 1970 by radionuclides introduced from fallout, as can be seen from Table 8-5, which compares the annual discharges from the Consolidated Edison reactors at Indian Point to fallout from weapons tests in past years. The presence of ^{137}Cs from such fallout was so predominant that the reactor contribution could be detected only by using the isotope ^{134}Cs as a tag. This nuclide is not present in bomb fallout but is formed in reactors by neutron capture in ^{133}Cs, the daughter of the fission product ^{133}Xe.

TABLE 8-5

Comparison of Discharges of the Principal Radionuclides from Indian Point I and Fallout from Weapons Tests (Measured in Curies)[a]

	^{90}Sr	^{54}Mn	^{137}Cs	^{3}H (tritium)
Annual discharge from Indian Point I (1968)	0.008	5.8	2.3	810
Fallout from weapons				
a. On Hudson River watershed (35,000 km²)	825	1236	1320	205,000
b. On Hudson River surface	3.7	5.5	5.9	920
c. On mixing zone of river, 16 km above and below plant	0.58	0.58	0.93	144

[a]For purposes of comparison, the year of heaviest fallout (1963) is compared to the year of maximum reactor discharge (1968). These data have been assembled from several sources. One Curie = 37 GBq.

Systems are now available by which the liquid wastes can be so thoroughly purified that the decontaminated water can be returned to storage tanks for reuse in the primary system. Of course, this does not apply to tritium, which cannot be separated by any practical means, but is returned with the treated wastewater to tanks where it can be stored awaiting reuse in the reactor.

REACTOR ACCIDENTS

Thus far the discussion of releases from light-water reactors has been limited to normal operating conditions. Malfunctions can of course develop, not all of which are abnormal any more than it is abnormal for an incandescent bulb to burn out or an electrical appliance switch to fail in the home. It is considered normal for the cladding of a small percentage of fuel elements to develop minor imperfections and for small leaks to develop in the PWR heat exchangers. Although, as noted earlier, it is possible to design for a certain amount of leakage and remove the radioactive impurities by systems of chemical decontamination, there are many advantages to minimizing the leaks. High concentrations of radioactive impurities in the coolant increase the gamma radiation emissions from the coolant system, which complicates maintenance and results in higher than necessary levels of worker exposure. Thus, the number of imperfections and the amount of radioactivity that passes to the coolant are performance indicators that have been constantly improved over the years.

Accidents that might result from relatively common failures of relays or valves can be avoided by redundancy in design, high standards of quality assurance during manufacture of the components, or stringent inspection procedures during plant operation. However, certain catastrophic failures can be postulated that could have severe consequences, but for which it is not possible to provide absolute assurance that the radioactivity can be contained. In general, it has been shown that the probability that a given accident will occur is inversely related to the severity of its consequences.

TYPES OF REACTOR MISHAPS

A sequence of events by which one can analyze possible reactor accidents is given in Fig. 8-7, which illustrates three important types of accidents: excessive reactivity, loss of coolant, and coolant-flow stoppage. Although various combinations of the blocks can be assembled to describe many possible accidents, not all the blocks will apply in any one mishap.

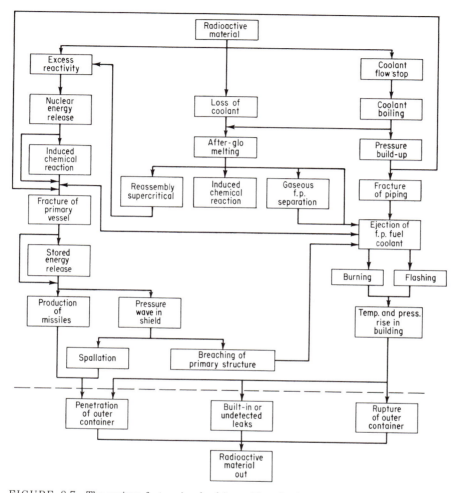

FIGURE 8-7   The various factors involved in accidental releases of fission products from reactors. Not all factors will be involved in any one type of accident. [From Brittan and Heap (1958).]

The more common types of accidents that must be considered in the course of risk assessment include stuck control rods, loss of coolant, loss of station pumping power, and the sudden addition of cold water to a core. Each of these and other contingencies must be studied and its effect on the permanence of the reactor evaluated. The loss-of-coolant accident may result in very serious consequences and will be discussed in detail.

RELATIVE HAZARDS FROM VARIOUS NUCLIDES

The fission fragments produced during reactor operation vary in mass number from 72 to 160 and include more than 80 important radionuclides produced in the frequency distribution shown in Fig. 8-8. It is seen that the percentage yields of the mass numbers, which are plotted on a logarithmic scale, range from about $10^{-5}\%$ to nearly 10% (Walton, 1961).

The inventory of fission products, activation products, and transuranic elements in the reactor core can be estimated by computer programs for any specified operating or shutdown history (Croft, 1980; Wenzel, 1994). Although nearly 800 nuclides are produced, not all of them are radioactive and others have such short half-lives that they are not significant in risk assessment. If stable nuclides and those with half-lives less than about 26 minutes are excluded, the list reduces to the 54 nuclides shown in Table 8-6.

In the event of a meltdown, the relative contribution of each nuclide to the dose received by the population at various distances from the plant depends on the quantity present in the core, its volatility, its chemical and biological behavior once it enters the general environment, and the effectiveness of containment.

It has been widely accepted that the radioiodines (and the nuclides of tellurium, which decay to the radioiodines) are responsible for most of the short-term dose received by persons downwind of the plant. It has been estimated that the nuclides of these two elements would be responsible for 83% of the dose received during the first day, as shown in Table 8-7. However, if the dose is integrated over periods of several years or longer, then ^{137}Cs dominates in many of the postulated accidents.

POTENTIAL CONSEQUENCES OF REACTOR ACCIDENTS

There have been many investigations of the consequences of the "maximum credible reactor accident." Some of these studies were made early in the history of the World War II atomic energy program and led to great conservatism in the sites selected for the location of the first reactors to be constructed. However, it must be remembered that when the first reactors were designed, their performance could be predicted only from theoretical considerations and time was required for operating experience that would demonstrate the validity of the underlying safety principles.

The early studies (Parker and Healy, 1956; Marley and Fry, 1956; AEC, 1957a) did, however, make it clear that the enormous inventory of radioactive fission products in a reactor core could cause catastrophic consequences if disseminated to the environment in a mishap. The most comprehensive

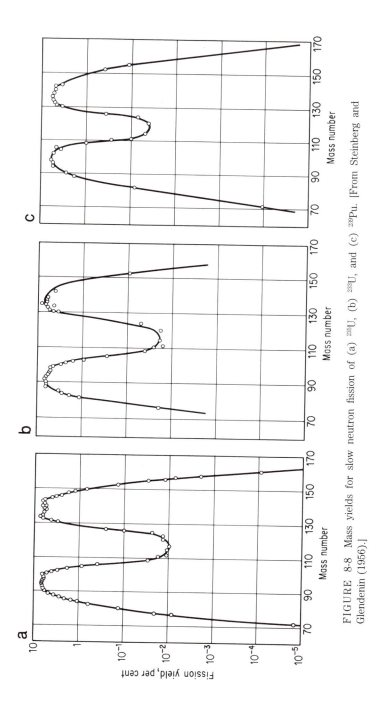

FIGURE 8-8 Mass yields for slow neutron fission of (a) ^{235}U, (b) ^{233}U, and (c) ^{239}Pu. [From Steinberg and Glendenin (1956).]

TABLE 8-6

Initial Activity of Radionuclides in the Core of a 3200-MWt Nuclear Reactor[a]

		Radioactive inventory	
No.	Radionuclide	Amount $(10^8$ Ci$)^b$	Half-life (d)
1	Cobalt-58	0.0078	71.0
2	Cobalt-60	0.0029	1,920
3	Krypton-85	0.0056	3,950
4	Krypton-85m	0.24	0.183
5	Krypton-87	0.47	0.0528
6	Krypton-88	0.68	0.117
7	Rubidium-86	0.00026	18.7
8	Strontium-89	0.94	52.1
9	Strontium-90	0.037	11,030
10	Strontium-91	1.1	0.403
11	Yttrium-90	0.039	2.67
12	Yttrium-91	1.2	59.0
13	Zirconium-95	1.5	65.2
14	Zirconium-97	1.5	0.71
15	Niobium-95	1.5	35.0
16	Molybdenum-99	1.6	2.8
17	Technetium-99m	1.4	0.25
18	Ruthenium-103	1.1	39.5
19	Ruthenium-105	0.72	0.185
20	Ruthenium-106	0.25	366
21	Rhodium-105	0.49	1.50
22	Tellurium-127	0.059	0.391
23	Tellurium-127m	0.011	109
24	Tellurium-129	0.31	0.048
25	Tellurium-129m	0.053	0.340
26	Tellurium-131m	0.13	1.25
27	Tellurium-132	1.2	3.25
28	Antimony-127	0.061	3.88
29	Antimony-129	0.33	0.179
30	Iodine-131	0.85	8.05
31	Iodine-132	1.2	0.0958
32	Iodine-133	1.7	0.875
33	Iodine-134	1.9	0.0366
34	Iodine-135	1.5	0.280
35	Xenon-133	1.7	5.28
36	Xenon-135	0.34	0.384
37	Cesium-134	0.075	750
38	Cesium-136	0.030	13.0
39	Cesium-137	0.047	11,000
40	Barium-140	1.6	12.8

(*continues*)

TABLE 8-6 *(continued)*

No.	Radionuclide	Radioactive inventory Amount (10^8 Ci)[b]	Half-life (d)
41	Lanthanum-140	1.6	1.67
42	Cerium-141	1.5	32.3
43	Cerium-143	1.3	1.38
44	Cerium-144	0.85	284
45	Praseodymium-143	1.3	13.7
46	Neodymium-147	0.60	11.1
47	Neptunium-239	16.4	2.35
48	Plutonium-238	0.00057	32,500
49	Plutonium-239	0.00021	8.9×10^6
50	Plutonium-240	0.00021	2.4×10^6
51	Plutonium-241	0.034	5,350
52	Americium-241	0.00017	1.5×10^5
53	Curium-242	0.0050	163
54	Curium-244	0.00023	6,630

[a]From U.S. Nuclear Regulatory Commission (1975), Appendix VI.
[b]One Ci = 37 GBq.

TABLE 8-7

Relative Contributions of the Major Radionuclides to First-Day Doses at 0.5 Mile

Radionuclide group	Curies (3000-MWt reactor)[a]	Relative dose[b]
Noble gases	3.44×10^8	0.8
Iodines	7.15×10^8	54.8
Telluriums	1.76×10^8	28.8
Cesiums	0.152×10^8	1.0
Ceriums	3.65×10^8	6.2
Rutheniums	2.07×10^8	1.0
Others	33.25×10^8	7.4
Total relative dose		100.0

[a]One Curie = 37 GBq.
[b]From American Nuclear Society (1984). Adapted from U.S. Nuclear Regulatory Commission (1975).

early study was that undertaken by the AEC in 1957, "Theoretical Possibilities and Consequences of Major Accidents in Large Nuclear Power Plants" (AEC, 1957a). This study undertook to analyze the consequences of destruction and volatilization of the core of a 500-MWt reactor in which the fission product inventory would be 4×10^8 Ci (15,000 PBq) measured 24 h after the accident. The reactor was assumed to be located about 3 miles from a major city.

The study was performed at a time when power reactor technology was in its infancy and the first commercially owned nuclear generating station was still several years away. By defining the enormous potential consequences of dispersing a reactor core of this size, if no mitigating features were incorporated into its design, the report served a useful purpose by pinpointing the kind of research that would be needed to determine which radionuclides would be disseminated in the event of an accident, the behavior of each nuclide in the environment, and the manner in which the releases could be contained.

In the 1957 study, there were essentially two boundary conditions: (1) the partially volatilized reactor core is fully contained and no contamination of the environs occurs and (2) 50% of the core is not contained and is volatilized. This assumption resulted in restrictions in the use of 150,000 square miles of land on which almost 4 million people lived. For the latter case, the cost estimates ranged up to $7.2 billion in 1957 dollars.

The assumption that 50% of all nuclides in the reactor core would be volatilized has been proven to be unrealistic, by both laboratory investigations and studies of accidents that have since occurred. There are great differences in the volatility of the various radioactive elements, and thus their availability for environmental dispersion. Because of the need to use a uniform method of risk assessment in the event of an LWR accident, the AEC in 1962 published the assumptions that should be made in the event of a loss of coolant accident associated with core damage. These assumptions made it possible to estimate the "source term," the quantities of radionuclides released. By methods of environmental pathway analysis, it then becomes possible to estimate the doses that would be received by people at various distances downwind of the reactor (DiNunno et al., 1962). The "source term" has a major influence on the doses received.

Beginning about 1962, all applicants for power reactor construction and operating licenses were required to estimate the doses that would be received by residents who live near the reactor in the event of a loss of coolant accident that was followed by destruction of the reactor core. That requirement exists up to the present. Based on work by DiNunno and associates, the regulations postulated that volatilization of the core contents would result in release of 100% of the noble gases, 50% of the iodines (of which one-half rapidly plates on surfaces within the reactor building), and 1% of the remaining fission products. One percent per day of the reactor building contents is assumed to leak to the outside atmosphere. These have been the basic assumptions that have been required by the NRC regulations contained in 10 CFR 100, and for many years have been used to assess the consequences of power reactor loss-of-coolant accidents. These assumptions were based on laboratory experiments with small quantities of irradiated fuel.

The assumptions about the source term came into question following the accident at Three Mile Island (see Chapter 12), when it was found that far less than 50% of the radioiodine escaped from the fuel. This resulted in a number of investigations that began to question the assumptions made concerning the release of the iodine nuclides from a damaged core. It was noted that the fraction of ^{131}I released from the reactor building was not 0.25, which is the assumption required by NRC, but about 3×10^{-9}. The releases of ^{131}I during other accidents were also known to be far less than assumed and further laboratory experiments have in fact also suggested that the assumed release of the radioiodines from a damaged core was much too high.

A major inquiry into the fractional releases of radionuclides from damaged cores has been completed by a Special Committee of the American Nuclear Society (ANS, 1984). It was concluded that the release of iodine and certain other elements has been overestimated by one or more orders of magnitude for many of the postulated accidents. A major reason for this discrepancy has been a failure to appreciate that the chemical conditions that exist within a reactor system during an accident should favor formation of CsI, a highly soluble nonvolatile compound. This greatly reduces the release of radioiodine in vapor form, with important consequences on the dose received by surrounding populations. The ANS report also concluded that the tellurium release was overestimated because reactions with zirconium and stainless steel were not previously considered. Subsequent to the ANS study, Soffer et al. (1992) reviewed the results of comprehensive studies initiated by NRC and reported that 95% of the iodine entering containment after an accident is in the form of the nonvolatile compound CsI. However, subsequent behavior of the iodine will be determined by the pH of the aqueous materials with which the CsI comes into contact, and without pH control a large fraction of the iodine will return to elemental form and be released to the containment atmosphere. If the reactor system can be designed so that the pH remains above 7 in the postaccident period, the amount of iodine available for release will be considerably reduced, and a substantial reduction in the number of expected health effects will result.

The findings, now more than a decade old, have not as yet resulted in any changes in the methods of risk analyses.

A considerable advance toward understanding the probabilities of occurrence and consequences of a major reactor accident was publication of the "Reactor Safety Study" in 1975, directed by Norman Rasmussen of the Massachusetts Institute of Technology and sponsored by NRC (1975). The study was a probabilistic assessment of the risks associated with reactor accidents and compared the probabilities of occurrence and consequences of accidents of varying degrees of severity. The report has been influential in determining the kinds of engineered safeguards that are required, including the area of the emergency evacuation zones around the plant. To provide perspective, the probabilities and consequences of reactor accidents were compared to other natural and man-made sources of risk. Those comparisons are shown in Figs. 8-9 and 8-10, where it is seen that at all levels of risk the probability of causing a given number of fatalities is orders of magnitude lower than that for other man-made hazards such as death from aircraft accidents, dam failures, fires, and explosions. The likelihood of a given number of fatalities due to natural hazards is similarly shown to be orders of magnitude greater than the likelihood of death from a nuclear

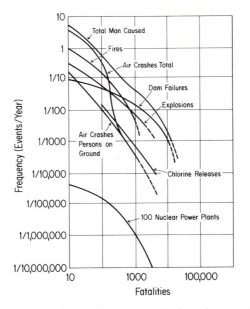

FIGURE 8-9 Frequency of accidents of varying severity due to human-caused events. [From NRC (1975).]

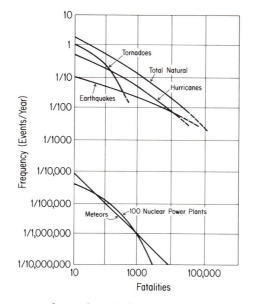

FIGURE 8-10   Frequency of natural events that cause varying numbers of fatalities. [From NRC (1975).]

power plant accident for all sources except meteorites. The risk of dying because of a reactor accident, assuming 100 operating power reactors, was estimated to be 5×10^{-10} per year compared, for example, to a risk of about 10^{-3} per year of dying as the result of other accidents.

The conclusions of the Rasmussen report were at once hailed by the proponents of nuclear power as convincing proof of the safety of nuclear power reactors, but were regarded with suspicion by their opponents. The NRC appointed an expert panel under the chairmanship of H. W. Lewis of the University of California at Santa Barbara (Lewis *et al.*, 1979), with the charge to review the Rasmussen report that "clarify the achievements and limitations" of the study. The Lewis committee lauded the Rasmussen report as a substantial advance over previous studies and for its sophisticated use of probabilistic methods of analyzing the consequences of reactor accidents. However, the committee concluded that the error bounds of the probabilities given in the report were probably wider than stated, although one could not determine whether the probabilities assigned in the Rasmussen study were overstated or understated.

The Lewis report, like its predecessor, was also both praised and condemned: from the pronuclear point of view it was thought to be supportive of the Rasmussen conclusion rather than a repudiation as claimed by others. It is of interest that a study conducted by the German government, similar in purpose to the Rasmussen report, came to conclusions that were quantitatively similar to those reported by Rasmussen (German Federal Minister of Research and Technology, 1979).

The severity of a reactor accident in which core damage occurs is dependent on the extent to which radioactivity, mainly as ^{131}I and ^{137}Cs, is released to the environment. In the 50-year period since the first reactor went into operation, there have been 14 accidents in which severe core damage occurred. These are listed in Table 8-8, where it is seen that only in the 1957 accident at Windscale in the United Kingdom and in the 1986 accident at Chernobyl were more than insignificant amounts of radionuclides released to the environment.

The Windscale accident was different from the others in that it was a plutonium-production reactor that was air cooled, graphite moderated, and fueled with metallic natural uranium. Part of the core was consumed in a fire, and 20,000 Ci (0.75 PBq) of ^{131}I was released to the environs via the cooling air. The Chernobyl reactor was also graphite moderated and had the disadvantage of having been built without some of the protective features

TABLE 8-8

Reactor Accidents That Involved Core Damage[a]

Year	Location	Name of reactor	Type	Extent of contamination
1952	Canada	NRX	Experimental	None
1955	Idaho	EBR-1	Experimental	Trace
1957	United Kingdom	Windscale	Military production reactor	20,000 Ci ^{131}I
1957	Idaho	HTRE-3	Experimental	Slight
1958	Canada	NRU	Research reactor	None
1959	California	SRE	Experimental	Slight
1960	Pennsylvania	WTR	Research	None measured
1961	Idaho	SL-1	Experimental	10 Ci ^{131}I
1963	Tennessee	Orr	Research	Trace
1966	Detroit	Fermi	Experimental power	No release outside plant
1969	France	St. Laurent	Power	Little, if any
1969	Switzerland	Lucens	Experimental	None
1979	Pennsylvania	TMI-II	Power	Slight
1986	U.S.S.R.	Chernobyl-4	Power	Extensive

[a]From Bertini (1980).

that are routinely adopted in Western countries. The more serious of the accidents listed in Table 8-8 will be discussed further in Chapter 12.

When it is announced that an accident has occurred, the media are often unable to evluate its severity, and relatively minor accidents frequently create far more anxiety than justified by the events. The IAEA (1990) has developed an International Nuclear Event Scale that should help to place announcements that an accident has occurred into proper perspective. The scale includes seven levels, as shown in Table 8-9, in which Level 1 refers to operational anomalies that do not present a risk of any kind, but may indicate the existence of procedural weaknesses. The accident classifications increase in severity to Scale 7, of which the Chernobyl accident has thus far been the only example. Under this system of classification, the accident at Three Mile Island would be rated as Scale 5, because severe core damage occurred, although environmental releases were minimal.

ENGINEERED SAFEGUARDS

A number of engineered safeguards are provided by the reactor designer to prevent release of the radioiodines and other radionuclides to the general environment. As noted earlier, every effort is made to design the reactor system so that the probability of accidents that can result in core damage is minimized. However, having designed the reactor to achieve this objective, the designer then makes the assumption that core damage will take place nevertheless and that exposure to nearby populations must be controlled by engineered safeguards. The most important of these safeguards are those provided to avoid a massive release of fission products in the event of sudden structural failure of the primary coolant system. This is the loss-of-coolant accident (LOCA).

A fundamental question is whether a massive failure of the piping or other components of the primary system is in fact credible (AEC, 1970b). There have been no such accidents in more than 50 years of experience with nuclear systems and, more important, massive failures are unknown in high-pressure central steam boilers, with which there is considerably more experience than is available in the nuclear industry. A study of some 500 boiler steam drums, designed for pressures over 600 psi, representing 4000 boiler years of operating experience, showed no failures of the steam drums themselves. Failures did occur in other parts of the high-pressure steam system, but they were not the massive type of failure that would cause a sudden release of coolant from a water reactor (Miller, 1966).

Should such a massive failure occur the coolant would flash to steam, and the core would quickly become subcritical from void formation due to boiling. In the absence of sufficient cooling, the fuel would overheat and

TABLE 8-9

The International Nuclear Event Scale for Prompt Communication of Safety Significance[a]

Level	Descriptor	Criteria	Example
Accidents			
7	Major accident	· External release of a large fraction of the reactor core inventory typically involving a mixture of short and long-lived radioactive fission products (in quantities radiologically equivalent to more than tens of thousands terabecquerels of iodine-131). · Possibility of acute health effects. Delayed health effects over a wide area, possibly involving more than one country. Long-term environmental consequences.	Chernobyl, USSR 1986
6	Serious accident	· External release of fission products (in quantities radiologically equivalent to the order of thousands to tens of thousands of terabecquerels of iodine-131). Full implementation of local emergency plans most likely needed to limit serious health effects.	
5	Accident with off-site risks	· External release of fission products (in quantities radiologically equivalent to the order of hundreds to thousands of terabecquerels of iodine-131). Partial implementation of emergency plans (e.g., local sheltering and/or evacuation) required in some cases to lessen the likelihood of health effects. · Severe damage to large fraction of the core due to mechanical effects and/or melting.	Windscale, UK 1957 Three Mile Island, USA 1979
4	Accident mainly in installation	· External release of radioactivity resulting in a dose to the most exposed individual off-site of the order of a few millisieverts. Need for off-site protective actions generally unlikely except possibly for local food control. · Some damage to reactor core due to mechanical effects and/or melting. · Worker doses that can lead to acute health effects (of the order of 1 Sv).	Saint Laurent, France 1980

Incidents

3 Serious incident
- External release of radioactivity above authorized limits, resulting in a dose to the most exposed individual off-site of the order of tenths of a millisievert. Off-site protective measures not needed.
- High radiation levels and/or contamination on-site due to equipment failures or operational incidents. Overexposure of workers (individual doses exceeding 50 mSv).
- Incidents in which a further failure of safety systems could lead to accident conditions, or a situation in which safety systems would be unable to prevent an accident if certain initiators were to occur.

Vandellos, Spain 1989

2 Incident
- Technical incidents or anomalies which, although not directly or immediately affecting plant safety, are liable to lead to subsequent re-evaluation of safety provisions.

1 Anomaly
- Functional or operational anomalies which do not pose a risk but which indicate a lack of safety provisions. This may be due to equipment failure, human error or procedural inadequacies. (Such anomalies should be distinguished from situations where operational limits and conditions are not exceeded and which are properly managed in accordance with adequate procedures. These are typically "below scale."

0 No safety significance

[a]Adopted from "Recommendations of International Atomic Energy Agency and OECD Nuclear Energy Agency" (IAEA Bulletin, April, 1990).

the zirconium cladding would react exothermically with steam. Without intervention, the fuel itself could melt. To limit overheating of the core, an emergency core cooling system would flood the core in the event of a LOCA and keep the temperature below that at which the cladding would fail.

An essential additional safeguard is the containment building designed to confine the steam and any entrained radioactive substances. The containment structure must be designed to withstand a variety of mechanical stresses, including the release of steam from the reactor system and the impact of missiles that could conceivably be produced (Gwaltney, 1969). Containment structures have become more complex since the reactors have become larger and have been built closer to centers of population. The first containment vessels were simple spheres, about 125 ft (7.5 m) in diameter, and fabricated from 1-inch (25-mm) welded steel plate. Subsequent containment vessels have been constructed of concrete, with linings fabricated from steel plate.

Means can be provided to condense the steam rapidly so that the pressure is greatly reduced, making it possible to use a less massive containment building. In one such system, designed by General Electric Company (Bertini, 1980), the steam would be vented into a water-filled pool that serves as a heat sink by condensing the steam. (Fig. 8-11). Another type of vapor suppression system, not widely used, employs an ice condenser in which the steam is condensed by venting through an ice-filled structure (Weems et al., 1970).

Additional protection is provided by a system of sprays that can wash the radioiodine and other fission products from the containment atmosphere into a sump. Adding sodium thiosulfate to the spray increases the efficiency of the radioiodine-scavenging process (American Nuclear Society, 1971; Parsly, 1971).

Finally, the containment structure itself must be designed so that leakage is minimized. It is conventionally assumed that the containment atmosphere leaks at a rate of 0.01% per day, and tests must be performed regularly to prove that the leak rate is smaller than this. Methods are available for testing the leak rates (Zapp, 1969).

Analysis of the so-called loss-of-coolant accident and several other types of accidents and the design of safeguards intended to mitigate the effects of such an accident comprise much of the material that an applicant must submit to the NRC for a power reactor license. During the past decade, the vast amount of technical information derived from laboratory study, pilot scale experiments, and theoretical analysis has been adapted to computerized methods of predicting the pressure and temperature transients involved in the various mishaps that might occur. This work has verified that the original assumptions are very conservative in almost every case. In a few

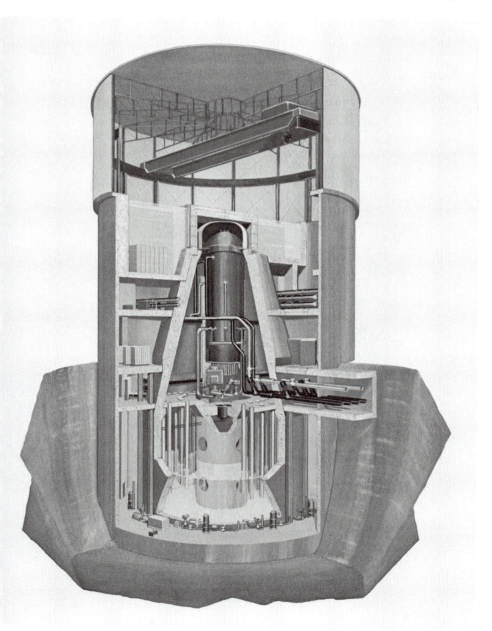

FIGURE 8-11 Boiling-water reactor conical concrete pressure suppression containment. (Courtesy of General Electric Company.)

cases in which the early practices were not conservative, new regulations were issued.

The conservatism of the methods used by the United States and other Western countries to protect against the effects of nuclear reactor accidents has been demonstrated by the contrast in the consequences of the accidents at Three Mile Island and Chernobyl. Both of these accidents are discussed in Chapter 12.

ADVANCED REACTOR DESIGNS

In the aftermath of the Three Mile Island accident, and during the many years that have passed without new reactor orders in the United States and some other countries, considerable effort has been devoted to increasing the inherent safety of the reactor systems. In the United States, the program has been sponsored by the DOE, the Electric Power Research Institute, and a consortium of utilities and nuclear power plant suppliers, the Nuclear Power Oversight Committee. Similar programs have been under way in the United Kingdom, Japan, Sweden, France, and Germany and have stimulated designs for advanced types of light-water, heavy-water, gas-cooled, and liquid metal reactors (Nuclear Power Oversight Committee, 1989; Golay and Todreas, 1990; American Nuclear Society, 1992).

The new reactor systems include improvements in inherent safety, economics, management, and programs of personnel training. In the United States, an important feature of the advanced reactor designs is that they will be standardized designs that can be licensed as a system by the NRC. A justified complaint about the past system of licensing has been that each plant has been different and has required individual licensing procedures. Standardization of design should shorten the time required for licensing and should also simplify personnel training and the supply of components.

One of the main advantages of the new designs is that simplicity is achieved by taking advantage of natural forces whenever possible to minimize the need for systems of pumps and pipes. Thus, in LWRs, the emergency core cooling system utilizes an overhead tank that drops its water into the pressure vessel by gravity, instead of having to depend on pumps, pipes, and a supply of electrical power. Overhead water storage also relies on gravity to provide the emergency sprays used to cool and depressurize the containment building. The passive safety features result in considerable simplification, as exemplified in one design by 60% fewer valves, 50% fewer large pumps, 80% less control cable, and 60% less pipe [Electric Power Research Institute (EPRI), 1989; Ahearne, 1993].

At this writing, four companies have submitted advanced designs to the NRC for generic approval. The first such approval was granted to the

General Electric Company in 1994. The first of two such plants in Japan began operation in 1996 (Nuclear Energy Institute, 1996).

SITE SELECTION CRITERIA

Selection of a site for a power plant, whether it is to be nuclear or fossil, is a complex procedure on which many constraints are imposed (Gifford, 1974; Okrent, 1981). These include economic factors, availability of rights-of-way for transmission lines, availability of a natural source of cooling water (or the feasibility of installing cooling ponds or towers), the geological and seismological characteristics of the region, the distance to the load center, and ecological or public health implications of atmospheric and liquid effluents from the plant. Until 1969, nuclear power plants were regulated exclusively by the AEC, whose authority was limited to control over the radiological hazards and who determined whether the plant could be built without jeopardizing the public health and safety only insofar as radiological hazards were concerned. Other environmental factors were not then subject to federal regulation. However, with the passage of the National Environmental Policy Act of 1970 the picture changed drastically, and a federal agency with the authority to issue a license, of any kind, was henceforth required to examine the full spectrum of environmental consequences of the proposed action. A comprehensive analysis (the Environmental Impact Statement) must now be submitted by the applicant to the licensing agency and be reviewed by other agencies of the government concerned with the various relevant environmental effects. This section will review that part of the licensing process that is concerned with radiological effects.

The detailed reactor siting criteria used by the NRC are contained in Part 100 of Title 10 of the Code of Federal Regulations. The rationale for these criteria is given in technical document TID-14844, "Calculation of Distance Factors for Power and Test Reactor Sites" by DiNunno *et al.* (1962), the purpose of which is to provide a uniform approach to site evaluation. A more detailed discussion of the evolution of reactor siting practices is given elsewhere by Okrent (1981).

According to Part 100, three zones surround the reactor:

1. An "exclusion area" in which the reactor licensee will have the authority to determine all activities including exclusion or removal of personnel and property from the area.

2. A "low population zone," which is the area immediately surrounding the exclusion area and which contains residents, the total number and density of whom are such that there is a reasonable probability that appropriate protective measures could be taken in their behalf in the event of a

serious accident. The guide does not specify a permissible population density or total population within this zone because the situation may vary from case to case.

3. A "population center distance," which is the distance from the reactor to the nearest boundary of a densely populated center containing more than about 25,000 residents.

Using these definitions, methods are proposed (Di Nunno *et al.,* 1962) for calculating the size of the required exclusion area, low population zone, and population center distance for a contemplated reactor design. The distances should be selected so that the following criteria can be met in the event of an accident:

1. The exclusion area should be of such size that an individual located at any point on its boundary for 2 h immediately following onset of a fission product release would not receive a total radiation dose to the whole body in excess of 25 rem (0.25 Sv) or a total radiation dose in excess of 300 rem (3 Sv) to the thyroid from iodine exposure.

2. The low population zone should be of such size that an individual located at any point on its outer boundary who is exposed to the radioactive cloud resulting from the release would not receive a total radiation dose to the whole body in excess of 25 rem (0.25 Sv) or a total radiation dose in excess of 300 rem (3 Sv) to the thyroid from iodine exposure.

3. The population center should be located at least 1.3 times the distance from the reactor to the outer boundary of the low population zone.

The NRC is careful to note that the 25 rem (0.25 Sv) dose for whole-body exposure and 300 rem (3 Sv) for thyroid exposure are not to be construed as being permissible in the event of accidents: "Rather, this 25 rem (0.25 Sv) whole-body value and 300 rem (3 Sv) thyroid value have been set forth in these guides as reference values, which can be used in the evaluation of reactor sites with respect to potential reactor accidents of exceedingly low probability of occurrence, and low risk of public exposure to radiation."

Document TID-14844 defines the basic ground rules for estimating the amount of radioiodine available for leakage to the environment and the emergency doses to the surrounding population that should serve as the design criteria. Ever since publication of these criteria, reactors designers have been devising safeguards, some of which have been previously described, that have the effect of greatly reducing the doses permitted by the NRC regulations. An example of the progress that has been made is shown in Table 8-10, which summarizes the calculated off-site doses under a variety of circumstances following loss of coolant in a large PWR. Whereas

TABLE 8-10

Summary of Off-Site Exposure Calculations for Loss-of-Coolant Accident[a]

Dose	2-h exposure at 520 m (minimum exclusion radius)	Total exposure at 1100 m (minimum low population zone radius)[b]
I. Thyroid dose (based on zero to 5% of airborne as CH_3I)		
Containment leakage terminated in 1 min by isolation valve seal water system	0.7 rem	0.36 rem
Gap release[c]—continuous leakage with 2 spray pumps and 5 fan filters operating	0.8–1.45 rem	0.42–0.68 rem
Gap release[c]—continuous leakage with 1 spray pump and 3 fan filters operating	1.7–2.7 rem	0.85–1.4 rem
Gap release[c]—continuous leakage with 5 fan filters operating	8.8 rem	4.8 rem
Gap release[c]—continuous leakage with 2 spray pumps	0.95–6.7 rem	0.55–13.9 rem
10 CFR 100 suggested limit	300 rem	300 rem
II. Whole-body dose		
Containment leakage terminated in 1 min by isolation valve seal water system	<1 mrem	<1 mrem
Gap release—continuous leakage	18 mrem	68 mrem
TID-14844 release—continuous leakage	3.8 rem	4.9 rem
10 CFR 100 suggested limit	25 rem	25 rem

[a]From USAEC Docket 50, Exhibit B.
[b]One rem = 0.01 Sv; one mrem = 0.01 mSv.
[c]TID-14844 initial iodine leakage inventory of 25% of core equilibrium quantity will result in thyroid dose 10 times value shown.

for design purposes 10 CFR 100 suggests a thyroid dose limit of 300 rem (3 Sv) in the low population zone following a loss-of-coolant accident, the calculated thyroid dose would be less than 1 rem (0.01 Sv) if a reasonable combination of safeguards is in operation.

Table 8-10 gives the dose estimates based on (1) the assumption that the radioiodine and noble gases related to the environment were contained in the gap between the fuel and the cladding and (2) the more pessimistic assumption of TID-14844 that 25% of the core inventory of radioiodine is released to the environment. Gap release would result from cladding failure and the TID-14844 assumption is based on failure of both the cladding and fuel.

LICENSING PROCEDURES

Whether a given reactor design is suitable for a given site is evaluated under the rules promulgated in Parts 50 and 100 of Title 10 of the Code of Federal Regulations. The first formal application to the NRC must be accompanied by a Preliminary Safety Analysis Report (PSAR) and (2) the Environmental Impact Statement.

The PSAR is a comprehensive document of several volumes that includes a complete description and safety assessment of the site, whose hydrology, geology, meteorology, and ecology must be adequately documented. The analysis must include a preliminary design of the reactor with an analysis and evaluation of the design and performance of structures, systems, and components of the facility. A description of the quality assurance program to be applied in design, fabrication, construction, and testing must also be included, as well as a preliminary plan for the applicant's proposed organization and personnel training program. To the extent that full technical information is not yet available, the PSAR must identify the research and development programs that must be completed before the design can be finalized. Finally, the PSAR must include a preliminary plan for dealing with emergencies.

The PSAR is reviewed by the NRC staff and by a statutory committee of outside experts known as the Advisory Committee on Reactor Safeguards (ACRS) (Okrent, 1981). The staff and ACRS reviews are accompanied by many conferences with the applicant and a massive amount of highly technical and voluminous correspondence. If the reviews are favorable, and depending on whether or not the public participates in the proceeding, the commission can elect to hold a public hearing by an Atomic Safety and Licensing Board (ASLB). In a few cases, public intervention at these hearings has been intensive and has delayed the licensing procedures. If the ASLB rules in favor of the application, then a license to construct the reactor is normally issued by the NRC.

Several years later, as the reactor is nearing completion, essentially the same procedure is followed for the operating license. The applicant submits a Final Safety Analysis Report (FSAR), which goes through the same reviews leading up to public hearings and a final decision as to whether the license will be issued. The licensing procedures have become overly complex and time-consuming, and modifications are badly needed that can streamline the procedures without sacrificing their effectiveness. This seems to have been accomplished with passage of the National Energy Policy Act of 1992, which greatly streamlined the licensing process so that all major issues, including design, siting, emergency procedures, and public concerns, can be settled before construction begins (U.S. Council for Energy Awareness, 1992).

DECOMMISSIONING

Nuclear power plants are designed to produce electricity for 30 to 40 years, and some of them have already been retired. The problems associated with decommissioning a reactor after a normal lifetime should be relatively simple in contrast to the complexities that will be required in the event of an accident that spreads radioactivity within the plant. The cleanup of Three Mile Island Unit 2 cost in excess of $1 billion dollars and took 10 years. The general problems that must be faced in the restoration of contaminated facilities are discussed in Chapter 15.

CHAPTER 9

Nuclear Weapons

The testing of nuclear weapons in the atmosphere has been responsible for the largest quantities of man-made radionuclides released to the environment. The radioactive debris from nuclear explosions divides into three fractions, depending on the height of burst and explosive yield. The first of these fractions consists of the larger particles that fall out within a few hours and that are intensely radioactive. The second fraction is dispersed into the troposphere, but may not produce fallout during the first day because the particles are sufficiently small to behave somewhat like aerosols and be subject to the laws of dispersion and rainout that govern small particles. The third fraction penetrates the stratosphere and will deposit worldwide over a period of many months. The tropospheric fallout tends to be distributed in bands at the latitude of detonation, whereas the stratosphere debris distributes itself globally, as will be described later.

Debris from bombs smaller than about 100 kilotons[1] tends to remain in the troposphere, whereas stratosphere injection is almost complete for detonations greater than 500 kilotons.

The physical and chemical characteristics of the particles have been observed to be highly variable in several respects (Crocker *et al.,* 1966).

[1]The explosive yield of nuclear weapons is conventionally expressed in units of the energy released by a ton (907 kg) of the explosive TNT, which equals approximately 4.2×10^{12} joules (J).

Depending on the temperature–time history of the particle, the radioactivity can be coated on the surface or distributed throughout.

The extent to which a given explosion will produce radioactive fallout in its immediate vicinity depends on the size of the explosion and its height above ground. The most fallout will be produced when the fireball touches the ground. If only 5% of the energy released by a 1-megaton bomb is spent in volatilizing soil with which the fireball is in contact, about 20,000 tons of debris will be added to the fireball (Glasstone and Dolan, 1977). Figure 9-1 shows the relationship between explosive yield in kilotons and the height above ground below which the fireball may be expected to touch the ground and produce heavy local fallout.

The fireball will rise to a height determined by the explosive yield, the height at which the detonation occurred, and the meteorological conditions existing at the time of detonation. Figure 9-2 shows that in temperate latitudes, fireballs from explosions under 100 kilotons stabilize below the tropopause (about 10 miles or 15 km in the midlatitudes), and detonations in the megaton range penetrate well into the stratosphere.

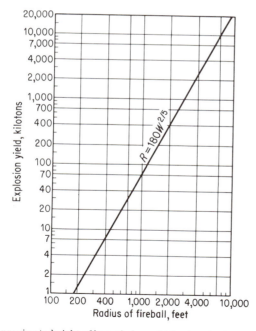

FIGURE 9-1 Approximate height of burst below which a bomb of a given yield will produce a fireball that touches the ground. [From Glasstone (1962).]

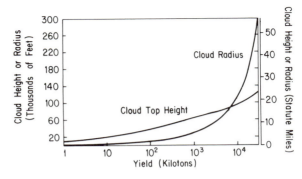

FIGURE 9-2   Approximate values of stabilized cloud height and radius as a function of explosion yield for land surface or low air bursts. [From Glasstone and Dolan (1977).]

Whether or not underground bursts produce fallout depends on whether the explosions vent through the surface. The radioactive debris from almost all deep underground explosions in Nevada have been contained below the surface. On the other hand, underground bursts that penetrate above ground have a very great potential for surface contamination.

Most of what we know about fallout from the explosion of nuclear weapons has been learned from studies of the effects of test explosions that have taken place in various parts of the world since the first atomic bomb was detonated on a New Mexico desert in July, 1945. In the intervening half-century, more than 400 test explosions in the atmosphere have been conducted, mainly by the United States and Soviet Union, but also by the United Kingdom, France, and the People's Republic of China. Table 9-1 summarizes the yields of atmospheric tests that have been announced by the various nations. The atmospheric tests have been conducted on land and sea, and hundreds of miles above the earth's surface[2] (Cochran *et al.*, 1989). In addition to the atmospheric tests listed in Table 9-1, a greater number of nuclear tests have been conducted underground, the United States alone having made 804 such tests in Nevada.

The first U.S. tests of nuclear weapons after World War II took place in the Marshall Islands in 1946. The Soviet Union conducted its first explosion in 1949, following which the United States accelerated its rate of testing and constructed a second proving grounds near Las Vegas, Nevada. Tests were also conducted by the United States in two other mid-Pacific locations, Johnson and Christmas islands. The latter is a British possession, where tests were conducted jointly by the two nations.

[2]These were small devices tested in a series named ARGUS to study the trapping of charged particles in the earth's magnetic field.

TABLE 9-1

Estimated Yields of Atmospheric Nuclear Weapons Tests[a]

	Period	No. of tests	Estimated yield (MT)	
			Fission	Total
United States	1945–1962	193	72	139
USSR	1949–1962	142	111	358
United Kingdom	1952–1953	21	11	17
France	1960–1974	45	11	12
China	1964–1980	22	13	21
Total		423	218	547

[a]From UNSCEAR (1982).

This chapter will be devoted primarily to the U.S. experience with 193 test explosions in the open atmosphere and many more tests conducted underground. Information about the USSR experience is beginning to emerge as a result of dissolution of the USSR and associated changes in public information policies. The Soviet tests begin in 1949, and from then until the 1963 test-ban treaty about 150 tests were conducted in the open air at sites in Kazakhstan, near Semipalatinsk, and on two Arctic islands in Nova Zemlya (Cochran et al., 1989).

Twelve tests in the kiloton range were conducted by the United Kingdom near Maralinga in Australia between 1952 and 1957. Nine tests, of which 7 were thermonuclear[3], were conducted on Christmas Island. France conducted 4 atmospheric tests in 1960 and 1965 in Algeria, and 44 atmospheric tests were also conducted on two uninhabited islands of the Tuamoto Archipelago in the South Pacific.

China has conducted more than 20 tests in the atmosphere at Lop Nor in Sinkiang Province beginning in 1964.

India has conducted only one test, underground, on a western desert. The test device was said to have been exploded for peaceful purposes.[4]

Testing nuclear weapons in the open atmosphere began to arouse world-wide concern in the mid-1950s, and fallout became a highly emotional and

[3]Thermonuclear bombs utilize the fusion reactions of light elements such as deuterium, tritium, or lithium.

[4]For a discussion of proposed peaceful uses of nuclear explosives see the second edition of this book published by Academic Press in 1973. This is a subject in which there was considerable interest during the period of open-air testing. Proposals were made to blast new harbors, to stimulate oil and gas production, and even to construct a new canal across the Isthmus of Panama. Interest has since waned, for reasons that will be obvious to the reader.

controversial subject. In response to worldwide pressure, in the fall of 1958 the United States, the United Kingdom, and Russia declared a moratorium on further weapons testing. By that time the three nuclear powers had conducted 38 separate series of tests that included a total of at least 227 detonations. France did not participate in the moratorium declaration and in 1960 became the fourth nuclear power by conducting a small series of nuclear tests on the Sahara Desert.

In 1961, without advance warning, the Soviet Union broke the moratorium agreement and exploded about 50 devices. The United States responded in kind, and two major powers began a frenetic competition that led to additional worldwide concern. The accelerated pace of weapons testing is illustrated in Fig. 9-3, which shows the rapidly increasing inventory of ^{90}Sr produced by nuclear explosions conducted up to 1962.

A nuclear weapons test-ban agreement was signed by the United States, the United Kingdom, and the Soviet Union early in 1963 and up to this writing it has succeeded in eliminating further testing in the open atmosphere by the three signatory powers (Seaborg, 1981; May, 1994). However, France did not sign this agreement, nor did China or India, all of which have conducted occasional tests in recent years. Nonetheless, these tests have not contributed significantly to the amount of radioactive fallout distributed around the world.

The test-ban agreement did not rule out underground explosions but stipulated that venting to the atmosphere must not be detectable beyond the borders of the nation that conducts the test. Accordingly, the limited test-ban agreement initiated a new era of nuclear weapons technology in which a variety of devices have been tested in underground cavities by methods that with only a few minor exceptions have prevented atmospheric pollution. Hundreds of underground tests have been conducted by the

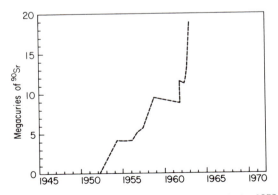

FIGURE 9-3   Inventory of ^{90}Sr by nuclear weapons tests, 1952–1963.

United States and the United Kingdom during the past 20 years. Whether the underground accumulations of radioactive debris will in time prove significant as a form of environmental pollution remains to be seen. The quantities of debris that remain underground after such tests are huge, but objective evaluation of possible long-range off-site risks has not been possible because little of the basic data has been made available. By the end of 1984, the United States had conducted a total of 635 tests at the proving grounds near Las Vegas. Of these, 105 were tests in the atmosphere and the remainder were conducted underground (DOE, 1985d). Underground testing has proceeded at a reduced rate in more recent years.

Fallout of radioactive debris from explosion of a nuclear weapon first occurred after the 1945 test in New Mexico in which a 19-kiloton device was fired from a 30-m steel tower. Some of the larger particles of radioactive debris fell on cows grazing about 20 miles (32 km) downwind and produced skin burns (Lamont, 1965). The finer particles drifted across the Middle West, and enough fallout occurred in Indiana (Webb, 1949) to contaminate cornstalks that ultimately found their way into a paper-making process used by the photographic industry. Radioactive particles in the finished product eventually caused damage to X-ray film that had been packaged with contaminated interleaving paper (Fig. 9-4). The extent to which fallout occurred at great distances from the test sites during the period of open-air testing is illustrated in Figs. 9-5 and 9-6.

Many thousands of people, mainly military, participated in the test programs. In the United States these included about 200,000 military personnel and concerns have been expressed about the health effects of their radiation exposure. This was prompted initially by a report of excess leukemia among participants in a test, named SMOKY, conducted in Nevada in 1957 (Caldwell *et al.*, 1983). A later report by a committee of the National Academy of Sciences confirmed that an excess of leukemia did exist among veterans who had participated in the SMOKY test but not among participants of other tests (NAS/NRC, 1985b). Deaths from all neoplasms were in fact lower than expected.

A study was also conducted of the health of about 22,000 participants in the test programs of the United Kingdom about 30 years after the tests were concluded. The health of the participants was found to be generally good, although a slight increase of leukemia and multiple myeloma was suggested by the data (Darby *et al.*, 1988).

PHYSICAL ASPECTS OF NUCLEAR EXPLOSIONS

Nuclear energy can be released from a bomb by means of either the fission or fusion process. As noted earlier, it has been customary to equate the

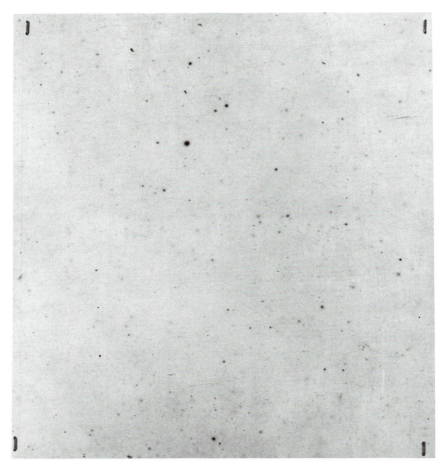

FIGURE 9-4   X-ray film marred by exposure to contaminated interleaving paper processed
August 6–10, 1945. (Courtesy of Eastman Kodak Company.)

explosive yields of nuclear and thermonuclear explosions to the equivalent
amount of TNT. Thus, a bomb having a yield equivalent to 1 million tons
(~900,000 metric tons) of TNT would be a 1-megaton bomb.

The pure fission (nuclear) bomb obtains its energy from either ^{235}U or
^{239}Pu. Fissionable material can be made critical either by quickly joining two
or more subcritical masses that then become supercritical or by implosive
compression of a subcritical mass. Both processes may be accomplished
with the aid of chemical explosives. The complete fission of about 56 g of

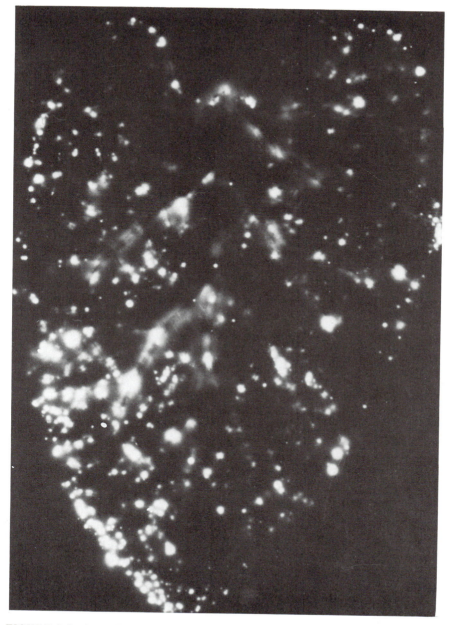

FIGURE 9-5   Autoradiograph of a leaf following a fallout of radioactive dust in Troy, New York. The dust originated from an explosion in Nevada about 36 h previously. (Courtesy of Herbert M. Clark.)

FIGURE 9-6   Autoradiograph of adhesive film exposed to the atmosphere for 24 h at a fallout monitoring station operated by the Health and Safety Laboratory of the U.S. Atomic Energy Commission in 1953. The sample was collected several hundred miles from the test site.

material will produce an explosion equivalent to 1 kiloton of TNT (approximately 4.2×10^{12} J).

Approximately 50% of the energy from a nuclear explosion is released in the form of blast, 35% as thermal radiation, and the remaining 15% as ionizing radiation. In considering the total consequences of a nuclear explosion, the effects of blast and fire may be of even greater importance than the effects due to ionizing radiations, but only the latter aspect will be discussed in detail in this text. The reader is referred elsewhere (Glasstone and Dolan, 1977) for a review of the effects due to blast and fire.

Of the ionizing radiations, one-third is prompt radiation produced within a few seconds after detonation and two-thirds, or 10% of the total energy released by the explosion, is in the form of delayed ionizing radiation produced by the decay of fission products and induced radionuclides.

The thermonuclear bomb utilizes the fusion reactions of light elements such as deuterium, tritium, or lithium. Several different reactions may occur, but all require that the nuclei have energies that can be obtained only with the aid of temperatures of several million degrees. Since this heat can be achieved in a fission bomb, such a device may be used as a trigger for a thermonuclear explosion.

The thermonuclear reactions are sources of fast neutrons that produce fission of ^{238}U and can thus be used to increase the explosive yield of a bomb by surrounding the fusion weapon with natural uranium. The fission of ^{238}U contributes a major fraction of the energy released by some thermonuclear weapons.

The terms "clean" and "dirty" have been sometimes used to describe the relative amounts of radioactivity produced by bombs. Those in which the energy is obtained primarily from fusion yield comparatively less radioactivity than weapons whose energy is derived entirely from the fission reactions.

Explosion of a nuclear or thermonuclear device produces a cloud of incandescent gas and vapor called the fireball, which is many times brighter than the noonday sun. Although the brightness begins to diminish after only 1 ms, the fireball continues to grow, reaching a final diameter equal to $D = 180W^{0.4}$ (Glasstone and Dolan, 1977), where D is in feet and W is the energy in kilotons.

In about 1 s, when the fireball from a 20-kiloton explosion has reached its maximum size, it will be about 440 m in diameter. In 1 min, the fireball will have cooled sufficiently so that it no longer glows, and by this time it will have risen to a height of about 7 miles (11 km). The size and height of stabilization are given as a function of yield in Fig. 9-2.

Convective forces initiated by the fireball result in enormous amounts of air and debris being sucked upward. Figure 9-7, which is a photograph of a relatively small nuclear explosion over the Nevada desert, shows how desert sands are being convected into the fireball, which has assumed a toroidal shape. Particles that enter the fireball sufficiently soon after its formation are vaporized and mixed within the fireball. Later, as the fireball cools appreciably, convected particles will no longer be volatilized but may serve as nuclei on which condensation of the radioactive constituents of the fireball can occur. Some of these particles may be as large as grains of sand, which posess sufficient mass so that fallout will occur in a matter of minutes. However, if the fireball is sufficiently high off the ground so that

FIGURE 9-7 Explosion of a nuclear weapon, showing the toroidal structure of the fireball shortly after its formation. The fireball did not touch the ground, but it is sufficiently low that dust dislodged from the surface by the blast waves is being sucked into the fireball. (Courtesy of U.S. AEC.)

large particles are not drawn into it, the vapors will condense into a fume in which the particles are very small and will fall more slowly.

The radioactive debris from a nuclear detonation originates in a number of ways. The principal source is the production of fission products in the relative amounts given by the fission-product mass yield curves in Fig. 8-8, and the fission-product yields shown in Figs. 9-8 and 9-9. The initial fission-product mixture contains more than 200 isotopes of 35 elements. Most of the isotopes are radioactive, and most of them have very short half-lives, so that the diminution in radioactivity is very rapid immedi-

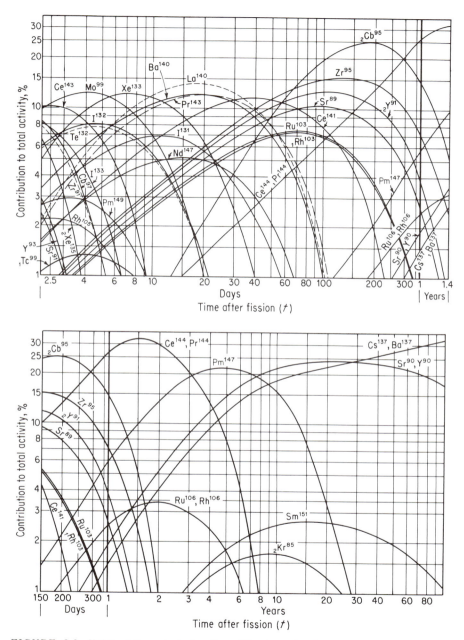

FIGURE 9-8 Yields of the principal radionuclides from the slow-neutron fission of ^{235}U. [From Hunter and Ballou (1951).]

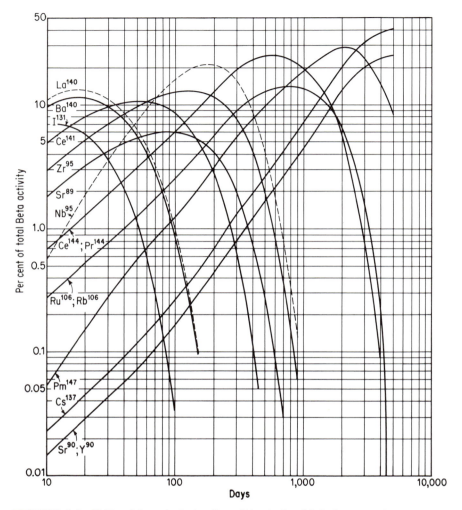

FIGURE 9-9 Yields of the principal radionuclides in the debris from megaton weapons. The differences from Fig. 9-8 occur because in such weapons fission occurs from both fast and thermal neutrons and in ^{238}U and ^{239}Pu as well as ^{235}U. [From Hallden *et al.* (1961).]

ately after fission. The yields of the principal fission products of concern are listed in Table 9-2.

The radioactivity A at any given time t after a nuclear explosion may be approximated if the radioactivity at unit time A_0 is known:

$$A = A_0 t^{-1.2} \tag{9-1}$$

TABLE 9-2

Approximate Yields of the Principal
Nuclides per Megaton of Fission

Nuclide	Half-life	MCi[a]
^{89}Sr	53 d	20.0[b]
^{90}Sr	28 y	0.1[b]
^{95}Zr	65 d	25.0[b]
^{103}Ru	40 d	18.5[b]
^{106}Ru	1 y	0.29[b]
^{131}I	8 d	125.0[c]
^{137}Cs	30 y	0.16[b]
^{131}Ce	1 y	39.0[b]
^{144}Ce	33 d	3.7[b]

[a]One MCi = 37 PBq.
[b]From Klement (1965).
[c]From Knapp (1963).

This equation provides valid estimates of the radioactivity for periods of time less than 6 months. When the radioactivity is decaying according to this law, the levels will diminish approximately 10-fold for every 7-fold increase in time since the explosion. This decay equation is only approximate, since the yields of the nuclides will vary from burst to burst, depending on the weapon design.

In addition to fission products, a number of induced radionuclides are produced by nuclear bombs, including substantial amounts of ^{239}Pu. Other radionuclides are produced by neutron interactions with nonradioactive elements of the bomb, with the atmosphere, and in some instances with seawater or soil. Libby (1958) estimated that the interactions of nuetrons with atmospheric nitrogen produce 3.2×10^{23} ^{14}C atoms per kiloton of yield. We will see that these reactions have been sufficient to produce marked increases in the natural background of radiocarbon and tritium.

When nuclear weapons are detonated close to the ground, a number of radionuclides may be produced by neutron reactions in soil. These are summarized in Table 9-3, which shows that ^{45}Ca and ^{55}Fe are the long-lived nuclides produced in significant amounts.

It is possible that under certain conditions fractionation of fission products will occur so that debris falling out in different places will be enriched or depleted in certain of the radionuclides. One reason for this is that among the fission products are noble gases such as xenon and krypton, which, though short-lived, exist sufficiently long so that the radionuclides to which they decay may be formed relatively late in the life of the fireball and, in

TABLE 9-3

Principal Radionuclides Induced in Soil[a]

Isotope	Half-life	Ci per megaton[b]
^{24}Na	15 h	2.8×10^{11}
^{32}P	14 d	1.92×10^{8}
^{42}K	12 h	3×10^{10}
^{45}Ca	152 d	4.7×10^{7}
^{56}Mo	2.6 h	3.4×10^{11}
^{55}Fe	2.9 y	1.7×10^{7}
^{59}Fe	46 d	2.2×10^{6}

[a]Adapted from Klement (1959).
[b]One Ci = 37 GBq.

some cases, after condensation of debris has already begun to take place. For example, there are several isotopes of krypton and xenon that decay to rubidium and cesium, respectively, and these in turn decay to strontium and barium. Thus, ^{90}Sr, which is the second decay product of ^{90}Kr, may exist in less than theoretical amounts in fallout close to the site of detonation, and there may be a corresponding enrichment in ^{90}Sr in the debris that falls out at a later time. This is explained by the fact that ^{90}Sr is not formed immediately, but is produced by the decay of ^{90}Kr, which has a half-life of 33 seconds.

Disequilibria may also occur because the different nuclides condense from the fireball at different temperatures. The more volatile elements, which condense last, are deposited on the surfaces of small particles, which settle more slowly.

From the data obtained during weapons tests, it is generally accepted that about 90% of the radioactivity produced in an explosion is located in the head of the mushroom-shaped cloud formed by the fireball as it cools. The remaining 10% is contained in the cloud stem.

The particle size of the debris is highly dependent on the type of explosion. Near-surface detonations will produce large glassy masses that are highly radioactive, as well as a log-normally distributed spectrum of smaller particles. Bursts that take place high in the atmosphere produce smokelike particles that remain suspended for considerable periods of time. It has been estimated on theoretical grounds that air-burst particles cannot grow to more than 0.3 μm by condensation, and that the larger particles are probably the results of coalescence of smaller ones or condensation of radioactive vapors on the surfaces of larger particles (Freiling and Kay, 1965).

RADIOACTIVE FALLOUT

The radioactivity of the debris produced in a nuclear explosion diminishes by a factor of about 20 from the first hour to the end of the first day. In addition, the radioactivity of the cloud becomes more diffuse during this period as a result of meteorological dispersion. Therefore, the potential hazards of radioactive fallout vary greatly, depending on whether the debris falls out within a few hours after the detonation or over a longer period of time. The basic factors that influence whether fallout will occur are the explosive yield of the device, the height of the explosion above the surface, meteorological conditions, and the design of the device. With respect to the latter, the ratio of fission to fusion is important because of its effect on the quantity of fission products produced per unit of explosive yield.

Our knowledge of fallout has been gathered from weapons tests and is limited by the fact that conditions were necessarily very different from those that would be faced in time of war. Tests within the United States were conducted in Nevada, where, for reasons of public safety, it was necessary to restrict the size of atmospheric tests to less than about 50 kilotons. Larger bombs were tested in the Marshall Islands, where much information about fallout was obtained, but the circumstances were somewhat atypical because of the enormous expanses of water over which fallout measurements were very difficult to collect. The Marshall Islands are thin coral reefs, and the effects of large land surface bursts could not be simulated. Many of the explosions took place on barges floating in lagoons. Other detonations were on reefs just barely above the surface of the water, with craters extending well below the waterline.

As a result of these limitations, many questions remain unanswered. Very little is known about the physical and chemical properties of the fallout that would be encountered under the wide range of conditions that might prevail during a nuclear war. Bombs of various sizes would be detonated at various heights, the terrain would vary from place to place, and it might be raining or snowing. The only thing about which one could be quite certain is that the conditions would be very unlike those experienced in either Nevada or the Marshall Islands.

Nevertheless, much has been learned. The sequence of events from the moment of detonation until the time of fallout is known, and a first approximation can be made of the probable extent of the area to be affected by fallout if conditions at the time of the explosion are known. However, any reliance on fallout predictions beyond the need to decide that fallout may occur in a given sector and that the public should seek the protection of evacuation or shelters would be extremely dangerous in view of the lack of precision of the fallout prediction methods even under the best conditions.

The only experience thus far has been from weapons tests for which there was far more exact knowledge of the explosive yield, the conditions of detonation, and the prevailing meteorology than would be available under war conditions, when it might be necessary to utilize meteorological data that are several hours old and that might have been collected at some location other than the target area. Even under the comparatively ideal conditions of the weapons tests, the fallout predictions were only an approximation of the actual fallout patterns.

Information as to exactly when, where, and how much fallout will occur would have to await actual observations. Figure 9-10 (Ferlic, 1983) compares the predicted and actual fallout from the test of a 43-kiloton device fired from a 500-ft (150-m) tower in Nevada in 1955. The prediction was made 8 hours before the test. The maximum cloud height was 42,000 ft (12,800 m). Although this is an extreme case in which the wind structure proved to be grossly different from what was predicted, Fig. 9-10 does serve to illustrate the difficulties involved in making fallout predictions. The problem would be even more complicated for the overlapping fallout patterns of several-megaton explosions. The complexity of the fallout pattern from a 5-megaton test on a coral reef in the Pacific is shown in Fig. 9-11. Note the irregularity of the isodose contours and the two "hot spots" 40 and 60 nautical miles (74 and 111 km) north of the hypocenter (Triffet, 1959).

In general, one may expect that relatively small detonations in the kiloton range will result in fallout patterns that more closely resemble the idealized cucumber-shaped patterns. However, fallout predictions following megaton bursts may be expected to be subject to great error because the greater cloud height subjects the debris to more wind shear.

Methods of Estimating the Dose from Fallout

There has been a need to develop both retrospective and prospective methods of estimating doses received from fallout. The retrospective estimates are used to estimate the doses received from fallout that occurred in the past during the era of weapons testing in the open atmosphere. Prospective dose estimates would be needed in the event of nuclear war, to serve as the basis for many civil defense and military decisions.

Retrospective dose reconstruction has been particularly important for epidemiological studies in the vicinity of testing grounds because of concern that increases in the rates of leukemia and thyroid cancer may have developed among the residents. This has necessitated use of transport models that are theoretically applicable for as long as four decades after the fallout occurred (Simon *et al.*, 1995; Till *et al.*, 1995). the conclusions of epidemio-

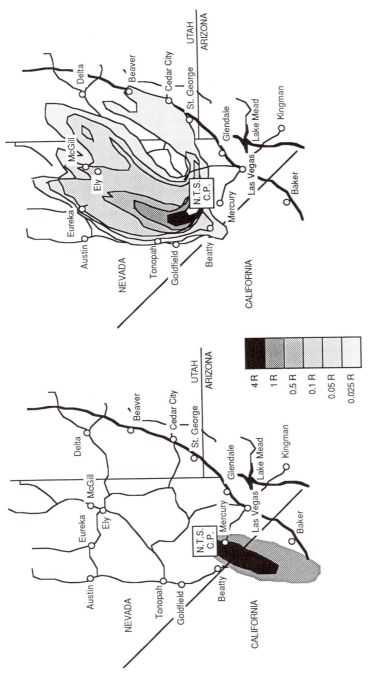

FIGURE 9-10 Predicted fallout (left) and actual infinite isodose contours (right) for TURK shot, Nevada, 1955. [From Ferlic (1983).]

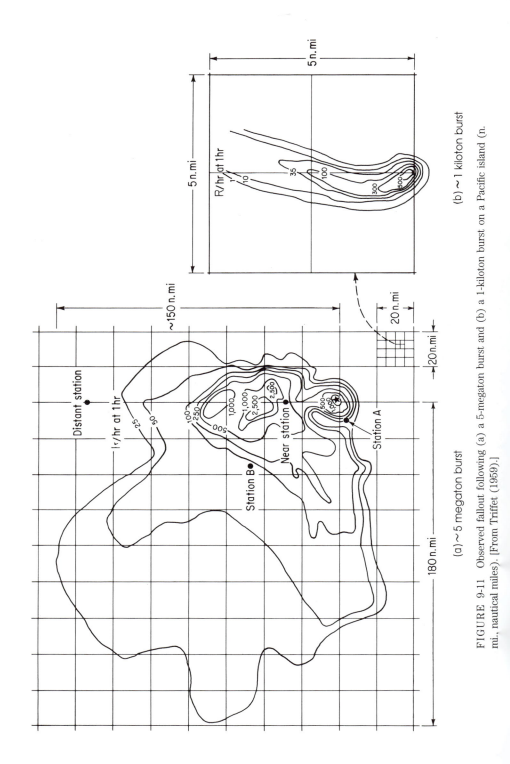

(a) ~ 5 megaton burst

(b) ~ 1 kiloton burst

FIGURE 9-11 Observed fallout following (a) a 5-megaton burst and (b) a 1-kiloton burst on a Pacific island (n. mi., nautical miles). [From Triffet (1959).]

logical studies conducted in the vicinity of the Nevada test site were discussed briefly in Chapter 2.

This section will discuss the methods used to estimate fallout patterns prospectively after explosion of a nuclear weapon. For the purpose of plotting data and standardizing the types of dose calculations that must be made to estimate dose rates and doses over various intervals of time, it is customary to forecast the fallout levels at a given point as though all the fallout occurred 1 h after the detonation. Thus, the fallout patterns in Figs. 9-10 and 9-11 are plotted in this way despite the fact that the contours cover such a large area that fallout at distant locations could not begin until 6 or 7 h after detonation.

The plots of radiation levels based on the activity at 1 h grossly overestimate the magnitudes of levels actually encountered, but they do serve the purpose of facilitating computation of dose rates and dose commitments during any given interval of time.

The dose calculations can be based on the approximation that a fallout of 1 MCi per square mile (14 GBq m^{-2}) will result in an exposure rate of about 4 R h^{-1} ($\sim$40 mGy h^{-1}) at 1 m above the ground (Glasstone and Dolan, 1977).

Given the dose rate at any time after a detonation and assuming that radioactive decay occurs at a rate proportional to $t^{-1.2}$, the various required calculations could be performed quickly with the older methods of nomograms, curves, and slide rules in the event that computer techniques are not available. Some useful examples of these calculations, including a specially designed slide rule, can be found in a U.S. government manual titled "The Effects of Nuclear Weapons" (Glasstone and Dolan, 1977).

SHORT-TERM RADIOLOGICAL EFFECTS OF NUCLEAR WAR

There have been many reports of the radiological effects of nuclear war, based on selected scenarios that describe the number of weapons, their size, heights of detonations, and meteorological conditions. In one 1959 congressional hearing, memorable because it was the first to receive the attention of the public, many experts presented their assessments of the consequences of an attack on the United States (U.S. Government Printing Office, 1959). A hypothetical attack was considered in which a total of 223 targets in this country were struck by nuclear and thermonuclear bombs, each of which had the explosive force of from 1 to 10 megatons of TNT. The total explosive yield of all bombs dropped on the United States was equivalent to 1453 megatons of TNT, and it was assumed that additional bombs delivered outside the continental United States had a total explosive yield equivalent to 2500 megatons of TNT. The fallout patterns in the continental United States 7 and 48 h after an attack are shown in Figs. 9-12 and 9-13.

This hypothetical attack, which may be presumed to be a realistic appraisal of the ability of a potential enemy to deliver nuclear and thermonuclear weapons within our borders in 1959, would have resulted in death to an estimated 42 million Americans and injury to an additional 17 million. Almost 12 million dwellings would have been so badly damaged that they could not have been salvaged, and an additional 8 million dwellings would have had to be evacuated for major repairs. Thus, more than 30% of the U.S. population would have been killed or injured, and more than 40% of the homes would have been destroyed or badly damaged. The stockpiles of nuclear weapons in the arsenals of the two superpowers increased greatly after the 1959 hearings and 10 years later (NAS/NRC, 1969) the assumed explosive yield of the attack was 12,000 megatons, an order of magnitude greater than the earlier assumption. Therefore, the fallout patterns in Figs. 9-12 and 9-13 must be taken as a lower estimate of the radiological problems that would be faced in the event of a massive nuclear attack. More recent scenarios [Office of Technology Assessment (OTA), 1979; Barnaby, 1982] have involved even greater explosive yields. Until the disintegration of the USSR, followed by major reductions in the numbers of strategic and tactical nuclear weapons, there were about 50,000 warheads in the stockpiles of the United States and the Soviet Union. Fortunately, the stockpiles are now being drastically reduced, but the process is slow and dangerously large stockpiles will continue to exist for many years.

The foregoing casualty estimates are based on an analysis of the radiation levels, the blast effects, and the effects of fires produced by the bombings. The extent to which additional casualties would be produced among survivors by infectious disease, starvation, deterioration of law and order, and exposure to the elements has been subsequently assessed by the Institute of Medicine (NAS/NRC, 1988). The vast numbers of injured and diseased people, the enormous destruction of production facilities and housing, interruptions of power communications, and disruption of the methods of distributing essential goods and services would be only a few of the factors that would interact in so complicated a manner as to place the true consequences of a massive nuclear attack beyond comprehension. In the hypothetical attack on the United States, about 96 million people lived outside the areas of likely blast damage. These people would be exposed to radioactive fallout at various times, from a few minutes to a few hours after the bombings, depending on their location. The principal immediate problem that would face this portion of the population would be the potentially lethal

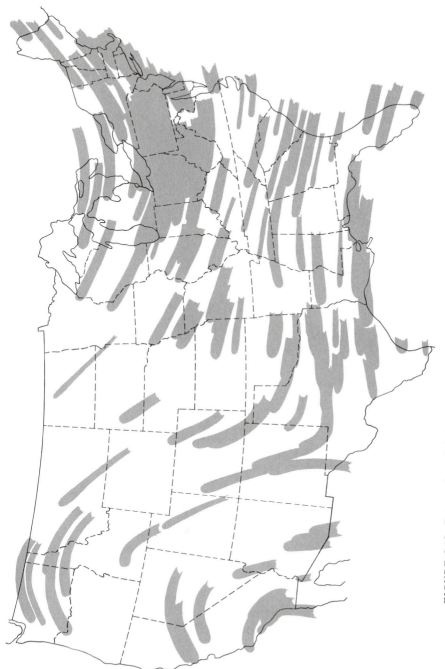

FIGURE 9-12   Pattern of near-lethal levels of radioactive fallout in the United States 7 h after a hypothetical attack with 223 bombs having a yield of 1453 megatons. [From Shafer (1959).]

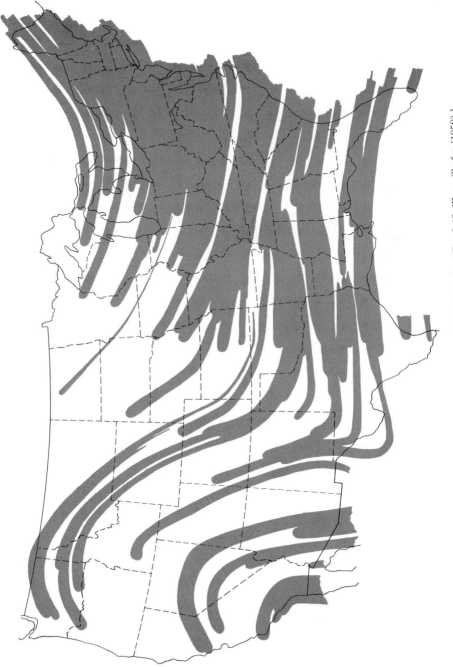

FIGURE 9-13   Fallout patterns 48 h after hypothetical attack of Fig. 9-12. [From Shafer (1959).]

levels of radioactive fallout. However, although they would be beyond the range of the effects of blast and thermal radiation, the problems of dealing with the radiation effects would nevertheless be complicated greatly by the enormous social stresses that would be produced throughout the nation.

The short-term biological effects to be expected from fallout would depend in a complex way on the rate at which the total dose was delivered and on whether there were any concomitant effects such as those from blast and thermal radiation. The effects of acute exposure were discussed in Chapter 2 and were summarized in Table 2-1.

In the immediate postattack period, there would be two principal dangers from contaminated food. Fresh foods that contain radioiodine could result in massive thyroid doses, and contamination of leafy vegetables or water by intensely radioactive fallout particles could result in high doses to the lining of the intestinal tract.

Of the several radioactive isotopes of iodine that are produced in nuclear explosions, the most significant is ^{131}I, with a half-life of 8.1 days. The principal pathway by which radioiodine reaches humans during the grazing season is by fresh dairy products, owing to the ability of dairy cattle to concentrate iodine deposited on forage into their milk and the short time that elapses between production and consumption of these products.

In wartime, the thyroid doses among the general population could be many tens of sieverts assuming that no countermeasures are taken, despite the fact that weathering and radioactive decay combine to reduce radioiodine deposited on foliage with a half-time of about 5 days. The requirement for restrictions and countermeasures might thus be limited to the first few weeks of the postattack period, so far as radioactive iodine is concerned. The subject is further discussed later in this chapter.

Individuals who survived the period of acute radiation danger would emerge into an environment that would be contaminated with radioactivity for the rest of their lives, but many of the fission products that have intermediate or long half-lives are relatively inactive biologically and would not appear as important contaminants of foods. These isotopes, which include ^{95}Zr, $^{106-109}$Ru, and $^{104-109}$Ce, would result in elevation of the external gamma-radiation background but would contaminate food only by being deposited on the surface of plants and would not be absorbed significantly from the gastrointestinal tract when eaten. However, these radionuclides could result in high doses to the gastrointestinal tract itself.

Considerably more significant would be the effects of ^{89}Sr and ^{90}Sr. ^{89}Sr has a half-life of only 59 days, but in fresh fallout the ratio of ^{89}Sr to ^{90}Sr may be as high as 200 to 1, diminishing to about 10 to 1 at the end of 1 year and to an insignificant fraction after 2 years. These strontium isotopes can enter human food supplies either by foliar deposition or by absorption from soil. The former mechanism is dominant when the rate of fallout is relatively high but becomes less important when the rate of fallout diminishes (see Chapter 5).

SOME PROBLEMS OF RECOVERY FROM NUCLEAR ATTACK

Although nuclear attacks may produce complete destruction by blast and fire over an area many miles in diameter, even a saturation attack would leave much of the country relatively intact except for the effects of fallout. Severe damage to wood-frame buildings would be limited to about 60 km from the explosion of a 1-megaton bomb, and fires would be produced up to about 30 km. Although the combination of blast, thermal, and radiation effects would make any discussion of countermeasures in the immediate target area of doubtful value, there is much that could be done to ameliorate the problems that would be faced by the large numbers of people outside the zone of blast and thermal damages but well within the region of potentially lethal fallout. During Operation Alert in 1959, a nationwide civil defense exercise in which a saturated attack on the United States was simulated, there were 5 million estimated fatalities in the state of New York (New York State, 1959). Of this number, approximately 1 million would have been killed by blast and heat, and 4 million fatalities would have been caused by fallout radiation. It is understandable that public officials should give attention to the need for fallout shelters.

The acute phase following attack would present the government with problems of unprecedented complexity. It would be necessary for officials to undertake complicated tasks at a time of enormous disruption of manpower, communications, and transportation and overwhelming destruction of the material resources of government.

The foregoing has been a superficial treatment that deals only with certain very obvious aspects of the radiological effects of nuclear weapons. Little or nothing has been said about the social, logistic, and medical implications of massive fallout. The problems that would face people at every level of the social structure would be so complicated as to defy meaningful analysis. The problems become more complex by additional orders of magnitude when we consider the concomitant effects of blast damage and fire superimposed on the radiological problems. The physical dimensions of the matter are illustrated by Fig. 9-14, which depicts the extent of the damage that would be produced if a 20-megaton bomb were exploded in mid-Manhattan. Stonier (1964) considered the consequences of such a catastrophe in vivid detail.

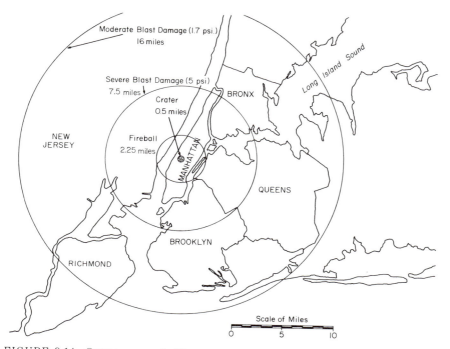

FIGURE 9-14   Consequences of a 20-megaton ground burst in mid-Manhattan. [From Stonier (1964).]

The radiation problems cannot be considered without understanding the social effects of blast damage and extensive conflagrations, but the effects of nuclear war that cannot be calculated are probably as great as those for which calculations have been attempted (OTA, 1979; NAS/NRC, 1964). A wide spectrum of individual factors, such as low water pressure, radiation damage to livestock (NAS/NRC, 1963), burned fields and forest subject to rapid erosion, acute food shortages, social disorganization, mass casualties, severe ecological disruption, disease, and pollution, would create a hundred new problems for every one foreseen. The disastrous immediate consequences of such a war would be followed by a prolonged period of social retrogression from which recovery would be slow and uncertain.

WORLDWIDE FALLOUT FROM NUCLEAR WEAPONS TESTS

The radioactive debris from a nuclear or thermonuclear explosion is apportioned among three fractions: large particles that deposit from the atmosphere within hours, smaller particles that remain in the troposphere from which they are removed on a time scale of days, and the fraction injected into the stratosphere, from which they are removed on a time scale of months. The first fraction, which includes the highly radioactive short-lived nuclides, is responsible for the patterns of lethal fallout that were discussed earlier in this chapter.

The importance of the fact that the radioactive debris is partitioned between the troposphere and the stratosphere was not appreciated until the advent of thermonuclear explosions equivalent to megatons of TNT. Prior to 1952, all the nuclear explosions were in the kiloton range, and after each series of weapons tests the atmospheric radioactivity diminished at a rate corresponding to the half-life of dust in the lower atmosphere, which was shown to be about 20 days (Stewart *et al.*, 1957). Essentially all the debris from kiloton bombs is deposited within about 2 months following injection into the atmosphere. The dust from such explosions is carried by the winds characteristic of the latitude in which injection takes place, and deposition ultimately takes the form of bands, as shown in Fig. 9-15.

The fallout from any one explosion is often spotty, because rainfall may coincide in some areas with passage of the cloud of radioactive dust. After one Nevada atmospheric test in April 1953, the highest fallout recorded in the United States was at Troy, New York, more than 2000 miles (3200 km) from the Nevada test site. Rain caused precipitation of the radioactive dust to the extent that it was estimated (Clark, 1954) that the cumulative gamma-radiation dose received by the inhabitants was about 100 mrad (1 mGy). This was a higher dose than was received anywhere else in the United States, except in the immediate vicinity of the test site, for the entire 1953 series of tests. As shown in Fig. 9-16, the tropospheric fallout from a thermonuclear explosion in the Pacific Ocean was distributed widely within 35 days, mainly in the Northern Hemisphere and to a lesser extent in the Southern Hemisphere. The debris injected into the stratosphere was deposited from pole to pole within 1 or 2 years.

Our knowledge of global fallout phenomena is largely the result of the collaborative efforts of many nations. In particular, the United Nations Scientific Committee on the Effects of Atomic Radiation (UNSCEAR), created by the General Assembly of the United Nations in 1955, has reported on the data accumulated by member countries. The periodic reports of UNSCEAR present excellent summaries of the subject of worldwide fallout from nuclear weapons tests (UNSCEAR, 1958–1994).

The cumulative yields of tests conducted in the atmosphere since 1945 are shown in Fig. 9-17. Of the many radionuclides produced in nuclear and thermonuclear explosions, the nine listed in Table 9-2 are the most

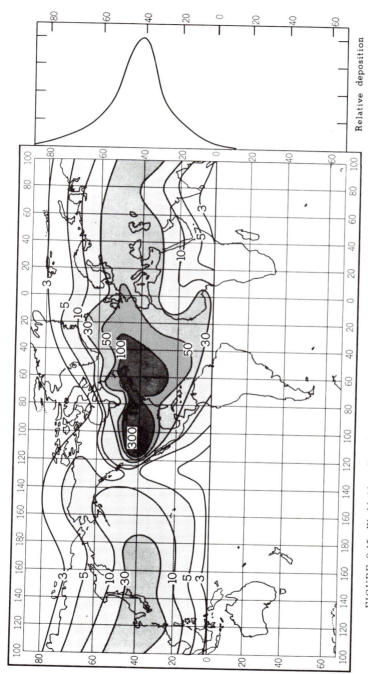

FIGURE 9-15 Worldwide radioactive fallout from nuclear weapons tests in Nevada in 1953. The explosions were in the kiloton range of yields, and debris was confined to the troposphere. The intensity of fallout is shown in relative units. [Reprinted with permission from Machta, L., List, R. J., and Hubert, L. F. (1956). World-wide travel of atomic debris. *Science* **124**, 474–477. Copyright 1956 American Association for the Advancement of Science.]

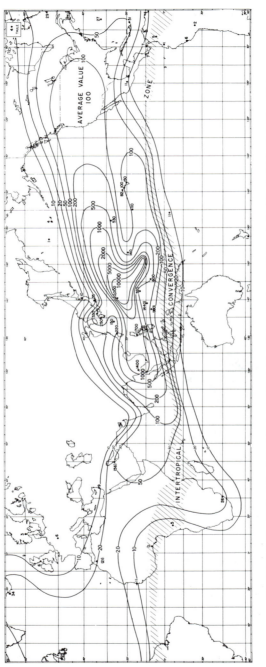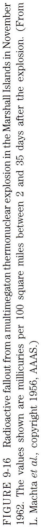

FIGURE 9-16 Radioactive fallout from a multimegaton thermonuclear explosion in the Marshall Islands in November 1962. The values shown are millicuries per 100 square miles between 2 and 35 days after the explosion. (From L. Machta *et al.*, copyright 1956, AAAS.)

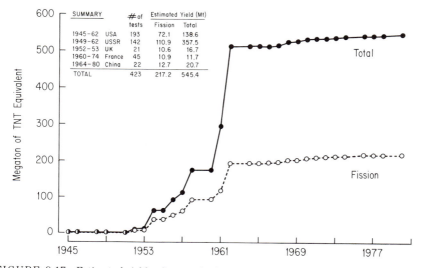

FIGURE 9-17   Estimated yields of atmospheric tests of nuclear weapons, 1945–1980. [Plotted from data provided in UNSCEAR (1982).]

important. Deposition from the stratosphere takes place so slowly that many radionuclides injected into that compartment decay before they settle to the lower atmosphere. However, since the average life of dust in the troposphere is about 1 month, many of the short-lived nuclides (such as ^{131}I) deposit on the surface of growing crops and thus enter the human food supply. However, the short-lived nuclides become relatively unimportant when deposited on soil because of the length of the growing time of most crops.

HOW RADIOACTIVE DEBRIS IS PARTITIONED AMONG THE THREE COMPONENTS OF FALLOUT

An understanding of the dose commitments that result from nuclear explosions requires knowledge of the manner in which the radioactive debris is transported by the atmosphere, the mechanisms by which it is deposited on the surface of the earth, and other pathways of human exposure. One of the first steps requires that estimates be made of how radioactive debris is apportioned among three fractions: (1) the fallout in the immediate vicinity of the explosion, (2) the debris injected into the troposphere, and (3) the debris injected into the stratosphere. Surprisingly, there is comparatively little reliable information on this subject. Devices in the megaton range that were exploded near the surface in the Marshall Islands are believed to have deposited as much as 80% of the fallout within about 100 km, but this is subject to great uncertainty and would depend on the size of the explosion, height above ground, type of terrain, and meteorological factors. The lack of adequate information can be explained by the operational difficulties that handicapped fallout studies in the Pacific, where the large U.S. test explosions have taken place.

The tropospheric component is estimated to account for no more than about 5% of the radioactive yield of megaton surface explosions (Machta and List, 1959) and less in the case of explosions well above the surface. The principal source of worldwide contamination by long-lived radionuclides from bombs in the megaton range of yield is the component of the debris that was injected into the stratosphere. Tropospheric fallout contains only a small fraction of the long-lived radionuclides from bombs in the megaton range but can be responsible for heavy exposure from the fallout of short-lived tropospheric debris, notably [131]I.

The first estimates of the amounts of radioactive dust injected into the stratosphere were made by subtracting the estimated amount of close-in fallout from estimates of the total amount of debris produced. The fraction of the debris that deposited close to the blast was approximated from field measurements that were often relatively crude. The amount of debris produced was estimated from knowledge of the type of device and its explosive yield. Local fallout was on average found to be 80% for land surface explosions, 20% for explosions on the surface of the water, and 10% for explosions in the air. The stratospheric inventory was not influenced significantly by the many detonations in the kiloton range since these did not penetrate appreciably into the stratosphere.

The stratospheric inventory was overestimated initially owing to underestimates of the amount of fallout in the vicinity of the explosions. Beginning in 1954, the United States began to collect stratospheric dust samples by balloons and high-flying aircraft (Feely, 1960). It was found that the inventory of ^{90}Sr was very much less than had been previously estimated, because the rate of fallout from the stratosphere was more rapid than had been believed. It was concluded that the half-time of fallout from the Soviet tests was less than 6 months and less than 12 months for the U.S. tests. These values would be equivalent to a mean residence time of about 8 months for Soviet debris injected at high latitudes and about 18 months for debris injected into the stratosphere near the equator. These residence times were very much less than the estimates of 5 to 7 years that had been made by earlier studies. The stratospheric inventory of ^{90}Sr from 1963 until 1980 is shown in Fig. 9-18 (Leifer et al., 1984).

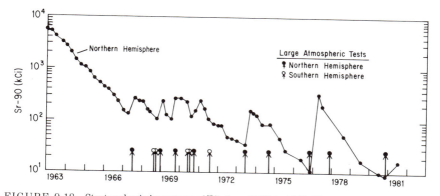

FIGURE 9-18   Stratospheric inventory of ^{90}Sr from 1963 to 1982. [From Leifer *et al.* (1984).]

BEHAVIOR OF INDIVIDUAL RADIONUCLIDES FROM FALLOUT

STRONTIUM-90

Investigations conducted since 1954 have resulted in the development of models that make it possible to predict the pattern of global contamination that will result from the injection of a given amount of ^{90}Sr into the atmosphere. It is also possible to estimate how deposition results in contamination of foods and, finally, given the level of contamination in food, it is possible to forecast human skeletal burdens.

Essentially all of the ^{90}Sr injected into the atmosphere during the period of weapons testing prior to the 1963 atmospheric test-ban agreement was deposited on the earth's surface by 1970. This is seen in Fig. 9-19, which shows the monthly deposition in New York City.

Worldwide deposition reached a peak of about 12.5 MCi (460 PBq) by late 1967, at which time Southern Hemisphere deposition was less than one-third that in the Northern Hemisphere. The amount of ^{90}Sr on the earth's surface has been diminishing at a rate of 2.5% per year because of radioactive decay, slightly offset by occasional tests by France and China (Volchok, 1970; UNSCEAR, 1982). In parts of the former USSR and some countries of central Europe, there was significant fallout of ^{137}Cs following the Chernobyl accident, which will be discussed in Chapter 12.

Strontium-90 deposition was far from uniform during the period of maximum fallout, as can be seen from Fig. 9-20. The band of relatively heavy fallout, 60 to 80 mCi km^{-2} (2.2 to 3 GBq km^{-2}), in the northern midlatitudes is conspicuous and is believed to be due to meteorological

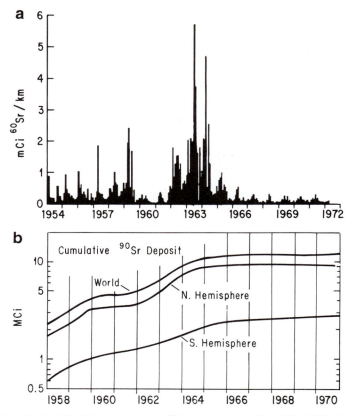

FIGURE 9-19   (a) Monthly deposition of ^{90}Sr in New York City, 1954–1971. (From U.S. Department of Energy, Environmental Measurements Laboratory.) (b) Cumulative ^{90}Sr deposition on the earth's surface, 1958–1970. Note that the cumulative deposition is plotted logarithmically. [From Volchok and Kleinman (1971).]

factors that result in increased stratospheric–tropospheric transfer during the spring months. This phenomenon was discussed in Chapter 4.

The dependence of fallout on the amount of precipitation is illustrated elegantly in Fig. 9-21, in which cumulative ^{90}Sr fallout is plotted against rainfall at five sites on the Olympic Peninsula in Washington State (Hardy and Alexander, 1962).

Strontium-90 in Food

As was discussed in Chapter 5, the ^{90}Sr content of plants can be due either to direct uptake from soil or to foliar deposition. The relative propor-

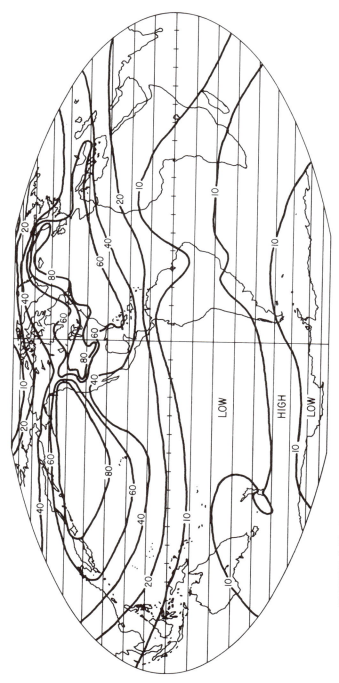

FIGURE 9-20 Isolines of cumulative ^{90}Sr deposits based on analyses of soils collected from 1965 to 1967 (in millicuries per square kilometer). [From UNSCEAR (1969).]

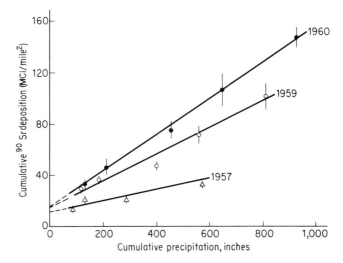

FIGURE 9-21   Dependence of fallout on precipitation on the Olympic Peninsula, Washington. [Reprinted from Hardy and Alexander (1962). Rainfall and deposition of strontium-90 in Callam County, Washington. *Science* **136,** 881–882. Copyright 1962 American Association for the Advancement of Science.]

tions of the two components depend on the rate of fallout in relation to the cumulative soil deposit, as well as the time of fallout relative to the growing season and the amount of plant surface area exposed to fallout (NCRP, 1991b).

The dietary sources of ^{90}Sr depend in part on the food consumption habits of the population, including the kinds of foods eaten and the way in which the food is processed or prepared. Because ^{90}Sr becomes part of the pool of Ca in the biosphere, the amount of this element contained in food is critical. Daily per capita consumption of calcium in the United States is about 1 g, but in other countries this may vary, as summarized in Tables 9-4 and 9-5.

For many years, beginning in 1960, the Environmental Measurements Laboratory of the DOE (formerly the Health and Safety Laboratory of the AEC) has reported on the ^{90}Sr content of typical diets in two cities, New York and San Francisco. There have been significant differences in the two cities, as shown in Fig. 9-22 (Klusek, 1984a), with the daily ^{90}Sr intake in New York being consistently higher than that in San Francisco. A comparison of the ^{90}Sr content of typical diets in the two cities during 1982 is given in Table 9-4. This difference has persisted for many years and is believed to be due mainly to lower annual rainfall in the regions that supply food to the San Francisco area (Harley, 1969).

TABLE 9-4
Strontium-90 in the Diet during 1982[a,b]

Diet category	kg y^{-1 c}	g Ca y^{-1 b}	Percent yearly intake Ca	New York City pCi ^{90}Sr kg^{-1}	New York City pCi ^{90}Sr y^{-1}	New York City Percent yearly intake ^{90}Sr	San Francisco pCi ^{90}Sr kg^{-1}	San Francisco pCi ^{90}Sr y^{-1}	San Francisco Percent yearly intake ^{90}Sr
Dairy products	200	216.0	58	3.2	641	32	1.0	200	21
Fresh vegetables	48	18.7		8.8	422		2.4	116	
Canned vegetables	22	4.4		5.4	119		2.9	64	
Root vegetables	10	3.8		3.4	34		3.8	38	
Potatoes	38	3.8		2.3	88		2.1	79	
Dry beans	3	2.1		15.9	48		7.9	54	
Fresh fruit	59	9.4	9	2.6	152	36	1.3	77	36
Canned fruit	11	0.6		1.1	12		0.8	9	
Fruit juice	28	2.5		1.7	48		1.4	40	
Bakery products	44	53.7	3	3.0	131	11	1.9	84	13
Flour	34	6.5		4.5	153		3.5	119	
Whole-grain products	11	10.3		6.2	69		2.9	32	
Macaroni	3	0.6		2.4	7		2.3	7	
Rice	3	1.1		0.6	2		0.8	2	
Meat	79	12.6	20	0.4	35	18	0.4	31	25
Poultry	20	6.0		0.3	6		0.3	5	
Eggs	15	8.7		0.6	10		0.6	8	
Fresh fish	8	7.6		0.2	1		0.1	1	
Shell fish	1	1.6		0.2	<1		0.7	1	
Yearly intake	370 g		10		1978 pCi	3		967 pCi	5
Intake per g of Ca					5.4 pCi g^{-1} Ca			2.6 pCi g^{-1} Ca	
Intake per day					5.4 pCi d^{-1}			2.6 pCi d^{-1}	

[a] From Klusek (1984a).
[b] One pCi = 37 mBq.
[c] From U.S. Department of Agriculture (1957).

TABLE 9-5

Dietary ^{90}Sr in the United States and Soviet Union, 1966[a,b]

	United States diet (New York City)				Soviet Union diet (country)[c]			
Diet category	Intake (kg y⁻¹)	^{90}Sr (pCi kg⁻¹)	^{90}Sr (pCi y⁻¹)	Percent of total intake	Intake (kg y⁻¹)	^{90}Sr (pCi kg⁻¹)	^{90}Sr (pCi y⁻¹)	Percent of total intake
Milk	200	13.4	2,970	42	110	14	1,540	12
Bread	89	17.4	1,580	22	220	33	7,250	55
Meat	114	2.8	320	4	60	10	600	5
Cereals	6	6.3	40	1	20	18	360	3
Fish	9	1.8	20	—	4	30	120	1
Potatoes and vegetables	219	8.3	1,800	25	220	12	2,640	20
Water	400	1.1	440	6	440	1.1	480	4
Yearly intake			7,170				12,990	

[a]From Harley (1969).

[b]One pCi = 37 mBq.

[c]Based on data given by Petukhova and Knizhnikov (1969).

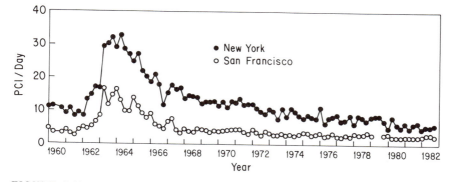

FIGURE 9-22   Results of quarterly analyses of ^{90}Sr in total diet of residents of New York City and San Francisco. [From Klusek (1984a).]

Table 9-4 shows that during 1982, dairy products accounted for 32% of the daily ^{90}Sr intake in New York compared to 21% in San Francisco. The ^{90}Sr content of cow's milk can be predicted accurately from knowledge of deposition rate (Fig. 9-23) and can be used to estimate the intake from all dietary sources (Bennett, 1972; Klusek, 1984a). This can be done from the following basic model:

$$M_n = p_1 F_n + p_2 F_{n-1} + p_3 \sum_{m=1}^{\infty} F_{n-m} e^{-m\mu} \tag{9-2}$$

where M_n is the average [90]Sr : Ca ratio in milk (pCi g^{-1} of Ca) in the year n, F_n is the deposition of ^{90}Sr in year n (mCi km^{-2}), p_1 includes the rate and deposition effects of the current year's deposition, p_2 is the lag factor

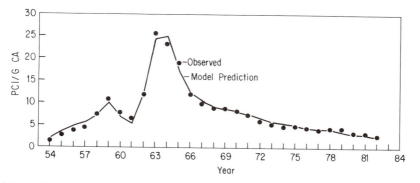

FIGURE 9-23   Observed and predicted average annual ^{90}Sr concentration in milk, 1954–1982, in New York City based on annual ^{90}Sr deposition. [From Klusek (1984a).]

for the previous year's deposition, and $p_3 e^{-m\mu}$ is the deposit factor with exponential removal owing to reduced uptake availability of the deposition for each of the preceding years.

Klusek (1984a) has evaluated the parameters of the model by fitting data from New York City through 1982. The p factors were found to be p_1 = 0.68, p_2 = 0.26, and p_3 = 0.16 and the value of μ was found to be 0.11. This latter value corresponds to a mean residence time of 9.1 years, after the initial deposition year, and includes both radioactive decay of 2.4% per year and reduced availability from soil of 8% per year. It is remarkable that this 1970 model and its parameter values have been found to be very stable with the addition of data for all succeeding years through 1982.

Milk is an important source of ^{90}Sr and a good index of exposure of growing children in many countries where dairy products are the main source of calcium. The longest series of ^{90}Sr measurements in milk is from the program of the Environmental Measurements Laboratory, which has been analyzing fresh milk from New York State since early 1954. The average annual ^{90}Sr content of liquid whole milk in New York City through 1982 is given in Fig. 9-23, from which it is seen that the milk reached a concentration of more than 25 pCi (0.93 Bq) of ^{90}Sr per gram of Ca in 1963. The tap water in New York City was also monitored during the period of weapons testing and reached a peak of 2.12 pCi L^{-1} (78 Bq m^{-3}) in 1963.

Figure 9-22 shows that the ^{90}Sr content of U.S. diets diminished for several years with a half-time of 3.5 to 4 years following the peak values of 1963–1964, when direct deposition on foliage was by far the major contributor. Dietary ^{90}Sr in recent years has originated almost entirely from soil. In subsequent years, the mean residence time of this nuclide in soil increases to about 9 years.

The models developed to predict the concentrations of ^{90}Sr in milk, total diet, and bone from rates of deposition on soil were among the first examples of environmental modeling that have evolved during the past 40 years, with major contributions by Knapp (1961), Kulp and Schulert (1962), Bennett (1972), and Klusek (1984a, 1984b, 1987).

The amount of ^{90}Sr contributed to the diet by grains is influenced by milling practices. In the United States, Canada, and the United Kingdom, the ^{90}Sr : Ca ratio in flour has been approximately one-third to one-half of that in the whole grain and one-quarter of that in bran (UNSCEAR, 1962). Milling reduces the ratio in rice grain to one-fifth to one-tenth of the value in whole rice. Harley (1969) attributed the relatively large contribution of bread to the daily intake of ^{90}Sr in the Soviet Union to the use of black bread, which includes deposition on the surfaces of wheat and rye (Table 9-5).

Strontium-90 in Human Bone

In parallel with the food sampling network, many countries have for many years reported on the ^{90}Sr content of human bone. It has been found (Klusek, 1984b) that a relatively simple model for the uptake of the nuclide can account for the amounts found in adult bone (vertebrae):

$$B_n = cD_n + g \sum_{m=0}^{\infty} d_{n-m} e^{-m\lambda} \qquad (9\text{-}3)$$

where B_n is the ^{90}Sr concentration in bone (pCi g^{-1} of Ca) in year n from m years of accumulation, D_n is the ^{90}Sr mean concentration in diet from midyear in year n-1 to midyear in year n (pCi g^{-1} of Ca), c is the fractional short-term retention of ^{90}Sr in bone, g is the fractional long-term retention, and λ is the rate constant for removal from the skeleton, including by radioactive decay.

The predicted and observed values for adult vertebral ^{90}Sr in New York City for 1954–1982 are given in Fig. 9-24. The model for children must be adjusted for changes in the various parameters with age. This has been accomplished by Klusek (1984b).

Because of the structural and physiological complexity of bone, calculation of the dose delivered is not a simple matter (Bjornerstedt and Engstrom, 1960; Spiers, 1966; UNSCEAR, 1969, 1982; NCRP, 1991b). Two tissues are of interest: the marrow and the bone cells lining the cavity within which the marrow is contained (the endosteum). The complexities arise from the fact that skeletal deposition is dependent on the variations of dietary intake with time, and with changes in the rate of calcium accretion and turnover with age.

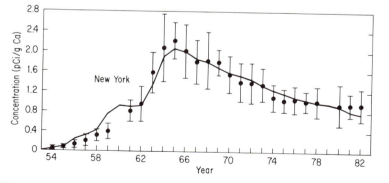

FIGURE 9-24   Predicted and observed ^{90}Sr in adult vertebrae in New York City, 1954–1982. [From Klusek (1984b).]

Taking all available information into consideration, UNSCEAR (1982) estimated that the average 50-year dose commitment from ingested ^{90}Sr by inhabitants of the North Temperate Zone has been 94 mrem (0.94 mSv) to the bone marrow and 210 mrem (2.1 mSv) to the endosteal cells. The dose commitments to residents of the South Temperate Zone were estimated to be about 25% of these values.

STRONTIUM-89

Strontium-89 behaves like ^{90}Sr in all respects, except that it is relatively more important as a foliar contaminant because of its short (50.5-day) half-life. It has been shown (Dunning, 1962) that ^{89}Sr may deliver as much as 3.4 times the dose from ^{90}Sr during the first year of a child's life. However, unless fresh ^{89}Sr is added during the second year, this isotope will have decayed almost completely, so that its contribution to the lifetime dose becomes unimportant.

The effective dose equivalent commitment from ^{89}Sr deposited in the Northern Hemisphere has been estimated to be about 0.2 mrem (2 μSv) (UNSCEAR, 1982).

CESIUM-137

Cesium-137 has a 30-year half-life compared to 28 years for ^{90}Sr, and it is produced somewhat more abundantly, at a rate 1.6 times greater than ^{90}Sr (in activity units). However, ^{90}Sr is more soluble than ^{137}Cs, so that in the years following the deposition of fallout on soil, the ratio of ^{137}Cs to ^{90}Sr would be expected to increase in soil near the surface. Because it is so tightly bound to soil (see Chapter 5), cesium is not readily incorporated metabolically into vegetation. During the first year following deposition, ^{137}Cs enters human food mainly by foliar absorption (UNSCEAR, 1982). During periods of active fallout, the ^{137}Cs content of land-grown food is generally dependent on the rate of fallout rather than cumulative deposition. This is not so for freshwater fish: in one lake system, the half-life of the ^{137}Cs concentration in fish was about 2.5 years (Gustafson, 1969).

Cesium-137 is a beta emitter, but can be most easily measured by the 0.662-keV gamma emission of its ^{137m}Ba daughter, which has a 2.5-minute half-life. This can be accomplished by gamma spectroscopy, but during most of the period of atmospheric weapons testing convenient methods had not yet been developed. However, when gamma spectroscopy became generally available in the early 1960s, its measurement became far simpler than the tedious technique of wet separations and beta counting for ^{90}Sr. The past deposition of ^{137}Cs could, however, be reconstructed from the constancy of its ratio to ^{90}Sr, the fallout of which had already been documented for more than a decade.

The distribution of ^{137}Cs in the stratosphere and the pattern of terrestrial deposition are similar to those for ^{90}Sr. Gustafson *et al.* (1970) used ^{90}Sr deposition at various stations in the United States to estimate the cumulative ^{137}Cs deposition and concluded that over most of the country the fallout by the end of 1965 ranged between 60 and 100 mCi km^{-2} (2.2 and 3.7 GBq km^{-2}) and that there was even less variation in the ^{137}Cs body burdens of the American population, owing to the fact that blending occurs because many important items of food originate from many different areas of the country.

It has proven useful to report the presence of ^{137}Cs in biological material as picocuries per gram of potassium, although it is known that the metabolisms of cesium and potassium are somewhat different from food to food and organ to organ (Yamagata and Yamagata, 1960). For purposes of dose calculation, ^{137}Cs is usually assumed to be distributed uniformly throughout the body.

The ^{137}Cs content of various foods from the Chicago area in 1968, by which time the maximum levels of fallout deposition existed, is given in Table 9-6, which shows that dairy products, grains, and meat products are the most important source of this nuclide. The data from 1961 through 1970 are summarized in Fig. 9-25, which shows that deposition reached a peak of about 130 mCi km^{-2} (4.8 GBq km^{-2}) during 1965 to 1966, and that this resulted in an annual dose of about 3 mrem (30 μSv) owing to internally deposited ^{137}Cs (Gustafson, 1969; Gustafson *et al.*, 1970). The dose commitment from ingesting ^{137}Cs in the Northern Hemisphere is estimated to have been about 17 mrem (0.17 mSv) on average (UNSCEAR, 1982). The dose is somewhat higher from external radiation. In 1969 the external radiation exposure due to fallout on an open field in Illinois was almost entirely from ^{137}Cs and was 13 mrem y^{-1} (0.13 mSv y^{-1}). UNSCEAR (1982) estimated that the 50-year external effective dose equivalent commitment from ^{137}Cs deposited in the North Temperate Zone was 40 mrem (0.4 mSv).

The biological half-life of Cs in humans is a function of sex and age. Representative biological half-lives are 2 days for the first 10% of the absorbed quantity and 110 days for the remaining 90% (ICRP, 1979–1988).

As was discussed in Chapter 5, the dose from ^{137}Cs can be increased markedly by unusual environmental factors. One example is food grown in potassium-deficient soils on the island of Jamaica, where the ^{137}Cs content of milk was 10 to 100 times that of milk from farms with normal soil (Broseus, 1970). In the subarctic, the body burdens of individuals who eat meat from moose or caribou have also been more than 10 times higher than the local population average (UNSCEAR, 1969). This is attributable to the consumption by the animals of lichens, which tend to concentrate many trace elements from the atmosphere.

TABLE 9-6

Cesium-137 in Chicago Diet in October 1968[a]

Diet category	Intake (kg yr^{-1})	Potassium			^{137}Cs[b]		
		g kg^{-1}	g yr^{-1}	Percent total intake	pCi kg^{-1}	pCi yr^{-1}	Percent total intake
Dairy products	200	1.4	280	21	18	3,600	29
Fresh vegetables	48	2.3	110		2	100	
Canned vegetables	22	1.3	29		9	200	
Root vegetables	10	2.9	29	29	7	70	8
Potatoes	38	4.5	171		17	650	
Dried beans	3	13.9	42		5	10	
Fresh fruit	59	1.9	112		4	240	
Canned fruit	11	1.2	13	13	18	200	10
Canned fruit juices	28	1.9	53		26	730	
Bakery products	44	1.2	53		21	920	
Flour	34	1.0	34		33	1,120	
Whole-grain products	11	3.5	38	10	30	330	20
Macaroni	3	1.8	5		19	60	
Rice	3	n.d.[c]	—		n.d.[c]	—	
Meat	79	3.3	261		26	2,060	
Poultry	20	2.7	54	25	15	300	20
Eggs	15	1.5	22		7	100	
Fresh fish	8	3.4[d]	27		194[d]	1,550	
Shellfish	1	n.d.[c]	—	2	n.d.[c]	—	13
Water	400	n.d.[c]	—		(0.05)[e]	20	
Yearly intake (rounded)			1330			12,300	

[a]From Harley (1969).
[b]One pCi = 37 mBq.
[c]Not determined.
[d]Based on 90% ocean, 10% freshwater fish.
[e]Number in parentheses represents estimated value.

CARBON-14

It has been seen in Chapter 6 that cosmic-ray reactions in the upper atmosphere result in the transmutation of atmospheric nitrogen to ^{14}C and that this nuclide has been in secular equilibrium in the biosphere in a concentration of 7.5 ± 2.7 pCi (280 ± 100 mBq) of ^{14}C per gram of total carbon. This equilibrium is believed to have been unchanged for at least 15,000 years prior to 1954,

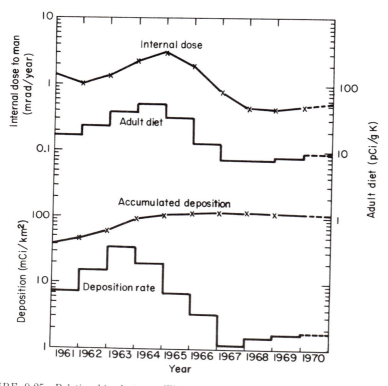

FIGURE 9-25 Relationships between ^{131}I content of fresh cow's milk and human thyroids in New York City during the Soviet tests of 1961. [Reprinted from Eisenbud *et al.* (1962). Iodine-131 dose from Soviet nuclear tests. *Science* **136**, 370–374. Copyright 1962 American Association for the Advancement of Science.]

when the advent of large thermonuclear explosions produced additional ^{14}C that perturbed the natural equilibrium. Carbon-14 in the atmosphere is believed to exist as $^{14}CO_2$. However, introduction into the atmosphere of CO_2 from combustion of fossil fuels tends to dilute the ^{14}C of the atmosphere.

Because the half-life of ^{14}C is 5730 years, the collective dose when it is introduced into the environment will be delivered for many generations. Since the genetic effects may be cumulative even in such small doses, there has been some concern about the effects of increasing the inventory of this nuclide in the biosphere (Totter *et al.*, 1958; Leipunsky, 1957; Zelle, 1960).

The best estimates of the amount of ^{14}C produced in weapons tests are those made by actual measurement of the ^{14}C activity of stratospheric air using high-flying aircraft and balloons. This quantity is estimated to have been 9.6×10^6 Ci (360 PBq) (NCRP, 1985b).

Carbon-14 distributes itself quickly among the major environmental compartments—the stratosphere, troposphere, biosphere, and surface ocean waters. Transfer among these compartments takes place with time constants on the order of a few years, but transfer to the deep ocean water proceeds more slowly. Following an injection of ^{14}C into the stratosphere, the biosphere will reach equilibrium after a relatively few years, and the ^{14}C will decrease slowly thereafter at a rate determined by transfer to deep ocean water and possibly humus (NCRP, 1985b).

By the end of 1967, the concentration of ^{14}C in the troposphere had increased about 60% above natural levels in the Northern Hemisphere and a little less in the Southern Hemisphere (Nydal, 1968). Because of the short time constants involved in transfer from the atmosphere to biosphere, the ^{14}C content of human tissues and foods increased rapidly following the heavy testing schedules of 1961 and 1962. The dose equivalent from ^{14}C in fallout is estimated to have reached a peak of 0.96 mrem y^{-1} (9.6 $\mu Sv\ y^{-1}$) in 1965 and had diminished to 0.37 mrem y^{-1} (3.7 $\mu Sv\ y^{-1}$) by 1984 (NCRP, 1985b). Because of its long half-life, bomb-produced ^{14}C will persist in the environment for many thousands of years.

PLUTONIUM

Plutonium injected into the atmosphere by nuclear explosions originates from both unfissioned plutonium and plutonium produced by neutron irradiation of ^{238}U. Several isotopes are produced, including ^{238}Pu (87.7 y), ^{239}Pu (24,130 y), ^{240}Pu (6570 y), and ^{241}Pu (14 y). The last nuclide decays by beta emission to ^{241}Am (430 y). Plutonium-239 and ^{240}Pu are the most abundant of these nuclides and because the pair cannot be distinguished by alpha spectrometry, the two nuclides are usually reported together as "plutonium." The ratio $^{240}Pu:^{239}Pu$ in global fallout has been reported to be 0.163 (Krey, 1976). It has been estimated that about 320 kCi (12 PBq) of Pu has been distributed globally, mainly from explosions of megaton-range weapons that took place before 1963. In addition, about 9 kCi (0.33 PBq) of ^{238}Pu has been injected to the atmosphere by nuclear tests, somewhat less than the 17 kCi (0.63 PBq) from the abortive reentry of a satellite in 1964 (see Chapter 12). The ratios of the various plutonium nuclides can be used to determine the sources of debris. The isotopic composition of plutonium produced by nuclear reactors is different from that in nuclear weapons, and the isotopic ratios can be used to distinguish weapons fallout from fallout from other sources (Krey, 1976; Burns et al., 1994). The isotopic ratio of the Pu in the bombs themselves is also different from the ratio found in fallout, which is influenced by the production of Pu from ^{238}U during the explosion.

During the atmospheric weapons testing era, the ratio Pu : ^{90}Sr in fallout was so constant, at a decay corrected value of 0.017, that estimates of global Pu deposition could be made from the known deposition of ^{90}Sr (Hardy, 1974b). Since deposition of ^{90}Sr in the North Temperate latitudes was about 80 mCi km^{-2} (3 GBq km^{-2}), one can in this way estimate the deposition of Pu to be about 1.4 mCi km^{-2} (50 MBq km^{-2}).

Using the constant ratio Pu : ^{90}Sr, and from historical ^{90}Sr measurements, Bennett (1978) estimated that the concentration of Pu in the air of New York City reached a peak of 1.7 fCi m^{-3} (63 μBq m^{-3}) in 1963.

The decay of the beta-emitting ^{241}Pu, which is formed in association with ^{239}Pu, produces ^{241}Am, an alpha emitter with a half-life of 432 years. The presence of this nuclide must be considered when calculating the dose commitment from ^{239}Pu. The lungs, osteocytes, and liver are subject to the highest dose commitment from inhaled Pu and Am. UNSCEAR (1982) estimated that the committed doses from inhalation of fallout plutonium in the North Temperate Zone are 80 mrem (0.8 mSv) to lung, 40 mrem (0.4 mSv) to osteocytes, and 100 mrem (1 mSv) to liver. In northern Finland, where the ^{137}Cs body burdens of the Lapps were very much larger than in southern Finland, there is no increase in the plutonium burdens (Mussalo-Rauhamaa et al., 1984).

RADIOIODINE

Although several species of radioiodine are produced in fission, only one, ^{131}I (half-life 8.2 days), is of major significance so far as worldwide fallout is concerned. This nuclide is produced copiously, at a rate of about 125 MCi (4.6 × 10^3 PBq) per megaton of fission. It was seen earlier that the fission yields of all nuclear weapons tests conducted in the atmosphere through 1980 totaled about 217 megatons. Thus, in excess of 27 GCi (10^6 PBq) of ^{131}I were released to the atmosphere, most of which was produced by explosions in the megaton range. The bulk of the ^{131}I was injected into the stratosphere, where it decayed substantially before transferring to the troposphere and depositing on the earth's surface. However, the ^{131}I produced by explosions having yields less than 100 kilotons remained in the troposphere, together with a small fraction of the debris from megaton explosions.

Van Middlesworth (1954) demonstrated the presence of ^{131}I in cattle thyroids throughout the United States during the period of nuclear weapons testing in Nevada in 1953. This was the first indication that this radionuclide might be present in the human food chain during periods of nuclear weapons testing.

As was true for ^{137}Cs, the paucity of information about ^{131}I during the 1950s, when extensive open-air testing of nuclear weapons was taking place within the continental United States, resulted from the fact that radioiodine was relatively difficult to measure with the instrumentation then available. Measurement is a simple procedure by means of gamma spectrometry, but this technique was not generally available prior to the first test moratorium of 1958, and no systematic measurements of radioiodine in food or human thyroids were made during the period from 1950 to 1958, when copious amounts of radioiodine were being released to the atmosphere from tests in Nevada. Radioiodine contamination of milk during that period may have reached higher levels in many parts of the country than at any time since.

On the basis of a study of the published values of ^{131}I in cow's milk, Lewis (1959) concluded that over the 5-year period prior to 1958 the average accumulated dose to the thyroids of children in the United States was 0.2 to 0.4 rad (2 to 4 mGy).

The thyroid doses were much higher among children who lived close to the test site. Following the weapons tests in Nevada in the early 1950s, it was known that relatively heavy fallout occurred at distances of a few hundred miles from the test explosions (Eisenbud and Harley, 1953, 1956; Eisenbud et al., 1962). More than a decade later, when it was realized that radioiodine was a significant source of exposure at considerable distances from the tests, efforts were made to reconstruct the thyroid doses in several communities where relatively heavy fallout was known to have occurred (Knapp, 1963; Pendleton et al., 1963, 1964). Pendleton and associates used air samples collected in a number of communities to estimate the ^{131}I content of cow's milk and the thyroid dose to infants. It was concluded that the exposures ranged from about 1 rad to as high as 84 rad (10 to 840 mGy) in the community of St. George, Utah, following a test on May 19, 1953. During the period 1965–1970, the U.S. Public Health Service conducted a study of residents who were children during the era of open-air weapons testing (Weiss et al., 1971). The children were presumed to have consumed fresh milk, which would have been the major source of radioiodine. The highest exposures were assumed to have been in Washington County, Utah, and Lincoln County, Nevada. Children who lived in Graham County, Arizona, served as controls because their exposure was considered to have been negligible. The results of the survey showed no elevation in neoplasms or other thyroid diseases among the more heavily exposed children. However, the studies were deficient in at least two respects: the data were gathered only 14 years after exposure, and there were great uncertainties in the dosimetry.

The study conducted by Weiss et al. (1971) was repeated by a team from the University of Utah in 1985–1986 (Stevens et al., 1992). The dose assessments were much more thorough than in the earlier study and showed that the mean absorbed doses to the thyroids were 5.0 and 17 rem (50 and 170 mSv) in the highly exposed groups and 1.3 rem (13 mSv) in those who were more lightly exposed. There were 10 subjects whose thyroid received more than 100 rem (1 Sv). The prevalence of thyroid carcinomas was of particular interest, and a slightly elevated number were found to be associated with the higher exposures, but the number of cases (eight) was too small to be meaningful. There were a total of 19 neoplasm's of all types, and these were found to be significantly associated with dose.

Residents of the Marshall Islands were much more heavily exposed to ^{131}I, and thyroid disease, including cancer, has been reported among them. This will be discussed in Chapter 12.

By the time the Soviet Union resumed weapons testing in 1961, gamma spectrometry had progressed to the point where *in vivo* measurements of ^{131}I in human thyroids were feasible (Laurer and Eisenbud, 1963). Measurements could also be made of human thyroid tissue that became available at autopsy, as well as cow's milk and other foods. Measurements made in New York City during weapons tests in the Marshall Islands and Siberia showed that during two 3-month periods in 1961 and 1962 the thyroid doses in children averaged 50 and 140 mrem, respectively (0.5 and 1.4 mSv) (Eisenbud *et al.*, 1963a).

MISCELLANEOUS OTHER NUCLIDES

Tritium (Moghissi and Carter, 1973; NCRP, 1979) is produced copiously in thermonuclear explosions, and its incorporation into precipitation caused the tritium content of surface waters in the North Temperate Zone to reach several nanocuries per liter in the mid-1960s (Wyerman *et al.*, 1970). Moghissi and Lieberman (1970) measured the tritium content of body water in children by means of tritium analyses of urine samples during the 2-year period 1967–1968, during which time the concentration in urine decreased from 1.5 to 0.2 nCi L^{-1} (55 to 7.4 kBq m^{-3}). The dose commitment from ^{3}H produced by nuclear weapons tests is estimated to have been about 5 mrem (50 μSv) on average in the North Temperate Zone (UNSCEAR, 1982).

Nuclides of the transitional elements are produced as fission or activation products, particularly in thermonuclear explosions. These include ^{55}Fe, ^{59}Fe, and ^{65}Zn. They are relatively short-lived nuclides that are easily detectable in the environment during and immediately following nuclear weapons testing, but they have not contributed significantly to human exposure. A possible exception is ^{55}Fe, which was first detected in fallout by Palmer and Beasley (1965) and was shown by Wrenn and Cohen (1967) to have delivered doses of about 1 mrem (10 μSv) to the erythrocytes of inhabitants of the New York City area. The dose to ferritin aggregates in hemoglobin was calculated to be 235 mrem (2.35 mSv) by including the energy of the short-ranged Auger electrons from ^{55}Fe (Wrenn, 1968).

LUNG DOSE DUE TO INHALATION OF DUST FROM WEAPONS TESTS

Radioactive dust particles that are too large to be respirable fall out quickly, but at distances beyond 200 miles (320 km) from near-surface explosions the mass median diameter of fallout particles has been found to be 2 μm, which is close to the optimum particle size for lung retention (Eisenbud and Harley, 1953). Shleien *et al.* (1965) found that 88% of the total activity of airborne dust was contained in particles less than 1.75 μm long. The radioactivity of the dust depends on the age of the debris and distance from the test site. In one city, a few hundred miles from the Nevada test site, the average beta activity of dust collected during a 24-hour period was 24 nCi m^{-3} (0.9 kBq m^{-3}) following a test in 1952 (Eisenbud and Harley, 1953). The extent to which radioactive dust was then present in the atmosphere is illustrated in Figs. 9-5 and 9-6.

There have been very few estimates of the dose to the human lung during the period of active testing. Wrenn *et al.* (1964) undertook a series of human lung measurements that showed that the principal dose was received from the ^{95}Zr–^{95}Nb pair. It was estimated that the lung dose was about 3 mrem (30 μSv) per day from lung burdens that ranged from 210 to 450 pCi (7.8 to 16.7 Bq). These lung burdens were associated with average air concentrations over a 6-month period of 2 to 4 pCi m^{-3} (0.07 to 0.15 Bq m^{-3}), several orders of magnitude lower than the peak concentrations reported. Fisenne *et al.* (1980) also estimated the lung dose to New York City residents from Pu in fallout.

EXTERNAL RADIATION

Many investigators have calculated the effect of delayed fallout on ambient gamma radiation using data on the quantities of the various radionuclides known to be present in fallout. These calculations give estimates of the gamma dose, assuming a deposition of a given mixture of radionuclides distributed uniformly on an infinite smooth plane. Among the difficulties that present themselves in calculations of this kind are the following. The dose integral is sensitive to the time of fallout, particularly if the fallout occurs within 2 or 3 days after the explosion. The time of arrival of the fallout is not always known. The surface of the earth is not an infinitely smooth plane but is highly irregular, particularly in inhabited areas. The natural and man-made irregularities tend to absorb gamma radiation and may modify the distribution of fallout to the extent that it is not correct to assume that the deposit is uniform. The effect of weathering is difficult to evaluate. Rain will in most cases lessen the dose received by washing the fallout from places of habitation, but in some cases the rain may tend to concentrate deposits and thereby increase the dose. Buildings provide an uncertain degree of shielding that depends on structural factors as well as the fraction of the day spent indoors.

Nevertheless, Beck (1966, 1972) showed that estimates of the gamma-radiation exposure rates based on rainfall deposition data are reasonably consistent with measurements made *in situ* using gamma-spectrometric techniques.

Gustafson *et al.* (1970) measured the external radiation dose from fallout near Chicago for many years beginning in 1960. He reported that the maximum open-field exposure was 56 mrad (0.56 mGy) per year in 1963, which diminished to 13 mrad (0.13 mGy) per year by 1969. Gustafson estimated that a more realistic estimate of the exposure would take into consideration the time a person spent indoors, in which case he estimated the annual exposure rates to have been 22 and 5.5 mrad (0.22 and 0.055 mGy), respectively. However, these exposures, as noted earlier, are due to dust from explosions many thousands of miles away. Localities closer to the test sites undoubtedly received very much higher doses, but very few measurements were made at the time. The dose commitment to inhabitants of the North Temperate Zone from external radiation due to tests conducted up to 1980 has been estimated to be about 100 mrem (1 mSv) (UNSCEAR, 1982).

The measurements of Beck included estimates of the contribution from the various nuclides present in fallout (Beck, 1980). In a total exposure of 76 mrad (0.76 mGy) during the period 1960–1964, the ^{95}Zr–^{95}Nb pair contributed about 37 mrad (0.37 mGy) and ^{137}Cs accounted for 13 mrad (0.13 mGy). The long life of ^{137}Cs results in a larger lifetime dose commitment from this nuclide. The NCRP (1987a) estimated that the annual dose received by the U.S. population from this source had declined to less than 0.01 mSv (1 mrem) by 1987.

CHAPTER 10

Various Other Sources of Exposure

The preceding chapters have been concerned with radioactivity in the environment from natural sources, the processing of uranium, the production of nuclear power, and the use and testing of nuclear weapons. There are a number of other ways in which exposure to radioactive substances can occur. These include their use in radioluminescent paints, as coloring agents in glass, ceramics, and dental prostheses, the use of various radionuclides in smoke detectors, power generators, welding electrodes, and other commercial and industrial devices, and as tracers and radiation sources in the biomedical sciences. Finally, transportation of radioactive substances must be considered.

In some publications (NCRP, 1977c; Moghissi *et al.*, 1978b; NCRP, 1987d), products such as drinking water, construction materials, fuels, tobacco, and phosphate fertilizers that contain natural radioactivity are included among consumer products in which radionuclides are used. However, the radioactivity in these materials is of natural origin and is inadvertently present in products that find their way to the marketplace. These examples were discussed in Chapter 6. This chapter will discuss products such as ceramic glazes and smoke detectors in which radioactive substances are deliberately used to achieve a desired objective that may or may not be related directly to the radioactive properties of the substance.

316

Before proceeding to a discussion of these and other uses of radioactive materials, the special case of radium will be reviewed because of the unique place this element occupies in the history of radioactivity and its relationship to luminous-dial painting, which was the first use of a radioactive substance in manufactured products.

THE EARLY HISTORY OF RADIUM

Radium was discovered by Marie and Pierre Curie in December 1898, but they did not produce the first 100 mg until nearly 4 years later. Radium in this context means ^{226}Ra. It is reported that about 1300 Ci (48 TBq) of radium was sold in the United States between 1912 and 1961, when the use of artificially produced radionuclides replaced radium to such an extent that new production had ceased [Stevens, 1963; U.S. Public Health Service (PHS), 1971]. Approximately 60 known deaths resulted from the use of radium in luminizing compounds and a historical epidemiological record of the surviving dial painters has been maintained (Thomas, 1992; Rowland, 1994). Luminous paints that contained radium were used not only in time-pieces, but also for compasses, religious articles, aircraft instruments, and even in luminous rings attached to the knobs of chamber pot covers that could thus be found more easily in the dark of night (Holm, 1978)!

Before the dangers of radium came to be appreciated, an unknown fraction of the total production was also used in quack medicine, as was discussed in Chapter 1. This practice resulted in additional cases of radium-induced bone cancer. The history of radium usage in industry and medicine is still of interest, particularly as it relates to remediation of processing sites and sites contaminated during its use (see, e.g., Steinhäusler et al., 1993). Remediation of radium sites is discussed in Chapter 15.

Of the 1300 g known to have been produced in the United States, only about 480 g were accounted for in 1971. The balance, about 820 g, may have been used largely for luminous compounds, static eliminators, or other ionization sources that were discarded as refuse after they had served their purpose (U.S. PHS, 1971). It was thought likely that some of the radium unaccounted for is unknowingly stored in safe deposit boxes or attics by survivors of early radiologists (Villforth, 1964; Villforth et al., 1969). This concern has been verified by a recent incident in which several members of a family were exposed to radium purchased unknowingly at a yard sale (Dunlap, 1995b). No overt injuries resulted from this incident, but the potential for serious injury from these uncontrolled sources remains.

From the time of its discovery and continuing up to the present, radium has been a source of many problems. In addition to the tragic misuses of radium in lumious-dial painting and quack medicine in which the radium was handled in loose form without precaution, problems developed because hydrogen and oxygen produced by radiolysis from the water of crystallization of the radium compound caused pressure to develop within the capsules into which radium was sealed for use in radiography. This resulted in occasional ruptures of sources, necessitating expensive building decontamination. In one case, rupture of a single 50-mg radium sulfate capsule used for instrument calibration caused such extensive contamination that abandonment of a building was necessary (Gallaghar and Saenger, 1957).

Another problem associated with radium sources, as well as sources using artificially produced gamma-ray emitters such as ^{137}Cs, is that they are small in size and are frequently lost in hospitals or in industry where they are used for industrial radiography. Every large city has had instances in which frantic searches of hospital incinerators, plumbing systems, municipal sewage treatment plants, or land fills were undertaken to find lost radium. Villforth *et al.*, (1969) analyzed 415 mishaps involving radium up to 1969.

EXPOSURE FROM RADIOLUMINESCENT PAINTS

It has been estimated that in an earlier period, 3 million radium-bearing timepieces were being sold annually in the United States (U.S. PHS, 1971), and that an additional 30,000 miscellaneous devices containing radium were sold annually. These included static eliminators, fire alarms, electron tubes, gauges, and educational products.

The radium content of men's wristwatches (Seelentag and Schmier, 1963) ranged from about 0.01 to 0.36 μCi (0.37 to 13 kBq). In Europe, a number of studies (Robinson, 1968) estimated that the per capita gonadal dose from the use of radium in timepieces ranged from 0.5 to 3.3 mrem y^{-1} (5 to 33 μSv y^{-1}). Others estimated the per capita gonadal dose from wristwatches containing 0.1 μCi (3.7 kBq) of ^{226}Ra to be 6 mrem y^{-1} (0.06 mSv y^{-1}) and the annual absorbed dose to the skin under the face of the watch to have been as much as 165 rem (1.65 mSv). Although clocks contain more radium than wristwatches, people are normally located at a greater distance from them. The annual whole-body dose equivalent from clocks containing about 0.5 μCi (19 kBq) of ^{226}Ra has been estimated to average about 7 to 9 mrem (0.07 to 0.09 mSv) among 10 million people who lived or worked in proximity to them (NCRP, 1977c). Although no wristwatches containing ^{226}Ra have been sold in the United States since about 1968, it was estimated in 1977 that 10 million such watches were still in use (NCRP, 1977c).

Radium has been largely replaced by less expensive, artificially produced radionuclides such as tritium and ^{147}Pm. Both radionuclides are regulated by the NRC, which limits the ^{147}Pm content per watch to 100 μCi (37 MBq) and tritium to 20 mCi (740 MBq). The minimum thickness of a watch case is sufficient to absorb the beta particles from both tritium and ^{147}Pm, but radiation exposure may result from inhalation or skin absorption of tritium that was volatilized (Fitzsimmons *et al.*, 1972). The wearers of ^{147}Pm radioluminescent wristwatches receive an average annual dose equivalent of about 0.4 μSv (0.04 mrem) and those wearing tritium radioluminescent watches receive about 1 μSv (0.1 mrem) (McDowell-Boyer and O'Donnell, 1978; NCRP, 1987d).

URANIUM AND THORIUM IN CERAMICS AND GLASS

Uranium has been a popular coloring agent in ceramic glazes in past years, producing colors that ranged from orange-red to lemon yellow. The dose rate to the hands in contact with these glazers ranges from 0.5 to 20 mrad h^{-1} (5–200 μGy h^{-1}) for glazes produced prior to 1944, but there is evidence that the dose from ceramics produced since 1944 may be less, by a factor of about five (Menczer, 1965). In 1961 the AEC ruled that a license would not be required for the use of uranium in ceramic glazes if the uranium content of the glaze was less than 20% for flatware and 10% for glassware (Simpson and Shuman, 1978).

Uranium has also been used at concentrations of a few hundred parts per million to enhance the appearance of porcelain teeth (Thompson, 1978). The principal dose is from irradiation of tissues within the oral cavity, estimated to be 1.4 rem y^{-1} (14 mSv y^{-1}) to a small volume of epithelial tissue. The presence of ^{40}K in the porcelain raises the dose to about 1.6 rem y^{-1} (16 mSv y^{-1}). A limit of 37 ppm has been proposed as a limit, which would result in a dose of about 1.0 rem y^{-1} (10 mSv y^{-1}). A program to replace the uraniferous teeth has not been considered to be warranted, because the risk is small. About 20 million bridges and dentures would require replacement, at a cost of many billions of dollars.

Thorium is incorporated into optical glass used in specialized instruments. Although there is no risk to the user from enclosed lenses, thorium has been found to be present in eye-pieces. Though the extent of this practice is not known (NCRP, 1987d), the dose to the 50-μm depth of the eye of a 20-hour-per-week user of a 0.05% thorium eyepiece is estimated to be 30 mSv y^{-1} (3 rem y^{-1}).

A more general source of exposure is the presence of thorium and uranium in ophthalmic lenses (Goldman and Yaniv, 1978). These elements enter the glassmaking process because of their natural association with rare earth elements used to tint the glass. In 1975, the Optical Manufacturers Association (OMA) adopted a voluntary standard to limit the radioactivity of ophthalmic glass to less than 0.45 alpha particles cm^{-2} min^{-1} from the lens surface (OMA, 1975). Goldman and Yaniv analyzed the dosimetric implications of the radioactivity in eyeglasses and concluded that the critical tissue was the germinal layer of the cornea estimated to be at a tissue depth of about 50 μm. On the assumption that an individual might wear spectacles 16 hours a day for a total of about 6000 hours per year, the maximum annual dose to the germinal layer of the cornea has been estimated to be about 500 mrem (5 mSv). Taking into account the quality factor for the alpha-radiation component, the NCRP (1987d) estimated an annual dose equivalent of 40 mSv y^{-1} (4 rem y^{-1}).

Thorium was also used in glass tableware and decorative glass to achieve a "true yellow" color in the glass (Sheets *et al.*, 1995). The investigators measured external radiation from 125 pieces of the yellow glass. Although concentrations were not reported, the external beta and gamma levels were comparable to uranium-glazed pieces.

DEPLETED URANIUM PROJECTILES

Depleted uranium (DU), a by-product from uranium enrichment, is used extensively for munitions by the military because of its high density and excellent armor-piercing qualities. Manufacturing and testing of the DU munitions have resulted in residual contamination at facilities, which in turn has required cleanup in some cases. Tens of metric tons of DU projectiles have been test-fired at numerous locations (Bernhardt, 1995). The basic clean-up criterion has been 1.3 Bq g^{-1} (35 pCi g^{-1}) of total uranium, but site-specific criteria have been developed in some cases (Bernhardt, 1992). Awareness of the environmental contamination problems from DU munitions has led to changed practices for testing (Oxenberg and Davis, 1993). Catch-boxes 30 feet (9 m) in depth filled with construction-grade sand have been built to improve recovery of DU projectiles and to limit the spread of contamination at the Aberdeen Proving Ground in Maryland. The catchboxes have arrested an estimated 85% of the projectiles since use began in 1989. Recovery efforts yielded only 5–57% of the DU projectiles prior to use of the catchboxes.

Concern has been raised by veterans that DU from projectiles used in Desert Storm may be partly responsible for the "Gulf War Syndrome," and the Department of Veterans Affairs has initiated research projects to address the question (Department of Veteran's Affairs, 1995). Results are not yet available.

THORIUM IN GAS MANTLES

Thorium has long been used for producing luminescence in gas mantles, of which an estimated 25 million are used annually, mostly by campers. Between 250 and 400 mg of Th is used in each mantle (O'Donnell, 1978). The dose to campers and other consumers is estimated to be less than 0.1 mrem y^{-1} ($1 \mu Sv$ y^{-1}), which is sufficiently low to permit these devices to be distributed to the public without regulation. More recently the NCRP estimated that 50 million people are exposed at an average annual level of 2 μSy (0.2 mrem). However, the doses received may vary, depending on the age of the mantle and its duty cycle. The thorium daughter products are separated in the manufacturing process and are thus not initially incorporated into the mantle, but build-in over a period of a few days. Radium-224, ^{212}Bi, and ^{212}Pb are volatilized in the process of burning but will again build-in if the mantle is not used for a few days (Mohammadi and Mehdizadeh, 1983; Luetzelschwab and Googins, 1984). Mantles made without thorium are now available and appear to be displacing thorium mantles at the retail level in the United States. This is a problem for science teaching laboratories, which have long relied on gas mantles as inexpensive, unregulated sources for radioactivity demonstrations.

THORIUM IN WELDING RODS

Tungsten electrodes containing about 2% thorium are used in the welding industry, technical schools, and home shops when high quality is required. Addition of thorium provides easier starting and greater weld stability. Approximately 1500 kg of thorium dioxide is used annually to manufacture about 5 million electrodes (NCRP, 1987d).

Thorium is released to the air during the welding process and during the periodic grinding of electrode tips that is required to maintain their shape. Breslin and Harris (1952) measured airborne concentrations of thorium during operations with thoriated electrodes and concluded that neither the external does rate nor the airborne thorium posed an unacceptable hazard to welders. McDowell-Boyer (1979) used the Breslin and Harris data, as well as interviews with welders and other data, to estimate doses to welders. Their estimates for a 50-year dose commitment ranged from 24 to 880 μSv (2.4–88 mrem). The NCRP estimates that there are approximately

300,000 welders using thoriated electrodes and that the average annual dose to these welders is 160 μSv (16 mrem).

Crim and Bradley (1995) have made new measurements of airborne thorium during welding and grinding operations. They did not express results in terms of dose but did conclude that airborne concentrations during normal operations were below the current derived air concentration (DAC) for thorium of 0.04 Bq m^{-3} (1×10^{-12} μCi mL^{-1}). They were able to cause airborne concentrations above the DAC only when electrodes were deliberately ground for times well in excess of normal.

USE OF SPECIFIC RADIONUCLIDES ("ISOTOPES") IN RESEARCH AND INDUSTRY

Radionuclides for use in medicine, research, and industry have become an important by-product of reactor operation and particle accelerators during the past 50 years. In general, these nuclides, which have come to be known loosely as "isotopes," may be employed for a wide variety of uses depending on their chemical and physical properties. The total quantities of the various radionuclides produced each year, the fraction of each that decays, the fraction that remains in use, and the amounts that go to waste-disposal sites are difficult to estimate.

A few radioactive isotopes that exist in nature among the heavy elements have been used as tracers, and early in this century it was shown that radium D (^{210}Pb) could be used as an indicator of lead in studies of the solubilities of lead compounds (von Hevesey, 1966). Subsequently, with the aid of particle accelerators such as the cyclotron and the Van de Graff generator, it became possible to produce many radionuclides that do not exist naturally in significant quantities. These included ^{14}C, ^{32}P, ^{24}Na, and ^{131}I, elements that are important in many biological processes. The potential importance of these artificially produced radionuclides as tracers was recognized in the 1930s, but the amounts then available were small and the instrumentation by which applications could be developed existed in very few laboratories.

The situation changed dramatically with the discovery of fission. In addition to the copious amounts of many new radionuclides that became available as by-products of the fission process, the high neutron fluxes available in reactors made it possible to produce new isotopic species by neutron irradiation. Transuranic nuclides such as ^{241}Am and ^{238}Pu could also be produced for application in smoke detectors and thermoelectric power generation. The Manhattan Engineering District (the wartime nuclear bomb project) was quick to recognize the potential value of these isotopes and took steps early in 1946 to make them available to science and industry. When the AEC succeeded the Manhattan Engineering District in 1947, the utilization of isotopes for peaceful purposes became a major program objective.

The importance of radionuclides as tracers and radiation sources is too well known to require elaboration. "Isotopes" are one of the standard tools of research workers. Equipment such as liquid scintillation and gamma counters with automatic sample changers to count samples efficiently is used in university, government, and industrial laboratories everywhere.

The sensitivity of the instrumentation used in tracer studies permits some isotopes to be used in such small quantities (designated "exempt") that their use does not require licensing by the NRC or the states. The NCR regulations pertaining to the use of the by-product materials are contained in Part 30 of Title 10 of the Code of Federal Regulations.

There are also many clinical procedures in which radionuclides are used for either diagnostic or therapeutic reasons. An estimated 100 million medical procedures using radioactive materials are performed in the United States each year (NCRP, 1989b). Most of these procedures use a small amount of radionuclide in tests performed outside the body. The numbers of diagnostic procedures performed in which radionuclides are introduced to the body were 6.1 million in 1950 and 7.4 million in 1990 (Metter et al., 1993). Radioactivity from patients and their excreta results in low doses to medical care workers and the public through direct contact (NCRP, 1996c). A fraction of radiopharmaceuticals administered to patients is excreted in urine and feces. In the United States, radionuclides in patient excreta are specifically excluded from the sewage disposal requirements that apply to radionuclides used in laboratory research or other applications (10 CFR 20.2003). As a result, radioactivity is readily measurable in the sewage from some medical institutions, particularly large institutions with active nuclear medicine therapy practice (Prichard et al., 1981; Piersanti et al., 1993). The isotopes can also be found in the solid sludge resulting from the treatment of sewage (Prichard et al., 1981; Kennedy et al., 1992; Ainsworth et al., 1994). Most isotopes used in nuclear medicine have a relatively short half-life. Rapid decay and dilution prevent any significant dose to the public from this practice.

AMERICIUM-241 IN SMOKE DETECTORS

The first smoke detectors, introduced in 1951, contained about 20 μCi (0.74 MBq) of radium, but the use of that material has been discontinued in favor of [241]Am (Johnson, 1978), which has made it possible to market inexpensive, efficient devices that are easily installed and can now be found in many homes and most hotel rooms, offices, and factories. Each unit contains a small ionization chamber in which the air between two electrodes is ionized by the presence of a small foil containing the radioactive source. Ionization produced by the radioactive source allows an electric current to flow across the gap between the electrodes. The current flow is reduced by smoke particles, thereby making it possible to actuate an electronic circuit and to sound an alarm. The americium is in the form of the dioxide, which is fabricated between a backing of silver and a front cover of gold or gold/palladium alloy that is sealed by hot forging (Wrenn and Cohen, 1979). Each detector now contains less than 1 μCi (37 kBq) of ^{241}Am (Stovall, 1989). However, smoke detectors of earlier design (circa 1970) contained an average of 79 μCi (3 MBq), which was gradually reduced by improved design. Twelve million units per year were being sold by the mid-1980s, containing a total of about 8.5 Ci (315 GBq) of ^{241}Am (Harris, 1986).

Americium-241 has a half-life longer than 400 years and emits both alpha and photon radiation. Because of the widespread use of these devices, a great deal of attention has been given to the exposure and risk to occupants of households in which they are used. The highest dose is from external radiation and is reported by Wrenn and Cohen (1979) to be 0.014 mrem y^{-1} (0.14 μSv y^{-1}) to an individual sleeping 6 ft from the detector for 8 hours per day. The NCRP (1987d) estimated that the average annual individual dose equivalent to a smoke detector user is 0.08 μSv (8 μrem).

Exposure of the general population to internal emitters disseminated by incineration of these devices, or their disposal in landfills, has been shown to be less than that from external radiation. A problem did occur when a company in Tonawanda, New York, that manufactured smoke detector sources disposed of ^{241}Am waste into sewers over a 6-year period (Kennedy *et al.*, 1992). Instead of dispersing and diluting, the americium concentrated somewhat in the sludge generated at the sewage treatment plant. The sludge had concentrations up to 100 pCi g^{-1} (3.7 kBq kg^{-1}), and concentrations up to 750 pCi g^{-1} (28 kBq kg^{-1}) were found in the dry ash resulting from sludge incineration. The ash was buried in a solid-waste landfill, and levels up to 160 pCi g^{-1} (6.0 kBq kg^{-1}) were found in landfill samples. Whole-body counting of sewage treatment plant and landfill workers detected no activity above background. The NRC now requires that any radionuclides disposed to the sewer be either soluble or contained in a readily dispersible biological material (see Chapter 11).

The exposures from smoke detectors involve minuscule risk in exchange for the great benefits that have been obtained. Historically nearly 9000 persons per year lost their lives from fires that occur in buildings and 90% of these fatalities occurred in private homes. Wrenn and Cohen (1979) concluded from an analysis of reports by others that 45% of these lives would have been saved had smoke detectors been available. In 1994 there were over 2 million building fires in the United States but fire deaths fell remarkably to 4275. The National Fire Protection Association attributes the decline to better safety standards and smoke detectors (*USA Today,* 1975).

RADIONUCLIDES AS SOURCES OF POWER

The decay heat of radionuclides has been used effectively to supply the electrical energy required for scientific instruments and communications equipment aboard space satellites. The power units are known by the acronym RTG for radioisotope thermal generator or SNAP for satellite nuclear auxiliary power. Twenty-three U.S. space missions have used RTGs through 1995 and the *Cassini* mission planned for 1997 will carry an RTG. The first of these devices was developed in the late 1950s, with the advent of miniaturized transistorized electronic circuits with low power requirements. They proved to be useful for supplying power to instruments and telemetering equipment in weather stations located in remote areas (Morse, 1963). In 1961, the first such device (SNAP 3) was launched into space to provide 2.6 We of power to a satellite used in the TRANSIT navigational system. The design life for the unit was 5 years, but the power plant was still operational 15 years later when the satellite was decommissioned (Bennett *et al.,* 1984). All of the SNAP devices used in outer space, except for SNAP 10A, have depended on the decay heat of ^{238}Pu, an alpha emitter with a half-life of 86.4 years. SNAP 10A utilized a nuclear reactor with a core containing 4.5 kg of ^{235}U. Thermal efficiencies of the systems have been gradually improved from 5.1% for the first TRANSIT satellite to 6.6% for the unit used for the *Galileo* mission. *Galileo* was launched by the space shuttle on October 12, 1989, for a 2-year trip to Jupiter, where it investigated that planet and its satellites. The *Galileo* mssion was equipped with two radioisotope thermal generators, with a total initial power of 570 We compared to 2.6 We for the generator used on the first TRANSIT mission.

When the SNAP program began in the late 1950s, the massive programs of testing nuclear weapons in the atmosphere that were then underway influenced the SNAP safety policies that were developed. The devices were then designed so that in the event of accidents during the launching procedures, or during the postlaunch period when the vehicle remained subject to the control of the range safety officer, or until the moment the vehicle achieved a satisfactory orbit, the integrity of the isotopic power unit would be maintained to prevent dispersion of the fuel. However, a conflicting requirement was also placed on the designers of the early SNAP units: if the generator reentered the earth's atmosphere after attaining orbit, it was required that the heat of reentry would result in total dispersion of the fuel in a fine particulate form.

The requirement that the fuel remain contained within its metallic canister was based on the desire to avoid massive exposure to ^{238}Pu in the event of an accident in the immediate environs of the launch site, and the need to avoid the partial disintegration of the canister that might occur if low-grade burnup occurred prior to entering orbit. Since the early flights contained on the order of 10^4 Ci (370 TBq) of ^{238}Pu, a total burnup in the upper atmosphere, with subsequent dispersion and gradual fallout, would have been a small increase above the fallout of ^{239}Pu, and other radionuclides that had been occurring for many years as a result of nuclear weapons testing. This philosophy was subsequently changed to require that the capsule retain its integrity under all circumstances, or at least until the ^{238}Pu has decayed. However, while the original policy was in effect, as will be discussed in Chapter 12, a SNAP device containing 17,000 Ci (630 TBq) of ^{238}Pu was volatilized by the heat of reentry about 150,000 ft above the Indian Ocean and resulted in a 50-year dose commitment to the respiratory lymph nodes of the world's population of about 36 mrem (0.36 mSv) (Shleien *et al.*, 1970). In contrast, a NASA spacecraft (*Nimbus-B1*) that was launched from California in 1968 was destroyed by the range safety officer at an altitude of 30 km because of a guidance error, and the radioisotope generators fell into the Santa Barbara Channel from which the generators were recovered intact (Bennett, 1981).

In 1970, the spacecraft *Apollo 13* was damaged on its way to the moon and the lunar module with its attached SNAP generator reentered the atmosphere over the South Pacific, where the generator landed in the 6-km-deep Tonga Trench. The package was never recovered, but subsequent surveys could find no evidence of radioactivity in either the atmosphere or the Pacific Ocean, leading to the conclusion that the containment remained intact.

Both the United States and the former Soviet Union have developed reactors for use in outer space, but only one such device has been used by the United States, the SNAP 10A, which was launched in 1965. It began operation after being placed in a 4000-year orbit, but shut down as a result of an electrical malfunction after only 43 days. The reactor remains in orbit. Its initial radioactive inventory contained 2×10^5 Ci (7.4 PBq) of fission products, but it has decayed by now to less than 100 Ci (3.7 TBq) and will diminish to less than 1 Ci (37 GBq) after 100 years (Bennett, 1981).

The former Soviet Union has launched more than 30 nuclear reactor-powered satellites. A satellite reactor in the *Cosmos 954* mission reentered the atmosphere and disintegrated over Canada in 1978. This incident is described in Chapter 12. Shortly after the first *Cosmos* accident, the United Nations established a Working Group on the Use of Nuclear Power Sources in Outer Space (United Nations, 1981).

A second satellite, *Cosmos 1402,* contained an enriched uranium reactor and was reported to have reentered the atmosphere on February 7, 1983. The reactor was designed to disintegrate on reentry but an undetermined fraction of the original core is believed to have entered the South Atlantic Ocean about 1600 km east of Brazil. Balloon samples of dust recovered from the stratosphere over New Mexico one year later showed the presence of enriched uranium that was attributed to the Russian reactor (Leifer *et al.,* 1984).

A third satellite (*Cosmos 1900*) containing a reactor lost communication on April 18, 1988. A safety system boosted the reactor into a 720-km orbit on September 30, 1988, and the remainder of the satellite reentered the atmosphere in October (Finck, 1990). On November 18, 1996, a failed Mars-bound Russian space probe that had a total of 200 g of Pu in four RTGs on board crashed into a 5000-m-deep portion of the Pacific Ocean 1000 km west of Chile. U.S. officials were of the opinion that the RTGs would still be intact.

The *Galileo* mission used isotope generators that contain 2.8×10^5 Ci (10.4 PBq) for the two heat sources. This is a huge quantity of ^{238}Pu that required maximum assurance that containment of the fuel would be achieved in the event of any of the many types of accidents that were hypothesized. These hypothetical accidents included explosions that might have led to plutonium release by the shock wave of the explosion, the thermal effects of the fireball, or projection of the generators at high velocity and their subsequent impact on hard surfaces in the immediate vicinity of the launch pad. Even in a successful launch there remained the possibility of an explosion of the rocket fuel prior to entry of the spacecraft into orbit or, having attained orbit, reentry into the earth's atmosphere as the result of a malfunction (General Electric Company, 1985). The successful October 12, 1989, launch of *Galileo* in now history, of course, and the plutonium power generator remains in space. Comparable quantities of ^{238}Pu have been used in more recently launced satellites. Emergency preparations similar to those for *Galileo* were made for the October 6, 1990, launch of *Ulysses,* a satellite designed to explore the polar regions of the sun with the aid of two 280-We generators.

The methods of probabilistic risk assessment referred to in Chapter 8, which have played an important role in the safety program of the nuclear power industry, were originally pioneered by the space program to provide estimates of the probabilities and consequences of mission failures at each step from prelaunch operations to postreentry recovery. The need for sophisticated methods of estimating the probabilities of mission failures was related only in part to the fact that some of the spacecraft would be equipped with radioisotope generators. These analytical techniques are required by the inherent characteristics of the space program: huge quantities of potentially explosive propellants are used and many of the missions are manned. The risks to populations in many parts of the world must be considered. The methods used to assure safety in the U.S. space program proved to be remarkably effective during the first two decades of its existence, but the explosion of the shuttle *Challenger* in January 1986 served to demonstrate the difficulty of identifying and quantifying all possible contingencies. As a result of the increased use of radioactivity in space and associated accidents, a small body of literature is developing concerning emergency preparedness (Finck, 1990; Lyman, 1993) and response to returning radioactive debris (Muck, 1993).

TRANSPORTATION OF RADIOACTIVE SUBSTANCES

There are currently about 500 billion shipments of all kinds made in the United States each year, of which about 100 million (0.02%) involve hazardous materials that are either flammable, explosive, toxic, or radioactive. The radioactive shipments number about 2 million per year and comprise about 2.8 million packages containing about 9 million curies (330 PBq), not including spent fuel (NCRP, 1977c; Wolff, 1984; McClure and Cashwell, 1992).

The rules and regulations that govern transportation of radioactive materials are complex owing to the varied types of shipments to which they are applicable. In the United States, the Department of Transportation (DOT) and the NRC have joint responsibility for regulating such shipments, most of which involve small quantities of radionuclides intended for use in research laboratories or medical facilities.

The first regulations for transportation of radioactive materials were drafted by the U.S. Postal Service after finding that photographic film was being fogged because of commingling with radium shipments (Pelletieri and Wells, 1985). The problems associated with shipments of radioactive materials became more complex as a result of developments during world War II, following which federal agencies overseeing hazardous shipments were concerned mainly with shipments of uranium and thorium ores, process residues, and radionuclides being shipped to and from research and medical institutions. There was as yet no nuclear power industry and significant shipments of potentially fissionable materials were preempted by the newly formed AEC. In 1961 the IAEA issued proposed regulations based on the U.S. rules, in which it was recommended that member states adopt those rules for the sake of uniformity. The IAEA-suggested regulations were modified in 1973, and the U.S. DOT revised its rules to be in essential conformity with IAEA. The DOT regulations are contained in Title 49 of the Code of Federal Regulations and, since they are subject to change, should be consulted directly by persons in need of detailed current requirements. Additional regulations that augment those of the DOT are published by the NRC (10 CFR 71) and the U.S. Postal Service (39 CFR 124).

International regulations and guides are promulgated by IAEA, the International Civil Aviation Organization, the International Maritime Organization, and the International Air Transport Association.

Under DOT regulations, a radioactive material is defined as one that has a specific activity in excess of 0.002 μCi g^{-1} (74 kBq kg^{-1}). This is a very conservative definition, as can be seen from the fact that it is only about 2.5 times that of elemental potassium.

The exact regulatory requirements depend on the kinds and amounts of nuclides involved and the type of vehicle being utilized. "Strong, tight packages" are used to transport material with extremely low levels of activity such as uranium ore, consumer goods such as smoke detectors, and much low-level waste. Small quantities of radioactive materials with a higher concentration of activity than allowed for the strong, tight packages can be shipped in Type A packages, which are fiberboard or wooden boxes, or steel drums, designed to withstand moderately rough handling conditions. The requirements for Type A shipments vary depending on whether or not the material is in a form that is capable of being disseminated in the environment. The maximum quantities that can be shipped in Type A containers vary from 0.002 Ci (74 MBq) of certain isotopes of californium, protoactinium, and plutonium (in any form) to 1000 Ci (37 TBq) of many of the less hazardous radionuclides in "special form." Examples of the latter include ^{3}H (tritium), ^{14}C, ^{35}S, and ^{147}Pm. All but 3.5% of radioactive shipments are made in strong, tight packages or Type A packages.

Type B shipments are required to be shipped in containers that have been tested to withstand more rigorous stresses than Type A containers. These include a 9-m drop to a hard surface, a fall of more than 1 m landing on the upraised tip of a 15-cm-diameter steel bar, and 30 minutes of exposure to a temperature of 1475°F for fissionable materials. A water immersion test requires that the package be submerged for not less than 8 h under at least 1 m of water. Type B packages account for only 3.5% of all radioactive shipments, but 90% of the radioactivity.

The applicable regulations (10 CFR 71) are highly detailed regarding methods of packaging and labeling and should be consulted directly for guidance on current requirements.

The shipments can include a wide variety of radioactive materials, from a few microcuries (about 100 kBq) of a relatively innocuous nuclide to high-level wastes or spent fuel. The types of shipments during a typical year are summarized in Table 10-1. The medical uses of radionuclides account for 62% of the shipments, followed by industrial uses, mainly in the form of radioactive sources. Table 10-1 does not include many classified shipments by the defense agencies.

The DOE sponsors a data base known as the Radioactive Materials Incident Report (RMIR), which is maintained by Sandia National Laboratories (McClure and Cashwell, 1992).

There were 329 accidents involving 3506 packages in radioactive shipments between 1971 and 1991. In addition there were 253 handling accidents and 924 lesser incidents. Strong, tight packages failed 65 times and Type A packages failed 55 times during that period, but the consequences were minor, being limited by the small amounts of radioactivity in materials that can be shipped in this way. Only 4.8% of the strong, tight packages and 2% of the Type A packages involved in accidents actually leaked contents. There were no failures of Type B packages during this period, although 85 such packages were involved in accidents. Two Type B packages were damaged but radioactivity was not released.

Millions of such shipments have been made and a solid actuarial base has been established that permits one to conclude that the precautions that have been taken are adequate to protect the public from either direct radiation from the packages or escape of the material to the environment. Shipments of high-level wastes, spent fuel, or large amounts of transuranic wastes occur with less frequency, but involve huge quantities of radioactivity and therefore require correspondingly greater degrees of protection. DOE

TABLE 10-1

Annual Unclassified Shipments of Radioactive Material[a]

End use	No. of shipments	Percent of shipments	Percent of radioactivity
Medical	1,730,000	62.2	34.3
Other	519,000	18.6	0.2
Industrial	213,000	7.6	63.1
Power	114,000	4.1	0.7
Waste	181,000	6.5	1.5
Research and teaching	17,100	0.6	0.1
Unspecified	7,550	0.3	0.1

[a]From Wolff (1984).

FIGURE 10-1   A spent fuel cask is struck by a locomotive traveling at 130 km h^{-1} as part of a testing program. Although the cask suffered slight external damage, its ability to safely contain spent fuel was not compromised. [From DOE (1995c).]

is advancing the concept of a multipurpose canister that could be used with various "overpacks" for storage, transport, and disposal of spent fuel.

The casks in which such shipments are made are huge in size but can be accommondated on either flat-bed trailers or railroad cars. Special features of design are needed to remove the heat that is generated during shipment. The required shielding and the need for structural strength add to the mass of the shipment, which may weigh as much as 100 tons. A shipping cask for spent fuel is shown in Fig. 7-1. The casks are subjected to severe tests for mechanical strength, fire resistance, and water resistance before their designs are accepted by the NRC. A particularly dramatic test is illustrated in Fig. 10-1 in which a cask is impacted by a railroad locomotive

TABLE 10-2

Estimated Annual Doses to the U.S. Population from Consumer Products and Miscellaneous Sources[a]

Source of exposure	No. of persons exposed in U.S.	Average annual effective dose equivalent (μSv)[b]	
		For exposed U.S. population	For total U.S. population
Television receivers	230,000,000	<10	<10
Video display terminals	50,000,000	<10	≤10
Ophthalmic glass	50,000,000	<4	<1
Gas mantles	50,000,000	2	0.4
Welding electrodes (thoriated)	300,000	160	0.2
Dental prostheses	45,000,000	0.7	0.14
Air transport of radioactive materials	14,000,000	2.4	0.13
Transport of nuclear fuel cycle materials	—	—	0.05
Luminous clocks and watches (tritium)	10,750,000	1	0.05
Electron tubes	230,000,000	0.04	0.04
Check sources	800,000	<10	<0.04
Smoke detectors	100,000,000	0.08	0.03
Luminous clocks and watches (promethium)	8,100,000	0.4	0.014
Airport X-ray baggage inspection	30,000,000	0.021	0.0027
Static eliminators	40,000	3.2	0.0005
Fluorescent lamp starters	50,000,000	0.0002	0.00004

[a]From NCRP (1987c,d).
[b]One μSv equals 0.1 mrem.

traveling at 130 km h^{-1}. The cask was hardly damaged despite the fact that the trailer on which the cask was mounted was destroyed and the front of the locomotive was badly damaged.

It is noteworthy that no member of the public has been significantly exposed to radiation from the millions of shipments of radioactive materials that have been made over many years in the United States.

SUMMARY OF DOSES RECEIVED FROM VARIOUS SOURCES

The NCRP has estimated doses to the U.S. population from many of the various sources discussed in this chapter in several of their reports (NCRP, 1987b–d). These doses are summarized in order of declining average individual dose in Table 10-2. Estimates of the kind included in this table are inherently difficult to make and are subject to great uncertainties. However, they do serve to illustrate the relative importance of the various sources.

CHAPTER 11

Radioactive
Waste Management

The technical aspects of radioactive waste management are perhaps less complex than the sociopolitical ramifications. The prospect of radioactive waste storage triggers fear in the general public, and this has been reflected in a "not in my backyard" (NIMBY) syndrome in which there is general agreement that there is a need to store radioactive wastes, but each locality wants them stored somewhere else. A more extreme view is NOPE or "not on planet earth," which holds that because radioactive wastes should not be disposed of anywhere, they should not be generated. A cynical view among politicians is NIMEY," not in my election year." Some of the sociopolitical issues of radioactive waste management are discussed in Chapter 16.

There are several kinds of radioactive wastes classified according to their physical and chemical properties as well as the source from which the wastes originate. For technical, legal, and political reasons, wastes that originate from the military programs are handled separately from wastes from the civilian program (DOE, 1983). A compendium on the technology of radioactive waste management has been assembled by Moghissi et al., (1986).

Among the physical properties that influence the manner in which a radioactive waste should be managed are the half-life of the nuclide and the chemical form in which it exists. The IAEA offers a waste categorization system to its member states (IAEA, 1993) but it is not followed uniformly. In this chapter, the following categories of wastes as employed in the United States will be discussed.

1. Low-level waste, defined by exclusion, is radioactive waste that is neither high-level waste, spent nuclear fuel, nor transuranic waste (see the following). It is not always "low level" and may include intensely radioactive gamma-ray sources used for medical or industrial irradiation or as accelerator targets that contain kilocurie quantities (hundreds of terabecquerels) of activity. Low-level wastes consist of residues from the nuclear power industry, laboratory research, industrial activities, and medicine. The physical forms of low-level waste also include contaminated protective clothing, paper, glass, plastic and other debris, biological materials, scrap metal, and even building materials. There are three categories of low-level wastes that may accumulate in huge volumes and require special attention. These are uranium mill tailings (see Chapters 7 and 15), wastes generated in the remedial action programs of the DOE, including residues from the Manhattan Project (see Chapter 15), and waste materials from decommissioning of nuclear power plants (see Chapter 8).

2. Spent nuclear fuel, or simply spent fuel, is nuclear fuel that has been withdrawn from a reactor following irradiation and has not been chemically separated into its constituent elements by reprocessing. Spent fuel includes all of the radioactive materials associated with fuel assemblies. For purposes of U.S. waste management regulation, the fuel must have undergone at least one year's decay since being used as a source of energy in a power reactor to be considered spent fuel.

3. High-level waste is reprocessing waste, the highly radioactive material resulting from the reprocessing of spent nuclear fuel, including liquid waste produced directly in reprocessing and any solid material derived from such liquid waste that contains fission products in sufficient concentrations. The definition of high-level waste in the United States has been broadened to include other highly radioactive material determined by the NRC to require permanent isolation (NRC, 1994).

4. Transuranic (TRU) wastes are mainly alpha-emitting residues from military manufacturing. The principal elements of concern are plutonium and americium. Waste that would otherwise be low-level waste is considered transuranic waste when the concentration of plutonium and americium exceeds 100 nCi g^{-1} (3.7 kBq g^{-1}).

Some radioactive waste may also contain hazardous chemical waste, in which case it is referred to as "mixed waste." In the United States, radioactive wastes and chemical wastes fall under different regulatory authorities, so mixed waste has been especially difficult to manage, for bureaucratic as well as technical reasons. However, there are current efforts to regulate mixed waste according to risk rather than by a combination of the existing, often conflicting regulations for radioactive and chemically hazardous wastes (Dragonette, 1995).

LOW-LEVEL WASTES

The early investigations of means for low-level waste (LLW) disposal involved possible disposal in the marine environment. In 1958 a panel of marine scientists of the National Academy of Sciences/National Research Council considered the impact of low-level radioactive waste disposal into the Atlantic and Gulf coastal waters and concluded (NAS/NRC, 1959) that relatively large quantities of radioactive wastes could be deposited safely in shallow coastal waters. Twenty-eight possible locations were selected that could be used for this purpose without limiting the areas for other uses. The total quantity of radioactivity that could be deposited in any one disposal area in any one year was estimated to be about 250 Ci (9 TBq) of ^{90}Sr or the biological equivalent of other isotopes. Compared to actual practice, this was a rather liberal recommendation. The calculated quantities of selected radioisotopes equivalent in hazard to 250 Ci (9TBq) of ^{90}Sr are given in Table 11-1. The practicality of disposing of low-level wastes into Pacific coastal waters was also examined by another committee of the NAS/NRC with similar conclusions (NAS/NRC, 1962).

The NAS/NRC studies seemed to justify practices that had already been used by the AEC and its contractors. The Atlantic and Pacific oceans and the Gulf of Mexico were used, beginning in 1946, for disposal of packaged low-level wastes originating mostly from research and development facilities. Fifty-five-gallon 18-gauge steel drums were used to contain mixtures of low-level wastes and cement, which was permitted to harden before disposal into the ocean. When this practice was discontinued in 1970, the total quantity deposited in the oceans by the United States totaled 94,600 Ci (3500 TBq), much of which were short-lived or relatively innocuous nuclides used in tracer applications and clinical practice. Additional use of the oceans for LLW disposal has been practiced by other countries (Holcomb, 1982; Sjöblom and Linsley, 1994). The quantities discharged in this way have been insignificant compared to the amounts lised in Table 11-1.

The practice of ocean dumping was discontinued in the United States mainly because of restrictions imposed by international agreements that resulted from mounting pressure against sea disposal of any kind. This was done despite the absence of evidence that there were damaging consequences from the existing practices. Ironically, during the same period when

TABLE 11-1

Quantities of Selected Radioisotopes Equivalent to 250 Ci of ^{90}Sr, Showing the Initial Quantities That Will Decay to 250 Equivalent Ci Allowing 1-Month and 1-Year Containment[a,b]

Isotope	Curies		
	No containment	1-month containment	1-year containment
^{24}Na	5.0×10^7	10^{24}	10^{183}
^{32}P	15.5	68.6	1.1×10^9
^{35}S	3.1×10^6	3.9×10^6	5.6×10^7
^{42}K	3.1×10^6	10^{14}	10^{226}
^{45}Ca	1.6×10^5	1.8×10^5	7.5×10^5
^{59}Fe	1.2×10^3	1.9×10^3	3.3×10^5
^{60}Co	6.2×10^3	6.3×10^3	7.0×10^3
^{64}Cu	5.0×10^4	10^{21}	10^{201}
^{65}Zn	1.4×10^4	1.5×10^4	3.8×10^4
^{90}Sr	250	250	250
^{131}I	9.3×10^2	1.2×10^4	10^{16}
^{137}Cs	9.3×10^4	9.3×10^4	9.3×10^4

[a] Reprinted with permission from Radioactive waste disposal into Atlantic and Gulf coastal waters. Publ. 655. Copyright 1959 by the National Academy of Sciences. Courtesy of the National Academy Press, Washington, D. C.
[b] Equivalence based on ratios of permissible seawater concentrations. One Ci = 37 GBq.

relatively small amounts of low-level waste were being introduced into the oceans, far greater quantities of radioactive debris from open-air nuclear weapons testing were being deposited in the oceans. From Table 11-2 we see that megacurie quantities of ^{90}Sr, ^{137}Cs, ^{14}C, and tritium were introduced

TABLE 11-2

Estimated Deposition of Radionuclides in the Pacific Ocean[a]

Nuclide	Activity produced by weapons tests, 1945–1972		Estimated deposition in Pacific Ocean	
	MCi	PBq	MCi	PBq
^{3}H	4,500	170,000	1,140	42,000
^{14}C	5.8	2,200	1.9	70
^{90}Sr	17	630	7.1	260
^{137}Cs	27	1,000	11	420
^{239}Pu	0.4	15	0.17	6.3

[a] Adapted from Eisenbud (1981a). Reproduced from *Health Physics* Vol. **40**, p. 435 by permission of the Health Physics Society.

into the Pacific Ocean. Despite the fact that the oceans have been the recipients of such enormous quantities of radionuclides from fallout, marine sources of food have not contributed significantly to the dose received from fallout in those countries of the world for which data are available (Eisenbud, 1981a). The UNSCEAR risk assessments have been based on contamination of the terrestrial food chains because most foods are derived from land sources. For example, in San Francisco, where representative diets have been monitored for ^{90}Sr for many years, fish and shellfish account for no more than about 0.2% of an annual per capita ^{90}Sr intake that has ranged between about 1000 and 3000 pCi y^{-1} (37 and 111 Bq y^{-1}). Similar findings have been reported from diet studies in New York City (Harley, 1969). Of course, there are places in the world where fish and shellfish form a much greater fraction of the diet, but there are no indications that the doses received in such plaecs have been significant

LLW PROGRAMS IN THE UNITED STATES

There are currently four accepted methods of LLW disposal: on-site decay, disposal to sewage, incineration, and shallow land burial. Compaction is often used to reduce volumes of buried wastes.

Decay is a logical way in which to deal with many of the short-lived radionuclides used in medicine and research. Iodine-131, with a half-life of 8 days, 14-day ^{32}P, and 80-day ^{35}S need not be shipped for off-site disposal if storage space is available to allow for decay. Decay in storage is permitted for isotopes with half-lives up to 65 days (10 CFR 35.92) and the NRC will allow longer half-life materials to be managed this way on a case-by-case basis. Once these isotopes have decayed for 10 half-lives, they may be disposed of as ordinary waste.

Small, dilute quantities of LLW are permitted to be disposed of into the sewer. The NRC has strict rules governing the amounts and forms that may be disposed of and records that must be kept (10 CRF 20.2003). It is important that the form of the waste be soluble or readily dispersible so that the activity is not concentrated in the sewage lines or treatment works (see Chapter 10).

About half of the wastes that originate from medical and research institutions contain small amounts of ^{3}H and ^{14}C generated in the course of liquid scintillation counting, a procedure that is in widespread use. The liquids are contained in 20-ml plastic vials, of which many millions are used in the United States each year (Roche-Farmer, 1980). These vials contained a total of about 500,000 L of organic solvent, frequently toluene, into which were dissolved a total of about 300 GBq (8 Ci) of ^{3}H and ^{14}C. The disposal problems were complicated by the flammable nature of the solvents used, and fires have been known to have occurred during transportation of these wastes. This amount of radioactivity is so insignificant that the vials could be incinerated at most biomedical laboratories without risk to the public. Any incinerator capable of properly burning animal carcasses and other biomedical wastes would be more than adequate to receive the scintillation vials (Eisenbud, 1980; Philip *et al.*, 1984).

In 1981, the NRC issued revised standards that would have permitted incineration of the scintillation vials (NRC, 1981). However, resistance to incineration of "radioactive waste" was so strong in, for example, New York City that the city council prohibited implementation of the NRC regulation despite an urgent resolution by the Committee on Public Health of the New York Academy of Medicine (NYAM, 1983) that emphasized the safety of the procedure. Nevertheless, incineration has been shown to be a safe option for the disposal of many types of wastes, often combined with the use of time to allow partial decay (Vetter, 1992; Emery *et al.*, 1992). Fortunately there has been a shift away from the use of hazardous solvents in liquid scintillation counting, so sewage disposal is often possible if the radioactive contents are below regulatory limits. Another development is the "minivial," which allows use of much smaller volumes of liquid scintillation fluid.

Shallow land burial of LLW is controlled by NRC specifications in 10 CFR 61. The basic objective of this regulation is that LLW facilities should be sited, designed, operated, closed, and controlled after closure so that human exposure will be maintained within regulatory limits. The site must be characterized and modeled, which means that its hydrogeological features can be described in quantitative terms. The direction and rate of transport of radionuclides released from the wastes must be predictable. This is not possible at some sites. For example, should the wastes be placed in fractured bedrock, the path taken by seepage might be indeterminate. At the other extreme, if the wastes are placed in a bed of homogeneous clay, the movement of groundwater could be well described, the exchange capacity of the clay would be known, and the rates of movement of the various radionuclides would be predictable. Design, operation, and closure of the facility must ensure protection for individuals who may inadvertently intrude on the site. The site must be well drained and not subject to flooding. The upstream drainage area must be minimized to avoid erosion. The groundwater must be at sufficient depth to avoid seepage into the repository. Subsidence should be avoided by minimizing voids between the packages of waste. Further, the regulations specify that liquid wastes must be solidified or packed with ample absorbing material. There are additional regulations that limit flammability and corrosion. Finally, there must be assurance of continuity to assure monitoring for at least 100 years after the site is closed. Engineered safeguards may be required to compensate for deficiencies in the natural properties of the site.

For many years, the United States has been heading for a crisis in the management of low-level wastes, not because the wastes have proved to be so hazardous, but because construction of disposal facilities has been halted by widespread public concern about the potential risks of shallow land burial, which is the most favored disposal option with regard to safety and cost.

In a special report to the President and the Congress, the National Advisory Committee on Oceans and Atmosphere (1984) recommended that the Congress and administration should revise the present policy of excluding the use of the ocean for low-level radioactive waste disposal, but that ocean disposal should not be initiated until it has been established that the fate and effects of such disposal will not result in adverse effects. This course of action is precluded by international agreements that severely restrict the use of oceans for waste disposal of any kind.

By far the largest volume of low-level radioactive waste has been disposed of by near-surface land burial. Until 1963, industry arranged to send such waste to AEC sites, but in that year AEC withdrew its offer of that service and began to license private companies for the operation of waste burial grounds. By 1971, there were six commercially operated licensed shallow land burial sites, as shown in Table 11-3 (Holcomb, 1980), at which a total of about 600,000 m^3 of low-level radioactive waste was stored. Al-

TABLE 11-3

Commercial Shallow Land Burial Sites[a]

Location	Year first licensed	Site operator	Licensing authority	Year closed
Beatty, Nevada	1962	Nuclear Engineering Company	State	1993
Maxey Flats, Kentucky	1962	Nuclear Engineering Company	State	1977
West Valley, New York	1963	Nuclear Fuel Services	State and NRC	1975
Richland, Washington	1965	U.S. Ecology	State and NRC	Open
Sheffield, Illinois	1967	Nuclear Engineering Company	NRC	1978
Barnwell, South Carolina	1971	Chem-Nuclear Systems, Inc.	State and NRC	Open
Clive, Utah[b]	1988	Envirocare	State and NRC	Open
Ford, Washington[c]	1996	Dawn Mining Co.	State	Open

[a]Updated from Holcomb (1980).

[b]Limited to wastes from the processing of uranium and thorium ores and contaminated soil, soil-like material, and debris.

[c]Limited to wastes from the processing of uranium and thorium ores.

though it has not been shown that any of the six burial sites created a public health problem, intense local opposition to operation of some of the sites began to develop, and three of the sites were closed.

In 1975 the burial grounds at West Valley, New York, were closed despite the fact that studies conducted by both private and public investigators demonstrated that continued operation posed no significant adverse effect on public health and safety (Giardina et al., 1977).

Low-level contamination of groundwater and milk supplies near a second burial site at Maxey Flats, Kentucky, caused the state to place a temporary ban on further operations, and in the following year the lease for the land on which the burial grounds were located was canceled. Studies by the EPA did find migration of tritium, ^{90}Sr, and a few other nuclides, but only in very small amounts (Montgomery et al., 1977). The tritium content of milk within about 3 km of the site ranged from 300 to 6500 pCi L^{-1} (11–240 kBq m^{-3}). An individual consuming 1 liter per day of milk containing the highest tritium level would receive an annual total body dose of about 0.4 mrem (4 μSv). The highest level of tritium-contaminated well water would deliver a dose of 0.1 mrem y^{-1} (1 μSv y^{-1}). A third site, at Sheffield, Illinois, closed in 1978 because it reached its licensed burial capacity and the state was opposed to any expansion.

For many years there were three commercial low-level burial sites in the United States, located in Barnwell, South Carolina; Hanford, Washington; and Beatty, Nevada. In late 1979, the governors of two of these states, Nevada and Washington, closed their sites temporarily because of poor packaging practices on the part of the shippers. This was done despite the absence of evidence that the practices were actually hazardous to either workers or the public, illustrating the seriousness with which compliance with regulations for safe practice is taken.

It was necessary to suspend the license of the Nevada burial site in March 1979, when it was found that employees at the site were removing contaminated hand tools, electric motors, and other items. As the result of an intensive effort, an estimated 25 pickup truckloads of radioactively contaminated equipment, as well as several loads of large items, were recovered and returned to the burial site (Wenslawski and North, 1979). The Nevada site closed permanently in 1993, leaving only two sites to cover the entire United States in North Carolina and Washington.

In addition to the burial sites that are operated commercially, there are five major shallow land burial facilities located at the major production and research centers operated by the DOE. These facilities serve the needs of the DOE and other government agencies.

Although no risk to public health has been shown to have resulted from the shallow land burial of low-level radioactive wastes, there have nevertheless been flaws in the management methods. These have included poor packaging, water infiltration, and insufficient compaction that resulted in creation of a "bathtub" owing to subsidence. All of the examples of mismanagement were of a type that could have been corrected. However, the various mishaps were so widely publicized that widespread public opposition developed that has jeopardized the nation's ability to manage its low-level radioactive wastes. All of this has resulted in considerable frustration on the part of users of radioactive materials, as exemplified by a New York State study group that expressed its concerns that low-level waste disposal services were likely to be severely disrupted unless the state acted to provide an in-state alternative to out-of-state shipments. At risk were the health value of the more than 1.9 million annual medical procedures using radio-pharmaceuticals, the multi-billion-dollar economic value of these procedures, the work of more than 5000 persons who perform research within the state of New York, and the operation of more than 300 industrial facilities (New York State Low-Level Waste Group, 1983).

In contrast to the objections to siting new waste facilities, there is usually no major concern on the part of the local communities about the dangers of living near the existing, operating low-level commercial sites. The facilities have been an important source of tax revenues and jobs in the rural counties in which they have been located. There is no public radiation exposure from the facilities. Nevertheless, they have became politically undesirable at the state level and the governors in the states in which the repositories are located appealed to the federal government to relieve their states from the perceived indignity associated with having become "dumping grounds" for the nation's wastes. As a result, in 1980 Congress passed the Low Level Policy Act that required each state to provide facilities for the disposal of wastes or join with other states to form compacts that would provide regional disposal facilities. It was specified that these facilities should be ready to receive wastes by January 1, 1993. The various arrangements made in conformity with the law are Illustrated in Fig. 11-1, but by early 1997, more than 3 years after the specified deadline, none of the new facilities had been built.

As a consequence, shallow land burial of LLW has been unavailable to many states for several years, although South Carolina recently opened its site to all states except North Carolina as part of a squabble within that state's LLW compact. In another recent development, a licensed disposal facility in Clive, Utah, is accepting bulk soil and rubble wastes containing artificial and natural radionuclides from all states. The facility began as a repository for uranium mill tailings wastes from Salt Lake City and has accepted remediation wastes from the Denver Radium Site and DOE's FUSRAP program (see Chapter 15). This facility is also permitted by the

FIGURE 11-1   Low-level radioactive waste "compacts" in the United States.

state of Utah to accept certain hazardous wastes for land disposal. The combination of the nuclear license and a hazardous waste permit allows the facility to accept some mixed wastes for burial.

Sources and Quantities of Low-Level Wastes

In general, the wastes generated by biomedical facilities have properties that simplify the disposal requirements. The half-lives tend to be short or, if not, the radiations are relatively weak, as in the case of ^{3}H or ^{14}C. The LLW produced by nuclear power plants tend to have longer half-lives, but much is in the form of activation products contained in stainless steel or other structural materials in which the nuclides are immobilized. Other LLW from the power plants, such as ion-exchange resins used to purify reactor cooling water, are potentially more mobile.

The capacity of the required LLW facilities can be descrbed in terms of volume or amount of radioactivity. Since the cost of disposal is influenced more by volume than by the amount of radioactivity, there is incentive to reduce the volumes of material for which disposal is required. This can be accomplished by incineration, which is done at a few large biomedical institutions, or by compaction. There have been substantial reductions in the volumes occupied by the waste materials owing to improved systems of compression and chemical separation of radioactive constituents. At the beginning of 1994, the total amount of LLW present at U.S. commercial sites contained an estimated 5.7 MCi (210 PBq) in a volume of about 1.5 million m^3 [Oak Ridge National Laboratory (ORNL), 1995]. The amount shipped for disposal in 1993 was 0.64 MCi (24 PBq) in a volume of 22,400 m^3 for an average specific activity of nearly 30 Ci m^{-3}. Volumes have declined in recent years and both activities and volumes are projected to decline over the short term, but the projections are recognized as uncertain.

In addition to the civilian sector, the DOE currently has 27 sites that generate, are projected to generate, or store LLW. At the beginning of 1994, the estimated DOE low-level waste volume was nearly 3 million m^3, twice the commercial inventory. The amounts for the seven sites managing the largest volumes were: 665,000 m^3 at the Savannah River site, 602,000 m^3 at the Hanford site, 458,000 m^3 at the Nevada test site, 442,000 m^3 at the Oak Ridge Reservation, 343,000 m^3 at the Fernald Environmental Management Project, 221,000 m^3 at the Los Alamos National Laboratory, and 147,000 m^3 at the Idaho National Engineering Laboratory (ORNL, 1995). Concentrations of radionuclides and chemicals in groundwater near LLW disposal facilities exceed applicable standards at several sites, although significant contamination of potable water has not occurred.

The DOE is currently evaluating its LLW management programs and is considering several alternatives (DOE Office of Environmental Management, 1995). These are decentralization, in which the waste is disposed of at 16 sites, seven regionalization alternatives in which wastes are treated and transported to a smaller number of sites for possible further treatment and disposal, and five centralization options in which treatment is performed at one or more sites but disposal is performed at only one site. "No action" is also considered as an alternative. Models indicate that the centralization options would be the most risky because of vehicle accidents and radiation exposure during transport. Estimated 20-year costs varied by less than a factor of two from $11.8 to $20 billion. The option with the lowest estimated cost was not the "no action" alternative but rather centralized disposal at the Nevada test site following minimal treatment. The most expensive alternative was a regionalization alternative involving treatment at 11 sites and disposal at 12 sites.

LOW-LEVEL MIXED WASTES IN THE UNITED STATES

Some LLW also contains hazardous waste as defined by the Resource Conservation and Recovery Act (RCRA). The hazardous wastes of concern are primarily chemicals that are hazardous because of properties such as flammability, corrosivity, reactivity, or toxicity. These wastes, called low-level mixed wastes, or LLMW, represent a management challenge not only because of their physical and chemical properties, but because the radionuclides and the chemicals fall under separate regulations and governmental agencies. In general, mixed waste management must meet regulations for both the chemical and radioactive properties. In some cases, regulations developed for hazardous wastes are not ideal for managing radioactive waste. One example is a RCRA requirement for frequent visual inspection of stored waste drums for leakage. When this requirement is applied to mixed waste, excessive radiation exposures to workers can result. Disposal of mixed waste is not generally available unless it is first treated to render the chemical portion nonhazardous or to separate hazardous and radioactive components. Exceptions are soil, soil-like material, and debris, for which a licensed mixed waste facility exists in Utah.

Because of the management difficulties discussed here, new generation of LLMW is now almost universally avoided. Most existing mixed waste in the United States is owned by the DOE and resulted from the development of nuclear technology. The DOE currently has 37 sites that generate, are projected to generate, or store an estimated 226,000 m^3 of LLMW (DOE Office of Environmental Management, 1995). The larger volumes of DOE LLMW are managed at the same locations discussed for LLW, plus the Rocky Flats Environmental Technology Site (formerly the Rocky Flats Plant). As with their LLW, the DOE is currently evaluating treatment, storage, and disposal options that include decentralized, regionalized, and centralized alternatives. Projected risks are smaller than for the much larger volumes of LLW and do not vary much among the alternatives. The most expensive option ($12.6 billion) is a regionalization alternative with treatment at 11 sites and disposal at 12 sites. "No action" is projected to be the least expensive alternative in the LLMW case ($5.2 billion).

HIGH-LEVEL WASTES

In the discussion of high-level wastes (HLW) it is useful to differentiate between "defense wastes," which have been accumulated during the past 50 years from military reactors and the production and use of plutonium at government installations, and "commercial wastes," which originate primarily from the production of civilian nuclear power.

DEFENSE HIGH-LEVEL WASTES

The principal accumulations of defense wastes are located at Hanford, Washington, and Savannah River, South Carolina, with lesser amounts at

Idaho Falls, Idaho, and West Valley, New York. The plant at West Valley reprocessed both civilian and government fuel until 1972 and is now a DOE waste management demonstration project. Most of the wastes originated from processing irradiated uranium for the purpose of plutonium separation. Additional defense wastes originated from the reprocessing of fuel from nuclear-powered naval vessels and government-affiliated research reactors. The DOE and its predecessor agencies have received, stored, and reprocessed approximately 100,000 MT (10^8 kg) of spent nuclear fuel (DOE, 1994h). Since the DOE stopped reprocessing reactor fuel in 1992, little additional high-level reprocessing waste is expected to be generated in the United States for the foreseeable future (ORNL, 1995).

At Hanford and Savannah River, the waste liquid from these processes has been stored in concrete-encased steel tanks a few meters below the surface of the ground. The original highly acid liquids were usually neutralized to reduce corrosion, causing a precipitate to settle to the bottom of the tanks, where in some cases it has attained a nearly rocklike hardness that makes it very difficult to transfer the waste to other containers when necessary. At the Idaho National Engineering Laboratory, the bulk of the liquid wastes from reprocessing fuels has been converted to calcine, a stable granular solid (ORNL, 1995). The calcine is currently stored in shielded, stainless-steel bins. The volumes and radioactivity of the accumulated high-level wastes in the United States are given in Table 11-4. Processes to convert the current waste inventories into forms sufficiently stable for permanent disposal are under development.

It was soon recognized that tanks develop leaks (AEC, 1973; Catlin, 1980) and it is generally recognized that an alternative to tank storage is required (NAS/NRC, 1978, 1985a; DOE, 1978). Suggestions include immo-

TABLE 11-4

Volumes and Radioactivity of Defense High-Level Wastes
in the United States as of 1994[a]

Location	Volume (10^3 m^3)	Activity (MCi)	Activity (10^6 TBq)
Hanford site	239	348	12.9
Savannah River site	127	535	19.8
Idaho National Engineering Laboratory	11	52	1.9
West Valley Demonstration Project[b]	2.2	25	0.9
Total (rounded)	380	960	35.5

[a]Adapted from ORNL (1995).
[b]Includes both defense and civilian reprocessing waste.

bilization of the waste in the present tanks, mixing the waste with concrete or other solidifying agents for injection underground as grout, and conversion to stable solids for burial. Currently the only plan under consideration for disposal of U.S. high-level military wastes is conversion to a stable solid-waste form and burial in a permanent geological repository under study. The site location, Yucca Mountain in Nevada, has been designated by the U.S. Congress and is discussed later in this chapter. The most likely solidification process entails conversion to borosilicate glass and encapsulation in stainless steel.

Whatever decision is made, the costs will be enormous. The Nuclear Energy Agency (1993) has compiled high-level waste and spent-fuel encapsulation and disposal cost estimates for several countries including the United States. The costs per cubic meter of waste in 1991 dollars ranged from $60,000 (Spain) to $560,000 (United Kingdom). The cost estimate for the United States was $110,000 per cubic meter. At this rate the cost of encapsulation and disposal of the 400,000 m^3 of accumulated U.S. reprocessing waste is estimated to be $44 billion. These costs do not include waste-disposal site selection, evaluation, research and development, or vitrification, nor do they include the enormous anticipated costs of "environmental restoration" at the current waste storage sites (see Chapter 15).

Although the high-level liquid-waste problems of all major defense program sites are complex, the problems of liquid-waste management at Hanford will be discussed in more detail because it is both the oldest and largest of the irradiated fuel-processing facilities.

Until 1973, and particularly during the 1950s, some of the liquid wastes were discharged directly into the ground using cribs, retention trenches, ponds, wells, and French drains (Wodrich, 1994). It is estimated that about 360,000 Ci (1.3 PBq) were introduced to the ground in these ways. We will see that this is about one-third of the radioactivity discharged to the ground as a result of leaks from waste storage tanks.

The liquid wastes at Hanford have been stored in 177 cylindrical underground tanks that have been constructed since 1943. The original design, used prior to 1964, contained a single carbon-steel liner. A total of 150 MCi (5500 PBq) is contained in 35 million gallons (132,000 m^3) of wastes stored in 149 of these tanks (Gephart and Lundgren, 1995). Another 28 tanks that were built between 1968 and 1986 have double liners that provide an added margin of safety by permitting early detection of leaks (Fig. 11-2). These double-lined tanks store an additional 82 MCi (3000 PBq) contained in 35 million gallons of waste. At the present time almost all of the radioactivity is due to ^{90}Sr and ^{137}Cs. Over the years, leaks in the single-lined tanks have released about 1 MCi (37 PBq) to the ground, but there have been no leaks from the double-lined tanks.

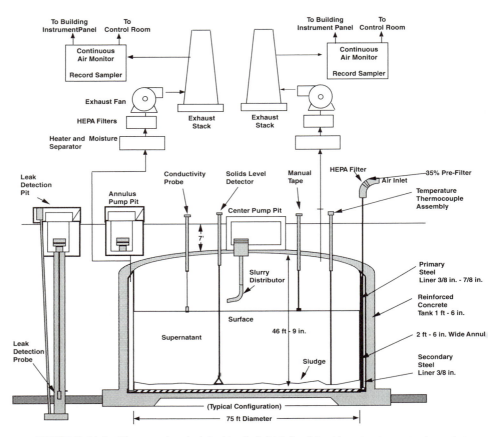

FIGURE 11-2   Diagram of typical double-shell, high-level liquid-waste storage tank used at the DOE Hanford site. (Courtesy of J. Lentsch, Westinghouse Hanford.)

Potentially serious radiological and chemical safety problems have been associated with the tanks. These include several corrosion mechanisms that can affect the integrity of the tanks (Anantatmula *et al.*, 1994), accumulations of hydrogen, nitrous oxide, ammonia, and methane (Lentsch *et al.*, 1995), and the presence of potentially explosive chemicals, such as ferrocyanide compounds that can react exothermically with other chemicals in the wastes (Babad *et al.*, 1995).

Radiolytic hydrogen and other flammable gases are generated interstitially within the sediments at the tank bottoms and from some of the tanks have been released in "burps," at intervals of 100 to 150 days, in concentrations above the flammable limits. Progress has been made in controlling these discharges by installing mixers that prevent accumulation of the gases within the sediments and permit them to be released at concentrations below the flammable limit (Lentsch *et al.*, 1995).

An array of about 800 sampling wells has been utilized in the ground-water monitoring program in the Hanford area (Dirkes and Hanf, 1995). Concentrations above the EPA drinking water standard have been found for some nuclides at a few locations within the reservation. In addition, it has been found that tritium is being discharged in seeps into the Columbia River in concentrations that exceed the drinking water standard at the point of discharge. However, as will be seen in Chapter 14, these usually have not resulted in more than trivial exposure to residents who live downstream from the reservation.

DEFENSE SPENT FUEL

In the United States, defense spent nuclear fuel is managed by the DOE and includes fuel from both the "production reactors" used to produce plutonium and tritium and fuel from Naval propulsion reactors. Although often unrelated to defense, government-owned nuclear fuel used in domestic and foreign research reactors is managed by the DOE and falls in this category. The core of the Three Mile Island reactor was made available for research at DOE laboratories and is also managed by the DOE (1994h). The quantities of defense spent fuel in the inventory are given in Table 11-5. The largest inventory is at Hanford and is primarily from the "N

TABLE 11-5

Inventory of Spent Nuclear Fuel Managed by the DOE as of 1995[a]

Generator or storage site	Mass (MT)	Percentage of total
Hanford site	2133	81
Idaho sites[b]	260	9.8
Savannah River site	206	7.8
Special case commercial reactors	41	1.6
Domestic research reactors	1.8	0.07
Oak Ridge National Laboratory	0.9	0.04
Other DOE sites	0.7	0.03
Total	2643	100.0

[a]Adapted from ORNL (1995).
[b]Includes Idaho National Engineering Laboratory, Naval Reactors Facility, and Argonne National Laboratory West.

Reactor," a large, graphite-moderated plutonium production reactor that is no longer in operation.

Based on current projections, the U.S. inventory is expected to grow only about 3% by the year 2035, primarily from naval propulsion and from research reactors at the National Laboratories, universities, the private sector, and in other countries.

DEFENSE TRANSURANIC WASTE

A special category of radioactive defense wastes are those contaminated with transuranic (TRU) radionuclides (Moghissi, 1983). These originate mainly in the production and fabrication of plutonium for military purposes, and ^{238}Pu, ^{239}Pu, and ^{241}Am are the most prevalent TRU contaminants. Most TRU waste is also contaminated with fission products but these have shorter half-lives than most TRU waste.

Prior to 1970, TRU waste was disposed of by shallow land burial in a manner still used for low-level waste. These wastes are referred to as "buried TRU." In 1970 it became a DOE requirement to separate TRU-contaminated wastes according to whether they contained more or less than 10 nCi (370 Bq) of TRU per gram of waste. Wastes that contained less than 10 nCi g^{-1} could be disposed of as low-level waste by shallow burial, whereas wastes containing greater amounts were placed in retrievable storage for eventual transfer to a permanent repository. This post-1970 waste is called "retrievably stored TRU." TRU waste is further categorized as "contact-handled" and "remote-handled," according to the external dose rate from the package. Packages exceeding 2 mSv h^{-1} at the surface are designated as "remote handled."

In 1982 the NCRP examined the TRU disposal practices and concluded that there was no basis for the 10 nCi g^{-1} criterion, that higher concentrations could be disposed of by near-surface land burial, and that what was required were site-specific limits based on geochemical and ecological conditions existing at individual sites (NCRP, 1982). Following publication of this report, the NRC changed the criterion to 100 nCi g^{-1} (3.7 kBq g^{-1}). Wastes with concentrations of TRU above 100 nCi g^{-1} (3.7 kBq g^{-1}) are, of course, still being held for ultimate disposal in a geologic repository. The quantities of retrievably stored TRU waste at the 6 locations where most of the material resides are given in Table 11-6. Smaller amounts of retrievably stored TRU are stored at 10 additional locations. The quantities of buried TRU are given in Table 11-7. For these tables, the total radioactivity in the TRU waste is given because separate data for TRU and fission-product waste in the containers are not available for all sites. TRU nuclides make up a substantial

TABLE 11-6

Quantities of Retrievably Stored Transuranic Waste at Various DOE Sites
as of the End of 1993[a]

Location	Contact handled			Remote handled		
	Volume (m³)	Total activity		Volume (m³)	Total activity	
		kCi	TBq		kCi	TBq
Idaho National Engineering Laboratory	64,800	365	13,500	80	7	277
Hanford site	15,600	213	7,880	201	36	1,340
Los Alamos National Laboratory	10,800	197	7,290	91	11	410
Savannah River site	8,930	561	20,800	0	0.0	0
Oak Ridge National Laboratory	2,020	68	2,520	564	288	10,700
Rocky Flats Plant	1,040	86	3,160	0	0.0	0
Other sites	890	7	260	10	0.3	11
Total (rounded)	104,100	1,500	55,400	950	340	12,700

[a]Adapted from ORNL (1995).

fraction of the total at sites for which data exist. A geologic repository in
Eddy County, New Mexico, is intended to provide for permanent isolation
of U.S. defense-related TRU wastes. This repository, known as the Waste
Isolation Pilot Plant, is discussed later in this chapter.

TABLE 11-7

Quantities of Buried Transuranic Waste at Various DOE Sites
as of the End of 1993[a]

Location	Volume (m³)	Total activity	
		kCi	TBq
Hanford site	63,600	600	22,200
Idaho National Engineering Laboratory	57,100	249	9,200
Los Alamos National Laboratory	14,000	n/a[b]	n/a
Savannah River site	4,870	34	1,260
Oak Ridge National Laboratory	176	0.24	9
Total (rounded)	140,000	883	32,700

[a]Adapted from ORNL (1995).
[b]n/a, Not available.

CIVILIAN SPENT-FUEL AND HIGH-LEVEL WASTES

The volume of waste from civilian nuclear power plants likely to become available in the foreseeable future has been greatly reduced by the slowdown in the development of nuclear power. In 1970, it was assumed that there would be 735,000 MWe (megawatts electric) of installed nuclear capacity in the United States by the year 2000 (ORNL, 1970), but 10 years later the estimate had been reduced to 180,000 MWe (ORNL, 1980). As of the end of 1992, however, the nation's installed capacity (excluding plants under construction) was only 99,000 MWe. Because no new plants have been ordered in the United States since 1979, the capacity is projected to remain fairly stable, growing to an estimated 101,300 MWe in the year 2000. Total spent-fuel estimates have undergone similar revisions. In 1970, it was estimated that the total inventory of spent nuclear fuel in the year 2000 would be about 90,000 MT (DOE, 1984a), but the year 2000 estimate had been reduced to about 42,000 MT by 1995 (ORNL, 1995).

By the year 2030, most of today's reactors will have reached the end of their design lifetimes. At that time the projected total mass of commercial spent fuel is estimated to reach 84,100 MT, an amount much larger than the 2675 MT from defense activities. The total radioactivity in the commercial fuel is projected to peak in the year 2014 at about 44,000 MCi (1.6×10^6 PBq). These estimates are based on the assumption that no new reactors are ordered but will obviously increase if new plants are constructed. These figures serve as the basis for national planning at the present time.

The spent-fuel elements from a commercial reactor, which consist mostly of zirconium-clad rods of uranium oxide, are placed immediately after removal in a large basin of water (the spent-fuel storage pool) located adjacent to a reactor. It was originally intended that the fuel rods would remain in these cooling basins for only about 6 months—to allow for reduction in their radioactivity and temperature—before being shipped to a reprocessing center. However, as mentioned in Chapter 7, the reprocessing of commercial spent fuel was halted in 1977 because of concern that the separated plutonium might be diverted for military or terrorist purposes. As it turned out, uranium economics changed drastically at about the same time because of the reduction in demand for nuclear power and fuel reprocessing could no longer be justified on economic grounds. As a result, the spent-fuel elements have been piling up on the nuclear power plant premises, in both the spent-fuel pools and, in some locations, dry cask storage. Efforts to site centralized temporary spent-fuel storage facilities (monitored retrievable storage, MRS) in the United States have failed despite efforts of the government and the nuclear utilities. MRS is discussed later in this chapter.

Although reprocessing of spent fuel is well established and used on a commercial scale in France and the United Kingdom (Nuclear Energy Agency, 1993), there are no plans to reprocess spent reactor fuel in the United States. Consequently the spent fuel itself is the waste material. However, should the economics of nuclear power change in the future, it may again become worthwhile to reprocess spent fuel.

OPTIONS FOR THE PERMANENT MANAGEMENT OF HIGH-LEVEL RADIOACTIVE WASTES AND SPENT FUEL

Several methods of isolating high-level wastes have been considered during the past 40 years, including on-site methods of solidification and disposal, use of the seabed and sub-seabed, injection as a grout in deep rock fissures, insertion in the Greenland icecap, and geologic isolation in deep mined cavities. Proposals have also been made that the wastes be lifted into outer space or into the sun by rocket, or be transmuted to more rapidly decaying elements in giant accelerators. The National Academy of Sciences has investigated the value of transmutation and has concluded that, although a major fraction of the transuranic actinide elements and some important long-lived fission products could be eliminated in this way, the benefits would not be sufficient to justify further consideration of this waste-disposal option (NAS/NRC, 1996).

Most of these disposal ideas must be considered infeasible for the foreseeable future because of economics, safety, and the current state of technology. Of these various options, only one appears politically and technically viable at this time: encapsulation and geologic isolation in mined cavities (IAEA, 1993). Nevertheless, some of the alternative options will be reviewed for historical interest.

On-Site Solidification and Disposal

At the Oak Ridge National Laboratory and in the former Soviet Union, moderately radioactive wastes have been mixed with cement to form a grout that is then injected through wells into rocks that have been fractured hydraulically (NAS/NRC, 1985a; Spitsyn and Balukova, 1979). Studies have suggested that this technique might be suitable for certain defense wastes at both Savannah River and Hanford (NAS/NRC, 1978, 1981), but it has also been shown at these sites that wastes with intermediate levels of activity can be hardened into concrete and buried in near-surface trenches. Solidification or other forms of stabilization followed by on-site disposal will likely continue to be useful for some lower-level wastes, particularly in remote environments.

The Marine Environment

The use of ocean waters for the disposal of low-level wastes has already been discussed in this chapter. The United Kingdom for some years discharged fission products into the Irish Sea from a fuel-reprocessing plant at Sellafield (formerly known as Windscale) near the coast of West Cumbria (Dunster, 1969). Experimental releases were first made in 1952, when about 10,000 Ci (370 TBq) of effluent were discharged over a period of about 6 months. Based on studies of the ecological behavior of the releases during this and subsequent experiments, the quantities released were increased beginning about 1969. The annual releases of ^{137}Cs from 1952 to 1984 are given in Fig. 11-3. The discharges of cesium, and other nuclides as well, peaked dramatically in 1976 and then decreased as a result of changes in operating practice (British Nuclear Fuels Limited, 1984; Black, 1984). Levels have continued to decline since the 1984 reports (Jones *et al.*, 1991).

By studying the dietary habits of the nearby populations in relation to radioecological factors, it was possible to derive working limits of permissible contamination of silt, sand, seaweed, and fish. The early studies identified contamination of local fish and a species of seaweed used in cooking as the critical pathway for human exposure (Dunster, 1958). Having defined the working limits of contamination of these vectors of human exposure, it was possible to calculate the permissible rates of release of radioactive aqueous wastes based on the public exposure standards in place at that time. It was found that the ^{106}Ru content of seaweed was limiting and that the permissible discharge rate of ^{106}Ru was 70,000 Ci (2600 TBq) per month. If it were not for the seaweed–food pathway, fish would have been limiting, in which case the permissible release could have been more than 1 million Ci (37 PBq) per month.

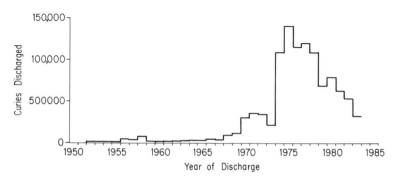

FIGURE 11-3   Discharges to the Irish Sea of ^{137}Cs from Sellafield. [From Black (1984).]

A formal inquiry into the public health impact of these releases was begun in 1983, in response to widespread concern that resulted from a television program that alleged an increase in the incidence of leukemia among young people in the vicinity of the plant. The report from the National Radiological Protection Board (1984) concluded that young people who lived in the vicinity of the plant from 1950 through 1970 would have received a total dose of about 350 mrem (3.5 mSv) during that 20-year period, which is about 13% of the dose received from background sources in the area (exclusive of radon).

In recent years the amounts of activity disposed into the sea at Sellafield have continued to decline as have the associated doses to the public. The harvesting of seaweed became uneconomic in 1970s and the critical population groups now are a few houseboat residents and a small group of people who are large consumers of fish and shellfish (Jones et al., 1991). Current maximum doses to these individuals are 150–200 μSv y^{-1} (15–20 mrem y^{-1}). Doses are projected to become insignificant by the year 2020 owing to reduced emissions and dispersal and decay of existing environmental concentrations (Hunt, 1991).

The former Soviet Union disposed of a variety of nuclear wastes directly into the Soviet sector of the Arctic Ocean and into rivers that flowed into the Arctic (Khodakovsky, 1994). The direct ocean disposals amounted to 2.5 MCi (93 PBq) as compared to a total of 1.7 MCi (63 PBq) by the other 12 countries that have practiced ocean disposal in the past. These Soviet wastes include 18 reactors from submarines and icebreakers, 6 of which still contain nuclear fuel. Fifteen of these reactors are resting in relatively shallow water of the Kara Sea, 20 to 50 m deep, and one rests at 300 m. Preliminary measurements of radionuclides near the disposal sites are stated to be low, suggesting that leaks have not yet occurred. Fortunately the Kara is frozen for 9 months of the year and has little biological activity. Fowler et al. (1994) confirmed that the ^{137}Cs concentrations of 4 to 12 Bq m^{-3} (0.1 to 0.3 pCi L^{-1}) in the Kara Sea are typical of those found in the open oceans and can be attributed to global fallout.

Sub-Seabed Disposal

The sediments of the deep ocean floor are uniform over a considerable area, show little evidence of mixing, have high ion sorptive capacity, and show virtually no movement of interstitial water (DOE, 1979b). The pressure is so high that the water will not boil if the sediments are heated by the decaying radionuclides. Because the sub-seabed has so many attractive features, a Seabed Working Group was established under the auspices of the Nuclear Energy Agency in 1977 to coordinate research on the possible use of the deep-sea sediments (at least 4000 m deep) as a repository for high-level nuclear wastes.[1] Several emplacement concepts have been proposed, including placing containers in drilled holes and the novel concept of a penetrator, a free-falling ballistically shaped container that could be released from a vessel and penetrate as far as 30 m into the sediments at the ocean bottom.

The sub-seabed concept is at a relatively early stage of development and further research would be needed to reduce the remaining uncertainties before actual disposal could occur. These uncertainties include pore water migration into the sediment, adsorption properties of sediments, and deep-sea biological activity. In addition to these technical and ecological considerations, there are legal and jurisdictional issues raised by the London Dumping Convention. Current interest in seabed disposal is quite low (IAEA, 1993).

Deep Geologic Repositories

The currently favored method of disposing of high-level wastes is in deep underground mined cavities and 11 countries are now in some stage of siting or developing these geological disposal sites.[2]

The two basic objectives of safe waste disposal are (1) to protect humans and our environment from the harmful effects of radioactive waste and (2) to dispose of waste in such a way that the transfer of responsibility to future generations in minimized (IAEA, 1993). The concept of waste isolation in mined cavities seeks to meet these objectives and depends on a multibarrier system that has the following component parts:

1. The solid form in which the wastes exist. For intact spent fuel, the waste form is the original clad fuel. For processed spent fuel, the radionuclides can be incorporated into solids such as glass or highly insoluble ceramics. A great deal of research on waste forms has been compiled by Lutze and Ewing (1988). The temperatures of the waste and host rock are important characteristics, which depend not only on the age of the waste but also the concentration of waste contained in the waste form, the spacing of the canisters, and the thermal properties of the rock in which the wastes are placed.

2. A canister that contains the waste and is designed to resist corrosion.

3. An "overpack" of highly adsorbent materials such as clay, which serves to reduce corrosion of the canister by retarding the migration of

---

[1]The member nations of the Seabed Working Group are Belgium, Canada, Federal Republic of Germany, France, Japan, Netherlands, Switzerland, United Kingdom, and United States.

[2]Belgium, Canada, Finland, France, Germany, Netherlands, Spain, Sweden, Switzerland, United Kingdom, and United States.

corrosive ions in groundwater. The overpack will also help to retard the migration of radionuclides that may escape from the canister.

4. The geochemical properties of the environment of the mined cavity repository, which govern the rate of migration of the radionuclides in groundwater.

5. The complex physical and biological pathways by which the radionuclides pass from the immediate vicinity of the repository to the biosphere and eventually to humans.

CRITERIA FOR SITE SUITABILITY

In the United States, general guidelines for determining the suitability of a repository site were published by the DOE in conformance with the requirements of the Nuclear Waste Policy Act (DOE, 1984b). An important requirement is that it should be possible to establish that the groundwater time of travel from the repository to the "accessible environment" should be no less than 1000 years. The nature and rates of hydrological processes during the past 2 million years should be such that, if continued into the future, the repository would not be affected deleteriously during the next 100,000 years. The guidelines then go into a great deal of detail concerning the required characteristics of the host rock, and specify that the repository should be placed at least 200 m below the ground surface.

An important criterion is that there should be no known natural resources at the site that might, in the foreseeable future, be of sufficient value to be commercially attractive. Geochemical and hydrological conditions should be such that the annual solubilization rate for the radionuclide inventory in the repository, after 1000 years, should be no more than 10^{-5}. Other required characteristics are tectonic stability, adequate rock mass, and suitable thermal conductivity, porosity, and permeability. An ideal rock mass would be large, homogeneous, dry, relatively free of fractures, and capable of sorbing or precipitating released radionuclides.

GEOLOGIC SETTINGS

Several types of rock are potentially suitable for repository sites, of which the following have received the most serious consideration.

Salt

Salt occurs in many localities throughout the world, either as extensive bedded deposits or in the form of domes. Salt has been considered a leading contender for repositories at least since 1957, when a National Academy of Sciences committee called attention to the many advantages of this type of rock (NAS/NRC, 1957a, 1970). A U.S. demonstration repository for TRU waste, the Waste Isolation Pilot Plant, is discussed later in this chapter.

Among the advantages of salt beds are their age (greater than 200 million years), which gives assurance of geologic stability and isolation from aquifers. Salt has high thermal conductivity, low permeability, and plastic characteristics that permit fractures to close. A disadvantage is that, because of its plasticity, the repository would become self-sealed after closure and wastes would thereafter not be readily retrievable. Another possible disadvantage is that the salt is frequently associated with deposits of potash or hydrocarbons and might result in human intrusion some time in the future. An obvious disadvantage is the solubility of salt, which requires that there be assurance that intrusion by groundwater will not occur.

A curious characteristic of salt is that brine inclusions are frequently present, which, in a thermal gradient, tend to migrate toward a source of heat. This phenomenon is the result of differential rates of solution at the walls of the inclusion, caused by the temperature gradient. The brine inclusions can thus move toward the canister and cause undesirable corrosion. This need not be a significant issue if the thermal gradients are properly specified in the repository design (NAS/NRC, 1983). Model development for salt deposits is being actively pursued (Nies, 1990).

Basalt

Basalt originates from volcanic flows that occur extensively in eastern Washington, Oregon, and other U.S. localities. The basalt beds near Hanford, Washington, were one of the five candidate sites originally selected by the DOE. One study has called attention to a number of critical geological problems related to basalt, including the effects of repository heating on the basalt, the abundance of faults and fractures, and uneven stresses in the basalt (NAS/NRC, 1983). Following designation of the Yucca Mountain site in Nevada as the U.S. high-level waste repository (discussed later), interest in characterizing basalt has waned.

Tuff

Tuff is a compacted volcanic ash, the accumulated glassy fragments from explosive volcanic eruptions, and occurs throughout the western United States and in other parts of the world. Tuff is characterized by relatively low permeability and porosity, but the negative aspects associated with tuff include its relatively high water content and the fact that the deposits are in areas of recent vulcanism and fault movement. A major advantage of tuff is that the deposits are usually above the water table. The proposed Yucca Mountain high-level repository site near Las Vegas, Nevada, is composed of tuff. Japan has also expressed interest in tuff as a potential repository medium (IAEA, 1993).

Granite

The granites and other crystalline "hard" rocks are a major class of rocks that were not included among the candidates for the first U.S. repository, but have been selected as the preferred rock type in both Canada and Sweden. Other countries considering hard rock repositories are Argentina, Finland, France, India, Japan, the Republic of Korea, Spain, Switzerland, and the United Kingdom (IAEA, 1993). Positive aspects are high thermal conductivity and structural strength, low porosity and permeability, and, usually, low water content. An important disadvantage is uncertainty about the degree of fracturing and its influence on the movement of groundwater. A large part of the characterization of potential hard rock repositories is study of fractures and their effect on groundwater movement.

METHODS OF RISK ASSESSMENT

To date no country has emplaced high-level waste in a geologic repository for permanent disposal. To assess the performance of a repository, it is necessary to model the expected behavior of the waste far into the future, based on current knowledge. There is no precedent in technology for the long periods of time for which risk assessments are required in radioactive waste management. Although there is emerging policy on the length of time for which safety must be assured and for the amounts of radioactive materials that should be permitted to enter the biosphere in future millennia, the policy is not universally accepted and is potentially subject to change. Approaches to the assessment and limitation of risk to individuals and populations potentially exposed to radionuclides from repositories have evolved and have even been codified in law or regulations, only to be remanded by courts and new approaches taken.

One approach to understanding the potential risk from high-level radioactive waste was the "water dilution volume" (WDV), an overly simplified indicator of risk that simply considered the amount of water required to dilute the radioactive material to the concentration that is safe for drinking water (NAS/NRC, 1983). The WDV does not consider the many ways in which the risk can be increased, such as by biomagnification, or decreased by immobilization in sediments.

The manner in which the radioactivity of high-level wastes diminishes with time is illustrated in Fig. 11-4, where the ADV is plotted against time for 10 million years. The units of water dilution are cubic meters of water per megagram of heavy metal (m^3 Mg^{-1} HM). It is seen that after a few hundred years, the WDV for fission products declines rapidly and the remaining nuclides are members of the actinide family of elements, except for ^{210}Pb, which grows in from ^{230}Th and peaks at about 10^5 years.

There have been a number of attempts to answer the question, "How long is long enough?" by comparing the WDV for nuclear wastes with that of the uranium ore from which the fuel was originally derived. Figure 11-5 does this by comparing the WDV for the high-level waste and uranium ore on the same time scale shown in the earlier figure

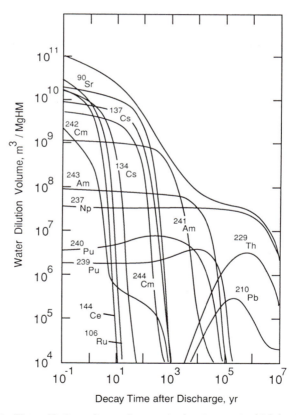

FIGURE 11-4   Water dilution volume of pressurized-water reactor high-level waste. [From NAS/NRC (1983).]

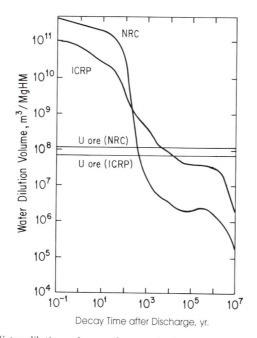

FIGURE 11-5 Water dilution volumes of pressurized-water reactor high-level waste and its parent uranium ore. [From NAS/NRC (1983).]

(NAS/NRC, 1983). Two comparisons are shown, using WDVs calculated by both the NRC regulatory limits and the International Commission on Radiological Protection (ICRP) recommendations of allowable annual intake in effect at that time. Using the NRC regulatory limit, the WDV for high-level waste crosses that for the quantity of uranium ore from which the fuel was obtained in less than 1000 years, but the crossover using the ICRP limits does not take place for more than 10^4 years. Figure 11-5 illustrates the sensitivity of such comparisons to changes in the underlying biophysical parameters used to calculate dose.

The ICRP and NRC values were the same for many years, but in 1980, ICRP published Part 2 of Publication 30 (1979–1988), which changed the absorption factors for transfer of a number of radionuclides from the intestines to blood, the net effect of which was to greatly increase the long-range potential hazard from ^{237}Np, and neptunium is an element of major concern. The difference between the two crossover times is mainly, but not entirely, due to that one change. The ICRP factors have since been adopted by U.S. regulatory agencies. Research in connection with the proposed U.S.

site at Yucca Mountain, Nevada, has shown that neptunium sorbs strongly
to calcite (DOE, 1995e). This could effectively retard the transport of neptunium through fractures.

Although the water dilution volume approach gives some insight into
the length of time it will take reactor wastes to decay to the equivalent
concentrations of uranium ore, it does not provide a quantitative estimate
of the dose or risk to persons potentially exposed to radionuclides from a

TABLE 11-8

Limits for Cumulative Releases to the Accessible Environment for 10,000 Years
after Disposal of Spent-Fuel or High-Level Waste[a]

Radionuclide	Release limit per 1000 MT of "heavy metal" or other unit of waste[b]	
	Ci	GBq
Americium-241 or -243	100	3,700
Carbon-14	100	3,700
Cesium-135 or -137	1,000	37,000
Iodine-129	100	3,700
Neptunium-237	100	3,700
Plutonium-238, -239, -240, or -242	100	3,700
Radium-226	100	3,700
Strontium-90	1,000	37,000
Technetium-99	10,000	370,000
Thorium-230 or -232	10	370
Tin-126	1,000	37,000
Uranium-233, -234, -235, -236, or -238	100	3,700
Any other alpha-emitting radionuclide with a half-life greater than 20 years	100	3,700
Any other radionuclide with a half-life greater than 20 years that does not emit alpha particles	1,000	37,000

[a]From Title 40, Code of Federal Regulations, Part 191.13, Appendix A.
[b]The release limits apply to the amount of wastes in any one of the following: (a) an amount
of spent nuclear fuel containing 1000 MT of heavy metal (MTHM) exposed to a burnup between
25,000 and 40,000 megawatt-days per metric ton of heavy metal (MWd/MTHM); (b) the high-
level radioactive wastes generated from reprocessing each 1000 MTHM exposed to a burnup
between 25,000 and 40,000 MWd/MTHM; (c) each 100,000,000 Ci (3700 PBq) of gamma- or
beta-emitting radionuclides with half-lives greater than 20 years but less than 100 years;
(d) each 1,000,000 Ci (37 PBq) of other radionuclides (i.e., gamma or beta emitters with half-
lives greater than 100 years or any alpha emitters with half-lives greater than 20 years); or
(e) an amount of transuranic wastes containing 1,000,000 Ci (37 PBq) of alpha-emitting
transuranic radionuclides with half-lives greater than 20 years.

repository. Current approaches to assessing repository performance are based on a process in which the release, transport, and uptake of radionuclides are modeled and doses to individuals calculated. Dose assessment, used also in other contexts such as historical dose reconstructions and preparation of environmental impact statements, is discussed in Chapter 14. Dose assessment for future repository performance is inherently more uncertain than that for past activities because it is necessary to make predictions of future geological, climatological, and social change. Thus the "scenario" postulated for the future is often the most important determinant of the calculated doses and risks. The following section discusses the criteria for high-level repository performance in the United States and a later section on Yucca Mountain gives the results of the current performance assessment of that proposed repository.

CRITERIA FOR OVERALL REPOSITORY PERFORMANCE IN THE UNITED STATES

The design of high-level waste repositories has been handicapped by the fact that repository performance requirements are slow to evolve and are still changing. In the United States, the EPA is charged with setting standards for high-level nuclear waste repositories. The agency followed a lengthy and public process to establish its first standard for deep geological disposal of a high-level waste in 1985. The standard was immediately challenged in court, rewritten, and ultimately published again in December of 1993 (EPA, 1993) with an effective date of January 19, 1994. This standard is based on a premise that the repository should be built so that it results in no more than 1000 hypothetical deaths worldwide in 10,000 years, using contemporary risk coefficients for development of fatal cancers. For the sake of implementation, however, the standard was expressed in terms of the probability of releases of radionuclides and classes of radionuclides from the repository. The standard also contains an individual dose limit of 15 mrem (0.15 mSv) per year and a requirement that there be a reasonable expectation that underground sources of drinking groundwater remain within the 1994 EPA drinking water standards for 10,000 years. Table 11-8 gives the basic EPA limits for release to the accessible environment.[3]

[3]Accessible environment is defined by the EPA in 40 CFR 191 as the atmosphere, land surfaces, surface waters, oceans, and all of the lithosphere beyond the controlled area. "Controlled area" is further defined by the EPA as (1) a surface location, to be identified by passive institutional controls, that encompasses no more than 100 km^2 and extends horizontally no more than 5 km in any direction from the outer boundary of the original location of the radioactive wastes in a disposal system, and (2) the subsurface underlying such a surface location.

For the expected case where a mixture of radionuclides is projected to be released, the limiting value is determined as follows. For each radionuclide in the mixture, the ratio of the projected release over 10,000 years and the limit for that radionuclide in Table 11-8 is determined. The sum of these ratios, commonly referred to as the "EPA sum," must have a likelihood of less than one chance in 10 of exceeding 1 and a likelihood of less than one chance in 1000 of exceeding 10. For the past several years, efforts to assess the expected performance of Yucca Mountain (discussed later in this chapter) have centered on this criterion (Wilson *et al.*, 1994).

Despite the efforts of the agency, this standard was overtaken by events. Before it was finally promulgated, Congress enacted the Energy Policy Act of 1992. This act required a separate standard specifically for the proposed Yucca Mountain repository and required the EPA to obtain an analysis of the scientific basis for the standard from the National Academy of Sciences.

The Academy completed its work in August of 1995 (Committee on Technical Bases for Yucca Mountain Standards, 1995), essentially rejecting the existing EPA standard. The Academy found no basis for limiting the assessment to 10,000 years, noting that the maximum releases projected were projected to occur tens to hundreds of thousands of years or even further into the future and that geological conditions could reasonably be assumed to be stable for a million years.

The Academy rejected the population risk basis of the EPA standard, preferring an individual risk standard and suggesting that the EPA use 5×10^{-4} per average lifetime as a starting point. They further argue that for Yucca Mountain, an individual risk limit would also provide adequate protection of the public at large, particularly if policy makers can accept that very low doses of radiation pose negligibly small risk. A dose-based standard is rejected because knowledge of the relationship between dose and risk is uncertain and subject to change.

The existing EPA standard requires consideration of human intrusion into the repository but the Academy concluded that it is not possible to assess the frequency of intrusion far into the future. The report endorses the critical group approach defined by the International Commission on Radiological Protection (ICRP, 1985). This approach protects the vast majority of the public while ensuring that repository decisions are not influenced unduly by risks to a few individuals with unusual habits or sensitivities.

The EPA has announced plans to work on a new standard for Yucca Mountain based on the recommendations of the National Academy of Sciences. The existing, extensive work on the performance assessment of Yucca Mountain (Wilson *et al.*, 1994) is based on the 1993 EPA standard and much will likely have to be redone if the EPA follows the advice of the Academy in preparing the new standard. The existing standard is still applicable to any other high-level or TRU repositories that may be proposed, including the DOE's Waste Isolation Pilot Plant repository, and to the management of high-level waste prior to disposal.

WHAT CAN WE LEARN FROM NATURE?

The models used to predict the performance of the repository are constructed by linking together the rates at which groundwater will seep into the repository, the rates of corrosion of the canister, solubilization of the waste form, and migration of the radionuclides through the backfill and rocks. After a thousand years or more the nuclides reach the biosphere, and then begins the task of modeling the pathways by which the nuclides reach human beings. Kocher *et al.* (1983) have estimated that the uncertainty in the dose estimates, based on transport only in the biosphere, may cover four or five orders of magnitude, and the uncertainty in the transport models used to describe the movement from the repository to the biosphere is indeterminate.

Most of the information on leaching rates, retardation factors, k_d, and other parameters of the dose models is obtained in laboratory measurements. A report by a panel of the National Academy of Sciences (NAS/NRC, 1982) noted that the leaching rates of solids under natural conditions are very much lower than those obtained by laboratory measurement and that the differences are frequently two orders of magnitude or more.

A number of investigators have used natural analogues to infer the behavior of a deep geological repository. The best known of these studies involves a natural fossil fission reactor in the Republic of Gabon, West Africa (see Chapter 6) (Cowan, 1976; IAEA, 1975). The reactor existed in what is now known as the Oklo uranium mine about 1.8 billion years ago, when conditions in the uranium deposit were such as to sustain criticality for a period on the order of 10,000 years, during which time huge amounts of fission products and transuranic elements were produced. The radionuclides have long since decayed to stable nuclides at the ends of the decay chains, and the presence of these nuclides, along with the stable nuclides produced in fission, has been detected in the environs of the fossil reactor. There is evidence that most of the nuclides of interest migrated for very short distances, on the order of meters, before they decayed. The fractional mobilization rates of the individual radionuclides from the reactor zones have been estimated to be on the order of 10^{-7} to 10^{10} years. The mobilization rates were so low that most of the radionuclides decayed before they migrated more than a few meters from their source.

A second natural analogue from which similar conclusions have been drawn is a highly weathered deposit of thorium and rare earth elements located near the summit of a hill in the state of Minas Gerais, Brazil. The hill, which is known as the Morro do Ferro (see Fig. 6-14), has been studied as an analogue for an ancient high-level waste repository that has been invaded by groundwater and eroded to the surface. The ore body contains about 30,000 MT of thorium, which is being used as a chemical analogue for Pu^{4+}. Lanthanum, a rare earth element, is being used as a chemical analogue for Cm^{3+} and Am^{3+} (Eisenbud *et al.*, 1984). As at Oklo, the mobilization rates for the two analogues for the transuranic actinide elements have been shown to be on the order of 10^{-9} per year. Thus, even in this near-surface, highly weathered, wet ore body, the mobilization rates are so low as to ensure *in situ* decay of the transuranic elements plutonium, americium, curium, and, probably, neptunium (Krauskopf, 1986).

One feature that Oklo and Morro do Ferro have in common is the presence of abundant quantities of clay minerals. At Morro do Ferro, the primary minerals in which the analogue elements were originally contained have long since been destroyed by weathering, and the analogue elements are presently immobilized in amorphous form associated with clays and iron oxides.

There are surely conditions in nature that would result in more rapid dispersal than has been seen at these two sites. On the other hand, the fact that some mineral deposits are stable in nature over geologic time is indisputable, and it would seem that long-term behavior of the wastes could best be predicted on the basis of the geophysical and geochemical conditions that tend to result in the natural stabilization of chemical elements.

PROPOSED REPOSITORIES IN THE UNITED STATES

At the present time, the United States is developing two deep geologic repositories, one for high-level waste at Yucca Mountain in Nevada and one for transuranic wastes near Carlsbad, New Mexico.

THE YUCCA MOUNTAIN REPOSITORY IN NEVADA (HIGH-LEVEL WASTE AND SPENT FUEL)

After 25 years of political indecision over how to deal with the high-level waste problem, the U.S. Congress enacted the Nuclear Waste Policy Act in late 1982, and it was signed into law on January 7, 1983, by President Reagan (Nuclear Waste Policy Act, 1983). The act established January 31, 1998, as the date when the first repository would begin operation. The law was highly specific as to how the U.S. program was to proceed. The process began with the issuance of guidelines for site selection, which were published in the summer of 1984 (DOE, 1984b) and enumerated the geological and other considerations that would qualify or disqualify a proposed site. The DOE then nominated five sites as suitable for investigation and carried out a comparative analysis. Following the comparative analysis, three sites were nominated for recommendation to the president. Table 11-9 lists the five preliminary sites that were recommended for characterization by the DOE as well as the three selected for detailed characterization (DOE, 1985c).

After proceeding along this path of development for a few years, it was clear that progress was slow and that selecting a site was more of a political than a technical decision. Just as serious site characterization was to begin at the Nevada, Washington, and Texas sites, Congress passed the 1987 amendment to the Nuclear Waste Policy Act in which the Yucca Mountain location in Nevada was designated as the one site to be characterized and developed (Nuclear Waste Policy Amendments Act, 1987). A site characterization plan was prepared consisting of 106 proposed studies in geology, hydrogeology, geochemistry, and engineering geology (DOE, 1988). These studies are to provide the necessary technical information to support the repository design and performance assessment (IAEA, 1993).

Although many agencies of the federal government will undoubtedly be involved with the repository, three are critical. The EPA sets the standards to be met by the repository. The DOE, through its contractors, characterizes, builds, and hopefully operates it. The NRC is charged with licensing the repository and verifying that the repository meets standards.

It is clear that a repository will not be available to accept waste from nuclear utilities in 1998, the date specified in the original act. The DOE is currently planning to have the repository open by 2010 for commercial spent fuel and by 2016 for DOE high-level waste and spent fuel (DOE,

TABLE 11-9

Sites Proposed for Nomination as Suitable
for Characterization[a]

State	Site	Geologic medium
Mississippi	Richton Dome	Domal salt
Nevada	Yucca Mountain[b]	Tuff
Texas	Deaf Smith County[b]	Bedded salt
Utah	Davis Canyon	Bedded salt
Washington	DOE Hanford site[b]	Basalt

[a]From U.S. Department of Energy (DOE, 1985c).
[b]Preliminary recommendation for detailed characterization.

1995d). The DOE has a statutory obligation to accept waste as of 1998 and there has been consideration given to storing it on existing DOE reservations. This plan has met with political opposition in prospective host states.

The development and operation of the high-level waste repository in the United States are funded by assessing nuclear utilities a fee of $1 per MW-hour of nuclear energy produced. The fee may have to be increased in the future (Nuclear Energy Agency, 1993). Approximately $8 billion has been collected to date and about $4 billion spent, but no fuel has been disposed of or transferred to a central storage location. This apparent lack of progress is disturbing to the nuclear utility industry (Denver, 1994).

Description of the Yucca Mountain Site

The proposed Yucca Mountain repository was described in the Site Characterization Plan (DOE, 1988), and other discussions can be found in IAEA (1993), in the performance assessment (Wilson *et al.*, 1994), and in a report of the National Academy of Sciences (Committee on Technical Bases for Yucca Mountain Standards, 1995).

The proposed repository site, northwest of Las Vegas and near the Nevada Test Site, is composed of welded tuff, a compacted volcanic ash. The site is characterized by relatively low permeability and 10% porosity. Less desirable aspects of this tuff include its relatively high water content and the fact that the deposits are in areas of recent vulcanism and fault movement. The Yucca Mountain tuff is underlain by layers of zeolites, which are known to be effective in retarding the movement of trace metals and radionuclides (DOE, 1995e). The repository will be in the unsaturated zone about 300 m below the surface and between 150 and 300 m above the water table. There are one to three fractures per meter but seals can be built into the repository.

The repository will consist of underground facilities, surface facilities, and ramps and tunnels connecting them. A conceptual sketch of the plan is given in Fig. 11-6. By law, the repository is being designed for 70,000 MT of waste; current plans indicate that 10% of the total would be defense high-level waste and 90% would be spent fuel from commercial nuclear power plants. Current conceptual waste container designs call for cylindrical metal containers with gas-tight seals. The original plan called for about 35,000 cylinders of waste placed vertically, but larger, horizontally placed, containers are now being considered. The projected waste could be accommodated by 8500 of these horizontal, "in drift" containers. Fifty to 100 years after the start of waste emplacement, the repository would be sealed by backfilling the tunnels, closing the opening to each emplacement tunnel, and sealing the entrances.

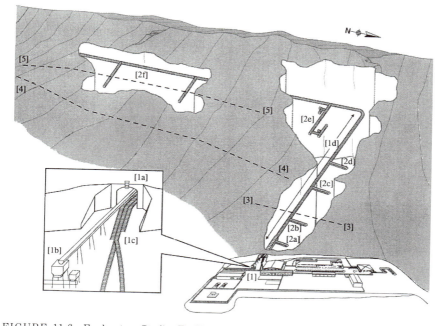

FIGURE 11-6 Exploratory Studies Facility (ESF) of the proposed Yucca Mountain geologic repository for high-level nuclear waste. [1] The North Portal is the main entrance to the ESF, where work began in April 1993; [1a] ventilation system; [1b] conveyor system; [1c] rail system used to move material and equipment; [1d] the north ramp is the main access to the potential repository level and to the main north–south tunnel; [2a–f] alcoves where various research projects are being conducted or proposed; [3] Bow Ridge Fault, the largest vertical displacement fault yet encountered in the ESF; [4] zone containing a series of small, near-vertical faults; [5] Ghost Dance Fault, the main fault found within the potential repository area. The fault is being studied to assess its influence on how water or gases might move from the surface through the mountain. (Courtesy of the Yucca Mountain Project.)

Expected Performance of the Repository

Work to date on expected performance of the repository has focused on modeling to compare with the existing probabilistic, cumulative release standards (EPA, 1993) and on assessing individual dose. As discussed earlier in this chapter, the U.S. Congress has required the EPA to prepare new environmental radiation protection standards specifically for the Yucca Mountain repository. Because the existing standards are expected to change, only a brief discussion will be given of progress and results to date. Modeling of the Yucca Mountain repository is the task of Sandia National Laboratory and they refer to their reports as "total-system performance assessments" (Barnard *et al.*, 1992; Wilson *et al.*, 1994). Model calculations were made for both the 10,000-year period specified in their current standard and for 1,000,000 years to study the impact of longer time periods on repository performance. An important factor affecting model outcome is how water is assumed to flow through the repository and two cases have been modeled.

For the first water flow case, referred to as the composite-porosity model, groundwater is assumed to flow according to Darcy's law (see Chapter 5) through a medium with porosity equivalent to the combination of matrix and fractures of the actual repository medium. The result of this model is relatively uniform flow that causes a large number of containers to be in a moist or wet environment, leading to aqueous corrosion and widespread early failure. The slow percolation in the unsaturated zone limits releases during the first 10,000 years but is insufficient to limit peak doses that could occur in a million-year period. For individual dose without a time limit, both dilution in the environment and release from the containers are important factors.

For the second water flow case, referred to as the weeps model, groundwater flow is restricted to locally saturated fractures (weeps) that contact the waste only at discrete points. Transport of the radionuclides is assumed to occur instantaneously through the unsaturated zone but container degradation and subsequent release of radionuclides is assumed to occur only when the containers and the weeps intersect. Contact between the water flowing in the weeps and the waste containers is relatively rare so that most containers stay dry and intact. The result of this modeling effort is that peak doses occur within the first 20,000 years and cumulative releases do not increase much thereafter.

For gas flow and $^{14}CO_2$ transport, a model was developed that utilized information about the site's topography and stratigraphy. Calculations indicated that transport times are fast enough that most of the gaseous radionuclides released from the waste containers will reach the surface.

Three disruption scenarios are included in the recent modeling: human intrusion, direct volcanic effects, and indirect volcanic effects. Human intrusion is assumed to involve exploratory drilling with present-day technology, waste container breakage or a near miss, and radionuclide release with the extracted drill core and drilling fluid. Direct volcanic effects are modeled as intrusion of a basalt dike directly into the repository. Indirect volcanic intrusion is modeled as volcanic activity near the respository that heats the waste containers and supplies corrosive volatile chemicals. Both heat and corrosive chemicals contribute to increased corrosion and failure of the waste containers. The probabilities of any of these events are low, but the consequences in terms of amount of released radionuclides are relatively high.

The repository models are probabilistic. The value of each of the important parameters is characterized by a probability distribution. The model is then run numerous times with each run called a "realization." For each realization, parameters are selected randomly from their individual probability distributions in accordance with the mathematical laws governing the distribution. The results from the realizations are combined to give overall probabilities, which are then expressed as functions of releases or doses.

For the composite-porosity model the repository does not meet the EPA 10,000-year performance criteria (see Criteria for Overall Repository Performance earlier in this chapter). The gaseous releases, primarily ^{14}C, are primarily responsible for the failure to meet the criteria. The reason for this is that there is a high degree of failure of containers due to the presence of moisture, and the geologic formation affords little retardation of the gaseous releases. With the weeps model, the EPA criteria are met because waste container failure is less probable in the weeps model than in the composite-porosity model and there is reduced probability of release through gaseous and aqueous transport.

When modeling was performed for one million years, over 90% of the realizations of the composite-porosity model and 1% of the weeps model realizations resulted in doses from drinking water exceeding the present-day total background of 300 mrem (3 mSv) per year. The composite-porosity model predicted higher doses because it predicts a higher rate of container failure than the weeps model. The doses are caused primarily by ^{237}Np, which is both long-lived, with a half-life of 2.1 million years, and fairly mobile.

Regulation is a critical factor in repository performance assessment. If the EPA follows the current advice of the National Academy of Sciences (Committee on Technical Bases for Yucca Mountain Standards, 1995), then individual dose over a longer term, rather than radionuclide release for 10,000 years, will become the measure of compliance. More emphasis will be needed on release rates in the models. Dose calculations require more information about the biosphere and more speculation about future conditions than do total release calculations. Uncertainties in doses will be greater than those for releases because of the additional assumptions required.

THE WASTE ISOLATION PILOT PLANT IN NEW MEXICO (TRANSURANIC WASTE)

The Waste Isolation Pilot Plant (WIPP) is a demonstration repository being constructed by the DOE near Carlsbad, New Mexico, but its use will be limited by law to TRU defense wastes that have been retrievably stored.

There is currently no provision for the "permanently disposed" TRU wastes that were buried prior to 1970. Although there is provision for demonstration of the feasibility of storing high-level wastes at WIPP, this seems unlikely at this time because the Yucca Mountain site in Nevada has been designated as the U.S. high-level waste repository.

The WIPP is a mined cavity in bedded salt 660 m below the earth's surface (NAS/NRC, 1985a). The underground layout is 776 by 1556 m and is served by two shafts, 3.6 and 1.8 m in diameter. About 4 km of tunnels and rooms have been constructed (IAEA, 1993). Costs to date have been $1.6 billion and total costs are expected to exceed $8 billion. Prior to permanent disposal of waste, an experimental program is planned to investigate thermal and structural interactions, plugging and sealing of the boreholes and shafts, and the interaction of the waste with the repository environment. Although the WIPP repository was physically ready by 1988, various environmental objections and legal and political maneuvers have prevented any wastes from being received, even for the experimental program. The DOE currently plans to begin receiving waste at WIPP beginning in 1998 (DOE, 1995d).

The DOE undertook initial risk assessments for the WIPP (DOE, 1979a). It was concluded that a person living near the storage site would be subject to a 50-year bone dose commitment of 1.1×10^{-8} rem (1.1×10^{-10} Sv) and a whole-body dose commitment of 8.7×10^{-11} rem (8.7×10^{-13} Sv). Even allowing for the uncertainties in dose assessment, it appeared then that isolation of transuranic radioactive wastes in geological repositories could be achieved.

As noted earlier in this chapter, the WIPP repository is now subject to the EPA standards for release of radionuclides to the accessible environment (EPA, 1993). The DOE is in the process of preparing a compliance certification application and has submitted drafts to the EPA for initial review (DOE Carlsbad Area Office, 1995). The methodology is to use information about the disposal system and the waste to simulate repository performance over the 10,000-year performance period. The process involves identification of features, processes, and events that might affect repository performance, and subsequent development of scenarios, scenario probabilities, and consequence models. As for Yucca mountain, complex probabilistic models are constructed and run on computers. The preliminary modeling for the WIPP repository includes the scenarios of undisturbed performance and exploratory borehole intrusion events. The preliminary results suggest that the repository will be able to meet the EPA release requirements, but the results will be undoubtedly subjected to critical review and further refinement.

Need for Interim Storage

Because spent-fuel storage capacity at the reactor sites is limited, an acute need is developing for some form of "away from reactor" storage. As part of the original Nuclear Waste Policy Act, the U.S. federal government was required to examine the need for construction of interim storage capacity and to make appropriate recommendations to Congress (DOE, 1985c). The costs of the facilities, if needed, would be borne by the electric utilities. The Nuclear Waste Policy Act as amended in 1987 provides that, pending development of a system for permanent isolation of high-level wastes, the government would construct monitored retrievable storage facilities. These MRS facilities were to have been built by the DOE in accordance with NRC licensing requirements, and the cost would have been borne by the users. To prevent the MRS from becoming a de facto permanent repository, the amended act prohibits fuel receipt at an MRS until after the permanent repository was licensed. Because Nevada was designated as the host state for the permanent repository, the MRS was to be built in another state.

Unfortunately, the U.S. government has been unable to site an MRS facility either through DOE siting processes or through the U.S. Office of the Nuclear Waste Negotiator. This small Office was authorized to provide study grants and negotiate with communities or Native American tribes who expressed interest in serving as a host to an MRS facility. Although the Office carried out its functions, it did not succeed in finding a host tribe or community and was abolished at the end of 1994 (Stallings, 1994). Thirty-one nuclear utilities in the United States have banded together to seek a host for a private MRS facility and are negotiating with the Mescalero Apaches in New Mexico. The situation in Europe appears to be better, with 12 operating "away from reactor" storage sites and 3 more under construction, all with international safeguards (Nuclear Energy Insight, 1995a).

THE SPECIAL PROBLEMS OF GASEOUS OR HIGHLY SOLUBLE LONG-LIVED RADIONUCLIDES

There are four radionuclides that, by reason of their long half-lives and either volatility or solubility, are accumulating in the general environment. These are tritium, krypton-85, carbon-14, and iodine-129, which are produced in both the generation of nuclear power and the explosion of nuclear weapons.

Tritium

Tritium (^{3}H) is present naturally in the environment and is also produced during the manufacture and explosion of nuclear weapons, as well as in nuclear reactors. The annual production of fission-product tritium by a 1000-MWe light-water reactor (Kouts and Long, 1973) is in the range of 15,000 to 25,000 Ci (560 to 930 TBq). When the tritium migrates from fuel during reactor operation, a fraction reacts with the zirconium cladding and becomes immobilized as zirconium hydride (NCRP, 1979). It has been found that about 13% of the total tritium contained in the fuel is in the form of zirconium hydride and about 87% remains as tritium in the fuel pellets. However, if the cladding has minor defects, the tritium can be released directly to the circulating water. Tritium production by neutron interaction with soluble boron can account for an additional 500–1000 Ci (19–37 TBq) per year.

Tritium remaining in the fuel will be released if the fuel is reprocessed, as a result of which the tritium emissions from a fuel-reprocessing plant are far greater than from reactors. Not only is the bulk of the tritium contained in the fuel being dissolved, but the reprocessing plant has a far greater throughput of fuel.

Tritium from all sources is disseminated in the environment as water and enters the hydrological cycle. The impact of tritium produced in weapons testing has been far greater than from all other sources, as can be seen from Fig. 11-7, which plots the contribution of the various sources to the tritium concentration of surface waters from 1960 projected to the year 2000. Assuming there are no further major programs of atmospheric weapons testing, tritium from fallout will continue to diminish with the radiological half-life of 12.3 years, and toward the end of this century the contribution of nuclear power will be about equal to the fallout residual. Nuclear power worldwide has not increased at the optimistic rates predicted in the 1970s, so these tritium predictions are likely to err on the high side.

Although copious quantities of tritium have been and will continue to be produced, the dose to humans will be small. It has been estimated, as shown in Fig. 11-8, that the dose to humans peaked in the mid-1960s at about 0.2 mrem y^{-1} (2 μSv y^{-1}).

KRYPTON-85

Krypton-85 is a noble gas nuclide that is produced copiously in fission. Krypton does not participate in metabolic processes, and the principal dose to an individual immersed in a cloud of this nuclide is to the skin, resulting from the 0.25-MeV beta emission.

Krypton-85 is produced in a 1000-MWe power reactor at an annual rate of about 500,000 Ci (18,500 TBq). Less than 1% of the krypton produced leaks through the fuel cladding during normal reactor operation, and for this reason the ^{85}Kr releases from normally operating reactors are insignificant

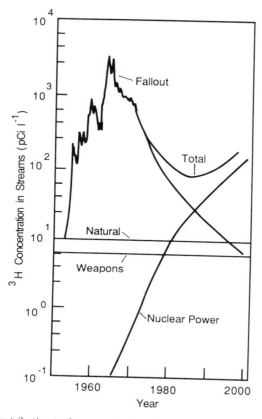

FIGURE 11-7   Contribution to the concentration of tritium in streams and fresh water from all sources. "Weapons" refers to nuclear weapons production activities. [From NCRP (1979).]

compared to the releases from fuel-reprocessing plants (NCRP, 1975b). However, about 50,000 Ci (1850 TBq) of ^{85}Kr was released as a result of the accident at Three Mile Island (NCRP, 1980). This will be discussed further in Chapter 12.

Because of its inertness, ^{85}Kr distributes uniformly throughout the earth's atmosphere within a few years after release. The ^{85}Kr concentration reached about 10 pCi m^{-3} (0.37 Bq m^{-3}) in 1970, mainly from nuclear weapons testing and plutonium production. It has been estimated that the skin dose to the world's population could reach about 2 mrem y^{-1} (20 μSv y^{-1}) early in the next century, but this was based on the somewhat larger nuclear power growth projections of the early 1970s and also assumed that all fuel would be reprocessed without control over the ^{85}Kr. Krypton

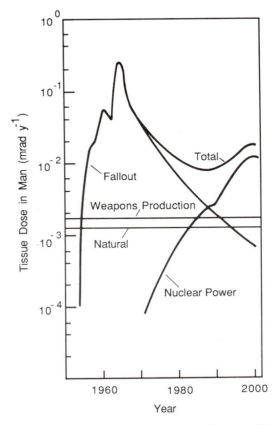

FIGURE 11-8   Projected tissue dose rate from tritium in humans. [From NCRP (1979).]

separation from exhaust gases is feasible using cryogenic and other methods. The EPA now requires that the krypton emissions be reduced from 500 kCi (18.5 PBq) per 1000 MWe to 50 kCi (1.85 PBq) per MWe, a reduction of 90%, from all processing plants built after 1983. As noted earlier, fuel reprocessing has stopped in the United States and its future is uncertain at best. If the spent fuel is stored, the ^{85}Kr will decay with its half-time of 10.7 years. If the fuel is reprocessed, 90% of the ^{85}Kr must be removed for extended storage until the gas has decayed sufficiently for release.

CARBON-14

The three sources of ^{14}C in the environment are from nature, nuclear weapons testing, and reactor operation. Carbon-14 from power reactors will accumulate in the atmosphere, but the dose to human beings will be insignificant for the foreseeable future. The contribution from fallout will continue to dominate the concentrations in the environment, about two orders of magnitude greater than that from reactor operation (NCRP, 1985b). In terms of compliance with now withdrawn EPA requirements, ^{14}C is the most troublesome gaseous release expected from the proposed Yucca Mountain repository (Wilson et al., 1994). It remains to be seen if ^{14}C will still be a problem once the required new EPA standards are developed for Yucca Mountain.

IODINE-129

Iodine-129 is one of the longest-lived nuclides produced in fission, with a half-life of 1.57×10^7 years. It is estimated that by the year 2000, about 2500 Ci (93 TBq) of ^{129}I will have been produced by power reactors. Iodine is such a soluble element, and the half-life of ^{129}I is so long, that the ^{129}I will eventually enter the stable iodine pool. The total amount of iodine that can be absorbed into the thyroid is under metabolic control and is limited to about 0.012 g. Iodine-129 cannot deliver a significant dose to the thyroid because this would require deposition of 34 g of ^{129}I, several thousand times the average normal value (NCRP, 1983). However, if one assumes that the 2500 Ci (93 TBq) of ^{129}I produced by the nuclear power industry up to the year 2000 will disseminate throughout the environment, it can be calculated that the collective dose equivalent to the world's population will be about 10^8 person-thyroid rem (10^6 person-thyroid Sv). Iodine-129 can be removed during fuel reprocessing, but whether this will significantly affect the long-range dispersal of ^{129}I in the environment is uncertain.

CHAPTER 12

Experience with Radioactive Contamination Due to Accidents

Contamination of the environment by radioactivity has occurred in a number of countries as a result of accidents or from the routine activities of industrial facilities. The nuclear weapons production complexes of the United States and former Soviet Union have been particularly important in this regard, although there are a number of other examples, some of which occurred early in this century in the processing of radium and other naturally occurring substances. Contamination from uranium mine tailings is discussed elsewhere (see Chapters 7 and 15). In Chapter 9 we discussed extensively the contamination that has occurred worldwide as a result of nuclear weapons testing. Waste management is discussed in Chapter 11, and the remediation of sites contaminated as a result of deliberate disposal practices or poor operating practices is discussed in Chapter 15.

The nuclear accidents reviewed here will be limited to those that have resulted in environmental contamination. Accidents that caused death or injury, but not necessarily environmental contamination, were reviewed by Lushbaugh *et al.* (1980, 1990). An example of an accident that caused severe plant damage but neither injury nor environmental contamination was the 1966 meltdown of the small Enrico Fermi reactor located on the western shore of Lake Erie near Monroe, Michigan (Scott, 1971).

Considering the size and complexity of the nuclear energy industry, the occupational safety record in the United States and most other countries has been excellent during its 50-year history. A total of seven deaths among employees of the nuclear industry have resulted from radiation accidents in the United States, the last of which occurred in 1964 in a privately operated industrial plant (Karas and Stanbury, 1965). Three of the seven fatalities occurred at a 1961 military reactor accident in Idaho that will be discussed later in this chapter. All but one of the fatal accidents in the United States were in connection with military nuclear research or development activities. With the exception of the Idaho accident, the workers were exposed only to external sources of gamma radiation with no dispersion of radioactivity. In addition to the accidents that took place in government facilities, there have been some that were due to the mishandling of industrial radiographic sources (Lushbaugh *et al.*, 1980). There have been about one dozen injuries from such accidents in the United States, and several more in foreign countries (Hubner and Fry, 1980; Ricks and Fry, 1990).

The number of accidents to be reviewed in this chapter has increased substantially in number and severity since the first edition of this book was published more than 30 years ago. Some are more important for our purposes than others. The accidents described have been selected either because of their historical importance or because of what we can learn from them. Thus, a minor incident in Houston, Texas, in 1957 is included to illustrate the importance of matching the response to the severity of the accident. That accident received sensational national attention because of the inexperience of the responsible government agency in judging the national hysteria that would result from its actions. At the other end of the severity spectrum is Chernobyl, the complexities of which make it difficult to summarize in just a few pages. One decade after this accident, hundreds of scientific papers have already been published on the biomedical consequences of the Chernobyl disaster (IAEA, 1996). At this writing, considerable information is being disclosed about other mishaps in the former Soviet Union that had not been previously publicized.

FALLOUT FROM THE THERMONUCLEAR WEAPONS TEST OF MARCH 1, 1954

Prior to March 1, 1954, there had been speculation among the very few people who were then informed about the subject of fallout as to whether surface and near-surface bursts in the megaton range could produce lethal amounts of fallout over large areas. The many doubts that existed even as late as March 1, 1954, were quickly dispelled when extensive fallout occurred following detonation of a 15-MT device (code-named BRAVO) mounted on a barge over the reef of Bikini Atoll in the mid-Pacific Ocean. The barge was located in shallow water and a large amount of coral was incorporated into the fireball. There was little wind shear, and it is estimated that from 50 to 80% of the radioactivity from the explosion fell out in the pattern shown in Fig. 12-1. This was the first test in the series known by the code name CASTLE (Conard, 1992; Eisenbud, 1990).

The first indication that radioactive fallout was occurring was an increase in a recording exposure rate meter located on Rongerik Island about 260 km east of Bikini. This island was inhabited temporarily by 28 American servicemen who were maintaining a weather station. The recording gamma-radiation detector was placed in their custody by the Health and Safety Laboratory of the AEC. The fallout began about 7 h after the explosion, and 30 min later the detector went off-scale at 100 mR h^{-1} ($\sim$1 mGy h^{-1}). Aerial reconnaissance confirmed that fallout had occurred on Rongerik and that even heavier fallout had occurred on Rongelap Atoll, about 170 km east of Bikini, as well as the atolls of Ailinginae, located about 120 km east-southeast, and Uterick, about 600 km to the east. Evacuation procedures were inexplicably delayed and not put into effect until about 30 h after the detonation, with the removal of the 28 Americans by air. The inhabitants of Rongelap, Utirik, and Ailinginae were evacuated after about 50 h.

Figure 12-1 shows the relative locations of these atolls. A number of unanswered questions about the fallout and the management of its consequences have been discussed elsewhere (Eisenbud, 1990, 1994a).

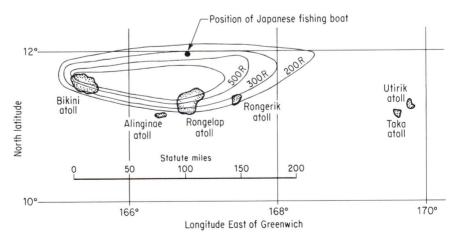

FIGURE 12-1 Fallout pattern from the thermonuclear explosion at Bikini Atoll in the Marshall Islands on March 1, 1954, as reconstructed from various sources.

It was later found that Rongelap Island was so heavily contaminated that the inhabitants would be unable to return for an extended period, and they remained on another atoll of the Marshall Islands for 38 months until July 1957. The 18 people evacuated from Ailinginae were actually Rongelapese who had been visiting the normally uninhabited island. They eventually returned to Rongelap with their kin.

The fallout on Utirik was insufficient to require prolonged evacuation, and the inhabitants were returned to their homes shortly after the incident (Conard *et al.*, 1980; Conard, 1992).

It was not known until some days later that a Japanese fishing vessel, the *Fukuru Maru* (Lucky Dragon), was located in the path of the fallout about 120 km to the east of Bikini. The boat had somehow not been observed during the aerial searches that preceded each test. When the explosion occurred, the men hauled in their lines and sailed away, but encountered visible fallout about 4 h later. The men lived on the vessel for 13 days until they returned to their home port of Yaizu (Kumatori *et al.*, 1980; Japan Society for the Promotion of Science, 1956).

Although the fallout was visible, neither the Marshallese nor Japanese took precautions to minimize their exposure. This is understandable in view of their ignorance of the subject. Thus, the Rongelapese and fishermen lived for 50 h and 13 days, respectively, in intimate contact with their contaminated environment. Large particles of fallout fell into their hair and came into direct contact with their skin. The Japanese fishermen described the dust deposit as similar to a light coating of snow (Japan Society for the Promotion of Science, 1956). Samples of fallout collected from the boat when it reached Yaizu were analyzed for $^{239+240}$Pu and ^{241}Am 20 years after the accident and were found to contain 12 pCi mg^{-1} (0.44 MBq kg^{-1}) of the two plutonium isotopes and 6 pCi mg^{-1} (0.22 MBq kg^{-1}) of ^{241}Am (Hisamatsu and Sakanoue, 1978). It has been estimated by Nishiwaki *et al.* (1956) and Tajima (1956) that the fallout had an original mass density of about 500 mg m^{-2} and a total radioactivity of about 50 Ci m^{-2} (1.9 TBq m^{-2}). The Pu and Am were thus present in the amounts of 6 nCi (220 Bq) and 3 nCi (110 Bq) per square meter, respectively.

The dust filled the air and fell on the bodies of the victims as well as into their food and water. The Marshallese sat on the contaminated ground, and the Japanese sat on the contaminated deck. No special precautions were taken in the consumption of food, and no special hygienic procedures were followed except that a number of Marshallese swam in the surf during the period of exposure. The whole-body and thyroid doses received by the several groups for whom estimates are available are given in Table 12-1.

TABLE 12-1

Summary of Doses Received Following Test Explosion of March 1, 1954

Group	Number exposed	Time fallout started (h after detonation)	Exposure duration	Whole-body dose (estimated rem)[a]	Iodine dose to thyroid (rem)[a]
Rongelap[b]	67	4–6	Evacuated in about 50 h, returned in 38 months	175	<10 y old: 810–1800 10–18: 334–810 >18: 335
Rongerik[b]	28	7	Permanently evacuated in about 30 h	78	50
Alinginae[b]	19	4–6	Evacuated in about 50 h, returned to Rongelap in 38 months	69	<10: 275–450 10–18: 190 >18: 135
Utirik[b]	163	22	Evacuated in 55–78 h, returned to Utirik	14	<10: 60–95 10–18: 30–60 >18: 30
Fukuru Maru[c]	23	4	Remained on boat for 13 days	170–590	296–1026

[a] One rem = 0.01 Sv.
[b] From Conard et al. (1960, 1980).
[c] From Kumatori et al. (1980).

MEDICAL AND ENVIRONMENTAL INVESTIGATIONS

Excellent reports of the American and Japanese studies of this accident are available (Japan Society for the Promotion of Science, 1956; Conard *et al.*, 1980; Kumatori *et al.*, 1980; Conard, 1992; Simon and Graham, 1994).

The Marshallese were evacuated to the island of Kwajalein, where an emergency hospital was established and teams of specialists from the United States were made available to provide medical care. Survey teams were sent to the islands for radiological measurement.

The Japanese fishermen did not enter their home port of Yaizu until March 14, by which time they were already suffering from the effects of radiation exposure, but it was another 2 days until it became known that their sickness was caused by the fallout. Although the men were showing signs of radiation sickness for many days prior to their landing in Yaizu, they maintained radio silence and it was not known that they had been exposed to fallout. It was later learned that the failure to communicate was due to the fact that the same crew had been poaching in Indonesian waters a year earlier, for which they were apprehended and jailed. There was evidently some uncertainty in their minds as to whether they had strayed into restricted waters, for which they might again be detained, this time by U.S. authorities.

The 23 men were at first hospitalized in Yaizu and later transferred to Tokyo. The early effects of the exposures of both the Marshallese and the Japanese fishermen can be described for skin, blood, and absorption of internal emitters.

1. *Skin effects.* Itching and burning sensations were experienced from 1 to 2 days after the fallout by all groups except the Utiriks. After various time intervals, ranging from 3 days in the case of the Japanese fishermen to 21 days at Rongerik and Ailinginae, skin lesions and epilation began to develop. The lesions became ulcerous in 70% of the Japanese fishermen and in about 25% of the Rongelapese. The investigators found that the severity of the injuries were related to obvious factors such as whether outer garments were worn and bathing habits. Rongelapese children who spent much time wading in the lagoon had fewer foot injuries, and the worst burns among the Japanese fishermen were on two men who did not wear hats.

2. *Hematological changes.* The diminution in white blood count was most marked in the fishermen, among whom the values diminished by as much as 50% in about 28 days, at which time slow recovery began. The values of the Marshallese varied from 55% of normal at 44 days in the case of the Rongelapese to about 84% of normal in the case of the less exposed Utiriks.

3. *Dose from internally deposited radionuclides.* Until they were removed from the contaminated environment, the Marshallese and Japanese existed under conditions that maximized the opportunity for inhalation and ingestion of the radioactive particles. Life on an atoll or fishing vessel in the tropics is largely an outdoor existence. The Marshallese drank from exposed cisterns and ate food that was exposed to the open air. The Japanese ate raw fish that were certainly in contact with the contaminated deck. Almost everything on the ship was found to be contaminated when it arrived in Yaizu 13 days later, and the contamination must have been even greater before some of the fallout was washed away by the frequent tropical rains.

Analyses of urine from both the Marshallese and Japanese were found to be positive for fission products. In the Marshall Islands, the first samples were collected 15 days after exposure. The principal absorbed nuclide was ^{131}I and two independent estimates of the initial thyroid deposits (Conard *et al.*, 1980) were 6.4 and 11.2 mCi (240 and 410 MBq). The dose from the short-lived nuclides, ^{132}I, ^{133}I, and ^{135}I, was estimated to be two to three times the dose from ^{131}I and is included in the thyroid dose estimates shown in Table 12-1.

Samples of urine were also collected from the fishermen and were sent to the AEC Health and Safety Laboratory (now the DOE Environmental Measurements Laboratory) for radiochemical analysis (Kobayashi and Nagai, 1956). Except for ^{131}I, only minimal urinary excretion of fission products was found, despite the fact that the men lived for 13 days on the contaminated fishing vessel. Urinary excretion of fission products was less than that reported for the Marshall Islanders. The beta activity of mixed fission products separated chemically from fallout samples collected on April 21, 1954, ranged from 10 to 110 pCi L^{-1} (0.37 to 4.1 kBq m^{-3}) compared with values as high as 37 pCi L^{-1} (1.4 kBq m^{-3}) obtained on five unexposed Americans and analyzed at the same time. All of these data were obtained before the availability of modern methods of gamma spectrometry, which permit more sophisticated analysis.

One of the fishermen died of a liver disease 6 months after the accident. The results of radiochemical analysis of his tissues as reported by Tsuzuki (1955) are given in Table 12-2. The amount of radionuclides in the tissues of this man was so low as to make it unlikely that injury would have resulted from absorption of intermediate and long-lived radionuclides. However, as noted earlier, the dose from radioiodine was certainly significant in all of these cases.

A second fisherman died in 1975, 21 years after the accident. Radiochemical analyses of lung, liver, spleen, kidney, pancreas, and bone were negative for two long-lived nuclides of concern, ^{90}Sr and ^{239}Pu (Kumatori *et*

TABLE 12-2

Distribution of Radioactivity in Deceased Fisherman[a]

Fraction	Probable nuclide	Liver[b] (pCi g^{-1})	Fresh tissue			
			Kidney	Lung	Muscle	Bone
Ru + Te	^{106}Ru + ^{106}Rh ^{129}Te	<0.1	0.9	<0.1	0.2	2
Zr + Nb	^{95}Zr + ^{95}Nb	1	1	0.4	0.3	2
Rare earth elements	^{144}Ce + ^{144}Pr	2	1	0.5	0.5	20
Sr	^{90}Sr + ^{90}Y	0.6	0.4	~0.1	<0.1	1

[a]From Tsuzuki (1955).
[b]One pCi = 37 mBq.

al., 1980). This is particularly noteworthy in view of the heavy deposition on the boat.

Follow-up studies of the surviving fishermen were negative for delayed effects 28 years postexposure, except that the frequency of chromosomal changes in blood cells remained abnormally high (Ishihara and Kumatori, 1983).

THE MARSHALL ISLANDS STUDIES

The more heavily exposed Marshallese were evacuated by plane and ship about 2 days after the fallout and were taken to Kwajalein, where they remained for about 3 months, during which time extensive medical studies were possible. The Rongelapese were then taken to a temporary village on Majuro Atoll, where they lived for 3½ years, by which time the radiation levels of Rongelap had been reduced to a point where they could return. The Utirik and Ailinginae inhabitants were allowed to return to their atolls following a 3-month stay on Kwajalein. Annual medical examinations of the Marshallese, undertaken under the direction of Robert A. Conard of the Brookhaven National Laboratory, have revealed a number of late effects of irradiation. The findings were reviewed in summary reports published 26 and 38 years after the fallout occurred (Conard et al., 1980; Conard, 1992).

The principal late effect of the fallout has been the development of thyroid abnormalities, including nodules, carcinoma, and hypothyroidism. There has probably been one death from radiation-induced leukemia. These effects were seen mainly in the Rongelapese, who were less than 10 years of age at the time of exposure. Of three persons who were exposed in *utero,* two have developed benign thyroid tumors. Several children became hypothyroidal, and growth was stunted in two boys who received thyroid doses of about 2000 rad (20 Gy) at one year of age.

During the first 4 years following exposure, the frequency of miscarriages and stillbirths was higher among exposed women, but no difference has been observed since that time.

Some low-level residual radioactivity remained on Rongelap when the natives from that atoll returned in 1957. It was found that the coconut crab, a much-favored food among the Marshallese, tended to concentrate radioactive Sr, and consumption of local crabs was banned. After 13 years, the concentration of ^{90}Sr in the flesh of these crabs was about 700 pCi g^{-1} (26 kBq kg^{-1}) of calcium. Whole-body counts of the natives indicated elevated levels of ^{137}Cs and ^{60}Co. By 1980, the body burdens of ^{137}Cs averaged about 190 nCi (7 kBq), having been reduced from about 680 nCi (25 kBq) in 1961 (Lessard *et al.,* 1984). A number of individuals who were then living on Rongelap were not present at the time of fallout. The ^{137}Cs burdens of the two groups were indistinguishable, implying that the nuclide was absorbed after the natives returned to the atoll. The ^{137}Cs body burdens increased from the time of their return in 1957 to about 1965, and then began to decline. It was estimated that the 50-year dose commitment from ^{137}Cs was 2 to 3 rad (20 to 30 mGy). Other nuclides contributed much less than the dose from radiocesium.

It was estimated from urine analysis that the mean ^{90}Sr burden of the Rongelapese increased to about 8 nCi (0.3 kBq) by about 1961. Analysis of bone samples from a Rongelap woman who died in 1962 corroborated this estimate (Conard *et al.,* 1970). The body burdens of both ^{137}Cs and ^{90}Sr began to drop after 1963, but it is not known if this has been due to lowered radioactive contamination of food from the island of Rongelap or because the diet is being diluted by an increasing amount of imported food.

With the passage of time the story became increasingly complicated by political and social factors. For many years after World War II the Marshall Islands were assigned by the United Nations to the United States to be administered as a Trust Territory. In March 1946, only 6 months after the war ended, the residents of Bikini were evacuated by the U.S. government to Rongelap to allow the United States to conduct a nuclear testing program called Operation CROSSROADS, in the summer of that year. One year later, the residents of Eniwetak Atoll were evacuated because of plans to conduct tests there in 1948. Thus began the first of the evacuations that have dislocated the people of the Marshall Islands until the present time. Although the last nuclear weapons tests were conducted in the Marshalls in 1958, some of the atolls were excessively contaminated and huge sums have been spent for remediation. As one example, $218 million was spent in the cleanup of Eniwetak (Simon *et al.*, 1993; Defense Nuclear Agency, 1981). In 1986 the Marshall Islands became an independent republic, and as part of the change from its trustee status was provided with a $150 million trust fund to compensate the residents for damage done by the testing programs. As of this writing, 40 years after BRAVO, and 50 years after the tests began, some of the evacuated people have not returned to their home islands. About 5000 claims for radiation injury have been made, involving $5.75 billion (Simon *et al.*, 1993). As part of the memorandum of understanding approved in 1992 by the newly formed Republic of the Marshall Islands and the U.S. government, guidelines were specified for the maximum doses that could be received by the Rongelapese, who were expected to resettle their atoll in 1995, before remediation should be considered. One mSv (100 mrem) per year was set as the upper limit, assuming a diet of locally obtained foods. A limit for transuranic elements in soil was also established at 0.2 μCi m^{-2} (7.4 kBq m^{-2}). This has been interpreted to mean 17 pCi g^{-1} (0.63 kBq kg^{-1}) averaged over the top 1 cm of soil with a density of 1.2 g cm^{-3}.

The gamma-radiation dose has been found to be well within this limit and could be reduced further by simple remedial methods such as the addition of potassium to soil, thereby reducing the uptake of cesium into food (Robison *et al.*, 1994). However, surveys conducted by a team assembled by the Republic of the Marshall Islands found the soil content of transuranic elements to be marginally above the specified limit. A fraction of the Rongelap population on the southern part of the atoll would be expected to receive a dose in excess of 1 mSv (100 mrem) per year (Baverstock *et al.*, 1994).

During the 5-year period 1989–1994, the Republic conducted its first comprehensive survey under the direction of Steven Simon. The final report of the study transmitted to the Republic president stated that "the current levels of radioactive contamination pose no risk of adverse effects to the present generation" and that "the risk of hereditary diseases was no greater than the . . . background of diseases characteristic of any human population" (Simon and Graham, 1994). However, this does not exclude the possibility that exposure to the radioiodines during the many nuclear explosions may have been the cause of thyroid injuries at other atolls that were not studied. A nationwide study of thyroid dysfunction is now under way but the results are not likely to be available for several years.

THE ACCIDENT AT THE WINDSCALE REACTOR NUMBER ONE, OCTOBER 1957

The Windscale (now called Sellafield) Works of the United Kingdom Atomic Energy Authority is located on a low-lying coastal strip in the northwest of

England. It comprised two air-cooled graphite-moderated natural-uranium reactors employed primarily for plutonium production. The core of one of these reactors was partially consumed by combustion in October 1957, resulting in the release of fission products to the surrounding countryside (Arnold, 1992). The Windscale accident resulted in the first major release of radioactive material from a reactor accident.

The interaction of neutrons with the crystalline form of graphite displaces carbon atoms from their normal positions in the molecular lattice. This has a number of effects on the physical properties of the graphite, including dimensional growth, which may occur linearly to the extent of about 3%. The ultimate effect of continued irradiation is disruption of the crystalline structure, with conversion of the graphite to carbon black. (Harper, 1961; Wittels, 1966).

If the crystalline structure has not been destroyed, the original molecular form can be restored by annealing. The disarray of carbon atoms represents stored energy that is released during the annealing process. As much as 500 calories may be released per gram of graphite, enough to raise the temperature of the graphite to a dangerous level.

The accident was caused by the release of this stored energy at an excessive rate during the regular annealing procedure, during which the temperature of the core was being raised using nuclear heat. This release of energy was excessive in portions of the core but went undetected because of insufficient core instrumentation (United Kingdom Atomic Energy Office, 1957, 1958). Failure of a fuel cartridge evidently resulted because of this. The metallic uranium and graphite began to react with air and, from the time combustion began on the morning of October 12, a substantial portion of the core was destroyed. The amount of radioactivity released to the environment was estimated originally by Dunster *et al.* (1958) and subsequently reviewed by Clarke (1974) and by Crick and Lindsley (1984). The data are in reasonable agreement, as can be seen in Table 12-3.

The original reports indicated that the reactor was being used only for production of plutonium at the time of the accident. However, in a reassessment of the accident published in 1983 it was reported that the reactor was also being used as an irradiation facility for the production of ^{210}Po by neutron bombardment of bismuth. This nuclide is used, with beryllium, to produce neutron sources in nuclear weapons. Table 12-3 does not include the estimate by Crick and Lindsley (1984) and Clarke (1990) that about 240 Ci (8.9 TBq) of ^{210}Po was released during the accident.

The first evidence that a mishap had occurred was the observation of elevated beta activity of atmospheric dust collected by an air sampler located in the open about a half mile from the reactor stack. A concentration of about 1.4 nCi m^{-3} (0.5 kBq m^{-3}) of air was observed, this being 10 times

TABLE 12-3

Principal Fission Products Released during
the Windscale Fire[a]

Isotope	Estimated releases (Ci)[b]	
	Dunster et al.[c]	Clarke[d]
^{131}I	20,000	16,200
^{137}Cs	600	1,240
^{89}Sr	80	137
^{90}Sr	9	6

[a]In addition, an estimated 240 Ci of ^{210}Po were released as discussed in the text.
[b]One Ci = 37 Gbq.
[c]From Dunster et al. (1958).
[d]From Clarke (1974).

the level normally present from ^{222}Rn and ^{220}Rn decay products. Air samples collected elsewhere in the vicinity of the reactor confirmed that a release of radioactivity to the atmosphere was occurring.

Visual inspection through a loading hole in the face of the reactor revealed that the uranium cartridges were glowing at red heat in about 150 fuel channels. Because of distortion that had already occurred, these cartridges could not be removed by the mechanical system normally used for this purpose, but the fuel could be removed from channels adjacent to the affected area, thereby creating a fire break that served to limit the extent of the mishap. For several hours, various schemes were devised for extinguishing the slowly burning core, but none was effective. On the following day, what must have been a most difficult decision was reached, and the graphite core was cooled by flooding the core with water. The reactor was cold by the afternoon of October 13.

ENVIRONMENTAL SURVEY PROCEDURES

When the Windscale mishap was discovered, procedures were implemented to determine the extent of exposure from external radiation, inhalation of radioactive dust or vapor, and contaminated food and water.

Vehicles equipped with radiation detection equipment were dispatched downwind from the stack and found the highest radiation level, 4 mR h^{-1} (~0.04 mGy h^{-1}), directly under the plume at a point about 1.6 km downwind. It was subsequently determined that the maximum dose of *external* radiation that would be received by a person remaining out-of-doors for 2 weeks following the accident would be in the range of 30–40 mR (~0.30–0.50 mGy).

During the period of release, about 12,000 air samples were collected on the site and about 1000 were collected in the environs. As would be expected, they revealed wide variations, and concentrations ranged as high as 0.45 pCi ml^{-1} (17 kBq m^{-3}). The average concentration during the period of the incident was approximately 4.5×10^{-3} pCi ml^{-1} (170 Bq m^{-3}), which was about 50% greater than the ICRP standard for permissible continuous exposure to ^{131}I (ICRP, 1960). As described by Dunster *et al.* (1958), the atmospheric contamination from the incident "rose on occasion to worrying but not damgerous levels on the site, while the dilution resulting from wind variations considerably reduced by hazard in the district."

Beginning on the afternoon of the first day, milk from cows in the vicinity of Windscale was found to be contaminated with ^{131}I. Up to that time no emergency criterion had been established for short-term permissible exposure to radioiodine in food, but after consideration of the problem the Medical Research Council promptly recommended that the maximum permissible concentration (MPC) should be 0.1 μCi L^{-1} (3.7 MBq m^{-3}) and that all milk containing more radioiodine than this be discarded. The manner in which the Council arrived at this figure is of some interest in illustrating the thinking of such groups nearly 40 years ago and the ingenuity with which workable guidelines can be arrived at on short notice when the occasion demands.

The Council started with the knowledge that cancer of the thyroid in children had been known to occur following X-ray doses greater than 200 rad (2 Gy). Although no cases were known to have occurred following exposure to smaller doses, the data were insufficient to permit the conclusion that 200 rad (2 Gy) was actually the threshold for tumor production. It was decided to limit the dose to children to a maximum of 20 rad (0.2 Gy). The amount of radioiodine in milk that would produce this dose in children became the permissible level for the entire population, children and adults alike.

Constants furnished in the ICRP tables were used to relate the concentration of radioiodine in milk to thyroid dose. One microcurie (37 kBq) of ^{131}I per gram of thyroid was calculated to result in a dose commitment of 130 rad (1.3 Gy). The mass of the child's thyroid was taken to be 5 g, and the thyroid was assumed to retain 45% of the ingested iodine. The limiting concentration in milk was computed to be 0.15 μCi L^{-1} (5.6 MBq m^{-3}), which was rounded off to 0.1 μCi L^{-1} (3.7 MBq m^{-3}).

Milk measurements were originally hampered by the fact that gamma-spectrometric equipment was not readily available, but beginning on the fifth day, such equipment was located in various laboratories in the United

Kingdom. Up to 300 samples per day were analyzed spectrometrically. The distribution of ^{131}I in milk on October 13 is shown in Fig. 12-2. Altogether the milk exceeded 0.1 μCi L^{-1} (3.7 MBq m^{-3}) in an area of approximately 500 km^2 stretching in a southeasterly direction from Windscale. The irregu-

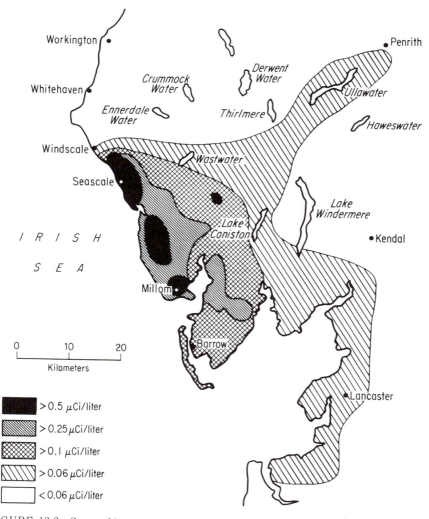

FIGURE 12-2   Geographic area surrounding the Windscale Works (now Sellafield) in northwest England, showing the ^{131}I concentration in the milk from various districts 5 days after the accident. (From Dunster *et al.*, 1958.)

larities can be explained by the changing meteorological patterns that existed during the period of emission (Crabtree, 1959). The highest concentration of radioiodine in milk was 1.4 μCi L^{-1} obtained from a farm located about 10 miles from Windscale. The amount of radioiodine in milk was found to correlate with the gamma-radiation levels in the area. The concentration of ^{131}I in milk exceeded 0.1 μCi L^{-1} (3.7 MBq m^{-3}) in pastures where the gamma radiation exceeded 0.035 mR h^{-1} ($\sim$0.35 μGy h^{-1}). The criteria for restricting the sale of milk remained unchanged throughout the episode. By November 4, the permissible level was exceeded only in a region extending about 12 miles southward from Windscale. This area remained under restrictions until November 23.

Drinking water samples were collected from reservoirs and streams and no concentrations of radioiodine or other radionuclides were found to exceed the limit permitted by ICRP. Children and adults living downwind as far as 24 miles from Windscale were scanned for iodine uptake. Among 19 children studied, the highest dose received was estimated to be 16 rad (160 mGy). The highest adult dose was estimated to be 9.5 rad (95 mGy).

Milk and other foods in the vicinity were analyzed for ^{89}Sr. The amounts found did not exceed those known to be present prior to the accident.

LESSONS LEARNED FROM THE WINDSCALE ACCIDENT

A number of significant conclusions were derived from this accident. In the slow oxidation of the Windscale fuel, radioiodine was released preferentially from the core. Because the stack effluents were filtered, other isotopes were not released to the environment in important amounts. The dose to individuals from inhalation of the iodine or from direct exposure to the plume or deposited radioiodine was negligible compared to the dose received from dairy products. The extent of iodine contamination could be readily estimated by scanning the gamma-radiation levels in the area. When the ^{131}I concentration in milk exceeds 0.1 μCi L^{-1} (3.7 MBq m^{-3}), the dose to the thyroids of children may exceed 20 rad (0.2 Gy). As a first approximation, it may be assumed that the concentration of radioiodine in milk will exceed 0.1 μCi L^{-1} (3.7 MBq m^{-3}) if it is produced by cows grazing in an area where the gamma-radiation levels exceed 0.035 mR h^{-1} ($\sim$0.35 μGy h^{-1}).

In addition, the accident at Windscale was the first to demonstrate the importance of detailed advance planning and the vast resources that must be made available. An incident of this kind taxes the health physics capacity of local organizations to the extent that only by pooling resources on a regional or national basis can the necessary technical assistance be brought to bear. More than 30 years later, the Chernobyl accident, which was orders of magnitude more severe, demonstrated that planning is even necessary on an international scale.

THE HOUSTON INCIDENT OF MARCH 1957

A relatively minor accident that occurred in Houston, Texas, in March 1957 and that received extensive publicity (*Look Magazine,* 1960) is worth discussing. This incident teaches us very little of a positive nature, but an examination of the details of the accident and subsequent events provides many illustrations of what not to do.

The Houston company was licensed by the AEC to encapsulate radioactive material for radiography sources known as "gamma cameras." In March 1957, two men were opening a sealed can containing 10 pellets of ^{192}Ir, each pellet being a ⅛ × ⅛-inch (3 × 3-mm) cylinder containing about 35 Ci (1.3 TBq). It was necessary to remove a plug at the end of the can to gain access to the pellets, and this was accomplished on a jeweler's lathe in a sealed Plexiglas box, which in turn was located in a hot cell. The operation was performed with master–slave manipulators through 33 inches (0.84 m) of concrete.

When the small can containing the pellets was removed from its larger container and opened, two of the pellets were found to be in a loose dusty form. Some of the dust escaped from the Plexiglas box and hot cell and was detected by an air monitor in the laboratory that alerted employees to the mishap. One of the two employees left the laboratory for home dressed in the street clothes he had worn at work. The second employee, who was dressed in work clothing and wearing a respirator, remained in the room for an unknown period of time.

The fact that the laboratory had become contaminated came to the attention of the plant management for the first time about 1 month after the incident occurred. It was then observed that contamination existed in the vicinity of the hot cell and that it had been spread to employees' street clothes, shoes, and even their homes and automobiles.

The AEC, which was then the licensing agency, was notified belatedly, about 5 weeks after the accident, and shortly thereafte the company retained the services of a private company to survey and decontaminate the affected areas. Of 19 private homes that were checked, 8 showed evidence of contamination; 7 out of 53 automobiles were found to have traces of contamination.

Nineteen employees and members of their families were examined by a physician. One neighbor of an employee was also examined. Except for minor radiation burns on each of the two employees who were present at the time of the incident, the medical findings were negative.

Very few quantitative data are available concerning the level of contamination encountered. However, it is reported that one of the employees received an exposure of 0.017 Sv (1.7 rem) during 1 week when decontamination of the laboratory was in progress. His 13-week exposure was 0.039 Sv (3.9 rem), which exceeded the permissible level of 0.03 Sv (3 rem) per quarter then in effect by 25%.

This incident became the subject of nationwide televised programs and nationally distributed magazine articles. One of the main themes of the publicity was that the affected individuals and their families were being ostracized socially because of their "radioactivity" and that serious disabilities were developing not only in the two employees but also in their families.

The incident occurred very soon after the licensing procedure of the AEC was put into effect, and the licensee was apparently ignorant of the need to notify the AEC that an accident had occurred. Moreover, the regulatory apparatus of the AEC had not yet been confronted with an incident of this type, and its handling of the incident indicated that the agency had much to learn about its role.

Many experts who might have been sent to Houston by the AEC to advise and assist the licensee to minimize the consequences of the incident were not made available, apparently because the regulatory function of the AEC in this particular instance was thought to forbid its participation in an advisory capacity. As a result, much potentially valuable information was never obtained. For example, urinalyses and whole-body gamma-radiation measurements would have certainly been useful for the two employees who were known to have been the most heavily exposed. Other employees might also have been so examined. Depending on these findings, it might or might not have been desirable to conduct similar examinations of the families and neighbors.

Decontamination procedures in the homes of the employees, including such drastic procedures as cutting out portions of rugs to remove measurable contamination, were undertaken without any basis for deciding that such measures were necessary. In summary, a valuable study that might have shed useful quantitative information regarding the consequences of an incident of this kind was not undertaken, and in the absence of reliable information, rumor and sensationalism swelled in uncontrolled fashion about the unfortunate principals for months afterward.

In 1961, after 4 years of exaggerated stories about the effects of the accident, some of the principals were examined at AEC expense by physicians at the Mayo Clinic (Atomic Industrial Forum, 1961). It was found that none was suffering from the alleged radiation effects. However, as is so often true in incidents of this kind, the factual announcement of the findings at the Mayo Clinic did not receive nearly as much public notice as the more sensational claims of injury.

This was a trivial accident that created anxiety all out of proportion to the potential dangers that existed. The total quantity of radioactive iridium that escaped to the environment probably did not exceed a few milicuries (a few hundred MBq). It would have been useful to have undertaken a risk assessment before corrective actions were ordered.

THE OAK RIDGE PLUTONIUM RELEASE OF NOVEMBER 1959

The radiochemical-processing pilot plant at Oak Ridge National Laboratory was built in 1943 and was used for processing irradiated fuel. In November 1959, a chemical explosion occurred in one of the shielded cells during a period when the process equipment was being decontaminated. No one was injured, and the monetary loss because of damaged equipment was relatively minor; however, the explosion resulted in plutonium contamination of the pilot-plant building, nearby streets, and building surfaces. The ensuing cleanup operations were costly, and the contaminated areas could not be used for many weeks while the cleanup was in progress.

The chemical explosion occurred during decontamination of an evaporator and is thought to have resulted from the formation of explosive compounds such as picric acid when concentrated hot nitric acid was mixed with a proprietary decontaminating agent that contained phenol. A small quantity of this solution had been left in the equipment because a normal water wash was omitted, and the explosion occurred when nitric acid was later introduced into the evaporator and was brought to the boiling temperature (King and McCarley, 1961).

The explosion breached a door leading from the cell directly to the outside of the building, and plutonium released through this doorway contaminated nearby streets and building surfaces. The adjacent air-cooled graphite-reactor building became contaminated when plutonium was drawn into the ventilation system. In addition, plutonium was forced through penetrations in the concrete cell walls into the remainder of the chemical-processing building. In all, an area of 350 m in diameter was contaminated. Quantities of ^{95}Zr and ^{95}Nd were also released, but were of minor importance.

Contamination within the cell in which the explosion took place was from 10 to 10^5 nCi per 100 cm^2 (3.7–37,000 MBq m^{-2}). A penthouse above the cell was contaminated from 2 to 20 nCi per 100 cm^2 (0.74–7.4 MBq m^{-2}). Contamination of streets and buildings in the immediate vicinity exceeded 50 nCi per 100 cm^2 (20 MBq m^{-2}) in spots, but was for the most part below 50 pCi per 100 cm^2 (0.02 MBq m^{-2}).

No employees were overexposed during the period of cleanup. Protective clothing consisted of two sets of coveralls, two pairs of shoe covers, two pairs of rubber gloves, an assault mask, and a hood. Wrists and ankles were sealed with masking tape. Employees leaving contaminated areas passed through monitoring stations where their protective clothing was removed. The men then showered and received a final examination for alpha contamination before being released.

Containment of the radioactivity was achieved initially by fixing the surface contamination in a number of ways. Washdown was not attempted because of the possibility that the plutonium might wash into inaccessible places and because the accumulation of wash water would have created a problem. Fixation was accomplished by resurfacing roads and by painting roofs, walls, and equipment. Paint was even sprayed on grass lawns and sidewalks that were found to be contaminated. Resuspension of the plutonium was greatly reduced and spread over long periods of time as the paint deteriorated. In addition, colored first coats of paint would give notice that repainting might be necessary.

The decontamination procedures consisted of brushing and sponging, scraping, grinding, and various other techniques that are enumerated in Table 12-4. The target levels for alpha decontamination are shown in Table 12-5.

When decontamination to this extent was not practical, but where the levels were less than 10 times the target levels, the surfaces were covered with brightly colored enamel and then covered with either paint or concrete. The brightly colored paint would serve as a warning in the future should the protective coating be removed.

The decontamination proceeded smoothly, but the incident resulted in costly interruptions to normal operations. The Oak Ridge National Laboratory graphite reactor was not in operation from November 20 to December 22. The processing cells directly involved in the incident could not be cleaned up for about 8 months until building modifications could be made to provide for improved containment of radioactivity.

It is estimated that only 15 g (about 1 Ci or 37 GBq) of plutonium was blown out of the evaporator subcell. This small amount of material resulted in contamination that could only be removed after an expenditure of hundreds of thousands of dollars and costly interruptions to an important research reactor and other activities in the immediate vicinity. All this occurred after 16 years of safe operation as a result of the rather subtle error of introducing phenol into a decontaminating agent that was later to be mixed with nitric acid. The incident serves well to illustrate the extraordinary care

TABLE 12-4

Summary of Decontamination Treatments for Various Surfaces[a]

Surface	Character of contamination	Primary decontamination treatment	Cleanup rate (ft^2 per man-hour)	Other decontamination treatment
All (walls, floor, ceiling)	Transferable	Scrubbed with detergent and water and brush or sponge	27	Dusty areas vacuumed
Painted metal (walls and ceiling)	Fixed	Paint removed with paint remover and scrapers, and surface scrubbed with soap and water	4	Outer layer of paint removed with sandpaper
Concrete (floor)	Fixed	Ground with terrazzo floor-grinding machine	5	"Hot" spots chipped out and vertical surfaces washed with dilute hydrochloric acid
Bare metal (SS[b] piping and tanks)	Fixed	Rinsed with dilute nitric acid and scrubbed with steel wool	—	Surfaces abraded with emery paper
Bare metal (other than SS)	Fixed	Abraded with emery paper or ground to remove pits	—	—
Lead shielding	Fixed	Rinsed with dilute nitric acid	—	—
Oily metal (pumps)	Fixed	Washed with Gunk, a commercial solvent	—	—

[a]From King and McCarley (1961).
[b]SS, stainless steel.

TABLE 12-5

Target Levels for Alpha Decontamination

	Direct reading	Transferable
Maximum	300 dpm per 100 cm^2	30 dpm per 100 cm^2
Average[a]	30 dpm per 100 cm^2	3 dpm per 100 cm^2

[a]At least 10 samples were used to derive an average sample, and there was at least one sample from each square meter of the projected surface area.

that must be exercised in the design and implementation of operations such as these.

Management of the accident would have been more difficult had the accident occurred anywhere but in a national laboratory that had unique experience with many minor incidents of this type. A well-trained technical staff was available, backed up by a well-indoctrinated labor force and services of a type that can only be found at large centers such as Oak Ridge.

THE ARMY STATIONARY LOW-POWER REACTOR (SL-1), JANUARY 1961

An explosion of the Army Low-Power Reactor (SL-1) in January 1961 resulted in the deaths of three military personnel at the National Reactor Testing Station in Idaho (currently named the Idaho National Engineering Laboratory, or INEL). The reactor was a direct-cycle boiling-water unit designed to operate at a level of 3 MWt and was fueled with enriched uranium plates clad in aluminum. After a little more than 2 years of operation, the SL-1 reactor was shut down on December 23, 1960, having accumulated an operating history of approximately 950 MW-days. A 12-day maintenance program was contemplated, and the reactor was scheduled to resume full power on January 4, 1961 (Buchanan, 1963; Horan and Gammill, 1963). On the night of the accident the crew consisted of three enlisted military personnel of whom two were licensed reactor operators and one was a trainee.

The first indication of trouble at the reactor was given by remote radiation and thermal alarms that caused the fire department and health physicists to respond. On arriving at the SL-1, they encountered radiation fields of 200 mR h^{-1} ($\sim$2 mGy h^{-1}) extending for a few hundred feet ($\sim$100 m) from the reactor building.

The emergency crews found neither a fire nor any of the three operators. A brief reconnaissance indicated radiation levels as high as 500 R h^{-1}

($\sim$5 Gy h^{-1}). After a search of about 1½ hours, the bodies of two of the personnel were found. A third man who was still alive was removed but died shortly thereafter. The two dead men were not removed for several days, during which time several hundred workers were engaged in recovery operations. Twenty-two personnel received radiation exposures in the range of 3 to 27 R ($\sim$0.03 to 0.27 Gy) (AEC, 1961). This is an excellent record, considering the complexity of the rescue and recovery operations and the fact that the radiation levels in some areas approached 1000 R h^{-1} ($\sim$10 Gy h^{-1}). According to Horan and Gammill (1963), the bodies had been saturated with contaminated water and penetrated by particles of fuel. The radiation intensities at the bodies were 100 to 500 R h^{-1} ($\sim$1 to 5 Gy h^{-1}) at 15 cm.

The accident evidently resulted when withdrawal of a single control rod caused the reactor to go into the prompt critical condition. Why the rod was withdrawn may never be known.

Despite the fact that the accident was a violent one and the reactor was not surrounded by a vapor container, the radiation levels outside the reactor building were minimal, about 5 mR h^{-1} ($\sim$0.05 mGy h^{-1}) at 1000 ft (300 m). Essentially all the radioactive material with the exception of ^{131}I was contained within a 3-acre plot. Thus, despite the fact that the reactor core contained approximately 1 MCi (37 PBq) of medium- to long-lived radionuclides, it is thought that less than 10 Ci (0.37 TBq) of ^{131}I was released to the environs.

Several months after the accident, it was decided to dismantle the reactor, an operation that presented many difficulties because of the high levels of radiation (Fig. 12-3).

Many years after the accident the Idaho National Engineering Laboratory performed a Historical Dose Evaluation related to activities at INEL (DOE-ID, 1991). The maximum effective dose equivalent to off-site individuals as a result of the accident was estimated to have been 3 mrem (0.03 mSv) for an adult, 4 mrem (0.04 mSv) for a child, and 7 mrem (0.07 mSv) for an infant. Maximum thyroid doses were estimated to have been 39 mrem (0.39 mSv) for an adult, 71 mrem (0.71 mSv) for a child, and 180 mrem (1.8 mSv) for an infant.

ABORTIVE REENTRY OF THE SNAP 9A, APRIL 1964

A navigational satellite launched on April 21, 1964, and carrying a ^{238}Pu power generator failed to reach orbital velocity and reentered the atmosphere at about 150,000 ft (46 km) over the Indian Ocean. The isoto-

FIGURE 12-3   The 13-ton core and pressure vessel of the Army Low-Power Reactor, located at the National Reactor Testing Station in Idaho, (SL-1) being removed from the reactor building 11 months after the accident. The top and one side of the building have been removed. The reactor is to be lowered into the plastic-covered concrete cask on the flatbed trailer. (From U.S. AEC.)

pic power unit was known as SNAP-9A and contained about 17,000 Ci (0.63 PBq) of ^{238}Pu (Krey, 1967).

Plutonium-238 was present in the upper atmosphere at the time as a residue from earlier nuclear weapons tests, but the system of high-altitude balloon sampling, mentioned in Chapter 10, made it possible to demonstrate the sudden appearance of a new source of this nuclide, first detected at an altitude of 34,000 m about 4 months following the abortive reentry. The stratospheric distribution of the debris during the next several years followed the predictions that had been made on the basis of transport models of the stratosphere developed through studies of the behavior of debris from weapons testing (Kleinman, 1971). However, the concentration of ^{238}Pu from SNAP 9A was somewhat lower in ground-level air than had been predicted (Shleien *et al.,* 1970).

About 16 kCi (0.6 PBq), representing 95% of the amount originally injected, was estimated to have been deposited by the end of 1970. The

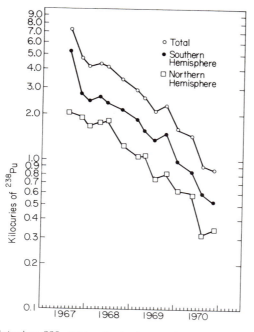

FIGURE   12-4   Plutonium-238 stratospheric inventories, 1967–1970. [From Kleinman (1971).]

estimated stratospheric inventories are given in Fig. 12-4, in which it is seen that the debris apparently took about 2 years in which to diffuse to the sampling altitudes, following which depletion of the stratospheric inventory proceeded exponentially with a half-time of about 14 months (Krey *et al.*, 1970).

The ^{238}Pu content of ground-level air was monitored by de Bartoli and Gaglione (1969) and by Shleien *et al.* (1970), on the basis of which they concluded that on average the 50-year organ dose commitment to the pulmonary lymph nodes was 36 mrem (0.36 mSv). More recently the UNSCEAR (1993) estimated the worldwide average individual effective dose from this event to be 0.4 nSv (0.04 μrem) and the collective effective dose to be 2.4 person-Sv (240 person-rem).

REENTRY OF THE SATELLITE *COSMOS 954,* JANUARY 1978

A Soviet satellite, *Cosmos 954,* powered by a nuclear reactor, reentered the atmosphere over the Canadian Northwest Territories on January 24, 1978, and spread radioactive debris over a 1000-km path stretching northeast from Great Slave Lake to Baker Lake (Tracy *et al.,* 1984). The satellite reactor was believed to have contained about 20 kg of highly enriched uranium. From radiochemical measurements of recovered fragments it was estimated that the burnup was 2×10^{18} fissions per gram of uranium. At the time of reentry, the core was estimated to have contained about 84 Ci (3100 GBq) of ^{90}Sr, 4900 Ci (180 TBq) of ^{131}I, and 86 Ci (3200 GBq) of ^{137}Cs. Searches directed by the Canadian government located about 65 kg of satellite structural components and equipment, some of which had dose rates of 5 Sv (500 rem) per hour on contact. More than 4000 radioactive particles that ranged in size from 0.1 mm to 1.0 mm were also recovered, but it was estimated that only 0.1% of the dispersed particles was found. Seventy-five percent of the original material is estimated to have remained in the upper atmosphere.

The recovered particles were found to have been highly insoluble in water and in 0.1 M HCl. No detectable contamination was found in samples of air, water, and food supplies. The Soviet government published no technical information about this incident.

THE ACCIDENT AT THREE MILE ISLAND UNIT 2, MARCH 1979

A relatively minor mechanical malfunction that occurred early in the morning of March 28, 1979, at the Metropolitan Edison Company (Met Ed) Three Mile Island Unit 2 (TMI) nuclear power reactor, initiated the world's first major accident at a civilian power plant. Radiation exposure to the plant workers and the general public was insignificant, but the financial consequences of the accident were enormous for Met Ed's parent company, the General Public Utilities Corporation (GPU). The accident resulted in extensive psychological trauma among nearby residents of TMI, and the sociopolitical implications of the accident impacted on the nuclear industry to a major extent and contributed to the halt in nuclear power plant sales in the United States during the 17 years since the accident occurred.

DESCRIPTION OF THE ACCIDENT

Three Mile Island Unit 2 was one of a pair of pressurized-water reactors of Babcock and Wilcox design. Each reactor had a generating capacity of 850 MW, and TMI-2 began commercial operation in December 1978. The plant is located outside of the town of Middletown, Pennsylvania, not far from Harrisburg, the state capital. The accident has been the subject of many investigations by governmental and private organizations (Kemeny, 1979; Rogovin and Frampton, 1980; New York Academy of Sciences, 1981).

At 4 A.M. on Wednesday, March 28, 1979, the feed water pumps that supplied the reactor's steam generators shut down, leading to automatic shutdown of the reactor and steam generating system. Heat from residual radioactivity in the core caused the temperature and pressure of the reactor coolant to rise and a relief valve atop the pressurizer to open (Nuclear Safety Advisory Center, 1981). The value should have closed in seconds, but was stuck in the open position and, since the reactor coolant water was draining, a loss-of-coolant accident was in progress.

The plant was equipped with three emergency feed water pumps and these started to operate automatically 14 seconds after the first shutdown. The control room operator was aware that these pumps were operating but did not notice two panel lights that would have told him the emergency feed water valves were closed and water was unable to reach the reactor.

The reactor safeguard systems responded properly to the emergency. Two minutes into the accident, high-pressure injection pumps began operating at a flow of 1000 gallons per minute (gpm) ($0.06 \text{ m}^3 \text{ s}^{-1}$), but 30 seconds later a reactor operator shut one pump down and reduced the flow of the second to less than 100 gpm ($0.006 \text{ m}^3 \text{ s}^{-1}$). As a consequence, steam bubbles began to form in the reactor coolant system, but the operators misdiagnosed the problem and, instead of restarting the emergency high-pressure injection system, they opened the reactor coolant letdown system, thereby draining additional water. By this incredible sequence of misjudgments, the water level continued to drop, the temperature rose, the fuel cladding failed, and fuel melting began. In the process, radioactivity that passed to the coolant water was leaving the reactor coolant system via the relief valve and was accumulating in the reactor building basement.

The core temperature rose to the point at which the zirconium alloy cladding began to react with steam to produce hydrogen, some of which escaped into the reactor building. Early in the afternoon of the first day, sufficient hydrogen had accumulated in the reactor building to result in a low-level explosion, but no damage was done.

A backup valve that could be operated manually was located between the pressurizer and its open relief valve, but it was not until more than 2 hours after the faulty valve stuck in the open position that the valve was closed. This stopped further loss of coolant, but the core continued to be partially exposed and further damage was occurring. By 10:30 A.M., 6½ hours after the accident began, the core was again fully covered with water. A full appreciation of the extent of the damage within the pressure vessel did not emerge until 1990, more than one decade later, when the debris was finally removed from the pressure vessel and it became possible to depict the extent of reactor damage shown in Fig. 12-5. An estimated 50% of the reactor core had melted during the accident (Electric Power Research Institute, 1990).

THE IMMEDIATE POSTACCIDENT PERIOD

By 7 A.M., 3 hours after the first indication of trouble, a site emergency had been declared because of the elevated radiation levels detected by radiation monitors in the reactor auxiliary building. About one-half hour later, a general emergency was declared, and by 8:30 A.M. the utility emergency response teams were finding levels of 1–3 mR per hour immediately off-site. Two survey teams were dispatched from the Pennsylvania Bureau of Radiation Protection at about 10 A.M., at which time assistance was requested from the network of emergency teams maintained by the DOE. A radiation monitoring helicopter and NRC inspectors began to arrive at the site.

The official agencies involved in the emergency management were (1) the Pennsylvania Emergency Management Agency, located in the office of the Governor, (2) the NRC, (3) the DOE, and (4) the Bureau of Radiological Health of the Food and Drug Administration.

The best information about off-site gamma-radiation levels of exposure during the first hours after the accident was obtained from 34 thermoluminescent dosimeters located around the plant to a distance of 15 miles. Additional dosimeters were emplaced by the Federal agencies, but not until the third day of the accident (Gerusky, 1981).

During the first few days postaccident, a considerable number of food samples were collected for analysis and whole-body gamma scans were completed on 700 people. All of these tests were negative.

In this situation, emergency management depended on the interpretation of measurements of many kinds. A serious misunderstanding of events occurred on Friday, March 30, in connection with the transfer of radioactive waste gases from the makeup water storage tank to a waste-gas decay tank located in the auxiliary building, outside of the containment structure. The operator was aware that this transfer would release radioactive gases to the outside atmosphere, and requested that a helicopter be assigned to make radiation measurements in the vicinity of the ventilation stack. The helicopter reported levels of 1200 mR h^{-1} ($\sim$12 mGy h^{-1}) at a point 300 ft above the stack.

Coincidentally, an NRC specialist had been calculating the ground-level exposures that would be experienced if there should be a failure of the waste-gas decay tank relief valve. A briefing was in progress at the time of the helicopter measurements, and the NRC specialist, literally within seconds of the 1200 mR h^{-1} ($\sim$12 mGy h^{-1}) report from the helicopter, stated that the

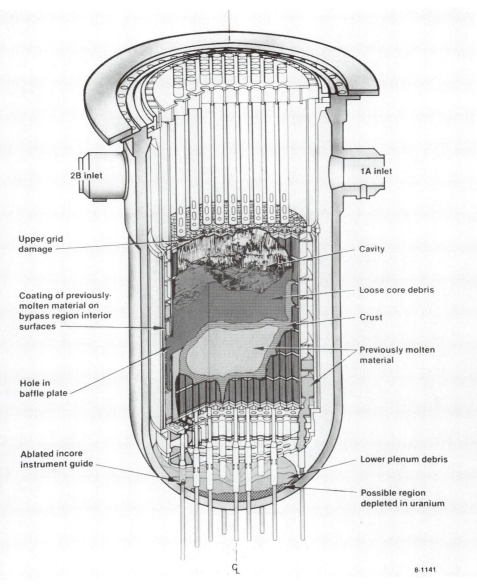

FIGURE 12-5 Postaccident conditions within the Three Mile Island Unit 2 reactor vessel, near Harrisburg, Pennsylvania. Note the extensive damage to the reactor core components and the solidified mass of previously molten material. (Courtesy of General Public Utilities.)

valve failure would result in a ground-level exposure of 1200 mR h^{-1} ($\sim$12 mGy h^{-1}). The officials erroneously assumed that the report from the field was of ground-level measurements, and therefore assumed that the waste-gas decay tank relief valve had failed. The gases, in fact, were not coming from the intact valve near ground level, but from minor leaks in the transfer system that were venting to the stack. On the basis of their misinterpretation of the reading, NRC officials recommended an evacuation of people as far as 10 miles downwind from the site. The head of the Pennsylvania Bureau of Radiation Protection knew about the helicopter reading and concluded that the evacuation was unnecessary, but communications were jammed and he could not reach the governor's office (Kemeny, 1979). In a telephone conversation later that morning, the governor talked with the chairman of the NRC, who assured the governor that no evacuation was needed. However, it was decided that the governor would urge everyone who lived 5 miles downwind of the plant to remain indoors for the next half hour. Shortly after noon, on the basis of another conversation with the NRC chairman, the governor recommended that pregnant women and preschool children leave the region within a 5-mile radius of Three Mile Island. They also decided to close all schools within that area.

Another serious misunderstanding developed Friday afternoon, when it was reported that a 1000-ft^3 (28-m^3) gas bubble that contained hydrogen had formed within the reactor vessel. Concern developed that an explosion might occur within the reactor and, as reported by the Presidential Commission, "that it was a groundless fear, an unfortunate error, never penetrated the public consciousness afterward, because the NRC made no effort to inform the public that it had erred." The hydrogen was produced as a result of high-temperature reactions between the hot steam and zirconium fuel cladding. However, the chemistry of the reactor system was such that excess oxygen could not accumulate. This was apparently known by the NRC experts, but during the weekend beginning March 31, no statement to this effect was made. Finally, on Monday, April 2, it was announced that the bubble had absorbed, and that perhaps the NRC calculations of oxygen generation rate were too conservative: no public statement was made to this effect.

Acting on the assumption that a major release of radioiodine was possible, the U.S. Department of Health, Education and Welfare (HEW) took steps to obtain supplies of potassium iodide that could be used by the surrounding population. Early Saturday morning, the Mallinckrodt Chemical Company of St. Louis agreed to provide 250,000 1-oz. bottles of the compound. The first shipment reached Harrisburg by early Sunday morning, and the balance was supplied by Wednesday, April 4. Fortunately, there was no need for it (Wald, 1980).

An incident that took place on the second day of the accident illustrates the difficulties of anticipating public and official reaction to actions that must be taken by those responsible on-site. About 400,000 gallons of slightly contaminated wastewater had accumulated in tanks by the second day and since the initial radioactivity of the water was well within the permissible limits for discharge into the Susquehanna River, it was decided to discharge the wastes in the usual way. When informed of the proposed action, the local NRC officials gave their approval and the utility also notified the State Bureau of Radiation Protection that the water was being discharged to the river.

The press, and many public officials and members of the general public, reacted adversely when they learned of the release. Although it was reasonable at the time for the plant personnel and NRC officials to consider it a routine matter since the amounts discharged were well within the limits permitted by the NRC license under normal operating conditions, the feeling developed that neither the government nor the plant management had dealt candidly with the public. The incident led to a requirement that TMI would not discharge any radioactivity to the river, a limitation that remained in effect for more than a decade, while cleanup operations were in progress.

Management of the contaminated water was the first major problem that had to be faced in the long-term program for decontaminating TMI-2. Hundreds of gallons of water accumulated in the reactor building sump with a concentration of 270 μCi mL^{-1} (1 GBq m^{-3}), mostly ^{137}Cs after decay of the short-lived components. Wastewater stored in various tanks contained a comparable quantity of water contaminated to only a slighly less extent. Ion-exchange systems were designed postaccident to process this water and performed well. Decontamination soon progressed to the point where tritium was the only remaining contaminant and the water could have been discharged to the Susquehanna River subject to the limitations normally imposed on operating reactors. The city of Lancaster downstream from the plant sued to prevent discharge of any TMI-2 water to the Susquehanna, and this resulted in an out-of-court settlement that prohibited discharge of any accident-generated water (Electric Power Research Institute, 1990). It became necessary to build holding tanks until a method of ultimate disposal of the tritiated water could be developed that would be acceptable to the NRC, the state of Pennsylvania, and the local community. The latter was highly influential as the result of intense and well-organized activism. Permission to discharge the water was never received. After about 10 years, during which there were many reports and hearings on the subject, it was decided that the water should be evaporated by construction of an expensive heating system. The dose in either case was insignificant, but evaporation was more acceptable to the public. This was only one of many expensive steps that were required as a result of public misconceptions of risk. These greatly contributed to the cost of cleanup.

Another example was the 1-year delay in entering the reactor building. Such entry was essential for the preliminary fact finding necessary to plan the recovery operations. However, entry could not take place until 57,000 Ci (210 TBq) of ^{85}Kr was vented from the building atmosphere. The various options for disposing of this gas were examined and it was decided that venting to the atmosphere under controlled conditions would be the safest method. A study by the NCRP (1980) concluded that exposure to the nearby residents would be trivial. The principal risk of exposure to ^{85}Kr is skin cancer, but the venting would result in a risk equivalent to the risk of about 20 minutes of exposure to sunlight. Nevertheless, a 1-year delay in the start of cleanup resulted from the indecision concerning krypton venting.

EXPOSURE OF THE NEARBY POPULATION

The task of estimating dose to the surrounding population was assigned to an Ad Hoc Population Dose Estimate Group (Battist *et al.*, 1979), which included representatives from all of the federal agencies involved in the accident investigation. Independent dose estimates were also made by a commission appointed by the president of the United States (Kemeny, 1979). It was concluded that the highest doses were received by a few people within a 2-mile radius who received doses between 20 and 70 mrem (0.2 and 0.7 mSv) (Gerusky, 1981). Four different methods of estimating population dose gave estimates ranging from 1600 to 5300 person-rem (16–53 person-Sv). The most credible estimate was thought to be 3300 person-rem (33 person-Sv). Using the risk coefficients discussed in Chapter 2, this would imply that perhaps one or two fatal radiation-induced cancers would occur in the lifetime of the exposed population. Beta radiation from the noble gases would have resulted in a skin dose as much as four times the gamma dose, but this would have been attenuated to an unknown extent by clothing. The effects of such low doses are of course subject to the uncertainties of estimating low-dose effects, and the true number of cases may in fact be higher, lower, or even zero.

The dose from ingested radionuclides was less than that from external radiation. Thousands of samples of air, milk, water, produce, and other environmental media were taken and analyzed by the various agencies. The highest doses would have been to a few people due to ingestion or inhalation of radioiodine and would have been no more than about 0.05 mSv (5 mrem) (Kemeny, 1979).

Psychological Effects on the Surrounding Community

The sociopolitical ramifications of the accident aftermath proved to be far more complicated than the technical problems. Public concern increased markedly on the third day of the accident because of apparent disagreements among the experts about the danger of the hydrogen bubble. Later surveys showed that 52% of the people living within 20 miles of TMI left the area, most of them on Friday, March 30 (Dohrenwend et al., 1981). The Governor's advice to pregnant women and preschool children was not lifted until April 9, and the schools within 5 miles of TMI did not open until April 11 (Flynn, 1981). Great distrust developed in the credibility of the electric utility company, the NRC, and other agencies of state and federal government. The tension was exacerbated by the presence of hundreds of reporters and other media representatives who descended on the small surrounding communities. It is reported that 1 year later, when the press was asked at a conference to comment on the possibility that the performance of key people involved in the accident may have been affected by the enormous pressure from the media, one reporter responded "that is one of the prices we must pay for the public's right to know" (Trunk and Trunk, 1981). It is estimated that 300–400 reporters were present in the small community of Middletown.

In the course of time, it became apparent that the only observable health effects of the reactor accident would be those due to the psychological trauma (New York Academy of Sciences, 1981; Hatch, 1991). The environs of TMI will be fertile laboratories for psychosocial scientists for years to come.

Despite the small doses that people received, there were many claims for damages. Diseases in farm animals and children that would have been normally accepted as due to the vicissitudes of living were ascribed to the accident, notwithstanding the absence of any evidence that the frequency of such occurrences was greater than normal after the accident.

THE CHERNOBYL ACCIDENT, APRIL 1986

Rumors of major radioactive contamination in the Soviet Union began to circulate as early as 1957, but state secrecy denied the world any information about those events until the period of political and social change that began in the late 1980s. In the midst of the changes that were taking place, in 1986 a catastrophic accident destroyed Unit 4 of the Chernobyl nuclear complex in the Ukraine. Although the severity of the accident made it impossible to keep the event a secret from nearby countries equipped with radiation detection equipment, the Soviet government made no announcement of the event for about 3 weeks, and no technical information was made available for about 4 months. However, under the new leadership, the centuries-old policies of secrecy that had characterized both the Czarist and Communist regimes were being changed to a new policy of openness (glasnost). This was first evident at the Post Accident Review Meeting sponsored in the summer of 1986 by the International Atomic Energy Agency (IAEA) in Vienna. Much to the surprise and delight of the world's scientists, their Soviet counterparts came to the week-long meeting prepared to discuss all aspects of the accident and its consequences as then understood (IAEA, 1986). This was a turning point in many aspects of East–West relationships concerning nuclear matters. Glasnost enabled the West not only to learn about the details of the Chernobyl accident but also permitted the release of information about other accidents as well as the consequences of a 50-year history of disregard for safety in the management of the Soviet nuclear program.

The 1000-MWe, water-cooled, graphite-moderated reactor at the Chernobyl power station in the Ukraine, about 90 km north of Kiev, was destroyed by an accident that occurred early in the morning of April 26, 1986. The reactor, Unit No 4 of the Chernobyl station, had operated since 1983. The Russian acronym for this reactor design is RBMK. The accident was by far the worst that has ever occurred to any reactor and was the first power reactor accident to result in radiation casualties[1] the first to result in extensive contamination of the environs, and the first to require evacuation of people from their homes. It was by far the most costly industrial accident in history. Although the accident would have caused worldwide concern in any case, the global reaction was exacerbated by East–West political and cultural differences, the fact that many days passed before it was announced that the accident had occurred, and the political changes that were beginning to take place in what was then the Soviet Union.

DESCRIPTION OF THE REACTOR

The core of the RBMK is a graphite cylinder 12 m in diameter and 7 m high, which is penetrated vertically by 1661 fuel channels and 222 control rod channels (Fig. 12-6) (IAEA, 1986). The fuel channels are housed within zirconium alloy pressure tubes. Within each tube are subassemblies of 18 zircalloy clad fuel pins. The fuel was 2% enriched uranium dioxide, but metallic uranium can also be used to facilitate plutonium production (American Nuclear Society, 1986). Light water passes vertically through each of

---

[1]The SL-1 reactor in Idaho that claimed three lives as a result of an accident was an experimental reactor rather than a power reactor.

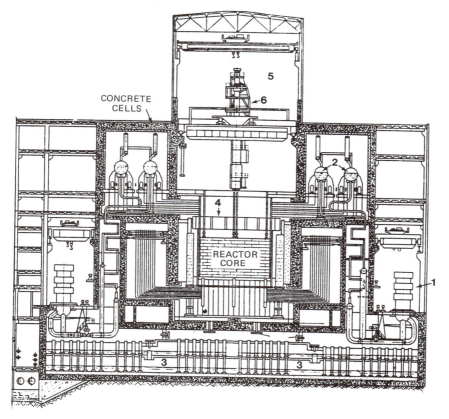

FIGURE 12-6 Cross section of the RBMK nuclear power plant at Chernobyl, Ukraine. The reactor core is shown at the center. The main circulating pumps (1) send the coolant through a complex system of piping into the fuel channels (not shown) that pass vertically through the reactor core. Water, with a steam (void) content of 14%, passes from the top of the reactor to the steam separators (2), from which the dry steam passes to the turbines. The coolant system is located within massive concrete cells designed to vent steam to vapor-suppression pools (3) in the event of a loss-of-coolant accident. Above the core is a massive concrete pad (4) assembled from removable blocks through which the coolant passes and penetrated by the fuel channels. The high bay above the reactor is of ordinary industrial construction and houses the fuel-handling machine (6).

the fuel channels, and the water, with a void (steam) fraction of 14.5%, then passes through steam separators to the turbine. Fourteen reactors of this design have been built since 1974.

The RBMK had been criticized in Western literature for many years prior to the accident because of design deficiencies. It has few of the engineered safeguards required in Western countries. The most serious defect is its positive void coefficient of reactivity. Thus, if liquid coolant changes to less dense steam during an accident, reactivity increases. Power reactors built in Western countries have a negative void coefficient that reduces reactivity as liquid coolant turns to steam. Another deficiency is the absence of a full containment structure. For many years, reactor experts who visited the Soviet Union had noted that the RBMK reactors were not fully surrounded by containment buildings of the type considered necessary in the West (Lewin, 1977). Instead, partial containment is provided by massive concrete walls that enclose most of the coolant system, but the region above the reactor floor is not enclosed and the roof of the reactor building is of ordinary industrial construction. This can be seen in Fig. 12-6. Another basic difference is that the RBMK reactor cores are not contained within steel pressure vessels as used in LWRs built in Western countries. Above the reactor is a high bay of conventional industrial construction that houses a tall and massive fuel loading machine. The roof of the building was not designed to contain a pressure surge, but the lateral components of a surge can probably be absorbed or deflected by the massive concrete shields also seen in Fig. 12-6. Beneath the reactor is a vapor-suppression pool, which would be useful in the event of sudden depressurization of the coolant circulating system.

CIRCUMSTANCES OF THE ACCIDENT

A routine shutdown of the reactor was scheduled for April 25, prior to which it was intended to test whether the momentum of the turbines would allow continued electricity production for start-up of emergency equipment in the event of an interruption in the steam supply (NRC, 1986). Similar tests had been performed previously without incident. Reactor power was gradually reduced from the full-power level of 3200 MWt in preparation for the test, which was to have been performed in the range of 700 to 1000 MWt. The emergency core cooling system (ECCS) was disconnected in accordance with the approved test procedure. However, because a need for power developed, the load dispatcher ordered the shutdown to be de-layed for 9 hours during which the reactor continued to operate at half full-power without the ECCS, in violation of operating rules. The automatic system for adjusting reactivity was also disconnected. During the 9-hour delay, core reactivity was greatly reduced because of the buildup of the fission product ^{135}Xe, which acts as a reactor "poison." The operator found the reactor difficult to control in the absence of the automatic system, and at one point the reactor power fell to 30 MWt. It then stabilized at 200 MWt but could not be increased further because of the presence of ^{135}Xe and insufficient reactivity reserve.

Standing instructions required shutdown of the reactor under these conditions, an action that would have avoided the accident. Instead, the operator took a number of actions, the net effect of which was to increase the void fraction and cause the power level to increase so rapidly that a steam explosion occurred. The steam released by the failure of pressurized tubes reacted exothermically with the zirconium, producing hydrogen that caused a secondary explosion. Parts of the core were scattered on the roofs of the reactor, turbine, and auxiliary buildings. Thirty fires were started and specially trained fire-fighting units from Chernobyl and nearby Pripyat responded within minutes. The building fires were extinguished about 4 hours after the accident, but the graphite continued to burn much longer, abetted by residual heat from the inventory of fission products. A source of radioactive emissions to the atmosphere was thus established that persisted for about 2 weeks. The remains of the power plant after the accident are shown in Fig. 12-7.

No announcement of the accident was made by the USSR at the time. Other countries soon learned of it from routine measurements made at nuclear power stations in Finland and Sweden, where monitoring equipment detected fission products in the atmosphere. It was soon determined that the source of the radioactivity was a reactor located in the USSR. However, inquiries made through diplomatic channels for several days elicited no information from the Soviet government. Finally, in response to worldwide pressure for information, an announcement that an accident had occurred was made on Soviet television on May 14, nearly 3 weeks after the accident took place. No quantitative information about the consequences was provided until late August at the IAEA conference in Vienna (IAEA, 1986).

The failure of the Soviet government to make a prompt announcement led to unfortunate rumors and speculation. A wire service dispatch from an Eastern European country stated that 2000 people in the Chernobyl area died from radiation exposure on the first day. This report was given credibility by a high State Department official. Since it was known that deaths from radiation exposure could only occur so quickly from doses in excess of about 50 Gy (5000 rad), speculation developed that a very much larger number of deaths would occur in subsequent days and weeks among those exposed to smaller doses. It later turned out that there were only two deaths on the first day, both of which were the result of trauma. In all, there were 28 deaths from the acute radiation syndrome (ARS) as a result of the accident (OECD, 1995).

Soviet investigators originally estimated that 81 MCi (3000 PBq) were discharged to the atmosphere between April 26 and May 6, by which time

FIGURE 12-7   Postaccident condition of the Chernobyl reactor building. The effect of fires can be seen on the turbine building roof in the foreground. (Photo by V. Obodzinsky.)

the emissions had practically ceased. During the past 10 years the source term has been refined and the estimated releases have changed considerably, increasing the total from 81 MCi to 295 MCi (11,000 PBq), a factor of nearly four. The 1986 and 1995 estimates of the principal nuclides that contribute to dose are compared in Table 12-6. About 6 tons of fuel were discharged in dispersible form (OECD, 1995). The percentages released varied from 100% for the noble gases down to 3.5% of the nonvolatile elements such as cerium and the actinide elements. Intermediate were 50–60% for ^{131}I and 20–40% for the cesium nuclides. The amount of ^{137}Cs discharged was about 3% of that released to the environment by all weapons tests conducted in the atmosphere. The quantity of ^{239}Pu released, about 700 Ci (26 TBq), was far less then the 350,000 Ci (13 PBq) known to have been distributed globally by weapons tests in the atmosphere. The rate of fission-product emission diminished from several MCi d^{-1} ($\sim$100 PBq d$^-$1) between April 6 and May 5 to less than 1 Ci d^{-1} ($\sim$30 GBq d^{-1}) by the end of May.

RADIOLOGICAL CONSEQUENCES

The radiological effects of the accident can be divided into four categories: (1) on-site personnel, including those mobilized for purposes of the required cleanup; (2) the environs within 30 km; (3) the European portion of the Soviet Union beyond 30 km from the reactor; and (4) European countries outside the USSR. Fallout was also detected in North America and Asia, but only in traces. Since the release was confined to the troposphere, transport to the southern hemisphere did not occur.

TABLE 12-6

Comparison of Estimates Made in 1986 and
1995 of Principal Dose-Producing
Radionuclides Released from Chernobyl[a]

Nuclide	Activity released (PBq)	
	1986 estimate	1995 estimate
^{133}Xe	1665	6500
^{131}I	270	1760
^{137}Cs	37	54
^{90}Sr	8	10
^{239}Pu	0.025	0.042

[a]The 1986 data are from IAEA (1986), and the 1995 data are condensed from OECD (1995).

On-Site Effects

Little information on the doses received by the workers was obtained from personal dosimeters. Either they were not used or became so badly contaminated that they were useless. In the absence of dosimeters, the Soviet scientists relied on biological indicators of dose based on earlier experiences with radiation injury in the USSR (Baranov and Guskova, 1990). Indicators such as the time of onset of nausea, vomiting, hematological changes, and chromosomal aberrations were used to estimate the dose received, the prognosis, and the methods of treatment. The scientists of the former Soviet Union report that they have had previous experience with about 1000 cases of the acute radiation syndrome.

A major unanticipated problem was severe "beta burns" of the skin because clothing became drenched with highly radioactive water. These effects were first evident within 1 or 2 days as a transient erythema that reappeared in greater severity in 2 or 3 weeks and sometimes resulted in blisters and severe ulceration. Eight workers with burns to 60–100% of the skin died during the first few weeks. Although severe bone marrow and intestinal injury were also present in these cases, it was concluded that it was the skin damage that caused death. Thus, in one case in which recovery of the bone marrow was occurring, death resulted on the forty-eighth day from endogenous intoxication arising from the extensive skin damage. Among the 28 early deaths from radiation sickness, the whole-body doses were as high as 1600 rad. The on-site personnel also sustained significant internal exposure from inhaled and ingested radionuclides, mainly [131]I. The thyroid doses were estimated to range up to 24 Gy (Guskova *et al.,* 1990). Potassium iodide (KI) had been administered to on-site personnel (250 mg twice daily), but not until about 90 minutes postaccident, too late to be fully effective. Some of the firefighters inhaled massive amounts of radioactive smoke and sustained severe beta burns to the lungs and nasopharyngeal region.

The task of cleaning up began on a heroic sale as soon as the fires were extinguished. About 800,000 workers that have come to be known as "liquidators" were recruited from the military and civilian sectors. Their duties included cleaning up around the remains of the reactor, destruction and disposal of contaminated structures, and construction of new roads. Some of the actions required great heroism. For example, during May and June a cooling system was installed beneath the reactor to control possible overheating of the damaged core material. In May, about one month after the accident, work began on an enclosure for the reactor, the "sarcophagus," that was completed in November[2] (Fig. 12-8). These and many other assignments were undertaken at great personal risk.

[2]The enclosure was not intended to be a permanent structure and soon began to show signs of deterioration. By 1990 it was estimated that 1200 m[2] of the external surfaces were open to the atmosphere. A permanent sarcophagus is now in the planning stage.

FIGURE 12-8   The "sarcophagus" built to enclose the remains of Unit 4 at the Chernobyl power station. The structure was built hurriedly under very difficult conditions. A secondary enclosure is being planned because of deterioration that has occurred. (Photo by V. Obodzinsky.)

During the first year, 200,000 workers were employed as liquidators and were allowed to accumulate 25 rem (0.25 Sv) before returning to their home bases. Some of the workers accumulated their allowable limit in a matter of minutes. It is estimated that the first wave of liquidators received doses that probably averaged 200 mSv (20 rem). However, the doses are uncertain because proper methods of physical dosimetry were either inadequate for the high radiation levels or simply were not available. Ten years after the accident there was a suggestion of slightly elevated cancer rates among the liquidators, but many more years of follow-up will be required before definite conclusions can be drawn.

The Population within 30 km

The main population center within a radius of 30 km was the town of Pripyat in which 45,000 persons lived only about 3 km from the reactor. Ambient gamma-radiation measurements, which began to be recorded about 3 hours postaccident, were less than 100 mR h^{-1} ($\sim$1 mGy h^{-1}) during the first day and then climbed to more than 1000 mR h^{-1} ($\sim$10 mGy h^{-1}) in some parts of the town. Evacuation of the village began about 36 hours after the accident, at which time the external gamma exposures ranged from 500 to about 1500 mR h^{-1} (~5 to about 15 mGy h^{-1}). In the meantime, residents were cautioned to remain indoors and distribution of KI began house-to-house at about 10 P.M. on the day of the accident. This was too late for suppression of iodine uptake to be fully effective, but the thyroid dose was reduced somewhat. Evacuation of Pripyat, which was completed in less then 3 hours, required the mobilization of 1100 buses. Evacuation of an additional 90,000 persons who lived within 30 km of the accident was completed by about the tenth day postaccident, bringing the total number of evacuees to 135,000. Among the many logistical problems that had to be dealt with was the presence of many cattle that could not be cared for in the absence of their evacuated owners. These were slaughtered, butchered, and sent to cold storage facilities. It has been reported that the only significant contaminant of the cattle meat was the short-lived ^{131}I, because accumulation of radioactive cesium in significant quantities had not yet occurred. Pripyat will remain evacuated for many years.

The mean whole-body dose received by the evacuees is estimated to have been about 1.5 rem (15 mSv), but there probably was wide variability and the estimated mean value must be regarded as very approximate. There is also uncertainty in the doses received by the thyroids of the evacuees because KI was administered at different times after exposure began. It is believed that the thyroid doses averaged 1 Sv (100 rem) for children under 3 years old and 0.7 Sv for adults (OECD, 1995).

Downwind fallout in this area ranged up to 4.8 PBq km^2 (130 kCi km^2), which was sufficient to kill about 1000 acres (400 ha) of coniferous forest. The conifers turned red in dying. The trees were so heavily contaminated that removing them was a major problem. They were cut and initially buried *in situ,* but then were disinterred and removed to one of the 800 high-level waste-disposal sites established in the vicinity. Two months after the accident, it was estimated that as much as 500 PBq (14 MCi), mainly ^{137}Cs, were on the ground within the 30-km zone. An extraordinary program of soil removal gradually reduced the amount of contamination, and a new village was built on the south edge of the 30-km zone to house the decontamination and power plant workers. Within a few years, some of the elderly farmers were permitted to return within the 30-km zone.

Effects in the Former Soviet Union beyond 30 km

The main sources of exposure beyond the 30-km zone were the radioiodines and ^{137}Cs, with a minor contribution from ^{90}Sr. Plutonium is present in the fallout but contributes less than 1% of the dose received. The radioactive iodines are mainly short-lived and were responsible for thyroid irradiation during the first few weeks after deposition, usually via the grass–cow–milk pathway. The dosimetry of radioiodine exposure under the conditions that existed at Chernobyl is made difficult by the fact that the nuclides are so short-lived and disappeared before measurements could be made on the massive scale required. Nevertheless, several hundred thousands of thyroid scans were completed.[3] Because ^{137}Cs has a half-life of about 30 years and remains fixed in the upper few centimeters of soil, it will remain a source of whole-body exposure for many decades. The dose from ^{90}Sr must be considered, since it is also absorbed via the grass–cow–milk pathway, but it was a minor problem owing to its lower volatility. The ratio ^{137}Cs : ^{90}Sr within the reactor core was reported to be about 1.4 compared to 8.5 in fallout, a depletion factor of nearly 6. After deposition, the ratio in near-surface soil will gradually increase because of the greater mobility of the ^{90}Sr.

During the 10-day period when emissions from the reactor were taking place, meteorological conditions changed several times. The plume drifted to the northwest on the first day and then veered in a northeasterly direction towards the Scandinavian countries, where its presence was detected by the routine monitoring programs at a Swedish nuclear power plant. The cloud then moved to the south, resulting in various amounts of fallout depending on the rate of cloud movement and the occurrence of precipitation. Although deposition over much of Europe was sufficient to result in considerable public and official concern, the greatest fallout by far occurred in the Ukraine, Russia, and Belarus. In 1986 these countries were part of the Soviet Union but soon became independent. We will see that contamination by radioiodine, particularly in Belarus, has caused thyroid cancer in children, and the levels of residual ^{137}Cs have resulted in restrictions of agricultural land use that have continued during the decade since the disaster occurred.

Effects Elsewhere in Europe

During the era when nuclear weapons were being tested in the open atmosphere, levels of fallout in many parts of Europe were not very different from those experienced in the less affected areas after the Chernobyl accident. However, that was a time when relatively few laboratories were equipped to measure environmental radioactivity. The Chernobyl accident occurred more than 20 years later, when many university laboratories, nuclear power plants, and national and provincial agencies possessed the

---

[3]One other radionuclide of iodine (^{129}I) is produced in copious amounts, but its 17-million-year half-life results in such a slow rate of decay that it delivers only a minor dose to the thyroid, compared to ^{131}I and ^{133}I. However, because iodine deposited in fallout remains in the top few centimeters of soil for decades, it could be of possible help in reconstructing the doses received.

equipment with which measurements could be made easily (Steinhäusler *et al.*, 1988; Bennett and Bouville, 1988). This resulted in a flood of information about the levels of fallout that was hard to interpret because the measurements often lacked intercalibration, and the data were reported in different units of measurement. Fortunately, the World Health Organization played an important coordinating role in the weeks following the accident and issued reports that summarized the data being accumulated by various countries (WHO, 1986). To further complicate matters, there were no uniform standards for instituting controls over consumption of contaminated food. This resulted in disagreements among both the federal and provincial levels of government, which served to emphasize the need for international standards to assist the authorities in making the decisions necessary to protect public health.

National regulations that controlled the permissible amount of ^{137}Cs in food restricted the sale of certain foods in some parts of Europe for several years after the accident. The restrictions on the distribution of sheep from certain parts of Cumbria, in the north of England, were not lifted until December 1991 (Ministry of Agriculture Fisheries and Food, 1991).

The effective dose commitments were higher on average in parts of southeastern Europe than in parts of the USSR itself. The doses received in Bulgaria, Austria, Greece, and Romania were among the highest, with average dose commitments in the range of 0.6 to 0.75 mSv (60 to 75 mrem). This was about equivalent to the dose received in many parts of the United States during the era of atmospheric weapons testing. In New York City, the thyroid dose to a 6-month-old infant as a result of the accident was less than 70 μGy (7 mrad) (Feely *et al.*, 1988). It was much higher during the nuclear weapons tests conducted by the Soviet government in the early 1960s. Anspaugh *et al.* (1988) estimated that the collective effective dose commitment to the 3 billion inhabitants of the Northern Hemisphere would be 930,000 person-Gy (9300 person-rem), with 97% in the western Soviet Union and Europe. Using risk coefficients derived from the linear–quadratic analysis of the Japanese studies, this leads to an estimate that the prevalence of cancer would be increased by no more than 0.02% in the European countries. This would be far too small an increase to be detectable.

HEALTH EFFECTS OBSERVED TO DATE

Thyroid Cancer

More than 200,000 ^{131}I thyroid measurements were made for children. In the Gomal region of Belarus, about 10% of the children less than 7 years of age received doses in the range of 2–40 Sv (200–4000 rem). Thyroid cancer, a comparatively rare disease, began to appear in greater than expected numbers as early as 4 years after the accident. In the years before the accident, the rate of childhood thyroid cancer in Belarus was comparable to that in England and Wales, about 0.5 cases per million children per year. By the years 1990–1994, the rate in Belarus had climbed to 92 cases per million children per year. By the latter part of 1995, a total of 463 children and adolescents had been treated surgically for thyroid cancer. This is about 50 times the normal rate (Konoplya and Rolevich, 1996). A surprising feature of the cancers is that they occurred so soon after exposure. It is notable that all cancers were in children less than 14 years of age. There is speculation that the short latency period and sensitivity of the children may result from the fact that the localities are noted for dietary iodine deficiencies. At this writing, this hypothesis can only be regarded as speculative but deserving of further study.

Leukemia and Solid Cancers

By 1996, 10 years after the accident, no difference has been found in the leukemia rates in regions in which the levels of fallout exceeded 550 KBq m^{-2} compared to the uncontaminated regions (WHO, 1995). Nor has there been an excess of solid tumors. However, because statistical uncertainties may obscure slight excesses, more time is required for study. It is possible that the leukemia risk coefficients, which have been derived from the experience in Japan, are too high for Chernobyl, where the dose was delivered over a much longer period of time compared to the Japanese bombings.

REPORTED HEALTH EFFECTS IN THE POPULATIONS OF THE THREE AFFECTED COUNTRIES

There has been widespread perception by the general public that increases in the prevalence of cancer began to appear almost immediately following the accident, despite implausibly short latency periods and, except for thyroid cancer, epidemiological evidence to the contrary. In addition, residents of both the contaminated and uncontaminated areas began to complain of an assortment of illnesses of types not accepted as resulting from radiation exposure.

An assessment of the health effects of the accident on the general health of the population was first undertaken in 1990 by the League of Red Cross and Red Crescent Societies. The survey found that in addition to widespread reports of increases in the prevalence of cancers, there were also reports of increases in morbidity due to hypertension, diabetes, chronic bronchitis, heart disease, and other disorders that are not considered to be plausible radiation-induced effects (League of Red Cross and Red Crescent Societies, 1990). Quantitative assessment was made difficult by the lack of good-quality baseline data. The report concluded that the only effects that could be definitely identified were those due to the anxieties of the general population. A similar conclusion was reached by the much more extensive study conducted by the United Nations (IAEA, 1991b; Ginzburg, 1993).

The anxiety of the people must not be dismissed lightly. The Chernobyl accident occurred at a time of extraordinary social and political changes in the USSR, which soon led to its breakup into the Commonwealth of Independent States. Long-standing animosities among the republics were quickly expressed. The Chernobyl reactors were located in the Ukraine, but control over them was maintained by Russia, and the heaviest fallout occurred in Belarus. The total secrecy with which government had been conducted for hundreds of years under both the Czars and the Soviets was suddenly changed to a policy of openness, and the public was not prepared to deal with the vast amount of often conflicting information presented by the media. Nor was the public prepared to deal with the rough and tumble tactics of political infighting. It has been no wonder that widespread anxiety developed in the general population (Shlyakhter and Wilson, 1992).

CONTAMINATION OF LAND BY LONG-LIVED NUCLIDES

Contamination of large tracts of land by ^{137}Cs has created by far the most serious consequences from the point of view of potential long-range health effects, social stress, and economic impact. Other nuclides have also attracted attention, most notably ^{90}Sr and Pu, but these have been shown to contribute only in a minor way to the doses received. The strontium nuclides contribute less than 5–10% of the dose received, and plutonium less than 1%, even for workers in the outdoors (Ivanova, 1995). When it became evident that standards would be required to govern land-use policies, the Soviet National Committee on Radiation Protection proposed that relocation of people should be required if the estimated lifetime dose exceeded 350 mSv (35 rem), a value corresponding to 70 years of exposure at 5 mS y^{-1} (500 mrem y^{-1}). However, opposition to the proposal became so intense that it was rejected by the Supreme Soviet and a special commission recommended limits based on levels of ^{137}Cs deposition. Remedial measures were required between 550 and 1480 kBq m^{-2}, and relocation was required for deposition above 1480 kBq m^{-2} (OECD, 1995). The remedial measures included relatively simple and practical practices such as avoiding consumption of mushrooms and washing vegetables before eating. At a contamination level of 550 kBq m^{-2}, the annual whole-body dose would be about 5 mSv, which declines as the ^{137}Cs decays. The Ukraine and Belarus were the provinces most affected. More than 10,000 km^2, mainly in Belarus, were contaminated above 550 kBq m^{-2} (15 μCi m^{-2}) and countermeasures were considered to be necessary.

Since the dose received is largely from ^{137}Cs deposited on agricultural land, the severity of the fallout in the European portions of the former Soviet Union will be discussed in terms of the deposition of that nuclide. In many localities, the ^{137}Cs deposition caused unacceptable contamination of agricultural produce. In Belarus, 30% of the milk and 5.7% of the meat produced in 1986 was confiscated. By 1989 this had been reduced to 2.8% and 0.1%, respectively (Alexakhin, 1990).

These assessments required the mobilization of laboratory resources in the former Soviet Union on a gigantic scale. About 7000 laboratories participated. About 200,000 children were thyroid-scanned and thousands of whole-body measurements for radiocesium were performed. Many tens of thousands of measurements were made of food and soil samples. The methods used by the investigators were examined by an international committee of experts nominated by the World Health Organization, the International Atomic Energy Agency, and other United Nations affiliates and were found to be of high quality. In fact, it was found that the methods tended to increase the estimates of dose because of conservatism in the assumptions made (IAEA, 1991b).

In addition to the 10,000 km^2 contaminated to levels greater than 550 mBq m^{-2}, about 3000 km^2 were contaminated to levels above 1480 mBq m^{-2} (IAEA, 1996). In 1987 in "all union" Chernobyl registry was established to provide the data base for a long-range followup system in which 270,000 people living on land contaminated to levels above 550 mBq m^{-2} would be required to receive annual physical examinations.

For the purpose of estimating the public health implications of land contamination, it is convenient to use collective dose. The 70-year collective dose commitment to the residents of these regions has been approximated to be 10,000 to 20,000 person-Sv (1 to 2 million person-rem) (Balanov, 1993). This would at most result in too small a number of radiation-induced cancers to be measurable, and a minor increment against the prevalence of cancers from other causes. One exception, discussed earlier, is thyroid cancer as a result of the high doses received by children. Another possible exception could be radiation-induced leukemia, which has a comparatively low spontaneous rate in the general population and a relatively short period of latency. No excess of leukemia has been identified after 10 years.

Attempts to reduce the effects of cesium contamination of agricultural land on food products have been undertaken on a large scale. The methods used, with more or less success, have included liming soils, adding potassium, deep plowing, removal of livestock from contaminated pastures before slaughter, and adoption of decontamination techniques in food processing. These measures have been reported to have reduced the effective dose commitment by about 60% (Fesenko *et al.*, 1995).

THE ECONOMIC COST OF THE ACCIDENT

The effects of the accident have been devastating to the agrarian economies of the three countries. It is estimated that 2.6 million km^2 of agricultural land has been affected. Fifty-four collective farms have been taken out of production, and restrictions on the use of extensive forest areas have been imposed. Mineral industries have also been affected: in Belarus, 91% of the available glass sands cannot be used, and 16% of cement production is in the highly contaminated zone (Konoplya and Rolevich, 1996). The government of Belarus has estimated that the 30-year program required to rehabilitate these areas will cost $235 billion in U.S. dollars, which is 32 times the nation's entire budget in the year of the accident. Belarus received the greatest amount of fallout and had had to deal with the most serious of the health and economic consequences. Although it is now more than 10 years since the accident occurred, it is still too soon to obtain realistic estimates of the economic cost of the disaster to that country.

SOME GENERAL OBSERVATIONS

The Chernobyl experience has served to identify many practical problems in the administration of protective action guides (PAGs). These had been established country by country and varied greatly. At the time of the accident, the PAG for ^{131}I in milk was 10 times higher in the United States than in neighboring Canada. In Europe, the spread in PAGs for protection of the adult thyroid varied about 30-fold among the various countries. There were comparable differences in the PAGs for ^{137}Cs in food. In 1991 the Food and Agriculture Organization and the WHO incorporated proposed PAGs into their food standards program (FAO, 1995). This may assist the adoption of uniform values on a worldwide basis.

    The relatively long time it takes to obtain reliable information from radionuclide measurements is a fundamental difficulty in the application of PAGs. The problem is particularly difficult with respect to ^{131}I, which has a short half-life and, because of its rapid uptake by grazing cows, can contaminate a milk supply in less than 2 days. This may be less time than it takes to collect representative samples of air, grass, or milk and transport them to a laboratory, complete the counting procedures, and report the results. The same problem exists with respect to cesium, for which time is not so important because of the long half-lives of the cesium nuclides and the much longer time during which the dose will be delivered. Data on radio-strontium are likely to be lacking totally in the first few days after an accident because it is a beta emitter and can be measured only after lengthy radiochemical procedures have been completed. However, the Chernobyl experience indicates that dose from the strontium nuclides is likely to be much less than that from cesium.

The time factor is even more important with respect to the use of thyroid-blocking agents. To be maximally effective, KI should be administered prior to exposure or, at most, within 2 hours postexposure. This will not be possible if the decision to administer KI is dependent on measurements. These difficulties will often make it necessary for the implementation of PAGs to rely on the use of models that depend on estimates of the source strength and on-line meteorological information to predict the times and places where protective action must be initiated.

Application of PAGs requires discipline on the part of local government, elected officials, scientists, and the media. The value of advance planning is greatly diminished when protective actions are taken (well meaning or otherwise) at lower levels of exposure than prescribed by the advance planning.

We discussed earlier that there were basic deficiencies in design that would have prevented the RBMK reactors from being licensed in any Western country. There was also an important institutional difference. The system of regulating nuclear power plants in the United States and other Western countries requires that all information about reactor design, construction, and operation be available to the public, and that provision be made for interveners in the licensing procedures. The intervention is frequently costly and time-consuming, but it is an effective way of probing the safety of the reactor design and the operating procedures. The complex system of licensing and the very threat of intervention by members of the public assure the best efforts on the part of the reactor designers, builders, and operators. The major lesson of Chernobyl may be the importance of a rigorous system of nuclear power reactor regulation, with unobstructed but reasonable opportunities for public participation in the licensing procedures.

OTHER CONTAMINATING EVENTS THAT OCCURRED IN THE FORMER SOVIET UNION

History will show that the Chernobyl accident was a turning point in the general policy of the USSR (now the Commonwealth of Independent States) concerning releases of information of many kinds to the open literature.

This policy was changed and replaced by openness (glasnost) by Mikhail Gorbachev during his term in power, just prior to the dissolution of the Soviet Union. Soon after, official reports began to be issued that revealed a pattern of other contaminating events that were the result of underlying factors similar to those responsible for the Chernobyl disaster. These included poor engineering, disregard for international standards of radiation protection, inadequate training of personnel, and just plain callousness on the part of officials. Events of which we are now aware caused thousands of workers and members of the general public to receive doses high enough to require long-term monitoring for delayed health effects. International collaboration is just beginning regarding such monitoring. In addition, contamination of land, fresh water, and marine environments will provide opportunities to refine the dose assessment models now used in the West to predict the behavior of radionuclides introduced into the environment.

The Soviet program for development of military uses of nuclear energy began in the late 1940s (Holloway, 1994). Thousands of workers were assembled at the research and production complex constructed in the Cheliabinsk district in the Ural Mountains, and a weapons testing site was established at nearby Semipalatinsk, where the first Soviet bomb was exploded in August 1949. Unfortunately, the Soviets failed to train a staff of radiation protection specialists similar to that relied upon by the U.S. program that was begun early in World War II. In contrast to the excellent safety record of the U.S. program, the Soviet scientists and engineers developed the plants and laboratories with insufficient attention to safety in both the workplace and the environment.

The main nuclear weapons production complex, called the Mayak Production Association, began operations in 1948 and has had a dismal record of worker safety, as well as exposure of three groups of off-site people: about 125,000 persons exposed to discharges into the Techa River, about 275,000 persons exposed to radiation following an explosion in a waste storage tank in 1957, and about 42,000 persons exposed to windblown dust from the contaminated bed of Lake Karachay in 1967 (Degteva, 1996). These sources of exposure are shown schematically in Fig. 12-9.

The basic statistics on occupational exposure during the early years of the program were not released until 1990 (Nikepelov et al., 1990). In one year, 1950–1951, almost half of the employees received doses of more than 1 Sv (100 rem). About 14,000 men and 5000 women were employed at this complex. Preliminary data suggest an excess of leukemia among the men and an excess of lung cancer among the women (Koshurnikova, 1996).

Little attention was apparently directed at isolation of large quantities of high-level wastes. Much of the wastes from the plutonium-production facility found its way into the Techa River and contaminated the shoreline

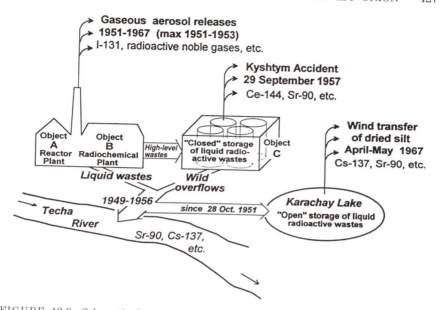

FIGURE 12-9   Schematic diagram of the radioactive releases from the Mayak Production Center, the main USSR nuclear weapons production complex, in the Ural Mountains. [From Degteva (1996).]

and drinking water (Trapeznikov et al., 1993; Balter, 1995; Degteva, 1996). It is estimated that from 1949 to 1956, 2.75 MCi (100 PBq) of wastes containing various nuclides (of unspecified age) was discharged to the river. Almost all of this was discharged between March 1950 and November 1951. In the summer of 1951, the exposure rate along the shore of a village downstream of the plant was 5 R h⁻¹ (~0.05 Gy h⁻¹). The streets and homes of villages had levels of 10 to 15 mR h⁻¹ (~0.1 to 0.15 mGy h⁻¹). The drinking water was also badly contaminated. After 25 years, examination of 12,000 residents of the riverbank area showed that nearly 1% had body burdens of ^{90}Sr greater than 2 μCi (74 kBq). It will take many years to reconstruct the exposure history of this population, and even longer to develop information about the effects of exposure on the health of the exposed people (Balter, 1995).

In 1976, Zhores Medvedev, a Soviet geneticist living in exile, first published accounts of a major release of radioactive material that took place in the Urals in 1957 (Medvedev, 1979). There had by then been no confirmation by the Soviet Union of such an occurrence, but the Medvedev claims received support from U.S. investigators (Trabalka et al., 1980), who undertook a comprehensive review of Soviet radioecological publications in the open literature and found a pattern of anomalous features consistent with the claims he made. The USSR maintained its silence on the subject until July 1989, when it submitted a report to the IAEA that described the principal features of the accident and its consequences (Nikepelov et al., 1989).

The accident occurred on September 29, 1957, and involved the failure of the cooling system for tanks in which high-level wastes were being stored as a nitrate solution. This resulted in an explosion that scattered fission products over an area of 16,000 km^2 inhabited by about 270,000 people. The effective dose equivalent received by about 10,000 people who were evacuated ranged up to 52 rem (0.52 Sv), and the collective effective dose equivalent totaled 115,000 person-rem (1150 person-Sv). There is every reason to believe that the report to IAEA was factual and, if so, it is clear from comparison with the previously published material that, in the absence of official information, the effects of the accident had become greatly exaggerated over the years. It was a serious accident with major economic consequences, but the public health consequences were minor. The total fallout of fission products beyond the facility itself was estimated to have been 2 MCi (74 PBq), compared to the estimated 325 MCi (12,000 PBq) released by the Chernobyl accident.

Large quantities of radioactive wastes, including reactor components, generated by the Soviet fleet of nuclear-powered ships have been disposed of in the northern waters of the open seas. The USSR operated 407 reactors, mostly on 235 submarines and other naval vessels, but also on nuclear-powered icebreakers (Office of the President of the Russian Federation, 1993). The 1993 report tallied about 10 MCi (370 PBq) of waste, expressed as ^{90}Sr equivalent curies.[4] This was done in knowing violation of international agreements that govern the use of the seas for radioactive waste disposal. The infractions will now offer the opportunity for long-range studies of the impact of such practices. Studies are needed and may lead to ways in which the marine environment can be used in the future without risk for controlled disposal of both chemical and radioactive wastes. At the present time, international rules deny nations the use of the seas for even well-planned waste disposal. As reported by the Russian Federation, the quantities disposed of are not large compared to the discharges by the British into the Irish Sea or the fallout into the Pacific Ocean from U.S. weapons testing practices (Eisenbud, 1981a). These practices, while best avoided, do not seem to have resulted in significant exposure of the human food chain.

[4]By using ^{90}Sr equivalents, the mixtures of radionuclides are normalized so that the total radiological impact is equivalent to the stated quantity of ^{90}Sr.

ACCIDENTS INVOLVING MILITARY AIRCRAFT CARRYING NUCLEAR WEAPONS

There have been several accidents to military aircraft carrying nuclear weapons. In two reported accidents involving U.S. aircraft, the bombs were destroyed and plutonium was dispersed, but no nuclear reactions took place. There are no reports of such accidents from other countries.

THE THERMONUCLEAR WEAPONS ACCIDENT IN PALOMARES, SPAIN, 1966

In January 1966, two U.S. Air Force aircraft crashed during a refueling exercise over the village of Palomares, on the southeastern coast of Spain. Of four thermonuclear weapons that fell with the wreckage, one was recovered intact from the Mediterranean Sea after a 3-month search, one was recovered intact from nearby fields where it landed by parachute, and two were destroyed on impact by detonation of the chemical explosives in the weapons. The pyrophoric plutonium metal was ignited, creating a cloud of oxide fume that was dispersed by a 30-knot wind from the west. Two hundred twenty-six hectares (about 0.87 square miles) of residential areas, farmland, and woods were contaminated. The accident occurred in an area where the fields were farmed intensively, primarily for tomatoes, the last seasonal crop of which was ready for harvest (Nuclear Defense Agency, 1975).

Initial Countermeasures

The initial levels of plutonium contamination are shown in Fig. 12-10. United States Air force and Spanish nuclear specialists cooperated in a radiological survey and cleanup of the area and decided to proceed according to the following criteria:

1. Soil contaminated above 1.2 MBq (32 μCi) per square meter was removed, barreled, and shipped to the United States for burial at the Savannah River plant. The area treated in this way totaled 2.2 ha, from which six thousand 250-L drums of soil were removed.
2. Arable land below 1.2 MBq (32 μCi) per square meter was mixed by plowing and harrowing to a depth of 30 cm. Seventeen hectares were treated in this way.
3. On rocky hillsides, where plowing was not practical, and the deposition was greater than 120 kBq (3.2 μCi) per square meter, the soil was removed with hand tools and drummed for shipment to the United States.

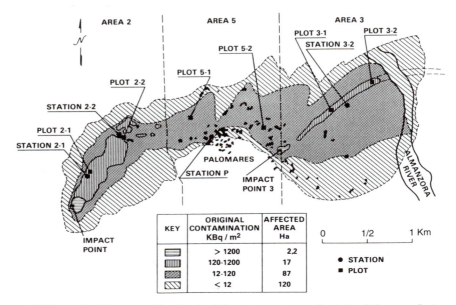

FIGURE 12-10 The postaccident levels of plutonium contamination in the Palomares, Spain, vicinity, showing the locations of sampling and experimental field plot stations. [From Iranzo and Richmond (1987).]

When these decontamination objectives were achieved, a surveillance program was established to monitor the air, soil, vegetation, farm animals, and people for plutonium. The surveillance program has now been in effect for about 25 years (Iranzo and Richmond, 1987; Iranzo et al., 1987, 1988, 1991, 1992).

Residual Contamination of Soil

The mean annual rainfall at Palomares is 20 cm. Studies of the plutonium content of the soil have been under way since 1968 in six 50 × 50-m study plots whose locations are shown in Fig. 12-10. Vegetation from the plots is also analyzed to measure plutonium uptake as well as the amount of contamination on the surface of the plants. From physical measurements and theoretical considerations, it has been concluded that soil conditions would favor maximum stability in the Pu^{4+} state. After 23 years, more than 99% of the plutonium in untilled soils has remained within a depth of 5 cm.

Contamination of Vegetation

Tomatoes, barley, and alfalfa have been the main crops since the time of the accident, with lesser production of melons, peppers, and corn. The analytical procedures are reported to differentiate between plutonium absorbed metabolically into the plants and that present as surface contamination due to dusting or rain splatter. As expected, the measurements show that contaminated food is insignificant as a source of human exposure. Nevertheless, the soil-to-plant concentration factors reported for Palomares are higher by about a factor of 10 than the values reported by investigators in other areas, which may be the effect of surface contamination that was not fully removed.

Plutonium Resuspension

Six air-sampling stations were originally established at Palomares, but for various reasons not all of the stations have been maintained during the entire period of the study. However, there is a full record for two of the stations, indicated as Stations 2-1 and P in Fig. 12-10. Station P is located in the village of Palomares, the population of which is currently about 760. Station 2-1 is located in a valley, the hills of which surround the region of maximum contamination.

The annual average concentrations for the two stations are summarized in Table 12-7 together with data from control stations in northwest Italy and New York City for purposes of comparison. Airborne ^{239}Pu has been detectable worldwide in tropospheric air as a result of nuclear weapons tests prior to the 1963 ban on atmospheric tests. Most of that plutonium was injected into the stratosphere, from which slow transfer to the troposphere has been occurring.

TABLE 12-7

Average Air Concentrations of Plutonium at Various
Locations (μBq m^{-3})[a]

Year	New York City, U.S.A.	Northwest Italy	Palomares, Spain	
			In town	Area 2-2
1966	3.1	2.6	14.8	44.8
1967	1.3	1.6	4.1	441.8
1968	2.0	2.7	2.6	21.8
1969	1.5	1.8	2.6	142.1
1970	1.8	1.8	2.2	5.9
1971	1.5	1.7	1.8[b]	2.2
1972	0.7	0.8	1.8[b]	10.4
1973	0.3	0.5	2.2	3.0
1974	1.1	1.4	4.1	8.1

[a]From Iranzo and Richmond (1987).
[b]Minimum detection level.

The concentration of airborne plutonium has been consistently elevated above "background" for all years at Station P and Station 2-1. The annual average concentrations at the two stations were elevated by factors of 2.5 and 48 during the period 1967–1974. The values within the town of Palomares showed a tendency to increase after about 7 years, which might have been the result of contamination of the streets with soil carried into town from the more heavily contaminated agricultural areas by farm equipment, people, and farm animals. The higher values at Station 2-1 are attributed by the Spanish investigators to the opening of new land to cultivation.

Because the data are obtained from fixed stations, they do not include exposures of people working in the fields. Operators of farm equipment when the fields are dry are probably exposed more heavily than is indicated by the data collected at the fixed stations. If, as was likely, the particles were originally formed as a fume from the burning metal, plutonium would be disseminated into the environment as fine particles (<1 μm), which would attach themselves to larger soil particles that would thereafter determine their aerodynamic behavior. About 90% of the plutonium has been found to be associated with particles larger than 10 μm, despite the fact that the measurements were made after the samples were subjected to maceration and ultrasonic dispersion (Iranzo et al., 1992). This would have increased the fraction of the very fine particles by separating them from large soil particles to which they were attached by surface forces.

Human Measurements

About 125 persons per year have been taken to Madrid for medical observation and radiological measurements, using urine analysis and whole-body counting. Of 1815 assays for plutonium, the urine from 124 people was above 0.37 mBq d^{-1} (10 fCi d^{-1}), the lower limit of detection. This sensitivity makes it possible to identify subjects for whom the 50-year effective dose equivalent (EDE) is greater than about 0.02 Sv (2.0 rem). It is estimated that the EDE exceeded 0.05 Sv (5 rem) in only 33 subjects, of whom the most heavily exposed (5 people) will receive a dose in the range of 0.15 to 0.2 Sv (15 to 20 rem) (Iranzo et al., 1988).

Whole-body measurements of Pu have also been made. All measurements have been below the lower detectable limit of 814 Bq (22 nCi).

The Palomares accident, unfortunate though it was, has provided a unique opportunity to study the uptake of plutonium in a contaminated rural environment. No medical findings of significance have been reported, but this is not surprising in view of the small population and the small doses received.

The Crash of a B-52 in Greenland in 1968

In 1968, a B-52 bomber of the U.S. Air Force crashed near Thule Air Base in northern Greenland, resulting in detonation of the chemical explosives in four nuclear weapons being carried aboard the aircraft (Aarkrog, 1971a). An unspecified quantity of ^{239}Pu and ^{241}Am was dispersed on the sea ice and, with its breakup in the spring, was transferred to the sea and sea bottom. Radioecological studies during subsequent years investigated the behavior of these elements in the sediments and benthic organisms (Aarkrog *et al.*, 1984). The marine sediments contained about 1 TBq (27 Ci) of ^{239}Pu in 1979, an amount similar to that found in 1968. From 1974 to 1979, horizontal transport was taking place at a rate of 0.4 km y^{-1}. The benthic community was found to contain only a small fraction, about 0.5×10^{-3}, of the sediment inventory. No Pu or Am was found in plants collected from surface water or in fish or sea mammals.

ACCIDENTS INVOLVING ABANDONED, LOST, OR STOLEN GAMMA-RADIATION SOURCES

There have been two serious accidents in which abandoned teletherapy units caused extensive contamination, acute radiation effects, and, in one case, fatalities. These occurred in Juarez, Mexico, and Goiania, Brazil.

There have been other instances in which misplaced or stolen gamma-ray sources have resulted in fatalities and injury, but these did not result in dispersed environmental contamination. In Morocco, eight members of one family died, and in Shanxi Province, China, a lost source resulted in 54 hospitalized cases of radiation injury and four deaths (UNSCEAR, 1993). Another accident involved the loss of a ^{192}Ir brachytherapy source from an Indiana, Pennsylvania, medical center (Jones, 1994). This accident contributed to the death of a patient and resulted in doses to nonradiation workers as high as 0.22 Sv (22 rem), but no distributed contamination. The most recently reported incident resulted in one death and several acute radiation injuries in Estonia (Dunlap, 1995a).

Recycled steel has also become contaminated in numerous incidents (Lubenau and Yusko, 1995). In one of the more extensive recycled steel cases, approximately 1000 apartments, offices, and classrooms in 85 buildings in Taipei, Taiwan, were constructed with reinforcing steel contaminated with ^{60}Co. Although no acute injuries were reported, doses to occupants ranging from 3 to 130 rem (0.03–1.3 Sv) over a 10-year period were given in a preliminary estimate, and more detailed work is in progress (Chang and Kau, 1993).

The Juarez and Goiania accidents are discussed in more detail in the following sections.

Cobalt-60 in Juarez, Mexico, 1983

In December 1983, a teletherapy unit that contained 450 Ci (16.7 TBq) of ^{60}Co was dismantled and sold as scrap in Juarez, Mexico, just across the Rio Grande from El Paso, Texas. Rough handling in the scrap yard ruptured the radioactive source and ^{60}Co pellets, each of which contained about 75 mCi (2.8 GBq), were scattered about the property (Burson and Lushbaugh, 1990). The pellets became mixed with scrap of all kinds, which was then transported to foundries, where some of them became incorporated into steel that was eventually converted to reinforcement bars, metal table bases, and electric motor parts that were shipped to the United States. In the course of events, several hundred pellets spilled into a truck that was left parked for more than one month on a street in Juarez. The scrap yard and the truck thus became the two main sources by which people were exposed.

The events might never have become known except that in mid-January 1984 a truckload of reinforcing rods stopped for information at the gate of Los Alamos National Laboratory in New Mexico. The truck triggered a radiation alarm that led authorities to a foundry in Mexico where the rods were fabricated. The DOE emergency response teams were alerted, and extensive aerial surveys were begun. In this way the truck parked in Juarez was found, and on inspection it was shown to be capable of causing exposures as high as 1000 R h^{-1} (~10 Gy h^{-1}). A radiation survey revealed that the exposures within houses in the area could have ranged between 50 and 200 R (~0.5 and 2 Gy). By that time the truck had been parked in the same place for about 50 days. Three people probably received doses in excess of 100 rem (1 Sv).

Conditions in the scrap yard were worse. The doses to 50 people who worked there ranged from 390 to 635 rem (3.9 to 6.4 mSv). Cleanup operations were begun, and 3000 table bases and 600 tons of bars were located. Total or partial demolition of 834 buildings was required. A waste-disposal depot was established, where 16,000 m^3 of earth and 5000 m^3 of other material were placed (Burson and Lushbaugh, 1990).

The estimated doses received by either the workers or members of the public are summarized in Table 12-8. The doses were received over an extended period of time and only mild symptoms of radiation exposure resulted. There were cases of mild stomach upsets and nosebleeds, and reduced sperm counts in the most heavily exposed men.

TABLE 12-8

Distribution of Doses Received by Workers
and the Public as a Consequence of the
Juarez, Mexico, Contamination Accident[a]

Number of people	Doses (Sv)	Doses (rem)
5	3 to 7	300 to 700
80	~0.25	~25
720	0.005 to 0.25	0.5 to 25
3200	<0.005	<0.5

[a]From Burson and Lushbaugh (1990).

Cesium-137 in Goiania, Brazil, 1987

A more serious accident, similar in its main features to the one in Juarez, occurred in September 1987 in the state of Goiania, Brazil. A teletherapy unit that contained 51 TBq (1375 Ci) of ^{137}Cs in the form of $CsCl_2$ powder was left in an abandoned medical clinic (Paschoa *et al.,* 1993, Oliveira *et al.,* 1991). The source assembly, which weighed about 50 kg, was removed from its shield and taken by wheelbarrow to the home of a scavenger. Several families lived in the house. Two men spent 3 hours attempting to disassemble the source, using a hammer and other crude means. On the fifth day the scrap was given to a junkman who placed it in a backyard dump and noticed its luminescence that evening. He brought it into the living room of his home, where for about 2 weeks it was an object of curiosity to family and neighbors, some of whom pried off pieces of the source as souvenirs. The source was then removed to a second junkyard, where an attempt was made to disassemble it with a power saw. On the fifteenth day, the wife of one of the junkmen became concerned because so many of their friends who had handled the source had developed gastrointestinal symptoms. She carried the source to the office of the local sanitary inspector, where on the following day a physicist determined that the object was highly radioactive. A team of specialists was dispatched to the scene by the National Nuclear Energy Commission and thereafter began an ordeal that for some of its members lasted 3 months, in which the technical difficulties were compounded by intense heat and a less than ideal geographical and social settings (IAEA, 1988; Health Physics, 1991).

The $CsCl_2$ was found to be distributed over an area of about 1 km^2, having been disseminated by the actions of people, heavy rainfall, wind, and foot and vehicular traffic. Decontamination methods were improvised, using solvents, sandpaper, and physical removal of the waste to a depot established nearby. The contamination limits used for the cleanup are listed in Table 12-9.

Four people ultimately died of acute radiation injury, and one person required amputation of an arm. One hundred twenty-nine persons were found to have measurable body contamination, of whom 21 were hospitalized because of serious skin burns or blood changes. Following the Goiania accident, Prussian blue was used in doses from 3 to 10 g d^{-1} to enhance the elimination of ^{137}Cs from the body (Melo *et al.*, 1994). Prussian blue caused internal dose reductions in adults in the range of 51–84%, with an average of 71%, independent of the dosage of Prussian blue. Total internal committed doses were in the range of 4.6×10^{-3} to 9.7×10^{-1} Gy (0.46 to 97 rad) for the most-contaminated adults.

Many tasks had to be undertaken (Rosenthal *et al.*, 1991) and had to be dealt with under highly adverse physical and social conditions. These included:

1. analysis of urine and feces;
2. construction of a whole-body counter;
3. whole-body counting of more than 200 people;
4. dose assessments for exposed people;
5. improvisation of decontaminating methods;
6. management of wastes;
7. patient management;
8. study of the efficacy of Prussian blue (ferric ferrocyanide to accelerate the excretion of cesium); and
9. communication with the media.

This accident serves as a grim reminder of the need to prepare for accidents that cause environmental contamination. More detail can be found in the two excellent reviews cited earlier (IAEA, 1988; Health Physics, 1991).

TABLE 12-9

Contamination Limits Used during Cleanup of the
Goiania, Brazil, Accident[a]

Quantity	Limits	
General surface contamination	0.1 nCi cm^{-2}	3.7 Bq cm^{-2}
Gamma radiation in homes	50 μR h^{-1}	~0.5 μGy h^{-1}
Gamma radiation outdoors	100 μR h^{-1}	~1 μGy h^{-1}

[a]From da Silva (1991).

The Goiania accident, like some of the others described, could have been prevented completely or its effects greatly ameliorated by more stringent regulatory enforcement and advanced planning. It is clear in this case that the use of water-soluble $CsCl_2$ contributed greatly to the severity of the accident. The source was registered by the responsible government agency, but it was a poor regulatory system that allowed the large source to be abandoned without control over its disposition. The regulatory system should provide supervision over such sources, literally from the cradle to the grave.

CHAPTER 13

Methods of Environmental Surveillance

Radiological environmental monitoring can be conveniently divided into effluent monitoring and environmental surveillance. Effluent monitoring is concerned with measuring the amounts and kinds of radionuclides released from radiological facilities, which are often called the "source term." Environmental surveillance is concerned with measuring radioactivity in the various environmental media such as air, water, biota, and foodstuffs. This chapter focuses on environmental surveillance and treats effluent monitoring primarily by reference. It is worth noting, however, that for diffuse sources such as radioactive landfills, the distinction between effluent monitoring and environmental surveillance blurs, and methods normally associated with environmental surveillance are used to characterize the source.

Broadly speaking, there are two basic types of environmental surveillance programs: those that are regional or global in nature and those that provide information about the vicinity of a particular facility from which radioactive substances may be emitted. Ideally, the individual plant-site monitoring programs should blend into surveillance programs operated by the local or state government to provide information on a county- or state-wide basis, and these programs, in turn, should blend into national and international surveillance programs operated by national governments or international organizations such as the Pan-American Health Organization, the World Health Organization, or the IAEA. Such international agencies have in fact assembled environmental radioactivity data, an important example being that of the United Nations Scientific Committee on the Effects of Atomic Radiation, which periodically publishes compendia of surveillance information from all parts of the world (UNSCEAR, 1958–1994).

Environmental surveillance programs for specific facilities may be further differentiated into the following: (1) preoperational design and surveillance to provide baseline radiological information against which any changes caused by a particular nuclear activity can be ascertained; (2) an operational surveillance program conducted during the operational life of a facility; (3) a surveillance program that can be held in abeyance pending an emergency in which extraordinary information is required; and possibly (4) a postoperational surveillance program to assess any residual effect of the facility on the radiological environment.

In surveillance programs, a primary objective is often to comply with governmental requirements for surveillance and to help ascertain compliance with relevant governmental radiation limits. Some types of surveillance samples, such as those of air, drinking water, and food, can provide information that is directly relatable to radiation dose to members of the public. Other types of surveillance samples are limited to verifying the integrity of the emissions controls of a facility. Examples of these kinds of samples are sediments, waters not used for drinking, and air samples in unpopulated areas.

It must be emphasized that surveillance programs can at best provide only an *estimate* of the radiation exposures to the off-site population. In fact the estimate is usually in the form of an upper bound. When operating within current regulatory and design limits (see Chapter 3), modern nuclear facilities' emissions are so small that facility-related radioactivity in the environment is frequently undetectable. Even when small amounts of facility-related radioactivity can be detected, the data are frequently accompanied by a great deal of uncertainty, even with state-of-the-art surveillance programs. The uncertainty in surveillance measurements depends on the kind and number of observations made, the type of instrumentation employed, the methods used in analyzing the samples and data, and basic physical limitations. As a consequence of the inherent limitations of surveillance programs, compliance with regulatory radiation dose limits during routine operation, particularly for airborne emissions, is usually demonstrated using a combination of measured emissions and environmental transport and dose models (see Chapters 5 and 14). Environmental surveillance data often provide only supplementary assurance that the regulatory limits have been met.

Arguably the most important aspect of a well-planned environmental surveillance program is the ability to provide data in case of an emergency that results in an unplanned environmental release of relatively large amounts of radioactivity. In the case of a large release, detectability is usually not at issue and the uncertainty of individual measurements is proportionally smaller. Also during emergency situations, the amount and temporal distribution of radioactive releases may not be known immediately and may never be known with desired accuracy. Thus the environmental surveillance data may well provide better and more timely information about the environmental consequences of an emergency than the use of models. In view of this, an environmental surveillance program should be flexible so that the needs of the routine program can be met while maintaining the ability to intensify and modify the program in response to potential emergency situations.

The basic principles of environmental monitoring were laid down by ICRP (1965, 1984) and by EPA (1972), and there is extensive literature in which refinements and experiences with a variety of surveillance programs have been described (IAEA, 1966; Reinig, 1970; NCRP, 1976; Kelly, 1980; Watson, 1980; Kathren, 1984; Keith, 1988; Kathren *et al.*, 1993c). The requirements for environmental surveillance programs for the various types of nuclear facilities have become increasingly prescriptive and guidance documents are available for the DOE and its contractors (DOE, 1991c) and for the nuclear power plants regulated by the NRC (Meinke and Essig, 1991a,b).

Ideally, surveillance programs should be designed to provide assurance that the dose to humans is below some prescribed level, usually specified in a governmental regulation or guideline. Because of the limitations discussed earlier, however, surveillance alone is often not sensitive enough to demonstrate compliance with today's restrictive environmental standards. Thus surveillance activities should be focused on pathways, media, and locations likely to contribute the greatest dose to a population. In practice, one usually finds that there are one or more critical pathways between the point of discharge of nuclides to the environment and human exposure, such as a particular food or perhaps airborne contamination. Within the critical pathway there will also be one or two critical nuclides that, in conjunction with a critical population, serve as the primary focus of the environmental surveillance program. Sometimes there may be secondary pathways involving nuclides of lesser significance that nevertheless require investigation.

In the past, a not uncommon assumption near nuclear reactors and fuel-reprocessing facilities was that radioiodine was the critical nuclide, and the critical pathway was deposition of this nuclide on pastures that nourish cows whose milk is furnished to children, the critical population. Today this pathway is still important for postulated accidents, but strict controls during routine operation have limited iodine releases. Now radioactive noble gases emitted to air are often the critical nuclides and pathway. An example of a secondary pathway might be the emission of tritium by the airborne pathway or discharge to rivers that serve as drinking water sources. Although one pathway may be critical in that the highest dose is delivered to humans by this route, it may also be desirable to monitor secondary pathways that might nevertheless be of potential importance, even though they result in lower dose to humans.

This chapter will focus on the requirements for environmental surveillance programs in the vicinity of nuclear facilities and surveillance for drinking water systems, but the basic principles can be applied to the design of regional, national, or international monitoring networks. Methods for monitoring for indoor radon can be found in NCRP (1988).

The subject can be treated under several headings as follows.

1. The design phase, in which regulatory requirements are identified and the environment is studied to identify the critical pathways in relation to: (a) the physical and chemical identity of the radioactive materials that will be discharged to the environment; (b) the distribution and living habits of the potentially exposed critical population; (c) transport mechanisms; and (d) the ecological mechanisms for dilution and concentration.

2. The preoperational phase, in which the surveillance program is operated, usually for a year prior to facility start-up, to obtain baseline data and to verify the performance of the equipment and design of the program.

3. The operational phase, the design and operation of which are based on information obtained in the preoperational phase, modified as necessary by the performance of the facility and continuing observation of critical pathway kinetics.

4. A surveillance plan to be used in emergencies, which is normally based on expected impacts of various hypothetical accident scenarios that are usually computer-modeled.

5. A postoperational surveillance program to assess the residual effect of the facility on the radiological environment. To take advantage of information gathered in the operational phase and advances in surveillance technology, design of the postoperational phase is best postponed until the facility is within a few years of completing its design lifetime. Postoperational surveillance is not treated further here.

The design of every surveillance program is based on five fundamental issues: (1) what should be measured; (2) what lower limit of detection, reliability, and accuracy is required (data quality objectives); (3) where should the measurements be made; (4) with what frequency; and (5) how are the data thus obtained converted to dose estimates.

DESIGN PHASE

Enough information must be obtained in the design and preoperational phases to identify the pathways by which humans will be exposed to the effluents from the proposed facility. In many cases, the critical pathway will be readily apparent from previous evaluations of similar situations, but one must always be on guard for unusual circumstances that may influence the radiation exposure of human populations. Examples of the latter were the use of uranium mill tailings as a building construction material (see Chapters 7 and 15) or consumption of seaweed that tends to concentrate ^{106}Ru (see Chapter 11). Figure 14-2 in the next chapter identifies the basic pathways by which humans may be exposed to radioactive environmental contaminants.

The design of the preoperational monitoring program will depend on the kind of facility being planned and whether sources other than natural ones are having an effect on the radiation background. The principal man-made source that affected background in the past was testing of nuclear weapons. Assuming that present bans on atmospheric testing are honored, it will be relatively easy to account for the existing, small residue from past fallout. More recently, the accident at Chernobyl produced rapid, temporary increases in atmospheric radioactivity detectable throughout the hemisphere and measurable residual contamination of land in Central Europe with ^{137}Cs. In some situations, one facility may be so close to another as to be affected by its routine radioactive releases. Surveillance programs for nuclear facilities that discharge liquid effluents to rivers need to account for potential contamination from upstream facilities. Medical institutions, for example, can contribute surprising amounts of ^{131}I to sewage and consequently to rivers, due principally to elimination of therapy-level doses in the excreta of patients (Prichard *et al.*, 1982). A discussion of environmental releases associated with nuclear medicine appeared in Chapter 10.

Thus, in principle, the preoperational surveillance requirement is for a system of measurements that will (1) determine if there are any natural radioactive anomalies in the region; (2) define the ambient levels of γ radiation apportioned between naturally occurring and man-made nuclides; and (3) obtain baseline data for some of the more important nuclides, both natural and man-made. The design of the program should allow for rapid identification of changes in background unrelated to the facility. These data can be accumulated in the year prior to start-up of the facility by techniques that will be described later in this chapter.

The preoperational study should also be used to gather the kind of environmental information that will be needed for critical pathway and critical population dose assessment. A basic objective of the preoperational survey should be to develop predictive techniques by means of which the effective dose equivalent and possibly the dose to the critical organ can be estimated per unit release of the critical nuclide, whether it be in the air or water. For example, what will be the dose to an individual consuming fish from a stream into which a given quantity of ^{137}Cs is annually discharged? Or what is the dose to a child's thyroid from a given release of ^{131}I discharged from a stack of given height? To enable one to make these calculations, preoperational meterological and hydrological observations are needed for a period of 1 year or longer. The climatological data should include a record of the joint fequency distributions of wind speed, wind direction, and class of stability. Additionally, the mixing characteristics of the receiving waters should be ascertained by dye studies, models, or other means. Approaches to dose assessment are discussed in Chapter 14. More detailed information on dose assessment and tabulations of default data[1] are readily available in the literature (NRC, 1977; Till and Meyer, 1983; NCRP, 1984a, 1996a,b).

OPERATIONAL PHASE

The information to be gathered in the operational monitoring program should be guided by the following principles (ICRP, 1965): (1) The information needed to assess the actual or potential exposure of populations can be averaged over extended periods, for example, 1 year. (2) Only the critical pathways need routine examination. (3) In addition to estimating exposures, it may sometimes be necessary to follow trends.

These are sound principles that, when adopted, greatly simplify the design of monitoring programs conducted in the vicinity of nuclear facilities. Over the past four decades the feeling has developed that extensive and expensive measurements, not necessarily dictated by the foregoing principles, are essential to satisfy a small but increasingly critical segment of the public. One result of this increased emphasis on environmental surveillance has been the development of detailed surveillance guides. The DOE Regulatory Guide (DOE, 1991c) contains nearly 150 specific requirements for DOE

---

[1]Default data can be used in models when site-specific data are lacking or are too costly to gather. Default data are usually "conservative" in the sense that their use results in larger values of calculated dose than would be likely if site-specific data were available. The most legitimate use of default data is in screening calculations; if a calculation with default data shows that doses are very low, it is unlikely that further refinement would result in increased dose and the associated costs to refine the modeling are unwarranted.

environmental surveillance programs. The NRC also issues detailed guidance in the form of reports and Regulatory Guides (NRC, 1979a,b, 1980a, 1983; Meinke and Essig, 1991a,b). These guides are a mixed blessing. On one hand, they provide concise sources of valuable information on environmental surveillance and a degree of standardization among programs; on the other hand, their prescriptive nature may tend to stifle advancements in the scientific bases of environmental surveillance.

The operator of a nuclear facility is normally responsible for monitoring the environs of the plant within the area influenced by its radioactive emissions. Governmental responsibility should involve sufficient monitoring close to the plant to validate the reports of the facility operator and should also include a monitoring program that covers the total geographical area for which the agency is responsible. This type of monitoring program can often be coordinated with air, water, and food surveillance programs conducted for more general reasons, such as analyses for trace metals, pesticides, or other chemical substances.

One of the most important factors that influences the design of a surveillance program is the required sensitivity of the procedures for dose estimation. Because environmental dose limits are fragmented among governmental agencies and environmental media (air, water, etc.), there is no single dose limit. In the United States, the various contemporary environmental dose limits for facilities range between 40 and 1000 μSv (4 and 100 mrem) per year (see Chapter 3).

It should be readily apparent that direct measurement of a small increment to a variable annual background (see Chapter 6) will rarely be possible. If quantitative estimates of doses are required at levels of a few millirems (a few tens of microsieverts) per year, then for the most part one must rely on "dose assessment." Dose assessment comprises calculations based on knowledge of radioactive emissions, results of environmental surveillance measurements in the critical pathways and quantitative understanding of critical pathways and critical populations (see Chapter 14). The main portion of this chapter will be concerned with the manner in which such measurements can be made, including measurements under emergency conditions.

RATES OF RADIOACTIVE EMISSIONS

Knowledge of the rates of radioactive emissions in air and water will frequently provide the principal assurances that population exposure criteria are not being exceeded. When the liquid and gaseous releases are below some predetermined limit, the requirements for environmental surveillance can be minimal, with sufficient sampling to confirm that the actual environmental concentrations do not exceed those predicted from knowledge of the release rates. For example, the DOE has graded requirements for air monitoring that provide for a minimal program if doses to the public can be shown to be less than 10 μSv (1 mrem) per year and an extensive program if doses are expected to exceed 50 μSv (5 mrem).

The methodologies for estimating the amounts of the individual nuclides being discharged in gaseous- or liquid-waste streams are beyond the scope of this chapter. The reader is referred to several useful documents for information on this topic (OECD, 1974; IAEA, 1978; American Public Health Association, 1985; EPA, 1990c: NCRP, 1986a; Chieco *et al.*, 1990; DOE, 1991c).

SAMPLE ANALYSIS

Many of the methods of environmental surveillance to be discussed involve analysis of samples in a radiochemistry laboratory. Although the sample media and sample preparation vary, the basic analytical methods are used for all types of samples. Many nuclides of interest in the vicinity of nuclear facilities can be ascertained by γ-ray spectrometry, which often requires little or no sample preparation. Radionuclides that emit only α or β radiation cannot be analyzed by γ spectroscopy. Examples of these nuclides are ^{239}Pu and ^{90}Sr and many of the radionuclides in the natural series. Measurement of gross α and gross β are often used to screen samples for α and β emitters. If predetermined screening levels are exceeded, then more complex analytical chemistry techniques are used to analyze specific nuclides. Tritium, which is a weak β emitter, is usually measured in a liquid scintillation counter. Several compendia of analytical methods for radionuclides in environmental media are available (Chieco *et al.*, 1990; IAEA, 1989; Bodnar and Percival, 1982; NCRP, 1986a).

ENVIRONMENTAL MONITORING STATIONS

Much of the environmental surveillance of major nuclear installations is conducted at fixed monitoring sites. A typical site is shown in Fig. 13-1. This site includes (1) a pressurized ionization chamber to measure direct radiation, (2) air samplers to measure both total suspended activity and particles less than 10 μm in diameter (PM$_{10}$), (3) a device to collect precipitation, (4) a desiccant cartridge to collect atmospheric moisture, and (5) a radio to transmit data to a central location. Not seen is a thermoluminescent dosimeter that measures total γ dose over a calendar quarter. Each of these sampling and measuring devices is discussed in greater detail in the following sections.

FIGURE 13-1 Modern environmental surveillance station. From right to left are a PM$_{10}$ air monitor, a total suspended particulate air monitor, a device to collect precipitation and water vapor, and the control unit and detector for a pressurized ion chamber. Not seen are a thermoluminescent dosimeter and the antenna for the radio that transmits data to the surveillance program headquarters. (Photo by David Myers, Idaho State University Photographic Services.)

DIRECT GAMMA RADIATION

Direct radiation sources from facilities can include radiation from passing airborne radionuclides, radiation from radionuclides previously deposited on the ground, sediments or vegetation, and radiation emitted directly from the facility or vehicles used to transport radioactive shipments to and from the facility. Radiation emitted directly from a facility is sometimes called "shine" radiation and can be important for high-energy accelerators, radiography installations, and facilities with large inventories of γ-emitting radionuclides. Exposure to shine radiation can occur along a line-of-sight path and by scatter from the atmosphere.

The ambient γ-radiation levels in the vicinity of a facility should be measured pre-operationally (new facilities) and throughout the operational life of the facility. The existing radiation levels can be ascertained very quickly by flying over the surrounding area with aircraft equipped with special γ-scintillation equipment having high sensitivity and a rapid response time (IAEA, 1991b). Any areas of elevated radiation will, in most cases, be due to outcroppings of geologic formations, such as granite, known to contain elevated natural radionuclides. However, the possibility of other causes of elevated radiation levels cannot be ruled out until verification is made on the ground. Aerial surveillance can also be useful in emergency situations. The availability of preemergency aerial surveillance data improves interpretation of aerial data taken during any emergency.

Aerial surveys provide information about a large area at a single point in time and are relatively expensive. However, for deposition over large areas there may be no alternative to aerial methods. Other methods are used to measure direct radiation over time at carefully chosen benchmark locations.

One method of making such estimates utilizes thermoluminescent dosimeters (TLDs), which can be exposed at these benchmark locations, typically for periods of several months (Cameron *et al.*, 1968; Becker, 1973). TLD materials store a small fraction of incident radiant energy in the form of trapped electrons. This energy is released as visible light when the TLD material is heated (thermoluminescence). Within limits, the thermoluminescence is proportional to the radiation received. The light is measured using a photomultiplier tube, which converts the light into an electrical signal. With proper calibration, the electrical signal produced by the thermoluminescence may be related to dose.

TLD elements are typically solid, granular, or powdered materials. Several elements are usually combined in a small package called an environmental dosimeter. The package provides for identification, protects against environmental insults, and supplies any necessary energy compensation. The sensitivity and accuracy of thermoluminescent techniques depend on the TLD material itself and the handling, transportation, and readout of the dosimeters. The presence of natural background radiation and its variation prevent TLDs from confidently measuring annual facility contributions of less than about 10 mrem. A performance standard for environmental dosimeters is available (American National Standards Institute, 1975), as well as articles describing their performance and use (Gesell, 1982; de Planque and Gesell, 1986; Klemic *et al.*, 1995).

Environmental TLDs are relatively inexpensive and are usually used at multiple sites around facilities. The NRC recommends 40 stations around nuclear power plants (Meinke and Essig, 1991a,b). At the large Idaho National Engineering Laboratory complex, approximately 150 locations are monitored, both on- and off-site.

Environmental TLDs can be valuable for after-the-fact documentation of direct radiation dose from accidentally released radionuclides or elevated radiation levels. The TLDs are always in place and their low cost means that many locations can be used. The entire period of the accident is covered because no time is lost between the inception of the accident and the deployment of emergency response instrumentation. Accidental releases from nuclear facilities can extend over days, if not weeks or longer. The temptation to exchange environmental TLDs too frequently in an effort to obtain immediate information on doses due to the accident can be offset with preplanning. Other kinds of instrumentation are more suited to real-time data collection. If environmental TLDs are to be used in an effort to follow the course of dose due to an accident, new TLDs should be deployed. The TLDs originally in place as part of the routine surveillance program should be left, if at all possible, so that the total direct radiation from the accident can be more accurately assessed.

Gamma radiation may also be measured instantaneously or continuously using several types of instrumentation, including pressurized ionization chambers, Geiger Meuller tubes, and scintillometers. The pressurized ionization chamber (PIC) is an accurate method of measuring dose and dose rate from ambient γ radiation and is widely considered to be the standard for determining environmental dose. Beck *et al.* (1971) developed the environmental application of the PIC. They used a stainless-steel-walled sphere, 25 cm in diameter, filled with very pure argon to a pressure of about 40 atmospheres. With a good electrometer it was able to measure hourly changes of about 0.1 μR per hour, equivalent to about 1 mR per year. PICs are now available commercially with a variety of readout and data acquisition, storage, and transmission options. Data from several units are commonly transmitted by radio to a central location. The normal fluctuations in background, owing to changes in soil moisture and other factors, can be documented by operating the PIC continuously in conjunction with a recording device. Alternatively, one can make monthly or semimonthly measurements over a period of 1 or 2 years. A modern, commercially available PIC is shown in the leftmost portion of Fig. 13-1. Exchange of a memory card in the control panel of the PIC is illustrated in Fig. 13-2. Memory cards are used to back up data normally transmitted by radio.

The contributions of natural and man-made radiation may be differentiated by analyzing γ spectra obtained with a suitable detector and pulse-height analyzer (ICRU, 1994). The techniques for separating the total γ-radiation dose rate into the various contributing components were described by Beck *et al.* (1971), who have used both thallium-activated sodium iodide scintillators (low resolution) and Ge(Li) detectors (high resolution). Figure 13-3 illustrates the excellent performance of which a high-resolution system is capable for *in situ* field spectrometry. The peaks due to fresh fallout are readily distinguishable from those due to natural radionuclides. Figure 13-4 shows a spectrum obtained with a low-resolution system and indicates the contribution to the ambient γ background from the principal sources

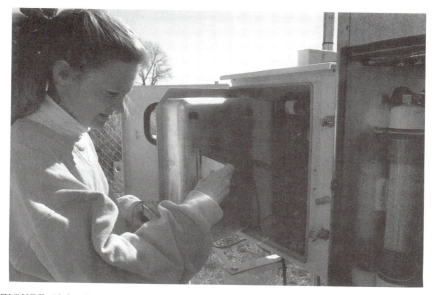

FIGURE 13-2   Control unit for a pressurized ionization chamber illustrating exchange of a memory card used to back up data that are routinely transmitted by radio to the surveillance program headquarters. (Photo by David Myers, Idaho State University Photographic Services.)

of background radiation, including the now ubiquitous ^{137}Cs. Ge(Li) detectors have now been replaced by the more durable intrinsic germanium detectors, which offer equivalent performance. These detectors must be operated at liquid nitrogen temperature but may be allowed to warm up for storage. Ge(Li) detectors, on the other hand, are destroyed if allowed to warm up.

SURFACE DEPOSITION

If one is concerned with the possibility of food-chain contamination, samples of radioactive particles settling to the surface of the earth as dry dust or in precipitation (fallout) are a convenient method of documenting the amount of contamination per unit area arriving at the ground surface. A simple method employed acetate film covered with a sticky substance and mounted horizontally on a frame. This material, known as gummed film, retains its adhesive properties when wet, and dust particles may be entrapped efficiently even though contained in raindrops. A major disadvantage of the technique is that the more soluble components can be washed off the film by rain. The gummed film method was developed to provide semi-quantitative data on the rates of fallout. Although the method has not been

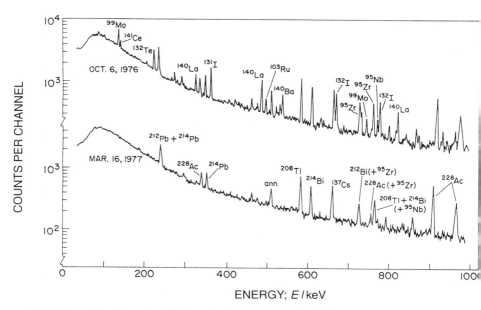

FIGURE 13-3   Example of high-resolution γ spectra taken following fresh fallout (October 6, 1976) and 5 months later (March 16, 1977) after most of the fallout nuclides had decayed. [From ICRU (1994).]

widely used for more than 30 years, the data collected during the period of atmospheric weapons testing have been useful in dose reconstruction efforts undertaken in recent years (Beck *et al.*, 1990; see Chapter 14). Several methods for collection of fallout can be found in the U.S. DOE Environmental Measurements Laboratory Manual (Chieco *et al.*, 1990).

SOIL SAMPLING

One method of obtaining a cumulative sample of deposited radioactivity is to collect surface soil for radiochemical analysis. Samples of deeper soil may also be appropriate when characterizing waste-disposal sites or when transport of radioactivity down into the soil column is suspected. Soil makes an excellent sampling medium, but it is, unfortunately, a difficult matrix from which to extract the radioactive substances for radiochemical assay.

Alexander *et al.* (1960) considered the problems of sampling soil in connection with an investigation of the worldwide distribution of ^{90}Sr from weapons testing. They recommended the selection of sites that have a good vegetative cover and are nearly level. Sites subject to overwash from higher ground or flooding should be avoided. Soils that pack when dry should also be avoided, as should sites that have a high population of worms, which might

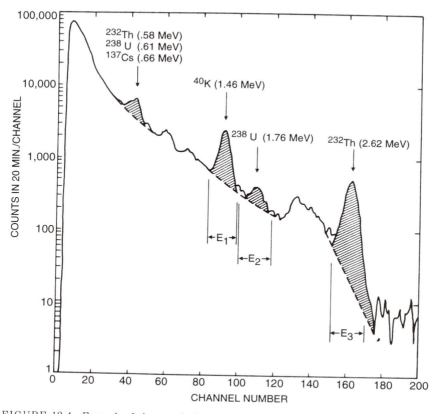

FIGURE 13-4   Example of a low-resolution γ spectrum taken at a background site illustrating the major contributors to terrestrial background radiation. [From NCRP (1976).]

affect the vertical distribution of isotopes. These criteria are particularly important where it is desired to obtain, by means of soil sampling, an estimate of the total amount of radioactivity deposited in a given area. These basic principles are still valid today.

Characterization of waste-disposal locations such as shallow land burial sites requires a different approach. Here the interest is often in the potential downward or lateral migration of radionuclides rather than fallout accumulation at or near the surface. Deep core samples may be essential. Another approach is to construct trenches and take samples at various depths and lateral locations along the walls of the trench.

An important consideration in any sampling program is the collection of samples intended to be representative of the environment. This is especially important in soil sampling because radionuclide distributions can be inhomo-

geneous and only small samples can be analyzed. The most accurate laboratory techniques cannot make up for a poor sample. The potential problem
of nonrepresentative samples from surface soils can be overcome by gather-
ing soil from a fairly large area, thoroughly mixing the sample, and then
removing an aliquot of a size suitable for analysis.

Whether or not one removes vegetation and organic debris from the
surface soil sampled depends on the purpose of the sampling. Where one
wishes to estimate the amount of fallout deposited on a given area, the
vegetation and debris should be analyzed along with the soil. However,
there should be no reason for retaining pebbles or twigs. Grass will trap
surface fallout and has been used as a sensitive indicator of fresh fallout.
Such a sample can be obtained and its spectrum determined in a few minutes
by means of γ-ray spectrometry. Detailed procedures for the sampling and
analysis of soil are given by Chieco *et al.* (1990). The NRC (1974) developed
a Regulatory Guide for soil sampling and analysis for plutonium based on
methods developed at the DOE Environmental Measurements Laboratory
and the Nevada Test Site.

Soil sampling has very limited application in the ongoing surveillance
programs of modern nuclear facilities. The relatively small releases coupled
with ongoing surveillance of air, water, and foodstuffs have reduced or
eliminated the need for soil sampling. The NRC does not list soil sampling
among its environmental surveillance recommendations for nuclear power
plants (Meinke and Essig, 1991a,b).

ATMOSPHERIC SURVEILLANCE

Air is a primary pathway for human exposure to radiation. Air contamination
may result from direct releases from facilities (or nuclear weapons) or from
resuspension of contamination on the surface of the ground. In addition to
direct inhalation, radionuclides in air can contribute to external dose from
passing clouds of contamination or by immersion in the contamination.
Airborne particulate matter also deposits on the surface of the ground and
water, contributing to those pathways. For the purpose of atmospheric
surveillance, radionuclides are divided into four categories based on physical
and chemical properties (DOE, 1991c). These categories are: (1) particulate
matter; (2) halogens, primarily iodines; (3) gases, primarily the heavier rare
gases such as krypton and xenon; and (4) tritium, which is often in the
form of tritiated water. Each of these four categories requires a different
strategy for sampling and analysis. For nuclear power plants the NRC recommends that a minimum of five sampling locations for particulate matter and
radioiodine be used (Meinke and Essig, 1991a,b). Three stations are to be
near the site boundary in sectors predicted to receive the maximum average

airborne concentrations. One station is to be near the community predicted to receive the maximum average airborne concentration and one station should be in a background location.

Particulate Matter

The general procedure used in sampling the atmosphere for radioactive particulate activity is to draw air through a filter at a known rate for a known period of time (IAEA, 1967; Lee, 1974; Hering, 1989). The radioactivity of the filters may then be analyzed and the activity per unit volumes of air ascertained. This basic principle is unchanged but both sample collection and analysis have evolved over the years.

Filters for particulate matter can be made from any fibrous material and many types are commercially available. Examples include cellulose, glass fiber, membrane, and polystyrene. Factors that should be considered in the selection of a filter include intended flow rate, retention, particle size selectivity, self-absorption of the α particles by the filter media, and compatibility with chemical analysis (DOE, 1991c). The efficiency of filter media is based on the ability to trap dioctylphthalate (DOP) particles with a mean aerodynamic diameter of 0.3 μm. DOP testing is a standard method for evaluating the efficiency of filters. Filters used in air monitoring should retain at least 99% of the DOP particles under normal operation conditions (American Conference of Governmental Industrial Hygienists, 1974). All the commercial filter media, when used properly, have efficiencies that are more than adequate to serve the purpose.

The measurement of both α and β activity is complicated by the fact that the atmosphere normally contains short-lived ^{222}Rn and ^{220}Rn decay products in concentrations that may be higher than the permissible concentrations of some of the long-lived nuclides for which one may be monitoring. For example, if one draws air through a filter paper at a rate of 1 cfm (28.3 L s^{-1}) and if the air contains 0.05 pCi L^{-1} of radon in equilibrium with its decay products, ^{218}Po through ^{214}Po, the α activity of the filter paper will increase to about 200 pCi (7.4 Bq) at the end of 2 hours (see Fig. 6-2). If this activity were assumed to be a long-lived α emitter, a concentration of 2.2 Bq m^{-3} (0.06 pCi L^{-1}) would be calculated. This value is many times the general public derived air concentration (DAC) for many α emitters. Thus, the natural α activity of the atmosphere tends to mask the presence of long-lived α emitters.

Fortunately, the natural α-emitting radionuclides in the atmosphere are short-lived. If one holds the filters for several days to allow the longest-lived species (10.6-hour half-life for ^{212}Pb in the thorium series) to decay, the natural nuclides will no longer interfere. If results are needed sooner, it is possible to compare the rate of decrease of the activity on the filter with that expected from natural airborne radioactivity. If the rate of decrease is less than expected, long-lived contamination is suspected.

Monitoring for particulate, atmospheric α activity in the vicinity of power reactors is unnecessary. Although it is theoretically possible for plutonium and other transuranic elements to be emitted to the environment, such emissions would be associated with overwhelming amounts of β- and γ-emitting nuclides. Although airborne β–γ measurements are recommended by the NRC in the vicinity of commercial nuclear power plants, airborne α measurements are not required (Meinke and Essig, 1991a,b).

Air samplers may be fixed or portable and may be devices that simply collect the particulate matter on filter papers for counting in a laboratory or they may be equipped with automatic counting and recording devices. Samplers are available in a range of flow rates. High-volume air samplers, $\sim 50 \, \text{ft}^3 \, \text{min}^{-1}$ ($\sim 25 \times 10^{-3} \, \text{m}^3 \, \text{s}^{-1}$), are used where a short sample collection time is necessary or where exceptional sensitivity afforded by a large sample volume is desired. Low-volume samplers, which collect at the rate of a few cubic feet per minute ($\sim 10^{-3} \, \text{m}^3 \, \text{s}^{-1}$), are normally used for continuous collection over a period of a week or two.

The shape and size of the intake orifice of an air sampler affect the distribution of particle sizes that can be collected on the filter. Many air-sampling devices used in the nuclear industry have air intakes intended to collect all airborne particles. In the simplest arrangement, the filter face is open to the atmosphere. The air sampler that collects all airborne particles has the best chance of detecting a release from a facility. However, the human respiratory system is effective in preventing particles greater than 10 μm in aerodynamic diameter from reaching the bronchial or pulmonary region of the respiratory tract. For this reason, the EPA has largely switched from measuring "total suspended particulates" to measuring PM_{10} as an indicator of air quality. A PM_{10} sampler has a specially designed intake maze that is designed to remove particles with aerodynamic diameters greater than 10 μm from the airstream before it passes through the filter. Following this logic, some organizations that monitor for airborne radionuclides have adopted the PM_{10} samplers. The choice depends on the purpose of the surveillance program. If the intention is to obtain an indication of a problem at the lowest possible level, then a total suspended particulate sampler would be the choice. However, if the primary purpose is to be able to use the results to calculate dose to humans, then a PM_{10} sampler is preferred. In some programs, both kinds of samplers are used. Figure 13-5 shows a filter being changed on a PM_{10} sampler.

The presence of the natural, short-lived α emitters makes rapid assessment of very low concentrations of long-lived α emitters impossible using

FIGURE 13-5   A PM_{10} air sampler illustrating exchange of the sampling head, which contains a filter to collect particulate matter and a charcoal cartridge to collect iodine. (Photo by David Myers, Idaho State University Photographic Services.)

simple techniques. Sophisticated instrumentation utilizing large-area filters, large air flow rates, energy discrimination, and "pseudo-coincidence" counting is being developed for timely determination of α emitters of concern such as plutonium (EG&G Berthold Company, 1992). Until such equipment is in widespread use, it will remain difficult if not impossible to provide prompt surveillance information about low levels of plutonium exposure in the event of accidental releases reaching the public. This would be particularly worrisome in the event of a suspected plutonium release from a launch pad accident (see Chapter 10).

Radioiodines

Radioiodine released from freshly irradiated fuel is likely to exist in vapor form, which will not be retained efficiently by filter media suitable for particulate matter. Activated charcoal and silver zeolite have been shown to be effective for radioiodine sampling (Cowser, 1964) and are normally used in series behind particulate filters. Where iodine in several chemical forms is suspected, several different kinds of absorbing cartridges can be arranged in series and analyzed separately. Cartridges used to collect iodine are normally analyzed by γ spectrometry (Intersociety Committee, 1972; DOE, 1991c).

Tritium

Atmospheric tritium usually exists as tritiated water (HTO) but may be present as molecular hydrogen (HT) under some circumstances. However, the dose potential from HT is about 25,000 times less than that for HTO (International Standards Organization, 1975), so surveillance is normally performed only for HTO. Sampling for HTO involves collecting water vapor from the atmosphere and analyzing the sample in the laboratory (Intersociety Committee, 1972; DOE, 1991c). One common sampling method is to pump air through a silica gel column at a flow rate of approximately 100 cm^3 min^{-1}. Essentially all of the water vapor in the sampled air is collected in the column. The maximum sampling duration for a given column size and flow rate depends on atmospheric humidity. If the method is to be quantitative, sampling must stop before the column becomes saturated with water. After sample collection, the silica gel is heated to drive off the water, which is condensed and analyzed by conventional liquid scintillation techniques. Molecular sieves have been used instead of silica gel. When intermittent sampling for short periods is desired, tritium can be collected by bubbling the airstream through a gas washing bottle. Figure 13-6 shows a cartridge used for the collection of water vapor for tritium analysis.

Gases

The longer-lived gases released from nuclear facilities, such as ^{85}Kr (1.08 y) and ^{133}Xe (5.3 d), can be collected for laboratory analysis by several methods. The shorter-lived gases such as ^{41}Ar (1.8 h) and ^{13}N (10 min) are best handled by measuring their external radiation directly (DOE, 1991c).

FIGURE 13-6 Two sampling devices in a single weather shelter. Precipitation is directed from the top of the weather shelter into the jug visible inside the shelter. Atmospheric water vapor is collected by pumping air through the desiccant column held by the individual who is servicing the station. (Photo by David Myers, Idaho State University Photographic Services.)

Grab samples can be collected by simply flushing enough air through a container to fully replace the air originally present with ambient air. Continuous sampling can be achieved by pumping air continuously into a large plastic bag specially designed for the purpose. The plastic must be carefully selected so that noble gases do not preferentially diffuse through the material. Once the sample is collected it may be compressed into a gas cylinder, such as those used for a self-contained breathing apparatus, before transport to a laboratory. Alternatively, the air sample stream may be continually compressed into a cylinder as it is gathered or collected with a cryogenic sampler. Once the sample is collected, the radioactive gases are separated in the laboratory using gas chromatographic columns and analyzed by liquid scintillation counting. Detailed methods for sampling and analysis are available (Grossman and Holloway, 1985; Trevathan and Price, 1985; Stanley and Moghissi, 1974).

AQUATIC SURVEILLANCE

If liquid wastes are being discharged into a receiving body of water, whether surface water or groundwater, the important pathways for human exposure should be thoroughly understood so that the sampling program can be designed with specific reference to the requirements for estimating human exposure (Foster *et al.*, 1971; DOE, 1991c). Samples may be required of the water, suspended particulate matter, sediments, benthic organisms, vegetation, shellfish, finfish, and waterfowl. Generally speaking, it is essential that water samples be analyzed if the water is a source for direct human consumption or irrigation. For surface water, particularly salt or brackish water, samples of shellfish or finfish will often serve to document the principal route of human exposure. Samples of sediments or vegetation may serve as indicators of potential exposure, since in many situations the vegetation and sediments will have higher concentrations of the critical nuclides than the food consumed by humans. For percolation or evaporation ponds used to manage low-level liquid waste, waterfowl can be an important pathway (Halford *et al.*, 1981; Morris, 1993). In practice, the design of the sampling program will probably be based on the preoperational and operational ecological studies and will result from consultation with regulatory authorities.

The proper selection of sample containers and preservative will guard against loss of the radioactivity to the container wall, algae, or slime growths and prevent volatilization, which may occur if iodides are oxidized to iodine. Acidification of water samples in the field is commonly used to prevent dissolved radionuclides from plating out on the surface of the sample containers, which are usually made of plastic.

Surface Water Surveillance

Surface waters can be divided into flowing bodies of water such as rivers and streams and relatively still bodies of water such as lakes and reservoirs. Estuaries affected by tides present an intermediate case between flowing and still water and can be particularly complex (see Chapter 5).

The concentrations of radionuclides expected to be found in surface waters under normal operating condition are sufficiently low that samples are normally analyzed in environmental radiochemistry laboratories. Thus the primary concern of surface water surveillance is to obtain representative samples and preserve their radioactive characteristics until they can be analyzed.

For still bodies of water, the selection of sampling points representative of background and the facility contribution may be difficult (DOE, 1991c). Flow pattern will likely not be obvious and may depend on the season. Dye or other environmental tracer studies will likely be needed. For small, still bodies of water, it may be impossible to locate a suitable background location. In this case, a nearby similar pond, preferably one that receives water from the same or similar source, may be used. Sampling points that are representative of the facility should be selected at the edge of the effluent mixing zone and at the nearest, and possibly more distant, points of withdrawal for use.

If the wastes are being discharged into a flowing stream, samples should be taken upstream for background assessment, at the edge of the effluent mixing zone, and at the nearest downstream point of withdrawal for domestic or other use. To ensure that sampling is representative, river and stream samples should be taken on a traverse across the stream and at more than one depth. Concentrations of radionuclides in rivers and streams arising from facilities can vary rapidly with time owing to the intermittent nature of discharges. Sampling should be coordinated with discharges or, preferably, be conducted continuously with an automatic sampler. Sampling at the intake of domestic water supplies should be coordinated with or even performed by the local water authorities.

It is often desirable to know if the radioactivity in water is dissolved or suspended as particulate matter. For this reason, samples are often filtered as they are collected. It is important to filter water samples before they are preserved with acid because introduction of acid can alter the partition of the activity between the aqueous and suspended phases. The filtrate and the particulate matter are analyzed separately. The data may be later combined to give the total activity per unit volume of the sample.

The NRC recommends that surface water be sampled upstream and downstream of nuclear power plants (Meinke and Essig, 1991a,b). Monthly composite samples should be analyzed for γ emitters and quarterly composite samples for tritium.

Groundwater Surveillance

Chapters 5, 11, 14, and 15 discuss groundwater contamination at several nuclear facilities and waste-disposal sites from direct injection of wastewater into aquifers and from seepage of contaminated fluids from wastes. Contamination in groundwater is of special concern because of its persistence. Much of the contamination introduced into rivers is quickly transported to the oceans, where it can become diluted to insignificant levels, but dilution mechanisms in most groundwaters are very slow. In addition to contamination of groundwater by artificial radionuclides from nuclear facilities, there is also a concern for elevated natural radionuclides such as radium, which occur in some groundwater.

Groundwater surveillance in connection with nuclear facilities is conducted to determine baseline conditions, to demonstrate compliance with applicable regulations, to provide early detection of contamination, and to identify potential and existing contamination sources (DOE, 1991c). As with other waters, groundwater samples are normally analyzed in an environmental radiochemistry laboratory, so obtaining and preserving representative samples until they can be analyzed is the major concern.

Where possible, existing wells can be used for groundwater sampling because new wells are expensive and each well can potentially serve as a conduit of pollutants from the surface to the aquifer. Nevertheless, new wells may be required based on knowledge of expected contaminant transport rate and direction. Great care should be taken in siting and constructing groundwater monitoring wells and applicable guidance such as the EPA's Technical Enforcement Guidance Document (EPA, 1986b) should be consulted. Depending on the applicable regulation, placement and construction of monitoring wells can be very prescriptive. Existing wells often do not meet these requirements. In some situations, the ability to sample at several depths from the same well may be desirable to characterize the depth distribution of the contaminants.

The low rate of flow of groundwater simplifies sampling. Grab sampling at intervals ranging from one to several months is usually sufficient and, in well-characterized, stable situations, intervals of 1 to 5 years may be adequate. Continuous sampling of the type used in river sampling is not necessary. In some cases, sampling frequency may be prescribed by a regulatory agency. If not prescribed, the normal approach is to sample frequently when a well is first included in a sampling program, and then to adjust the sampling frequency based on the history of the rate of change of the measured concentrations of the contaminants of interest.

"Thief sampling," in which a sample is collected from water standing in the well bore by a bailer introduced from the surface, is undesirable. Thief samplers do not assure a sample that is representative of the aquifer and may introduce contamination from the surface. Despite the disadvantages, thief sampling may be necessary in wells of limited production capacity or where cost is an overriding factor. Portable pumps may also be used for sampling. These pumps allow the well bore to be cleared, affording a representative sample, but still present the risk of introducing contamination from the surface. Potential contamination from thief and portable pump samplers can be minimized with scrupulous cleaning of the apparatus prior to introducing it into the well. Far preferable to thief sampling or portable pumps is the use of submersed pumps permanently installed in the wells.

Sampling from a pumped well is conducted by pumping and discarding water from the well until at least three standing well volumes of water have been removed. Temperature, specific conductance, and pH of the pumped water should be continuously monitored (Wood, 1981). When these parameters stabilize, it can be assumed that the sampling train is well rinsed and that water representative of the aquifer is being drawn. At this time a sample can be drawn into a sample vessel. Collection of a groundwater sample at a remote site is illustrated in Fig. 13-7, where one individual monitors temperature while another collects the sample. However, recent research indicated that large purge volumes may be unnecessary in some circumstances (Powell and Puls, 1993).

FIGURE 13-7   Collection of a groundwater sample at a remote location. The individual on the left is collecting a filtered sample for laboratory analysis. The individual on the right is continuously monitoring the temperature of the water as an indicator that the sample is representative of the aquifer. Note the portable generator that supplies power to the dedicated pump in the well. (Photo by David Myers, Idaho State University Photographic Services.)

The size and type of the sample container and any pretreatment or preservation required depend on the analysis to be performed, and this information can normally be obtained from the analytical radiochemistry laboratory. In some programs the analytical laboratory provides the sample containers and the preservatives.

The NRC recommends that groundwater that may be affected by nuclear power plants be grab-sampled quarterly and analyzed for γ emitters and tritium (Meinke and Essig, 1991a,b).

Surveillance of Aquatic Organisms

Because the concentration factors for most aquatic organisms and radio-nuclides are greater than one, the radioactivity of aquatic organisms will often be higher than the water itself. With a few exceptions, aquatic plants are not a part of the normal human diet, at least not in the United States. On the other hand, fish and shellfish are important diet items. Species of some aquatic organisms, as well as sediment, may also be sampled as a useful indicator of radioactivity in the environment rather than because they are consumed as food. Studies of aquatic organism populations and local dietary habits should be reviewed (or performed if not existing) to help decide which organisms to sample.

As with other kinds of environmental surveillance, it is important to establish background sampling locations. The considerations discussed earlier in regard to surface water background locations are applicable to sampling for aquatic organisms.

Just as for water samples, analysis of aquatic organisms is performed in an environmental radiochemistry laboratory; the laboratory should be consulted for information on sample size, containers, and preservation. Freezing samples and transporting them in insulated containers with dry ice is common in order to prevent the otherwise rapid spoilage. Collection of many species for monitoring purposes requires special permits from state or federal fish and game authorities.

The NRC recommends that each aquatic species of commercial or recreational importance near the outfall be sampled once during the season or twice per year if harvesting of the species is not seasonal (Meinke and Essig, 1991a,b).

Drinking Water Surveillance

The Safe Drinking Water Act (SDWA) and its implementing regulations require radiological (and nonradiological) surveillance of drinking water supplies. The concern of the SDWA for radioactive contaminants is almost exclusively for natural radioactivity, including uranium, radium, and radon. Although some monitoring is required for ^{3}H and ^{90}Sr, isotopes usually associated with human activities, EPA has estimated that no systems will be found out of compliance for these isotopes, whereas approximately 28,000 systems are expected to be found out of compliance with the standards for naturally occurring radionuclides (EPA, 1991a).

The schedule of sampling has been dependent on numerous factors, such as the source (ground or surface) of the water, the size and other demographic characteristics of the community served, the vulnerability of the system to contamination, the prior history of water contamination, and state discretion (EPA, 1990c). Recent trends in proposed rulemaking are to eliminate some of the complexity and variability and to standardize monitoring requirements nationally (EPA, 1991a). Table 13-1 summarizes the frequency of drinking water monitoring required under the EPA July 1991 proposed rule. About 80,000 water supply systems are expected to be affected and EPA estimates the annual costs to be $7,000,000. These figures imply an average annual monitoring cost of $87 per water supplier and appear to be incremental costs above existing requirements. Most of the cost is associated with radon and radium. This cost estimate seems remarkably low if all the costs of administering a program, collection of samples, analysis by a qualified laboratory, quality assurance, and review and reporting of data were included in the estimate.

The sensible approach to all environmental surveillance is to start with inexpensive screening techniques and then advance to more specific, and usually more costly, tests if the screening tests show the presence of contaminants above predetermined action levels. This approach is used by EPA in its requirements for surveillance of drinking water supplies. Increasingly complex and costly tests (isotope-specific) are applied if the simpler tests (gross α and gross β) exceed certain levels.

Operators of nuclear facilities that discharge liquid wastes to receiving bodies of water that are sources of drinking water usually include drinking water surveillance in their overall environmental surveillance program. The surveillance may be done as part of the facility's program or performed by the water supplier under agreement with the facility.

FOOD SURVEILLANCE

It is sometimes desirable to sample foods, including game animals, to evaluate the extent to which humans are exposed. Food, like drinking water, is directly consumed by the population and measurements of radioactivity in food can provide a direct measure of the intake (or lack of intake) of radionuclides. Food sampling may be required by regulatory agencies or

TABLE 13-1

Example of Minimum Radioactivity Surveillance Frequency (Samples per Indicated Period) Proposed for Drinking Water Systems[a]

	Year	1993	1994	1995	1996	1997	1998	1999	2000	2001	2002	2003	2004
Monitoring cycles and periods		First 9-year cycle									Second 9-year cycle		
		Monitor for 1990 maximum contaminant levels			Initial period			Repeat period			Repeat period		
Gross beta	Ground and surface water	1990 requirements			4	4	4	4	4	4	4	4	4
³H, ⁹⁰Sr	Ground and surface water	1990 requirements			1	1	1	1	1	1	1	1	1
Radon	Groundwater	1990 requirements			4	1	1	1	1	1	1 per 3-year period		
	Mixed ground/surface systems with waiver	1990 requirements			4	1	1	1 per 3-year period			1 per 3-year period		
Gross alpha, ²²⁶Ra, ²²⁸Ra, uranium	Ground and surface water systems	1990 requirements			1	1	1	1 per 3-year period			1 per 3-year period		
	Mixed ground/surface systems with waiver	1990 requirements			1	1	1	State discretion			1 per 9-year cycle		

[a] Adapted from EPA (1991a).

may be voluntary on the part of the facility. The scale of such a sampling program will vary, depending on the circumstances. If it is established from knowledge of the kinds and amounts of radionuclides discharged that significant food-chain contamination is not possible, no food-sampling program should be necessary. This will, in fact, be the case for most places in which only minimal if any environmental contamination is occurring. In other instances, the food-sampling program may be limited to the produce of relatively few farms. However, if the opportunity for widespread contamination exists, the samples may be collected from very large areas, namely, hundreds of square miles in the case of a major spent-fuel-reprocessing plant or a whole country as was the case when fallout from the explosion of nuclear weapons was occurring.

Food surveillance can also be desirable for its community relations value. The results of testing food directly for radioactivity are simpler to interpret for the public than the complex atmospheric transport and food-chain models used to relate activity in diet to facility releases.

In some instances, food may, in fact, provide the most meaningful of all possible environmental samples. For example, it is desirable to sample shellfish harvested from an estuary into which radioactive corrosion products such as ^{60}Co or ^{65}Zn are being discharged. Such radionuclides are known to concentrate in shellfish, samples of which will yield more meaningful information than other biota.

If ^{131}I is being released into the atmosphere in proximity to cow pastures, the ^{131}I concentration in the milk produced will provide far more meaningful information than air samples, deposition samples, or samples of the forage for the purpose of estimating potential dose to populations through the milk pathway. Significant contamination of pastures by ^{131}I will result in detectable contamination of milk within 24 hours.

Under any given circumstances, the need for food sampling and the methods by which such sampling should be conducted should be based on a thorough understanding of agricultural practices in the area, and the program should be guided by competent specialists.

The needs for radiological analyses of food are apt to be highly variable, depending on the numbers and types of nuclear facilities in the area. Nuclear accidents, such as the one at Chernobyl, or a resumption of atmospheric testing of nuclear weapons could place a heavy demand on a food-sampling program. Thus, regional or national monitoring programs should be flexible so that the extent of sampling can be varied as the needs require. Because radiochemical analysis involves relatively expensive procedures, careful attention should be given to the design of the sampling program when one is required to sample a large geographical area such as a state or country. Apart from economic considerations, there is the need to avoid overloading radiochemical laboratory facilities, which typically cannot accommodate sudden, large analytical loads that may be imposed by poorly designed sampling programs.

Basically, there are four questions that must be answered before starting a food-sampling program: (1) What foods should be sampled? (2) Where should they be collected? (3) How many samples are required? (4) What kinds of analyses should be performed, and with what sensitivity, accuracy, and precision?

The foods one selects as part of a large-scale sampling program should depend on the characteristic behavior of the significant radionuclides in food chains and on the dietary habits of the population. Although it is of scientific interest to follow as many of these isotopes as possible through the various steps in the food chain, not all foods or all radionuclides need be studied for the purpose of surveillance. Some of the radionuclides present will be of little public health significance because they are present in such small amounts or because they are chemically inactive and do not enter metabolically into food chains. Since the basic purpose of a regional monitoring program will be to estimate the dose equivalent among the general population, only those foods should be sampled and only those nuclides analyzed that contribute significantly to population exposure. We have seen from previous chapters that in most cases the important nuclides are ^{90}Sr, ^{137}Cs, and ^{131}I. In many countries, fresh milk is the food of choice for obtaining samples of a population's total intake of ^{90}Sr, ^{137}Cs, and ^{131}I. This greatly simplifies the sampling problem, because the very nature of the milk distribution system in most large cities results in considerable blending from various sources. The extent of the network and the frequency of sampling depend entirely on the circumstances and the objectives of the program (Eisenbud et al., 1963b; Stein, 1971).

The NRC recommends milk and other foodstuff sampling in the vicinity of nuclear power plants. Specifically, recommendations include semimonthly milk samples during the pasture season in three locations, samples of each principal class of food product irrigated with water in which effluents have been discharged, fish in the vicinity of the discharge area, and three different kinds of broad leafy vegetables (Meinke and Essig, 1991a,b).

DIRECT HUMAN MEASUREMENTS

The purpose of food analysis is to enable one to estimate radionuclide intake in exposed populations. Another method of estimating intakes in living individuals is by γ-ray spectrometry, usually using sodium iodide detectors in well-shielded rooms called whole-body counters.

Whole-body counting is particularly useful for measuring incorporated nuclides having penetrating γ radiations (Meneely and Linde, 1965), but it has also been applied (Laurer and Eisenbud, 1968; Ramsden, 1969) to measurement of nuclides such as ^{241}Am, ^{239}Pu, and ^{210}Pb, all which have low-energy emissions (13–60 keV) and require special techniques specifically applicable to this portion of the spectrum. Whole-body γ-ray spectrometers are sophisticated instruments, and relatively few exist in the world. Whole-body counters are useful for research and the determination of incorporated radionuclides in radiation workers but have little place in a program of routine surveillance of the general population except under unusual conditions, such as the accident in Goiania, Brazil (Chapter 12).

ENVIRONMENTAL SURVEILLANCE UNDER EMERGENCY CONDITIONS

Radiological emergencies potentially include a broad spectrum of circumstances. At one end is the innocuous incident in which there is no opportunity for significant public exposure but that generates fear by the knowledge that an accident has occurred involving radioactive materials. At the other end of the spectrum might be a serious accident involving a nuclear weapon or major nuclear facility, such as the accident at Chernobyl in the Ukraine, in which there is substantial potential for human injury and where quick action is required to obtain the environmental information on which decisions will be made regarding countermeasures. Intermediate between these examples would be an accident, such as that at Three Mile Island Unit 2 in Pennsylvania, where the amounts of material released to the environment are in excess of those approved for routine circumstances and where a quick appraisal of public risk is required.

Experience has shown that regardless of the severity of the accident, these incidents have in common the need for an emergency plan designed to deal with the full spectrum of accidents that might occur at a given facility. The design of the emergency plan should take into consideration the characteristics of the site with respect to geography, population density, governmental organization, and the information media. All radiological incidents that involve the general public are capable of producing misunderstanding, fear, and frequently a degree of press attention that may be all out of proportion to the true importance of the incident. It is essential that the emergency plan be designed to provide prompt and accurate information that can be dispensed candidly and efficiently to the public agencies and the media.

Since it is not possible in this volume to cover all of the possible contingencies that may arise, this section will deal with certain general principles that should be observed, and foremost among these is that the emergency plan should be in writing and have the concurrence of all key individuals both within the organization operating the facility and among the various governmental organizations that might be involved. In many states, responsibility for coordinating the off-site surveillance activities, digesting the data, preparing public information releases, and making decisions about required countermeasures is that of the state health or environmental authority. The state police, the local police, and numerous other governmental agencies will also be involved. The interrelationships of these various organizations should be clearly documented and understood by everyone concerned.

The full course of the emergency can be separated into two phases. Initially, the organization having operational responsibility for the facility involved in the accident must rely on its own resources, pending arrival of outside assistance. This phase may last a few minutes or several hours, depending on the geographical situation and the degree of advance preparation. In the second phase, the somewhat limited facilities of the local organization will be augmented by outside assistance, and some governmental organization, such as a state authority, will usually assume responsibility for managing the off-site aspects of the emergency.

EMERGENCY PREPAREDNESS

An essential first step is to analyze the kinds of accidents that can take place and their radiological consequences in a Safety Analysis Report. This has been done for many years as part of the NRC licensing procedure and for nuclear reactors, fuel-processing plants, reprocessing plants, and other major facilities operated by the DOE. Accident analysis is an essential part of the Safety Analysis Report and provides the basic descriptions of the contingencies with which the plan must deal.

An early step in development of the plan should be consultation with the various governmental organizations that would become involved in the event of an accident. This will vary from place to place, but most states have standby emergency procedures to which the radiological emergency plan should be adapted. Thus, in many localities, the state police, an organization whose staff normally provides continuous emergency readiness, can be relied on to receive information about emergencies and to place into effect various standard operating procedures and contingency plans. The police organization might also have responsibility for controlling traffic in the vicinity of the accident, arranging for detours, and arranging emergency transportation for certain key people.

As mentioned earlier, it is likely that a state health or environmental authority will be responsible for evaluating the significance of off-site radiological data and deciding what actions should be taken. The response to any emergency should be graded according to its severity (see Table 8-9), and the actions that are taken to reduce the dose to the off-site population should be determined in advance and be included in the emergency plan that is developed jointly by the facility operator and the governmental organization that has responsibility for dealing with the emergency.

For large facilities, an important part of the plan is liaison with supporting radiological assistance teams. Accidents of the severity of the Windscale production reactor in England, the Three Mile Island reactor, and the Chernobyl reactor, as described in Chapter 12, require relatively large numbers of skilled personnel to gather the necessary information. In the United States, the DOE and other governmental organizations maintain a network of emergency teams located at all major facilities. These teams are prepared to offer assistance to others upon authorized request.

The written plan should describe the kinds of radiological monitoring equipment that is available locally for emergency use. The equipment should be stored in a place that would be accessible regardless of the kind of accident and where opportunities for radioactive contamination are minimized. It requires great discipline to maintain such standby equipment in proper operating condition, and the emergency plan should include the necessary maintenance and inspection schedules.

The plan should identify the principal pathways of human exposure so that the radiological monitoring efforts can be properly directed. By the time the facility goes into operation, enough radiological information should be available so that computer programs, tables, or charts can be developed that make it possible for the plant operator and health physicist to estimate the potential dose commitment to the surrounding population from any postulated release. Computer software or tables can be prepared so that the dose commitment can be calculated from whatever mixtures of radionuclides are released, by either air or water. This will require identifying the locations of water intakes, dairy farms, and other exposure pathways, which should be listed in the original emergency plan.

CHAPTER 14

Radiological Assessment and Its Application to Dose Reconstruction

This chapter will describe how the methods of radiological assessment are used to reconstruct doses received by individuals when exposure has occurred as much as several decades previously. The bulk of the chapter addresses recent dose reconstructions for residents in the vicinity of some of the major facilities of the U.S. nuclear weapons complex. However, the practice of dose reconstruction has a much longer history. Perhaps the first such effort was by R. D. Evans, who in the 1930s estimated the doses delivered to the skeletons of radium dial painters, based on the rate at which radon is eliminated in the expired breath of individuals who have accumulated skeletal deposits of radium. One of the most complicated efforts at dose reconstruction has been undertaken over a long period of time in the cities of Hiroshima and Nagasaki, where estimates of the doses received by all survivors of the World War II atomic bombings have been made (Auxier *et al.*, 1966; Roesch, 1987). Another monumental effort was required to estimate the doses received by about 220,000 military personnel who participated in the atmospheric nuclear weapons tests (NAS/NRC, 1985b). Dose reconstruction on a large scale has also been conducted for the inhabitants of the Marshall Islands (Conard, 1992; Simon and Graham,

1994, 1995) and for residents near the weapons testing site in Nevada (Voillique and Gesell, 1990; Stevens *et al.,* 1992). Yet to be accomplished are the various dose reconstructions to be undertaken in the former USSR, as a result of exposure from both the Chernobyl accident and the poor practices (by Western standards) that characterized its weapons production program.

RADIOLOGICAL ASSESSMENT

In their authoritative work on the subject, Till and Meyer (1983) defined radiological assessment as "the quantitative process of estimating the consequences to humans resulting from release of radionuclides to the biosphere." In this chapter we have broadened the definition somewhat to include assessments of occupationally exposed groups such as the radium dial painters and impacts from direct radiation such as those from the atomic bombings in Japan. Radiological assessment has a myriad of uses other than dose reconstruction. These other uses can be grouped according to whether the doses are occurring at or near the time of the assessment, or are expected to possibly occur in the future.

One use of radiological assessment in the present time frame is the interpretation of results of an ongoing environmental surveillance program (see Chapter 13) in terms of radiation dose to the population. Annual reporting of estimated doses to the public is required of most nuclear facilities (40 CFR 61). A particularly important use of radiological assessment is rapid forecasting of doses in the event of a nuclear accident. This requires that potential accidents be identified ahead of time and that most of the information on pathways and populations be known in advance. A computer model is normally in place and ready to run, although as a backup crude estimates can be made with manual techniques. In the event of an accident, it is necessary only to estimate the actual radionuclide releases and input current meteorological information to assess dose. Rapid identification and communication of potential risk to the public are crucial to managing an accident effectively (see Chapter 12).

An important use of radiological assessment in the future time frame is to support the process required by the National Environmental Policy Act (NEPA), which mandates that the environmental impact of major actions by the public and private sectors be used as a part of the decision-making process. Accordingly, environmental assessments, or the more detailed environmental impact statements, are prepared for each new project or significant modification, depending on their potential impact. For nuclear facilities, this includes estimating the effects of projected routine and accidental releases of radionuclides. Up-to-date safety analyses are also required of existing and planned nuclear facilities, and radiological assessment is an important tool in their preparation.

ELEMENTS OF RADIOLOGICAL ASSESSMENT

The elements of radiological assessment include the specific radionuclides released ("source term"), their transport, bioaccumulation, and uptake by humans, doses resulting from the uptakes, and an estimation of the risk due to the dose (Till and Meyer, 1983; NCRP, 1984a, 1996a,b). The process is illustrated in Fig. 14-1. In most cases, radiological assessment is performed using models, because environmental data are obviously not available when predictions for the future are being made, as in the case of NEPA studies, and because environmental data are frequently sparse or unavailable for historical studies. For ongoing routine releases from modern facilities, levels are usually so small as to make reliance on environmental surveillance measurements impractical.

The source term is ideally derived from current or historical data, when available, but may have to be modeled or even be hypothetical, as when applying radiological assessment to future cases. For radionuclide release, the basic source term information is the kind and quantity of activity released per unit time. The chemical and physical properties of the release are also very important because transport and accumulation depend on these properties.

The source may be released to air or water, which then transport and disperse the radionuclides in the environment. Transport and dispersion by air, surface water, and groundwater are complex but have been studied extensively. The basic principles of transport and dispersion of radionuclides have been reviewed for air in Chapter 4 and for water in Chapter 5. Transport models are developed based on the basic behavior of radionuclides in these media.

As the radionuclides move through the atmosphere and hydrosphere they are absorbed (and eliminated) by plants and animals. When concentrations in air or water become relatively stable, such as during continuous releases, a state of equilibrium is sometimes reached between the amount of activity in a given species of plant or animal and its immediate environment. For intermittent or accidental releases, equilibrium is usually not reached between biota and the environment and more complex, dynamic modeling is required. The fundamentals of movement of radionuclides from the air and water to biota were discussed in Chapter 5. Food-chain models such as the PATHWAY model (Whicker et al., 1990) are developed using these fundamental principles, as well as experimental and observational

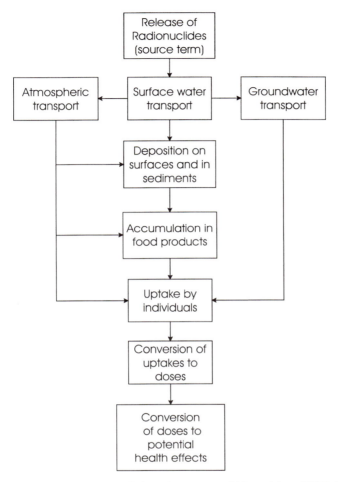

FIGURE 14-1   Major steps in radiological assessment [Adapted from NCRP (1984a).]

data concerning uptake and elimination by plants and animals. Figure 14-2 illustrates the many pathways by which radionuclides can travel through the environment and reach humans.

Once radionuclide concentrations in air, drinking water, and food species have been modeled, information about human metabolism, consumption patterns, and food sources are used to calculate human intakes. These intakes can then be converted to dose using established conversion factors such as those promulgated by the EPA (Eckerman *et al.*, 1988). In some cases, such as external dose from a passing cloud of radioactivity or direct

Transport Pathways Leading to Risk

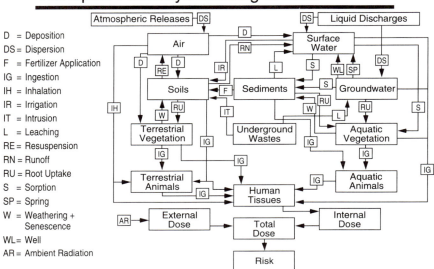

D   = Deposition
DS = Dispersion
F   = Fertilizer Application
IG = Ingestion
IH = Inhalation
IR = Irrigation
IT = Intrusion
L   = Leaching
RE = Resuspension
RN = Runoff
RU = Root Uptake
S   = Sorption
SP = Spring
W = Weathering +
       Senescence
WL= Well
AR = Ambient Radiation

FIGURE 14-2   Pathways by which radionuclides can be transported through the environment to humans [From Whicker (1995).]

radiation from the detonation of a nuclear weapon, intake is not required to cause dose.

Even a brief review of dose assessment is incomplete without a discussion of uncertainties. Nearly every aspect of radiological assessment is inherently uncertain, including the data, assumptions, and scientific judgments used in the development and running of the models. Not only are data uncertain, but they may be missing or only partly relevant. Uncertainties are usually greater for the biological rather than the physical components of models. They can be estimated for each element of the assessment and then propagated through the calculations to arrive at a final uncertainty, which should be expressed in probabilistic terms. Modern uncertainty calculations are usually performed using Monte Carlo techniques. Uncertainties of the individual components of the model are expressed in terms of probability distributions that are sampled each time the model is run. Each run of the model is called a "realization" and the results of hundreds of realizations of a model can be used to construct a probability distribution of the final result. This probability distribution is the source of the best estimate of the dose and its uncertainty (Hoffman *et al.*, 1995).

DOSE RECONSTRUCTION

Dose reconstruction is an application of radiological assessment to estimate past exposures, often those that occurred many decades prior to the study. It may be defined simply as the processes of estimating radiation doses to the public from past releases of radionuclides. (Dose reconstruction may also be performed for chemical releases, but this discussion is limited to radionuclides and direct radiation.) On the basis of more than a decade of work on the nuclear weapons complexes of several countries, the NAS/NRC (1995) reviewed the status of dose reconstruction and made a set of recommendations for future studies. Dose reconstructions are conducted because of public concern over possible health effects from past releases, and it is expected that some reconstructions may lead to epidemiological studies if doses are sufficiently high. Ideally, epidemiological studies should not be initiated unless justified by the results of risk assessment. However, the level of public anxiety may be such that decisions are made to proceed independently of risk assessment.

To be fully useful, dose reconstructions should meet the two basic criteria of scientific and public acceptance. First it is important that the *process* of dose reconstruction, not just the final product, be subjected to ongoing scientific peer review and public participation. Generally it is public suspicion that leads to the need for dose reconstructions in the first place and it is difficult to allay suspicions once formed. Inclusion of the public helps to ensure that their concerns are addressed, builds confidence in the study, and provides information that might be otherwise unavailable to the investigators. Public participation in the planning and conduct of the reconstruction may also help reduce suspicions that the studies are less than objective. The requirements for peer review and public input can be accomplished in part by an advisory or steering panel that reviews reports from the scientists conducting the reconstruction and provides advice or even makes decisions concerning the ongoing conduct of the study. This advisory or steering panel should include scientists and public representatives.

A simple study, referred to as a scoping study, is usually performed once a decision is made to consider a dose reconstruction. The scoping study can help to determine if a full dose reconstruction is warranted, or even feasible, given the data available and will usually identify the important source terms, pathways, and exposed populations. Simple models can be used to make either realistic or bounding calculations of the source term and dose, and it is preferable to do both so that the range of possible exposures can be evaluated. A full dose reconstruction, if conducted, provides refined estimates of the same quantities. Dose reconstructions at U.S. nuclear weapons sites are conducted by the Centers for Disease Control and Prevention and their contractors.

UNCERTAINTIES

Dose reconstructions differ from other applications of radiological assessment because the data needed for modeling and other reconstruction efforts are often difficult to locate and interpret and may be unavailable. Also, historical measurements may have been made with instruments and methods that were not as accurate as today's methods. As with any application of radiological assessment, it is necessary to estimate the uncertainty in the individual components and in the final result. Because of the uncertainties in the historical data used in dose reconstructions, the final uncertainty may be expected to be larger than that for radiological assessments performed for contemporary facilities.

Bioassay may eventually play a role in reducing uncertainty in dose reconstruction for some cases. For the plutonium-fabricating facility at Rocky Flats, in Colorado, it was possible to obtain urine samples from the residents in the area as well as samples of human tissues that became available from surgery or at autopsy. Plutonium is long-lived, highly insoluble, and can be detected in attocurie amounts, if necessary (1 aCi = 37 nBq). Strontium-90 also has a relatively long radiological half-life (29 years), and once deposited in bone is excreted so slowly that effective dose equivalents can be estimated even many years after exposure. However, bioassay cannot be used for nuclides such as ^{131}I, which has a half-life of only about 8 days.

THE SOURCE TERM

There are three possible methods for estimating the source term from releases in the past. These are reports made at the time of the releases, engineering calculations based on knowledge of the processes leading to release, and estimates based on historic environmental surveillance. Ideally the estimates are made using at least two of the methods in order to develop scientific and public confidence in the source term. A critical task is to identify and locate the pertinent historical records. Searches may be comprehensive or selective. Selective searches are usually quicker and less costly, but comprehensive searches are less likely to miss important data and are often more satisfying to the public. When possible, original data in the form of logbooks or charts should be used rather than summary documents. Any gaps in the data should be examined carefully and consideration given to filling in the gaps using other sources or interpolation if necessary. It is important to recognize that releases may have occurred from points other than stacks or liquid-discharge points. These other sources could include waste-disposal sites and miscellaneous diffuse sources such as windows, fires, or outdoor storage areas. Episodic releases resulting from accidents or special tests should be identified separately from chronic releases, especially if they resulted in much higher discharges. These episodic releases can then be treated with improved accuracy in the modeling process if warranted. The process of finding and reviewing the hundreds of thousands of records that have accumulated over several decades has in some cases cost many millions of dollars.

Environmental Pathways

Environmental pathways were discussed briefly in the radiological assessment section of this chapter and in more detail in Chapters 4 and 5. A critical analysis of environmental pathways is necessary for any reconstruction of doses resulting from environmental releases. Since most releases have occurred to the atmosphere or to surface water or groundwater, the resulting pathways to humans are often similar. Nevertheless, site-specific variability in climatology, waters resources, and agricultural practices must be considered and the problems of limited or poor quality data must be confronted. If good historical environmental surveillance data are available, they can be used to help validate the pathway models or, if sufficiently abundant, can be used directly in estimating intakes for some of the population. In general, models must be relied on to estimate doses to the entire affected population.

Doses and Risk

Once the pathway analyses are complete and concentrations in air, water, and foodstuffs are determined (and any direct radiation), doses can be calculated as for any radiological assessment and potential risks estimated from the doses. A set of intake-to-dose and dose-to-risk conversion tables should be agreed upon in advance. A graded approach may be taken by first estimating effective doses to representative populations. If the levels are high enough to warrant more specificity, organ doses and doses to specific individuals may be estimated. If doses are low, resources can be saved or redirected to sites with higher doses.

Epidemiological Considerations

Depending on the outcome of the dose assessment process, the public may be satisfied with simply knowing the doses and the risks (if small), or they may request that epidemiological studies be performed. These are likely to be monitoring studies to inform specific populations of the risk rather than formal studies to increase general knowledge of radiation effects. The possibility of follow-on epidemiological studies can be considered at each step of the dose reconstruction process so that the dose data will be useful in the event that an epidemiological study is performed (NAS/NRC, 1995). For example, organ doses are more desirable for epidemiological studies than is effective dose. The expected limitation of any epidemiological study based on environmental doses from facilities is illustrated in Table 14-1, which lists the size of the studied population sample required to identify an increase in cancer with 80% probability as a function of equivalent dose, using the potential risks developed by the NAS/NRC (1990). As can be seen, unrealistically large populations are required for the smaller doses, and very high doses are required for realistic population sizes. If the smaller risk rates of the ICRP (1991) were used and if a more stringent criterion was required for the probability of observing an excess, for example, 95%, the required populations would be even larger.

TABLE 14-1

The Required Sizes of Exposed Population Samples to Be Studied with Lifetime Followup to Detect an Increase in Cancer with 80% Probability as a Function of Dose[a]

Mean dose[b]		Required exposed population sample size		
mSv	rem	All cancers	Leukemia	Respiratory cancer
2.5	0.25	32,000,000	74,000,000	>100,000,000
5	0.5	7,900,000	19,000,000	44,000,000
10	1	2,000,000	4,700,000	11,000,000
20	2	500,000	1,200,000	2,700,000
30	3	220,000	520,000	1,200,000
40	4	130,000	300,000	680,000
50	5	80,000	190,000	440,000
60	6	56,000	130,000	310,000
70	7	41,000	99,000	220,000
80	8	31,000	76,000	170,000
90	9	25,000	61,000	140,000
100	10	20,000	49,000	110,000
120	12	14,000	25,000	77,000
150	15	9,100	11,000	50,000
200	20	5,200	3,900	28,000

[a] Reprinted with permission from "Radiation Dose Reconstruction for Epidemiologic Uses." Copyright 1995 by the National Academy of Sciences. Courtesy of the National Academy Press, Washington, D. C.
[b] Dose means effective dose for all cancers and target organ equivalent dose for leukemia and respiratory cancer.

EVENTS LEADING TO THE NEED FOR DOSE RECONSTRUCTIONS

The facilities of the weapons production complex have had a complicated history that originated in the laws that established the DOE and its predecessor agencies. The Atomic Energy Acts of 1946 and 1954 specified that the AEC should have responsibility for regulating both occupational and environmental safety at its facilities. This stipulation was originally due to the fact that the responsibility would require highly specialized personnel and equipment to oversee safety, and these were not available outside of the AEC in the early post-World War II years. Moreover, neither the EPA nor the NRC had as yet been established, there were no other major federal regulatory agencies, and there was only minimal regulation by some of the states.

For many years the AEC and its successor agencies (hereafter referred to as DOE) regulated their activities by means of internal directives and, as will be seen in our final chapter, succeeded in maintaining a commendable industrial safety record. In 1984 the primacy of DOE in regulating environmental matters was challenged in the courts, which ruled that DOE should be subject to EPA regulations (Dorian, 1991). This required that the DOE contractors comply with two complex federal laws, PL-98-616), The Hazardous and Solid Waste Amendments of 1984 (RCRA), and Pl-96-510, the Comprehensive Environmental Response, Compensation, and Liability Act (CERCLA), which took effect in 1980. The latter is known as the Superfund Act, which is the basic law that requires cleanup of contaminated sites. Other laws, including the Clean Water Act and the Resource Conservation and Recovery Act, have also had major impacts on the DOE. In retrospect, it is difficult to understand why the DOE did not adopt the EPA regulations earlier, even though it was not legally bound to do so.

Until the 1984 court decision, DOE confined its regulatory activities to the ionizing radiations[1] and, in general, followed the guidance provided by the NCRP. However, there was no such guidance for chemical wastes, which were handled in a manner considered to be satisfactory at the time. Thus, about 40 years after its major facilities were built, the DOE in 1984 suddenly

---

[1]The one notable exception was the regulatory system developed for control of disease associated with the use of beryllium. When knowledge about this subject developed in the 1940s, AEC decided to establish standards for workroom and community air. Although intended only for use by AEC and its contractors, the standards have been adopted by regulatory agencies in the United States and abroad. The reason for this exception was the fact that AEC became a major user of this metal, and its toxicological properties were poorly understood at the time (Eisenbud, 1982).

became subject to strict laws that they had not previously believed applied to its activities.

It was soon evident that the DOE weapons facilities were not in compliance with EPA requirements. Admiral James Watkins, who became Secretary of Energy during this period of turmoil, was formerly an officer of the nuclear navy, which maintained high standards for chemical and radiological safety. He set about trying to rectify the compliance situation using principles that had worked well in the nuclear navy.

Beginning in the early 1950s, under Admiral Hyman Rickover and his successors, the U.S. Navy had developed a superb and deserved reputation for the high standards of its operations (Hewlett and Duncan, 1974), but the environmental problems of the Navy were not necessarily an appropriate model for the far more complicated and vastly larger problems that faced the weapons production complex. Nevertheless, as we will see in the final chapter, DOE had conducted its operations for half a century with an exemplary record of occupational safety and no member of the public was known to have been injured by the operations of the nuclear weapons production complex. In such a vast complex there were certainly defects, but the overall performance deserved high marks. Nonetheless, DOE facilities were not in compliance with modern regulations, particularly the ones promulgated by EPA, which were more generally concerned with chemical wastes than radioactive substances. To achieve compliance will take well into the next century, at a cost estimated to be upward of $150 billion (NAS/NRC, 1989).

This development coincided with a period of heightened public concern about DOE facilities. Unprecedented class action suits and other litigation were brought against DOE contractors by residents living near the facilities who believed they were injured as a result of deficiencies in operation and management. Vast amounts of money were involved in these lawsuits, and both the plaintiffs and defendants began to undertake massive efforts to estimate the doses received by people who were exposed to the effluents from the facilities. A demand also developed to bring the facilities into compliance with the EPA regulations. Thus was born the new field of "environmental remediation," a subject that will be discussed in the next chapter.[2] The remainder of this chapter will be concerned with the methods by which the doses received over a period of many decades were estimated. The fields of dose reconstruction and remediation have become growth industries in the United States, requiring expenditures of hundreds of millions of dollars for the former and hundreds of billions of dollars for the latter.

---

[2]The term "environmental restoration" seems to be used interchangeably with environmental remediation.

DOE sites can be grouped into two major classes of facilities: the chain of production plants built to satisfy the requirements for uranium that developed during World War II, and the complex of nuclear weapons research and production plants built shortly after World War II.

THE WORLD WAR II URANIUM PRODUCTION PLANTS

The World War II uranium-production plants were originally intended to operate for only a few months to produce enough uranium to meet wartime needs, but they remained in operation for several years. There are 46 such sites,[3] and they have been subject to a major decontamination program known as the Formerly Utilized Sites Remedial Action Program (FUSRAP), which is discussed in Chapter 15. An important law, adopted in 1978, was PL-95-604, the Uranium Mill Tailings Radiation Control Act, which dictated the practices that should be followed to control exposure of the public to the radiation from mill tailings. This is discussed in Chapters 7 and 15. Although these uranium sites are being remediated, they have not been subjected to the formal dose reconstruction efforts that are occurring at the major weapons materials production sites.

THE NUCLEAR WEAPONS PRODUCTION COMPLEX

The U.S. nuclear weapons production complex includes 17 major sites located in 12 states. Some of the sites, such as Hanford, Oak Ridge, Los Alamos, and Savannah River, occupy large areas that were selected during and following World War II to provide both the required security and a means of limiting exposure of people in the event of accidents. A list of these sites is given in Table 14-2. With the demise of the Soviet Union and cessation of nuclear weapons production in the United States, there has been much concern about residual contamination by radioactive and chemical substances at these sites.

The halt in weapons production occurred when some of the facilities were reaching the end of their useful lives. Some had been built during World War II under conditions of great urgency and no longer met the modern standards of design and operation. Many of the facilities were cloaked in secrecy and neither the physical plants nor the methods of operation were subject to outside scrutiny. In addition, the emphasis on waste disposal was almost entirely focused on radioactive wastes. Chemical wastes were managed throughout the period in the manner customary prior to the development of the far more restrictive methods of the past two

[3] A few of the 46 predated the World War II effort but were included by Congress in the FUSRAP program because of the similarity of the problems encountered.

TABLE 14-2

Significant Facilities in the DOE Weapons Complex[a]

Weapons laboratories	Materials production	Weapons and components production	Testing sites
Lawrence Livermore National Laboratory	Feed Materials Production Center (Fernald, Ohio)	Burlington assembly plant	Amchitka Island test site
Los Alamos National Laboratory	Hanford Nuclear Reservation	Kansas City plant	Bikini Atoll test site
Sandia National Laboratory	Idaho National Engineering Laboratory	Mound plant	Eniwetok Atoll test site
	Oak Ridge Reservation	Pantex plant	Nevada test site
	Paducah Gaseous Diffusion Plant	Pinellas Plant	
	Portsmouth Gaseous Diffusion Plant	Rocky Flats plant	
	Weldon Spring Site		

[a]Compiled from DOE (1995a).

decades. For many years, DOE was not required to comply with the standards of EPA or other regulatory agencies that came into existence in the late 1960s. As we have seen, this position was changed by the courts in 1984. Although there were undoubtedly sites where problems had developed that posed avoidable risks to workers, no studies have indicated conditions of imminent danger to the general public. In fact, we will see that except for World War II radioiodine releases from Hanford, there has been no evidence that any member of the public in the vicinity of these sites has been exposed in excess of current recommended limits.

The failure of the 50-year-old weapons complex to meet the environmental standards of the 1990s became a subject of considerable national interest despite the lack of evidence that the deficiencies were hazardous to the public health (NCRP, 1991c). A major complication has been the litigious atmosphere that has existed in the United States in recent years. Major lawsuits have been brought against the uranium-production center in Fernald, Ohio, the Rocky Flats plant in Colorado, and the Hanford Nuclear Reservation in the state of Washington.

The Fernald case was brought by property owners surrounding the plant. They claimed, among other things, that emissions from the plant had diminished the value of their property and increased the risk to their health.

In 1989, the government settled the case for approximately $80 million, without going to court. The settlement provided for the establishment of a medical monitoring program for area residents who lived within 5 miles of the plant, as well as payments to property owners and other payments based on claims of emotional distress.

The plaintiff's attorneys in the Fernald case received an award of nearly $20 million. Shortly after the Fernald case settled, the same plaintiff's attorneys, acting on behalf of other citizen groups, filed substantially similar suits against the operators of other nuclear weapons plants, including Rocky Flats and Hanford. These lawsuits are still pending.

It is not clear why the U.S. government or its contractors should be considered responsible for whatever emotional distress existed in the Fernald case. For three decades the community had been bombarded by sensational stories in the media. For example, it was stated in the weekly magazine *TIME* (1988) that the name Feed Materials Production Center was selected to mislead the public into thinking that the plant was producing cattle food. There was ample evidence, however, that the workers at the plant knew what they were producing, as did local government. In fact, prior to start-up of the plant in 1951, the Ohio Health Department was informed in great detail of expected amounts of radioactive materials that would be discharged to the Miami River (Eisenbud, 1991). The *TIME* magazine article also reviewed the tragic experience of a child who developed bone cancer, allegedly because of emissions from the plant. The allegation was supported by a report that bone from the child contained 10 times the normal amount of uranium. If so, this would have resulted in a dose of about 0.3 mSv per year (3 mrem), which is about the dose received to the whole body from a single cross-country airplane flight. Much of the anxiety caused by the existence of the plant was likely due to the constant barrage of negative information from the media. Public concern and the pending litigation have resulted in the need for massive studies designed to reconstruct the doses received by the public over the entire period of plant operations.

SUMMARY OF DOSE RECONSTRUCTION PROJECTS AT U.S. NUCLEAR WEAPONS PLANTS

The largest dose reconstruction studies at the U.S. nuclear weapons production plants are the Hanford Environmental Dose Reconstruction (HEDR) project (Farris *et al.*, 1994a,b) and similar studies of the environs of the Rocky Flats plant (ChemRisk, 1994) and the Feed Materials Production Center (Meyer *et al.*, 1994). These studies are summarized in the following sections. Although appearing too late to be reviewed here, the reader is also referred to a special issue of *Health Physics* devoted to dose reconstruction (Little, 1996).

THE HANFORD NUCLEAR RESERVATION

Hanford, located in central Washington State, was selected in 1942 as the site for plutonium-production reactors and the plants required to separate plutonium from irradiated fuel. Within a few months the U.S. Army Corps of Engineers began to assemble a 500,000-acre tract that is roughly circular, with a diameter of about 40 km (Hewlett and Anderson, 1962). In one of the most spectacular achievements in engineering history, and without any more prior experience than with the research reactor operated briefly in the squash court under the stands of the stadium at the University of Chicago in late 1942, three plutonium-producing reactors were built and placed in operation at Hanford by early 1944. These were followed by six more reactors constructed between 1949 and 1963. A series of four plants for the separation of plutonium were also constructed, the first of which began operation in 1944.

The HEDR project was massive in scope and complicated in design. The reader is referred to the excellent summaries by Farris *et al.* (1994a,b) of the two major pathways of exposure, the Columbia River and the atmosphere. Most of what follows is drawn from the two Farris reports.

The Columbia River Exposure Pathway at Hanford

Eight of the reactors used natural uranium, were moderated by graphite, and were cooled by single pass-through of Columbia River water. This cooling water accumulated activation and fission products before being returned to the river. Some of the radionuclides released to the river were produced by neutron activation of trace substances in the cooling water. The river water took only one to two seconds to pass through the reactor core, but it was subjected to intense neutron radiation at a flux of 10^{13} to 10^{14} neutrons cm^{-2} s^{-1}. The most notable of the nuclides produced were ^{24}Na ($T_{1/2} = 15$ h), ^{32}P ($T_{1/2} = 14$ d), and ^{65}Zn ($T_{1/2} = 243$ d). Hold-up basins were provided to permit decay of the shorter-lived nuclides. The short half-lives of the nuclides limited the extent to which they could accumulate in the river and ocean. This is shown in Fig. 14-3, which shows the annual releases to the river from 1944 through 1971, together with the accumulation in the river and ocean. Because of the short half-lives of the radionuclides released, the accumulation is relatively modest.

In addition to activation products, fission products also escaped because of failures of the aluminum cladding that encased the natural uranium fuel.

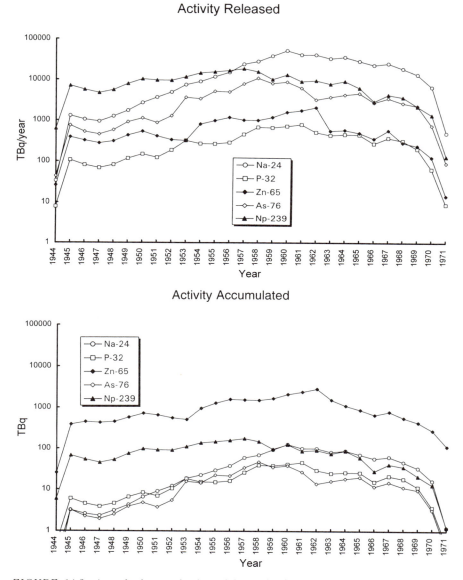

FIGURE 14-3 Annual releases of radionuclides to the Columbia River from 1944 through 1971, together with the accumulation in the river and ocean. [From Farris *et al.* (1994b).]

Nearly 2000 fuel failures were known to have occurred during the operating history of the reactors (Farris *et al.*, 1994a).

It was known early in the program that waterfowl, fish, and mollusks were accumulating radionuclides from the Hanford aquatic releases, and reports on the subject were published as early as 1946 (Parker and Norwood, 1946). The amounts being found were not believed to represent risks to either the wildlife or humans, but the studies provided the first opportunities under field conditions to obtain environmental bioconcentration factors of the kind that have proved essential to the development of transport models.

With the passage of time the original judgments have been shown to be correct. Most people received no more than 2 or 3 mrem (20 or 30 μSv) per year from the radioactivity released to the river water. The principal source of exposure for most people was consumption of drinking water. The maximally exposed individuals were those who consumed large quantities of fish (assumed to be 40 kg y^{-1}) from the river. During the nearly 50 years from 1944 through 1992, the accumulated effective dose equivalent (EDE) to such an individual would have been about 1500 mrem (15 mSv). Ninety-three percent of the dose from the Hanford aquatic releases would have been received during the period 1950–1971 and was due to the presence of ^{65}Zn and ^{32}P. The total number of people so exposed has not yet been estimated, but is likely to be no greater than a few hundred.

These dose estimates were generalized using great quantities of data obtained from the operating history of the facilities between 1944 to 1971. To start, estimates of the monthly releases of 11 nuclides were made based mainly on monitoring data. Where gaps in the data existed, the estimates were completed using statistical analysis of effluent monitoring data with Monte Carlo uncertainty analysis (Farris *et al.*, 1994b). This was done for the eight reactors that were in operation during that period.

From the point of release to the river, a model of the Columbia River was used to calculate dilution of the radionuclides. This model had been under development for many years and had been verified frequently by comparison with the measured amounts found in the river.

The amounts of the various nuclides present in fish and waterfowl were estimated by applying bioconcentration factors (BCFs), the ratio of the concentration of a substance in biota to that in water, or to the assumed concentrations in river water. Here the investigators had an important advantage in that the Hanford environmental science staff had for many years been studying BCFs in the Columbia River. The availability of site-specific data eliminated one of the major sources of uncertainty in dose assessment. The highest BCF was a median value of 1500 for ^{32}P in omnivorous fish. The lowest BCF was for the transuranic actinide ^{239}Np, which had a median value of 21. Only two radionuclides proved to be present in significant amounts of waterfowl, ^{32}P for which the median BCF was 290 and ^{65}Zn for which the median BCF was 44.

Shellfish have long been known to be strong concentrators of many radionuclides and for that reason measurements among the important oyster beds at the mouth of the Columbia River had been made for decades. It was found (Heeb and Bates, 1994), that the concentration of ^{65}Zn in oysters could be estimated ($R^2 = .83$) from the linear equation

$$WBO = 0.0019 \ C$$

in which

$$WBO = \text{concentration of } ^{65}\text{Zn in oysters (pCi g}^{-1})$$
$$C = \text{curies of } ^{65}\text{Zn released per given year.}$$

Anadromous fish, such as the various species of Columbia River salmon, spend part of their lives in fresh water and part in the ocean. The salmon are spawned in the far reaches of the Columbia River system and then migrate downstream for several months before they enter the ocean, where they spend most of their adult lives. They return to the Columbia River as adults and begin their hard journey to the headwaters for spawning and death. Historical measurements existed of the amounts of the various radionuclides accumulated in the flesh of the salmon at various periods in its life cycle.

All of the information about river flow, radioactive emissions, BCFs, other biological factors, and rates of human consumption was used to compute doses for hypothetical individuals at various distances downstream from Hanford. Also included were numerous other sources of exposure, such as river use by boats or swimming and drinking untreated water.

With the foregoing information, the doses received by representative individuals were then computed. The maximally exposed individual was assumed to be a person who resided in the town of Ringold during the period 1944–1992. The estimated EDE was calculated to have been 1.5 rem (0.15 Sv) during the entire period. Uncertainty was estimated by 100 realizations of the Monte Carlo technique discussed earlier in this chapter. The ratio of the maximum to minimum estimated doses was less than a factor of 10. Doses received by the maximally exposed individual at other locations on the river would have been lower than at Ringold. By way of comparison, the EDE from natural radioactivity would have been about 14 rem (0.14 Sv). It was estimated that the dose received by the average individual would have been lower by a factor between 10 and 40.

Atmospheric Pathways at Hanford

The most important atmospheric release from Hanford was ^{131}I from the chemical separations plants, the first of which began operation in December 1944 (Farris *et al.*, 1994a). The atmospheric pathways study covered the period 1944–1992, but the period of particular importance was 1944–1945, when more than 500,000 Ci (18.5 PBq) were released. The 1945 releases were greater than those for all other years combined.

The quantities of radioiodine released could be estimated from the radiation history of each batch of fuel and the chemical-processing schedules. There were also stack monitors that maintained a record of the gaseous discharges and extensive environmental monitoring and meteorological records. During most of the period of interest, direct measurements of ^{131}I were not possible because convenient methods of gamma spectrometry did not become available until the early 1960s. Until then the radioiodines could be assayed only by laborious chemical procedures. What was measured in the stack and environmental monitoring procedures was either total beta or total gamma radiation and the amount of ^{131}I was inferred from the operating history of the reactors and separations plants.

The availability of computers now makes it possible to go far beyond what is necessary technically for risk assessment, but a comprehensive review of all possible contributors to dose helps credibility in the view of the public. By 1994, when the studies were being undertaken, it was known that only a few of the hundreds of radionuclides produced in fission would be significant. Nevertheless, 1200 radionuclides were screened as possible contributors to the dose received. This list was narrowed down to 150 nuclides that seemed to be of most interest (Napier, 1991). Further study led to the conclusion that ^{131}I accounted for more than 90% of the dose, with marginal contributions by [103/106]Ru. (Farris *et al.*, 1994a).

The dose computations required to digest the vast amount of available information were made possible by linking numerical models that described the source term, atmospheric transport, and environmental accumulation. The source term was evaluated from hourly releases of ^{131}I during the period of highest release but less frequently during most of the nearly 50-year period. The complexity of the calculations is illustrated by the fact that the ^{131}I doses to 12 representative individuals, differentiated by sex, age, and pattern of food consumption, were prepared for 1102 locations within the nearly 200,000-km^2 area included in the study.

As might have been expected, it was concluded that the most heavily exposed persons were infants who consumed milk from grass-fed backyard goats or cows. The median thyroid dose for such an infant in 1945 was estimated to be 192 rad (1.92 Gy) with a range of 45 to 824 rad (0.45 to 8.24 Gy). At the time of this writing there are no estimates of the number of people who might have been so exposed, but the highest exposures were in small towns with only a few if any infants. The dose was markedly influenced by seasonal factors and whether the cows were fed stored feed.

The HEDR study will presumably be followed by an epidemiological investigation to determine if an excess of thyroid cancer has occurred among the more highly exposed residents of the area.

THE ROCKY FLATS PLANT

The Rocky Flats Plant, near Boulder and Denver, Colorado, has been engaged in the processing of plutonium for nuclear weapons since 1953. A major fire occurred at the plant on May 11, 1969 (AEC, 1969), and raised suspicions that plutonium contamination of the environment might have resulted. This possibility received national attention in 1970 as a result of analyses of soil samples taken under the aegis of a group of independent scientists (Colorado Committee for Environmental Information, 1970). It was also known that a less severe fire had occurred in 1957.

Traces of plutonium were found in soil by the independent scientists. A subsequent study by the AEC Health and Safety Laboratory (Krey and Hardy, 1970) confirmed that, although plutonium was indeed present, its distribution was not consistent with wind directions recorded at the time of the fires. This study concluded that the main source of contamination was an outdoor area on which drums of machine oil containing plutonium particles had been stored and leaked.

The report of the independent scientists led to allegations by the study sponsor that "the continuation of plant operations at this site represents a serious threat to the health and safety of the people of the Denver area." Investigators from the Colorado Department of Health confirmed that plutonium was present in soil beyond the plant boundary but reported that "no public health hazard existed within the time encompassed by this report" (Shapley, 1971).

During the 1970s, court activity began that lasted 10 years, during which the parties to the lawsuit conducted extensive area surveys. After 10 years the court ruled that there was no basis for a lawsuit but in the meantime the state of Colorado specified a Pu limit of 2 dpm g^{-1} of soil[4] or 0.01 μCi m^{-2}. Since there was off-site contamination in excess of that limit, DOE undertook some modest remedial actions (such as plowing where feasible) and also expanded its landholdings.

Opposition to continued operation of the Rocky Flats Plant continued to increase during the 1980s based on employees' statements about deficiencies in safety-related plant operating practices. Similar complaints were being made about other components of the weapons complex. At Rocky

---

[4]dpm g^{-1} means disintegrations per minute per gram of soil, equivalent to 0.9 pCi g^{-1} (33 Bq kg^{-1}). The dose from inhalation of soil contaminated at this level is of the same order of magnitude as that from natural radionuclides present in ordinary soil.

Flats this resulted in an unannounced raid by 70 FBI agents and EPA staff for the purpose of seizing documents and other evidence needed to determine compliance with hazardous waste regulations. The raid was authorized by a court based on allegations of criminal violations of hazardous waste regulations. In addition, a so-called "Tiger Team" was sent to the Rocky Flats Plant by the Secretary of Energy to provide an independent assessment of conditions.[5]

Soon the DOE shut the plant down until it could be upgraded. It may have been no coincidence that at the time global politics were such that there was no longer any need for plutonium production or bomb fabrication. In fact, the name of the Rocky Flats Plant was soon changed to the Rocky Flats Environmental Technology Site. Government reaction to the allegations of the public was bizarre to say the least and tragic because of the staff demoralization that resulted and the unnecessary anxiety caused among people living nearby.

The tiger teams were a phenomenon of the late 1980s during which a demand developed in the office of the Secretary of DOE for a new DOE "culture" that emphasized safety rather than production. We will see in the final chapter of this book that the DOE and its predecessor agencies maintained an excellent safety record throughout its history. Threats of criminal indictments for malfeasance were made at the time, but no such action has been taken during the intervening years, although fines were levied against the operating contractor of the Rocky Flats Plant.

Residents living in the vicinity of the Rocky Flats Plant brought a class action suit against its operators. This stimulated a series of additional, extensive investigations of the extent of off-site contamination during the late 1980s and early 1990s.

Subsequently, the emission records of the plant were incorporated into environmental transport models, from which the amounts of plutonium deposited on the ground and resuspended by wind action have been estimated. Plutonium fixes firmly to soil and tends to remain in place. Even if distributed originally as a fine fume (less than 1 μm) the particles become relatively immobile because they tend to attach to much larger particles in the soil that are not readily resuspended.

Dose reconstruction by such methods has been completed (ChemRisk, 1994). As expected, inhalation proved to be the main source of exposure, from 10 to 1000 times greater than other routes of exposure. The highest

---

[5]The Secretary of Energy and his immediate staff were made aware of the raid so that there would not be a dangerous confrontation between the raiders and the plant security force, whose primary mission was to safeguard the weapons-grade nuclear materials in the facility. However, most of the plant staff were unaware of the impending raid. The "Tiger Team" was assembled and moved on site within a few hours of the start of the FBI/EPA raid.

effective dose commitments were downwind of the drum storage pad and were found to total between 9.1 and 72 μSv (0.91 to 7.2 mrem) during the period when plutonium was being released from the pad. For purposes of comparison, the average effective dose from indoor radon is 2 mSv (200 mrem) per year. Thus one year of normal indoor radon exposure results in a dose 28 to 220 times higher than the range of estimates for all of the Rocky Flats plutonium contamination.

Samples of human tissue obtained from nearby residents at autopsies or during surgical procedures would be a good indicator of the extent to which plutonium exposure has taken place. About 700 such samples were obtained and analyzed for plutonium. The amounts of plutonium found in the tissues were no greater than that found in tissues of other people who live at great distances from the plant and have been exposed only to the worldwide fallout from weapons testing (Cobb *et al.*, 1983).

Another measure of plutonium exposure can be obtained by urine samples, since it is possible by the methods of fission track analysis (Wrenn *et al.*, 1994) to detect amounts as low as a fraction of a femtocurie (10^{-15} Ci; 37 μBq). If the rate of urinary excretion is known, the body burden and the dose received can be estimated using metabolic models developed by the ICRP and others. The plutonium content of urine collected from people who live near the plant was found not to be significantly different from the amounts found elsewhere in the United States (Whicker and Ibrahim, 1994). The 50-year effective dose commitment based on the results of urine analysis and other methods of dose reconstruction has proven to be, at a maximum, on the order of 1 mrem y^{-1} (10 μSv y^{-1}).

There have been a number of estimates of the total amount of plutonium deposited on the site and its environs. The Colorado State University Radio-ecology Group performed the most extensive study to date and concluded that the total amount was 4.2 Ci (155 GBq), but that most of this was deposited on-site. The deposition off-site was estimated to be 1.0 Ci (37 GBq), but includes the fallout present worldwide from weapons testing. When this is subtracted, the deposition of off-site plutonium due to the Rocky Flats Plant is estimated to be 68 mCi (2.5 GBq) (Whicker and Ibrahim, 1994). Plutonium from Rocky Flats can be distinguished from plutonium from weapons testing fallout using differences in isotopic ratios, which can be obtained by mass spectrographic analysis (Krey, 1976).

Thus by the methods of dose reconstruction that utilize emissions from the plant, analysis of human tissue, and analysis of human urine, the doses received by people who lived near the plant were far less than that received ordinarily from radioactivity naturally present in the environment (see Chapter 6).

THE FEED MATERIALS PRODUCTION CENTER

The Feed Materials Production Center, now known as the Fernald Environmental Management Project, is located near Fernald, Ohio, about 18 miles northwest of Cincinnati and a few miles east of the Indiana border. From 1951 to 1989 it processed uranium concentrates and uranium compounds recycled from other stages of nuclear weapons production, as well as some uranium and thorium ores. In July of 1989 the production mission was suspended to focus on environmental restoration. In December of the same year EPA added Fernald to its National Priorities List ("Superfund") of federal facilities in need of remediation. In June of 1991, DOE closed the plant insofar as any future production was concerned and dedicated the site and its workforce to cleanup programs. The Feed Materials Production Center (FMPC) had released particulate matter, primarily uranium, and radon and its decay products from two large silos containing radium.

The Fernald Dosimetry Reconstruction Project is currently evaluating doses received by the public from radionuclides released to the environment by FMPC between 1951 and 1988 (RAC, 1996). A review of historic records was the foundation for reconstructing routine operations, documenting accidents, and evaluating unmonitored emission sources. Information was traced to original sources whenever possible. Former and current employees and residents provided information throughout the process.

The quantities of materials released to air, surface water, and groundwater have been estimated. Airborne release estimates were made for dust collectors, scrubbers, waste silos, unmonitored sources, and accidental releases. Radon releases from the silos remained elevated through most of the 1970s, whereas uranium releases to air decreased during that time. The quantity of uranium released to surface water was much less than that released to air. Best estimates of releases are reported as median values, with associated uncertainties calculated as an integral part of the estimates. The methodology to describe the environmental transport of the materials and mathematical models to calculate the resulting radiation doses have been developed and environmental monitoring data have been evaluated to verify that estimates of releases and transport are reasonable.

Although screening calculations showed that atmospheric pathways dominated the total dose from FMPC release, other routes of exposure were also considered. Accordingly, the investigators considered groundwater characteristics and local farming practices in addition to atmospheric parameters.

To undertake the dose reconstruction, 37 years of plant records were located and studied. A great number of interviews were held with former employees. The annual releases of uranium, thorium, radium, and radon were estimated and used as source terms for pathways analysis. Among the kinds of information obtained for the dose reconstruction were the following:

Plant operating procedures
Effluent sampling procedures
All records of atmospheric and discharges
Original analytical worksheets
Plant processes logbooks
Daily sump discharge logbooks
Reports on ventilation system tests

Over the life of the plant, it is estimated that about 300,000 kg of uranium was discharged from the plant to the atmosphere. For nine scenarios, representing people living within 10 km from the plant, the effective doses due to uranium over the 37 years of plant operation ranged from 0.99 to 56 mSv (99 to 5600 mrem). For all but a few of the nearby residents, the effective dose was less than 10 mSv (1000 mrem). Most of the dose was received during the early years of plant operation (RAC, 1996). These doses were relatively small compared to those near the Hanford site discussed previously.

The doses from radon were higher than those from uranium and were by no means insignificant. The radon originated mainly from process residues known as K65, which were raffinates from the extraction of radium from high-grade ores, as described briefly in Chapter 7. The K65 residues have been stored in silos for decades and have been a constant source of radon release to the atmosphere. The median cumulative equivalent doses to the tracheo-bronchial epithelium ranged from 0.3 to 3.4 Sv (30 to 340 rem), compared to about 1 Sv (100 rem) from radon normally present in the atmosphere (RAC, 1996).

The dosimetry reconstruction projects frequently lead to estimates of the number of predicted premature deaths expected from the estimated doses. In almost all such reports there is an absence of accurate demographic information for the period when the highest doses are delivered. Thus, while one can calculate the additional individual risk from the imposed dose under various scenarios, it is not possible to translate the individual risk into probable numbers of illnesses or deaths unless the number of individuals actually exposed at various levels is known. A major philosophical question is whether the object of dose reconstruction should be individual doses, the collective dose to the exposed population, or both. In many cases the dose to the individual is so small that the risk might be considered to be negligible. But, as we saw in the discussion of worldwide fallout in Chapter 9, if a small risk is imposed on a large enough population, the numbers of health effects may not be insignificant.

CHAPTER 15

Remediation of Contaminated Sites

Contamination of land and water by chemical and radioactive materials is a subject that has received widespread media attention and caused much public concern. Although chemicals have caused many of the serious environmental contamination problems, radioactive contamination seems to receive special attention, perhaps because its relationship to the nuclear weapons program and the associated public fears.

The processes of lessening the contamination or its impact by removal, treatment, or stabilization of the contaminated materials have come to be called "environmental remediation" or "environmental restoration." When contaminated buildings or other structures are involved, the process is usually referred to as "D&D," for decontamination and decommissioning (or dismantlement). The disposition of radioactive wastes, including those generated as a result of environmental remediation work, is discussed in Chapter 11. This chapter treats the contaminated sites themselves.

The most important radioactively contaminated sites in the United States can be divided for convenience of discussion into four major groups.[1] The first comprises sites that predate the nuclear era, which began during

---

[1]There are also examples in which contamination has resulted inadvertently, and in some cases unknowingly, by above-background concentrations of natural radionuclides. These are discussed in Chapter 6.

World War II. These sites are generally small and are the legacy of nonnuclear industrial activities involving naturally occurring radionuclides such as radium and thorium. The second group includes sites at which uranium ore was processed during and after World War II for nuclear weapons and power programs. Included in this group are numerous sites throughout the country where high-grade uranium ores were experimented upon and processed, though mostly in the arid West, where lower-grade uranium ore was mined and milled until recently. The third group includes the plants and laboratories that make up the nuclear research and weapons production complex operated by the DOE today (the "DOE complex"). Most of the contamination and the anticipated costs of its remediation will be seen to be associated with the development, production, and testing of nuclear weapons (the "weapons complex").

The fourth group comprises sites that have been licensed by the NRC or the states and that require decontamination of the facilities when they cease operation. Many will also require cleanup of contaminated environments. Examples of these licensed sites include commercial nuclear power plants, fuel-manufacturing sites, laundries that clean protective clothing, source manufacturers, and licensed uranium mills. The requirements to decontaminate sites at the end of their useful lives and for financial assurance are normally conditions of obtaining operating licenses for the facilities. Because of these requirements, licensed sites typically do not have the problem of enormous public costs presented by the weapons complex sites and are not discussed further in this chapter.

Contaminated sites are known to exist at U.S. military facilities and in other countries that have sponsored military or civilian nuclear energy programs, but they are beyond the scope of this chapter.

EXAMPLES OF U.S. STANDARDS FOR CLEANUP

The standards by which radiation is regulated have been reviewed in Chapter 3. However, the established standards that limit exposures of the public are often difficult to apply to environmental remediation because (1) the standards may not be acceptable to members of the public or their representatives, (2) the site may be located in a controlled area such as a government reservation for which it is not possible to predict future patterns of use, and (3) the ICRP (1991) recommends against applying standards developed for "practices" to situations for which intervention is the only possible remedy. Remedial action may then be based on the doses that may be received by a hypothetical future person who has no knowledge of the contamination. Usually the hypothetical person who would receive the greatest dose is the "resident farmer," who will grow much of his or her food on the site and obtain water from a well drilled into the contaminated site.

The cleanup standards applied to specific classes of site remediation depend on the type of site and the cognizant regulatory authority. Even a cursory review of the standards and their application to specific sites reveals a pattern of inconsistency, driven primarily by public opinion and politics. Standards can be broadly classed as activity-based or risk-based. Activity-based standards specify maximum concentrations of radioactive concentrations that are permissible. These may be activities per unit mass or volume in the case of contaminated soil or water, or activity per unit area in the case of surface contamination. Risk-based standards may be specified in terms of numerical probabilities of fatal cancer or in terms of radiation dose. Cleanup standard are changing rapidly and both the EPA and the NRC have prepared draft standards that would limit residual doses from remediated sites to 15 mrem (0.15 mSv) per year. The fate of these draft standards has not been decided as this book goes to press. Examples of some existing remediation standards follow.

COMPREHENSIVE ENVIRONMENTAL RESPONSE, CLEANUP, AND LIABILITY ACT (CERCLA)

CERCLA, also known as "Superfund," is invoked at most of the larger, unlicensed sites, including much of the DOE weapons complex. Superfund encompasses chemicals as well as radionuclides. Money from the government superfund is used to pay for cleanup of contaminated sites when the company or party responsible for the contamination no longer exists or does not have the wherewithal to pay. However, restitution is usually sought from "potentially responsible parties." For federal facilities, the government bears the costs, which are normally funneled through the agency responsible for the contamination.

Three approaches to CERCLA cleanup standards are specified in 40 CRF 300.430. If available, "applicable or relevant and appropriate requirements" (ARARs) under federal or state environmental or facility siting laws are invoked. Examples of ARARS include numerical concentration limits promulgated for water under the Safe Drinking Water Act and the Clean Water Act and radioactivity standards promulgated under the Uranium Mill Tailings Radiation Control Act (UMTRCA). For known or suspected carcinogens, acceptable exposure levels are generally concentration levels that represent an excess upper bound lifetime cancer risk to an individual of between 10^{-4} and 10^{-6} using information on the relationship between dose and response. The 10^{-6} risk level is considered the point of departure for determining remediation goals when ARARs are not available or are not sufficiently protective because of the presence of multiple contaminants at a site or multiple pathways of exposure. For systemic toxicants, such as uranium, acceptable exposure levels are those to which the human population, including sensitive subgroups, may be exposed without adverse effect during a lifetime or part of a lifetime, incorporating an adequate margin of safety. A recent trend is to consider also "ecological risks" and "cultural risks" in establishing cleanup standards for specific sites.

STANDARDS FOR REMEDIATION OF URANIUM MILL TAILINGS

Specific numerical standards have been established for the cleanup of uranium mill tailing piles. According to EPA standards, control afforded by the remedial action should be effective for up to 1000 years, to the extent reasonably achievable, and for at least 200 years.

Average annual atmospheric releases of ^{222}Rn from residual material should not exceed 20 pCi m^{-2} s^{-1} (0.74 Bq m^{-2} s^{-1}) or increase the annual average concentration of ^{222}Rn in air outside the disposal site by more than 0.5 pCi L^{-1} (19 Bq m^{-3}). The concentration of ^{226}Ra in land (above background and averaged over 100 m^2) is limited to 5 pC g^{-1} (185 Bq kg^{-1}) for the top 15 cm of soil and 15 pCi g^{-1} (550 Bq kg^{-1}) for deeper layers.

For habitable buildings the original guidelines were issued by the Office of the Surgeon General in 1970 but are now included in the EPA regulation. The annual average radon decay product concentration, including background, should not exceed 0.02 WL[2] and must not exceed 0.03 WL. Gamma radiation in such buildings is limited to 20 μR h^{-1} ($\sim$0.2 μGy h^{-1}) above background.

Standards for radionuclides in groundwater have been established at 5 pCi L^{-1} (185 Bq m^{-3}) for combined ^{226}Ra and ^{228}Ra, 30 pCi L^{-1} (1100 Bq m^{-3}) for combined ^{234}U and ^{238}U, and 15 pCi L^{-1} (550 Bq m^{-3}) for "gross alpha" excluding uranium and radon (EPA, 1995). Provision is made for alternative concentrations if protection of human health can be demonstrated, but they require specific approval of the EPA.

RISK ASSESSMENT, RISK MANAGEMENT, AND THE PUBLIC

Site remediation is costly and frequently controversial. Risk assessment is but one of a number of steps in the decision-making process. It is considered to be a science and is distinguished from risk management, which must

---

[2]One WL (working level) corresponds to approximately 50 pCi L^{-1} (1.85 kBq m^{-3}) of radon in typical indoor environments. See Chapter 6 for a complete definition.

also take into account technical resources, social, economic, and political values, and control or response options to determine means of reducing risk. Public participation and economic benefit of remediation efforts within the local community are also factors (NAS/NRC, 1994b).

Risk assessment can be conducted in many organizational settings. Although a risk assessment group outside the organization responsible for the contamination might have more credibility, the gain in credibility might come at a cost in time needed to organize the effort and to obtain background information.

For the weapons sites, the entire process of risk assessment is so strongly influenced by lack of trust in the DOE, regulatory agencies, and even the scientific community that it is important, where practical, to involve the public in a formal way. If respected members of the affected community participate in the process, the government and its representatives will develop a better understanding of the concerns of the public and the public representatives will develop a better understanding of the issues involved.

The risk assessment process used by DOE, specifically at Hanford, has been criticized in a report commissioned by the U.S. Senate Committee on Energy and Natural Resources (Blush and Heitman, 1995). The authors conclude that risk assessments are being done with few or no data, that data from the nuclear power industry are being used inappropriately, that the risks of "no-action" alternatives are not realistic, that different programs use differing methodologies, that uncertainties are rarely accounted for or even discussed, and finally that integrated risk assessments (of the whole site) are not being done.

THE REMEDIATION PROCESS

The remediation process should begin after a risk assessment has been completed. Ideally the remediation plan should be based on the risk being imposed by the contaminated site. This will involve site-specific considerations depending on whether the contamination is in an urban or suburban area, in the middle of a desert, or on a large tract of government-controlled land. Swindle (1990) has identified six major phases of the remediation process that may be applied to specific sites as warranted; the less complex sites frequently require fewer steps.

1. Preliminary assessment and site investigation
2. Maintenance and surveillance
3. Remedial investigation and feasibility study
4. Technology demonstrations

5. Corrective measures
6. Facility decomissioning or closure

Remedial actions are normally conducted under the supervision of one or more regulatory agencies. In the United States, the responsible regulatory agency is usually the EPA or an approved state environmental regulatory program. Often the federal and state regulatory programs work in cooperation. The first step in the process is to make an inventory of the specific sites to be remediated and to seek regulatory approval of the inventory. Based on evaluation of the hazard at each site, the preliminary assessment establishes whether further action is required and, if so, identifies the activities required to carry out the remedial investigation and feasibility study. During this step, and required maintenance or surveillance is also identified. Routine maintenance is performed in an effort to contain residual contaminated materials at the site and surveillance is used to determine whether or not the containment efforts are successful.

The remedial investigation and feasibility study is used to fully characterize the nature and extent of the contamination and to define the remedial actions to be taken. Efforts are made to involve states, Native American groups, and the public at this stage. The investigation process usually involves analysis of environmental samples to characterize the contamination and to identify any releases of contaminants to the surrounding environment. The data required to evaluate alternative cleanup strategies are gathered through feasibility studies. These may include bench scale or pilot studies of proposed methods. Potential cleanup alternatives are screened for applicability to each site and the most promising alternatives are given a detailed technical evaluation and a recommended approach is developed. At this stage the recommended approach must be approved by the regulatory authority. Remedial actions performed in the United States under the Resource Conservation and Recovery Act (RCRA) can be approved with an "Operating Permit" or a "Consent Order/Compliance Agreement." Actions performed under CERCLA are approved in a "Record of Decision." Sites deemed to have high risk according to a ranking system are listed on the National Priorities List and these are referred to as Superfund sites.

Technology demonstrations may be performed at sites to evaluate or demonstrate the suitability of technologies for the task at hand. Technologies may also be evaluated at a small site with a view toward broader application if they are found to perform well. Past corrective actions may be evaluated as part of a technology demonstration.

Interim corrective actions may be performed rapidly when the remedial investigation or other process identifies a site that poses an unacceptable risk to humans or the environment. Interim action can be in the form of engineered barriers to prevent the spread of contamination or removal, treatment, or interim storage of the contaminated materials. These actions maintain safety during the often lengthy process of arriving at a final decision.

Decommissioning of facilities or "closure" of a contaminated site results in the long-term containment and control of the contamination. Even when the primary remedy for a site is removal of the contamination, residual contamination can exist at low concentrations and containment on site may be required. Decommissioning or closure is performed in accordance with the decisions made and recorded under the remedial investigation/feasibility study phase and with priorities and schedules negotiated with the regulatory authority. Inspections and monitoring of the site by a third party or the regulator are usually performed at this closure stage.

The complex technical, regulatory, and public issues that surround environmental remediation require a strong infrastructure. This support activity is typically involved in managing the data, integrating the information developed in the six phases described here, interfacing with regulators, conducting community relations, and preparing documentation.

Surprisingly, risk assessment was not explicity identified by Swindle as part of the process, but it may be considered to be implicit in the studies conducted in support of the remediation process. Risk assessment is central to the decision-making process under CERCLA and the DOE supports its use in remediation prioritization and decision making.

REMEDIATION MEASURES AND TECHNOLOGY

The three basic options for contaminated sites are (1) no action, (2) containment, and (3) some combination of retrieval, treatment, storage, and disposal (DOE, 1989). One conceptually simple method for remediation of contaminated soils is to excavate the material and transport it to an approved disposal site. This approach is often favored by nearby residents who usually prefer that the disposal site be in some other state. Excavation of soils with low-level contamination can be accomplished with conventional construction techniques. Excavation of heavily contaminated materials requires advanced remote or robotic techniques. As discussed in Chapter 11, however, establishing new radioactive disposal sites for excavated materials is extremely difficult, and some formerly available sites are now closed. Costs for transport and disposal, especially when large volumes are involved, can be very high. For these reasons, there is need for technologies that can be used to reduce waste volumes by concentrating the radionuclides in a fraction destined for disposal and to stabilize residues for possible on-site disposal.

Selection of the no-action alternative may be possible when risk assessment demonstrates that there are no unacceptable risks associated with leaving the contamination in place. The role of risk assessment and risk management in remedial action decisions was discussed earlier in this chapter. The no-action alternative, however rational in some situations, may nevertheless be unacceptable to a community that can only be satisfied by removal of the contamination.

Containment can be accomplished with a variety of techniques (Swindle, 1990). Options include capping to prevent mobilization of the contaminants by precipitation, vertical barriers, and collection of water that percolates through or runs off the site. Some treatments, referred to as *in situ*, can be performed within the contaminated material at the site. Dynamic compaction can be used to reduce voids and volumes of waste but the presence of liquids in containers that may rupture during the process is a limitation. Soil grouting may be used to reduce contaminant mobility by decreasing soil permeability. For the grouting process, a stabilizing material such as polyacrylamide is mixed with the waste and the mixture is injected into wells, where it flows into voids and fractures, hardens, and immobilizes the waste. *In situ* vitrification is an experimental process in which electrodes are inserted into waste or contaminated soil and current is used to heat the material (Jacobs and Spalding, 1988; Alexiades *et al.*, 1994). Volatile materials are driven off, collected, and treated and the residue is melted in place (vitrified) to form a glasslike substance that retains the nonvolatile contaminants in a stable form.

Soil washing is a treatment process that removes contaminants from excavated soils using physical separation, chemical extraction, or a combination of the two (Parikh *et al.*, 1995). Physical separation uses screening and water to separate soils into several size fractions. Chemical soil washing is removal of contamination by mixing contaminated soils and extractants in a continuous vat leaching process. Heap leaching involves percolating the extractant through a fixed bed of soil. The physical separation method is most effective on contaminated soils having higher concentrations of contaminants in the finer fractions. Chemical extraction may be more effective if the contaminants are bonded to coarse soil particles. Successful soil washing results in reduced volumes of materials that require disposal as radioactive waste and larger volumes of soil that are sufficiently decontaminated that they can remain at the site.

The plasma hearth process is a high-temperature vitrification process that has potential for stabilizing excavated wastes and contaminated soils (Geimer *et al.*, 1994). Whole drums of waste can be processed without pretreatment. The arc of a plasma torch mounted in a refractory reaction chamber destroys organic materials and melts the inorganic residue into a nonleaching glassy waste form. Waste gases produced in the process are treated and monitored. This process has a long history of use in the metallurgical industry but has only recently been applied to the problems of waste treatment.

Groundwater contaminated with radionuclides may be remediated by a process known as "pump and treat," in which water is pumped from a contaminated aquifer, treated at the surface, and returned to the aquifer. Pump and treat is widely used for nonradioactive contaminants such as volatile organic chemicals, where it has met with some success, particularly where the stratigraphy of the aquifer is well known. Combined treatment using air stripping and carbon adsorption for volatile organic compounds and resin ion exchange for ^{90}Sr was attempted for an old injection well on the Idaho National Engineering Laboratory site (Cotten, 1955). The water treatment system, designed to treat contamination levels found during sampling, had to be shut down when the actual concentrations were found to greatly exceed concentrations found during the sampling phase. Apparently the much greater pumping rates used during attempted remediation dislodged contaminants that had remained in place during the sampling period.

Conventional methods of waste treatment (Moghissi *et al.*, 1986; Berlin and Stanton, 1988) may also be of use in treating wastes resulting from remedial actions. However, the DOE recognizes that current technology is inadequate to the remediation tasks at hand and is currently supporting technology development at a cost of about $400 million per year (DOE, 1995c).

SITES PREDATING WORLD WAR II

Not all of the problems of radioactive contamination are traceable to the nuclear industry. Waste products that contain radioactive substances from other industries can also present vexing problems. Among these are the facilities that processed radium or thorium for medicine, research, luminous paints, and other uses in the early part of the twentieth century. Other plants have engaged in the extraction of rare earth elements from monazite or zircon sands that also contained thorium.

THE DENVER RADIUM SITE

Prior to 1913, low-grade ore (~1–2% uranium) was shipped from Colorado and Utah to Europe for extraction of radium (Sowinski, 1996). There was no domestic U.S. extraction industry and radium was bought from European suppliers at prices ranging from \$120,000 to \$180,000 per gram, even though production costs were only \$40,000 (Landa, 1987). In 1912, the National Radium Institute, a joint venture between private industry and the Bureau of Mines, developed and began operating a radium-processing plant in Denver. The location was chosen because of its proximity to the supply of ore, and nine other radium-processing operations subsequently opened in Denver along with companies that manufactured consumer products containing radium. The Denver radium industry remained economically viable until pitchblende ores (containing ~50% uranium) were discovered in the former Belgian Congo in the 1920s. These richer ores were processed more economically by the Belgians, forcing most of the Denver companies out of business. The last radium company in Denver ceased production in 1941 but continued with molybdenum and uranium processing until 1984.

These operations and their potential contamination were largely forgotten until references to the National Radium Institute were noted many years later in a 1916 Bureau of Mines Report. Following this lead, the state of Colorado and the EPA found 31 radioactive locations within the Denver Metropolitan Area, which were included on the EPA National Priorities List in 1983 and collectively named the Denver Radium Site. Potential risks included radon exposure from direct radiation and ingestion. Gamma-radiation exposure rates above local background ranged from a few μR h^{-1} (a few tens of nGy h^{-1}) to about 2 mR h^{-1} (~20 μGy h^{-1}). Exposures to individuals would be expected to increase if the materials were misused or if the areas were redeveloped without remediation.

All but one of the sites had been contaminated by companies that were long out of business. These sites were cleaned up at taxpayer expense by removal, storage, and ultimate disposal of the 175,000 m^3 of contaminated material at a licensed radioactive landfill in South Clive, Utah. The last site to be remediated was owned by a company still in business and therefore responsible for costs under CERCLA. The company fought the high-cost option of out-of-state disposal, and eventually on-site immobilization of the 38,000 m^3 of contaminated material was chosen. The contaminated material is to be treated with a mixture of cement and fly ash, compacted to create a concrete monolith, and covered with a synthetic liner, clay, dirt, and riprap. Although the city of Denver sued in an attempt to force removal of the materials, EPA's decision for a local remedy was upheld and is expected to be implemented in the near future.

MONTCLAIR, NEW JERSEY

At another radium site, the former location of the U.S. Radium Company in Montclair, New Jersey, soil in the environs was found to be contaminated with radium-containing wastes generated in the 1920s (Eisenbud, 1991). Decades later, 12 homes since built on the site of the plant were found to be resting on radium-contaminated soil, which caused elevated concentrations of radon within the homes. Decontamination of the site required the removal of 15,000 fifty-five-gallon drums of soil (3100 m^3) that contained about 1.5 Ci (56 GBq) of ^{226}Ra at an average concentration of 300 pCi g^{-1} (11 kBq kg^{-1}).

The state of New Jersey first made several unsuccessful proposals to dispose of the material. Among the options considered were ocean deposition (ocean water has a natural radium content of about 1 GBq km^{-3}), dilution with uncontaiminated soil, and storing the soil in abandoned mines. None of these methods was permissible under existing regulations. The problem was finally solved by shipping the barrels to the Oak Ridge Reservation where the slightly contaminated soil was used to dilute wastes that had a higher level contamination before shipping to Hanford, Washington, for storage. The cost of decontaminating the 12 homes was $8.5 million (Cole, 1989; New Jersey Department of Environmental Protection, 1987). This relatively high cost of decontaminating only 12 private homes illustrates the funding problems of decontaminating the numerous plants and laboratories of the DOE weapons production complex.

SITES RELATED TO THE DEVELOPMENT OF NUCLEAR ENERGY AND NUCLEAR WEAPONS

Fifty years of the development of nuclear technology and the manufacture and testing of nuclear weapons at 137 locations in 33 states and one territory have left an enormous legacy of contaminated sites. Altogether, these locations cover an area of 3300 square miles (8600 km^2). The DOE has identified nearly 10,000 individual contaminated sites at these locations that it believes may have to be remediated (DOE, 1995c). Almost a million cubic meters of radioactive waste have to be managed until adequate treatment and disposal are available. More than 7000 contaminated buildings are no longer needed and more waste is generated as they are decontaminated and dismantled. However, the problems of contaminated sites in the weapons complex were not unanticipated. An early report of the AEC stated that "the disposal of contaminated wastes in present volumes and by present methods (in tanks or burial grounds or at sea), if continued for decades, presents the gravest of problems" (AEC, 1948).

A complete listing of the contaminated sites is given in DOE (1995a) and their locations throughout the country are shown in Fig. 15-1. Included are (1) industrial sites contaminated early in the nuclear era, primarily with

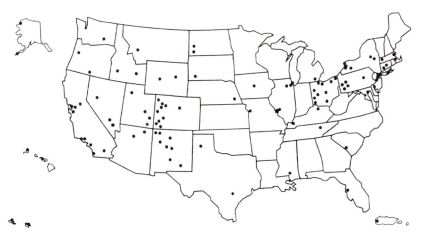

FIGURE 15-1   Locations of contaminated sites related to the development of nuclear energy for both defense and nondefense purposes. [From DOE (1995a).]

high-grade, imported ores; (2) uranium manufacturing , mill, and tailings sites; and (3) the sites of the nuclear research and weapons complex. The progress and plans for remediation of some of these sites are discussed in the following sections.

INDUSTRIAL SITES OF THE EARLY NUCLEAR ERA

From the early 1940s through the 1960s, work was performed at many private and a few publicly owned industrial and laboratory sites throughout the United States as a part of the early atomic energy programs. Some sites were involved with the raw materials used in the development of the first atomic bomb, whereas others were involved in later peaceful development of nuclear energy. Much of the waste remaining at these materials processing sites is a sandlike by-product of the processing of uranium and thorium ores and materials contaminated with these by-products. The important radionuclides are the natural isotopes of uranium, thorium, and radium and their decay products. Sites contaminated in the early years were generally cleaned up to the standards in effect at the time. The early cleanup standards were not as stringent as those of today, and radioactive material at levels above today's standards remained at many of the sites.

In 1974 the DOE initiated the Formerly Utilized Sites Remedial Action Program (FUSRAP) to study the nonfederal sites and take appropriate cleanup action (DOE, 1985b, 1995f). Over 400 potentially contaminated sites were examined for possible inclusion in the program. During this process, several contaminated sites not associated with nuclear energy activities were added to the program by congressional action because DOE had developed the expertise to deal with them. Ultimately, 46 sites were accepted into FUSRAP. The larger sites, in terms of waste volumes, are listed in Table 15-1. Some of the contamination originally confined to the sites was spread to other locations ("vicinity properties") by deliberate movement, by demolition of buildings, or by wind and water, resulting in a total of about 1,800,000 m³ of waste materials. There are now 309 vicinity properties requiring cleanup associated with the 46 sites. A unique feature of the FUSRAP sites is that they are mostly privately owned industrial sites that are currently in use for purposes unrelated to radioactive material. This may be contrasted with the uranium mill tailings piles, which are mostly in remote locations, and with the nuclear materials processing sites, which are controlled on government-owned reservations (discussed later in this chapter).

Although materials exist at most FUSRAP sites at levels that exceed current cleanup guides, they are believed to pose no immediate health risks given current land use. Cleanup is being based on hypothetical future use and the fact that the mere presence of contamination reduces property values and restricts use. Contamination that presents no risks at an industrial site today may pose a risk if the land were to be converted to residential or resident farmer use. Cleanup guidelines have been adopted from those developed for uranium mill tailings (see the next section), but efforts to develop site-specific standards based on risk assessment are ongoing.

The amounts of waste material range from <1 m³ for the Granite City Steel site in Illinois to about 300,000 m³ for the Maywood site in New Jersey. Limited cleanup under FUSRAP began in 1979 and major remedial action started in 1981. Twenty-one of the 46 sites (46%) have been completed as of this writing, but these sites account for only 9% of the vicinity properties and 4% of the total materials. Partial cleanup and stabilization has occurred on 11 sites. To date, cleanup has consisted of excavation and removal at the smaller sites and stabilization and limited removal for the larger projects. Even though 96% of the material awaits final disposition, much of that is in a stable form and is being monitored. Future options being considered include various combinations of institutional controls, capped disposal cells on or near the sites, disposal at commercial or DOE radioactive landfills, treatment, and beneficial reuse. Some of the waste materials have economic concentrations of rare metals, but the associated radioactivity has minimized interest in their recovery.

Clearly much remains to be done. About $425 million has been spent from inception of the project through 1995. The work is expected to be completed in the year 2016 at a total cost of $2.5 billion. However, this

TABLE 15-1

Locations and Status of the 18 Largest Sites in the DOE FUSRAP Program, Listed in Order of Waste Volume[a]

Site name	City	State	Number of vicinity properties	Waste volume (m³)	Clean-up completed?	Notes[b]
Maywood site	Maywood	NJ	83	300,000	No	NPL, DOE
St. Louis airport site	St. Louis	MO	0	191,000	No	NPL
St. Louis downtown site	St. Louis	MO	6	188,000	No	
Latty Avenue properties	Hazelwood	MO	6	161,000	No	NPL, DOE
Niagara Falls storage site	Lewiston	NY	26	157,000	No	DOE
St. Louis airport vicinity prop.	St. Louis	MO	78	149,000	No	NPL
Ashland 1	Tonawanda	NY	0	92,000	No	
Seaway Industrial Park	Tonawanda	NY	0	89,000	No	
Wayne site	Wayne	NJ	23	83,000	No	NPL, DOE
Middlesex Sampling Plant	Middlesex	NJ	0	68,000	No	DOE
Linde Air Products	Tonawanda	NY	1	54,000	No	
Painesville site	Painesville	OH	0	53,000	No	
Colonie site	Colonie	NY	56	41,000	No	DOE
Ashland 2	Tonawanda	NY	0	40,000	No	
Niagara Falls vicinity prop.	Lewiston	NY	26	38,000	1986	
W.R. Grace & Co.	Curtis Bay	MD	0	28,000	No	
Luckey site	Luckey	OH	0	26,000	No	
Middlesex Municipal Landfill	Middlesex	NJ	2	24,000	1986	
Remaining 28 sites			2	38,000		
Total			309	1,820,000		

[a]Adapted from DOE (1995f)

[b]NPL. Site listed on the EPA National Priorities List. DOE: Site is owned or leased by DOE.

estimate assumes that cost savings will be realized by innovative technology and by acceptance of on-site disposal for much of the high-volume, low-specific-activity wastes. Both of these assumptions are problematic at this time, and if excavation and off-site, commercial disposal are used for all sites, the cost will increase to at least $5 billion.

MINING, MILLING, AND PROCESSING OF URANIUM

Mill Tailings Piles

The mining and milling of uranium in the United States and the impact of mill tailings on the environment were discussed in Chapter 7. This section is concerned with the remedial action programs for mill tailings (DOE, 1995b,g). Most of the uranium ore produced in the United States between 1943 and 1970 was mined and milled under contract to the government. After completing the contracts, many of the mills shut down, leaving large quantities of mill tailings at the sites. At many of the sites, no effort was made to prevent the spread of the tailings by wind or water or to preclude their use in construction, which resulted in contamination of vicinity properties.

The concern about gamma radiation and radon from uncontrolled mill tailings, particularly from their use in construction, arose originally in Grand Junction, Colorado. This concern led to a 1978 law specifically requiring cleanup of structures in Grand Junction and, shortly thereafter, to an inclusive law, UMTRCA[3], addressing 24 inactive, abandoned mill tailings sites in 10 states. Approximately 3×10^8 m^3 of tailings and contaminated material is present at these sites, which are located mostly in the U.S. West. Uranium mill sites that were still in operation in 1978 are licensed by the NRC and remain the responsibility of their private owners. These private sites make up about 75% of the total tailings generated, leaving 25% to be cleaned up under UMTRCA.

The uranium mill sites and contaminated vicinity properties designated under UMTRCA are being cleaned up by the DOE under EPA regulations (40 CFR 192) and with consultation by the NRC. Once remedial actions are completed by DOE, the sites are licensed by NRC and subjected to long-term surveillance to verify that the remedial measures were adequate. The numerical cleanup standards were discussed earlier in this chapter.

---

[3]UMTRCA is the Uranium Mill Tailings Radiation Control Act. The original requirement to clean up Grand Junction is found in Public Law 95-338, Authorization, Appropriation: Arms Control and Disarmament Act (22 U.S.C. §2576 *et seq.*). The other sites were included by UMTRCA, Public Law 95-604, Uranium Mill Tailings Radiation Control Act (42 U.S.C. §7901 *et seq.*). Both laws were passed in 1978.

The sites are being remediated in two phases, the surface phase and the groundwater phase. The three approaches to surface remediation of uranium mill tailings piles are to stabilize in place or on-site (on-site disposal) or relocate the tailings (offsite disposal). The tailings from the 24 sites are being consolidated into 16 disposal cells. Stabilization is achieved by capping designed to hold the tailings in place and limit radon emissions.

Properties in the vicinity have been remediated by removing contaminated materials, which are then disposed of with tailings from the piles. Tailings had their greatest use for construction in Grand Junction and account for about 85% of the contaminated properties. Seasonal swelling associated with moisture changes in the local bentonite clay soils would damage basement foundations unless sand or a sandlike fill was used between the soil and basement. Tailings had the required mechanical properties and were readily available.

As of August 1995, surface remediation was complete at 14 sites, work is under way at 8 sites, and planning is in progress for the remaining 2 sites. Ninety-six percent of the vicinity properties have been cleaned up and 80% have been independently certified as clean. Most of the surface remediation work is expected to be completed in 1997, with final closure of the disposal cells scheduled for 1999. The volumes of tailings, contaminated areas, and status of the neighboring properties are given in Table 15-2 for the 10 largest sites.

A decade after it was passed, Congress amended UMTRCA to include groundwater contamination.[4] Studies have now shown that radionuclide contamination at levels above the recently established standards exists at 22 of the 24 sites and that all except the Lowman, Idaho, site have some form of chemical or radiological contamination above the standards. Risk assessments for half of the sites were performed by the end of 1994, but no site-related human health risks were found. It is too early in this project to know how much or what form of groundwater remediation will be proposed or funded, but planning is ongoing (DOE, 1995b).

The mill tailings remediation program is estimated to cost $2 billion. The surface phase is nearly complete and the 1.5 billion estimate for that phase is probably accurate. The groundwater remediation has just begun, however, and actual costs may be quite different from the estimate of $540 million.

Government-Owned Uranium Processing Sites at Fernald, Ohio, and Weldon Spring, Missouri

Government-owned uranium-processing sites were operated near Weldon Spring, Missouri, from 1957 through 1966 (Green, 1993) and near

[4]The amendment is contained within Public Law 101-616; 42 U.S.C. §7922 (a).

TABLE 15-2

Remediation Data for the 10 Largest Sites Falling under the Uranium Mill Tailings Radiation Control Act, Listed in Order of Tailings Volume[a]

Location	State	Volume of contaminated material (1000 m³)	Area contaminated (ha)[b]	Number of vicinity properties identified (1994)	Number of vicinity properties completed (1994)
Falls City	Texas	4,400	240	13	13
Grand Junction	Colorado	3,600	46	4,349	4,255
Ambrosia Lake	New Mexico	2,900	250	5	0
Maybell	Colorado	2,700	87	6	0
Rifle (new)	Colorado	2,700	96	110	91
Mexican Hat	Utah	2,200	100	11	11
Salt Lake City	Utah	2,100	52	118	118
Durango	Colorado	1,900	51	130	130
Riverton	Wyoming	1,400	57	42	41
Shiprock	New Mexico	1,200	53	15	15
Remaining 15 sites		4,900	568	420	376
Total		30,000	1,600	5,219	5,050

[a]Adapted from DOE (1995b, g).
[b]One hectare = 2.47 acres.

Fernald, Ohio, from 1951 through 1989 (Janke *et al.*, 1992). Both of these sites have been shut down and renamed to emphasize that their mission is limited to environmental remediation. The Weldon Spring site is now the Weldon Spring Site Remedial Action Program (WSSRAP) and the Feed Materials Production Center near Fernald is now the Fernald Environmental Management Project.

Uranium and thorium processing at Weldon Spring contaminated the 44 buildings, surrounding land areas, and waste pits. Raffinate, the waste product resulting from the chemical separation of uranium from the feed material, was piped to four surface impoundments, occupying 10 ha (25 acres). Approximately 170,000 m³ of the raffinate sludge has settled to the bottom of the impoundments and is covered by standing water (Hillman *et al.*, 1994). The primary radioactive contaminants in raffinate are [230]Th and [226]Ra.

The buildings have now been dismantled and WSSRAP is designing a chemical stabilization and solidification plant for treatment of the raffinate. Mixing of the raffinate during recovering and treatment is expected to release significant amounts of radon, which is to be stored in high-pressure tanks for decay. Once stabilized, the raffinates will be retained in an on-site disposal cell. A major issue at WSSRAP is a 9-acre (3.6-ha) water-filled quarry that was used for waste disposal. Use of the quarry began for chemical disposals in 1942 and then for radioactive wastes when uranium processing began there in 1957. Water from the quarry has been removed and treated and the remainder of the waste is being excavated and transported to the main site (WSSRAP, 1995). The cost of the cleanup at Weldon Spring is estimated to be $370 million (DOE, 1995d).

The DOE Fernald site in southwestern Ohio, about 18 miles northwest of Cincinnati, was used for the processing and storage of uranium metals and compounds (Janke *et al.*, 1992). There were about 200 structures on the 425-ha (1050-acre) site. Activities there resulted in chemical and radioactive contamination on and near the site, and in 1989 the site was included on the EPA National Priority List. Waste is stored in six pits, three silos, and thousands of drums and other containers. Some of the radioactive waste is mixed with hazardous chemicals. The total waste volume has been estimated to be 2.4×10^6 m³, more than half of which is contaminated soil.

It is estimated that as much as 450,000 kg of uranium was released to air and water over the life of the plant (Fernald Citizen's Task Force, 1995). Groundwater and surface water have become contaminated with above-background concentrations of uranium, but current doses are small. In 1993, the annual dose was estimated to be 0.7 mrem (7 μSv) from drinking well water from the area and 0.01 mrem (0.1 μSv) from eating locally caught fish. Bottled or municipal water has been supplied to residents whose wells draw from the contaminated water. Radon released from materials stored on site currently contributes an estimated annual effective dose of 450 mrem (4.5 mSv) to the "fence line" dose, a little over two times background (Environmental Protection Department, 1994). A historical dose reconstruction for the Fernald site is nearly complete; the preliminary results are discussed in Chapter 14.

Studies of the site are essentially complete and remediation began in 1995. The planned remedial actions include vitrification of materials with the highest concentrations, off-site disposal of some materials, and on-site disposal of the lightly contaminated soils in an engineered facility (DOE, 1995c). At one stage, cleanup was expected to take about 25 years and cost $5.7 billion, but the current DOE estimate is $4.2 billion. Area citizens are arguing for an accelerated schedule, which they believe could reduce costs even further (Fernald Environmental Management Project, 1995).

SITES CONTAMINATED BY FISSION, ACTIVATION, AND TRANSURANIC RADIONUCLIDES

The sites discussed in the previous sections associated with the milling and processing of uranium and other source materials have generated huge volumes of waste materials that require control and large contaminated areas that require remediation. Actual radiation exposures occurred to individuals from close contact with the materials, particularly where the waste materials were used in construction. However, the radionuclide concentrations and associated hazards were low. By contrast, the potential hazards from the high concentrations of radionuclides in wastes and contamination resulting from the production of tritium and plutonium for nuclear weapons are very great. Yet these wastes and contaminated areas exist mostly on government reservations and have been much more closely controlled than mill tailings and uranium- and thorium-processing residues. Current radiation exposures of the public from wastes and contaminated sites in the nuclear weapons complex are minimal, but the potential for future risk is great, particularly if institutional control of the government reservations were to be lost as a result of political and societal change. Remediation itself will present risks to workers and the public as wastes and contaminated soils are disturbed, treated, and transported.

A useful way to illustrate the magnitude of the weapons complex cleanup is through projected budgets. Up to 1996, about $28 billion had been spent and estimates for the entire job ranged from $100 to $300 billion. In 1996 the DOE compiled estimates from site-specific data and estimated that remediation of the DOE complex would take 75 years and cost between $189 and $265 billion with a midrange estimate of $227 billion (DOE, 1996). Three-fourths of this amount is related to past nonweapons research and production and the remainder for ongoing programs and past nonweapons activities. Approximately $63 billion is expected to be spent on actual environmental remediation and the remainder on related issues of waste management ($111 billion), nuclear material and facility stabilization ($21 billion), landlord costs and program management ($20 billion), and technology development ($12 billion). The midrange estimate assumes that significant savings will result from use of new technologies and improved efficiency. These overall figures include the relatively small amounts being spent for remediation of uranium-production and -processing sites discussed in previous sections. The 1996 budget for all environmental management activities by DOE was $6.5 billion, of which $2 billion was earmarked for remediation.

Figure 15-2 illustrates how the costs are expected to be allocated among sites. It is clear that the major production sites at Hanford, Savannah River, Rocky Flats, Idaho, and Oak Ridge account for three-fourths of the costs, whereas the weapons research laboratories at Los Alamos and Oak Ridge have much smaller requirements. Figure 15-3 illustrates the expected budget requirements over the planned 75-year period. It is important to note that these figures exclude Naval propulsion facilities, currently active facilities such as the Pantex plant (nuclear weapon assembly and disassembly), and contamination for which no feasible cleanup technology exists. Examples of the latter include most groundwater contamination and the nuclear weapons test sites. As the currently excluded sites are included, total costs can be expected to rise.

Remediation activities at a few of the more important sites in the DOE complex and one example of remediation taken to excess are discussed in the following sections. Remediation at the major DOE sites is being carried out under Federal Facilities Agreements that typically involve the DOE, the EPA, and the appropriate state regulatory authority. This discussion is not intended to be comprehensive, nor could it be. However, the DOE has been very forthcoming with information in recent years and has published a directory of information resources to which the reader needing more information is referred (DOE, 1994g). Also, the major DOE sites have Site-Specific Citizens' Advisory Committees that can be contacted for independent information on DOE's progress with remedial action. Additional background information on some of the sites may be found in Chapter 14.

The Savannah River Site

The Savannah River site is located on 840 km² along the Savannah River near Aiken, South Carolina (DOE, 1994e). The historical mission of the site was to produce tritium and plutonium for the weapons program in large nuclear production reactors. The tritium and plutonium were separated and

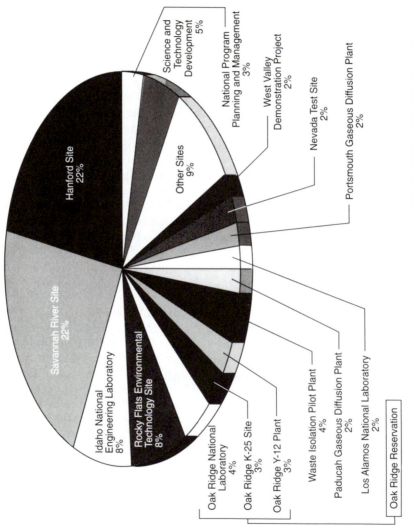

Science and
Technology
Development
5%

National Program
Planning and Management
3%

West Valley
Demonstration Project
2%

Nevada Test Site
2%

Portsmouth Gaseous Diffusion Plant
2%

Hanford Site
22%

Other Sites
9%

Savannah River Site
22%

Idaho National
Engineering Laboratory
8%

Rocky Flats Environmental
Technology Site
8%

Oak Ridge National
Laboratory
4%

Oak Ridge K-25 Site
3%

Oak Ridge Y-12 Plant
3%

Waste Isolation Pilot Plant
4%

Paducah Gaseous Diffusion Plant
2%

Los Alamos National Laboratory
2%

Oak Ridge Reservation

FIGURE 15-2 Cost estimates by site for environmental management within the DOE complex, including environmental restoration. [From DOE (1996).]

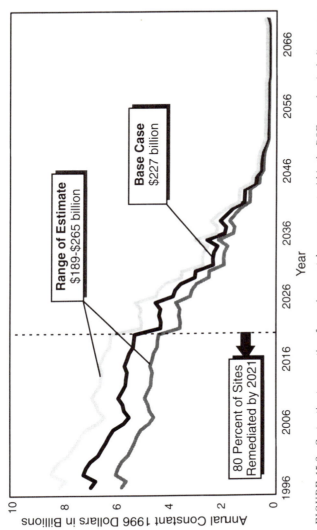

FIGURE 15-3 Cost estimates over time for environmental management within the DOE complex, including restoration. [From DOE (1996).]

refined from targets irradiated in the reactors. The processes generated high-level, low-level, transuranic, and mixed wastes that were managed in seeping and settling basins, unlined disposal pits, waste piles, burial grounds, and underground storage tanks. Contamination was obviously anticipated when wastes were placed on or in the ground and the storage tanks have leaked, resulting in extensive environmental contamination.

More than 400 individual contaminated sites and 659 contaminated buildings have been identified at Savannah River, including 82 settling or seepage basins, 99 rubble or burning pits, 6 areas of groundwater contamination, 85 spill sites, and 85 miscellaneous units. Remediation options are site-specific and most involve some combination of stabilization, capping, removal, and grading. Vitrification is planned for some liquids. Cleanup is proceeding under a Federal Facilities Agreement that recognizes the regulatory authority of the EPA and the South Carolina Department of Health and Environmental Control. Although a few locations have been fully cleaned up, most of the work remains to be done.

The Hanford Nuclear Reservation

The Hanford site is located on 1450 km^2 of semiarid land near the towns of Hanford, Pasco, and Kenewick in southeastern Washington State (DOE, 1995c). The Columbia River flows through the site and was used for cooling and process water for the reactors and other operations. The principal original mission of the Hanford site was the production and refining of plutonium for the weapons program, an activity that began 1943. The nearby Pacific Northwest Laboratory was responsible for much of the early research on radiation protection and the fate and transport of radionuclides in the environment. A reconstruction of the off-site doses due to releases in the early years was discussed in Chapter 14.

There are about 200 inactive, contaminated facilities that require decontamination and dismantlement, including eight surplus reactors. The reactors are to be removed from the buildings and buried on site (DOE, 1994c).

Activities at Hanford have generated high-level, low-level, and mixed wastes and resulted in soil or groundwater contamination at about 1100 locations, ranging in size from less than one square foot ($\sim$0.1 m^2) to 1800 acres (720 ha). Much of the environmental contamination originated from storing wastes in direct contact with the ground and disposing of liquid wastes to soil columns, a practice known locally as cribbing.

The most notorious waste issue at Hanford is the high-level waste tanks, which were discussed in Chapter 11. From approximately 1943 to 1987, large underground storage tanks were used to store millions of gallons of high-level radioactive liquid wastes. The DOE estimates that 67 of the 149 single-shell tanks have leaked in the past, releasing approximately 1 million gallons of high-level mixed waste into the dry sediments under and surrounding the tanks (Butherus *et al.*, 1995). These leaks have contributed to the 1100 contaminated sites.

The Hanford cleanup is being conducted under a Federal Facilities Agreement ("Tri-Party Agreement") among DOE, EPA, and the Washington Department of Ecology. Relatively minor contamination in an area comprising 46% of the Hanford site, known as the North Slope and Arid Lands Ecology Study Area, has been cleaned up. Other areas will clearly require much further study and effort before remediation is complete. Management of the Hanford cleanup has been severely criticized in a report to the U.S. Senate Committee on Energy and Natural Resources (Blush and Heitman, 1995). the authors conclude that the Hanford cleanup is the largest public works project in world history, that the process is not working effectively, that the site is spending more money on the cleanup than it can justify, and that no reliable data exist for completing the cleanup.

The Oak Ridge Reservation

The Oak Ridge Reservation (ORR), an area of 240 km² along the Clinch River in eastern Tennessee, includes a national laboratory where several research reactors and two significant former production facilities were sited (DOE, 1994b, 1995c). The Oak Ridge National Laboratory (ORNL), originally the X-10 Site, was built beginning in 1942 as a facility to produce samples of plutonium and to develop methods for its separation and purification for early nuclear weapons development and later evolved into a research facility. The production facilities were the "K-25" plant, where uranium hexafluoride was enriched in ^{235}U by gaseous diffusion, and the "Y-12" plant, where enriched uranium weapons components were manufactured.[5] The K-25 plant is now the principal waste storage facility on the Oak Ridge Reservation and houses an incinerator. The Y-12 plant has become a center for handling, processing, storage, and disassembly of enriched uranium weapons components but the reduced mission has resulted in 155 surplus facilities. The approximate areas occupied by the three major facilities are ORNL, 3000 acres (1200 ha), K-25, 300 acres (700 ha), and Y-12, 800 acres (325 ha).

Contaminants in the environment include radionuclides, especially uranium, and chemical hazardous wastes such as polychlorinated biphenyls and mercury. Cesium-137 was released in liquid wastes from chemical separation activities at X-10 and traveled over White Oak Dam to the Clinch and

---

[5]The Y-12 plant was originally built to enrich uranium by electromagnetic separation but this process was found to be less effective than gaseous diffusion, and the plant was converted to the manufacture of weapons components.

Tennessee rivers. Mercury was used between 1950 and 1963 in processes to enrich lithium for use in thermonuclear weapons. Two million of the 24 million pounds of mercury handled at Y-12 were reported lost or unaccounted for (Widner, 1994) and mercury continues to enter East Fork Poplar Creek. ORNL has more than 200 individual contaminated sites on the reservation. In addition to the contamination on the ORR, there is off-site contamination in the Clinch River, East Fork Poplar Creek, Watts Bar Reservoir, the Oak Ridge Institute for Science and Education, and a few private sites.

The ORR was placed on the National Priorities List under CERCLA effective December 21, 1989. As a result of this listing, the DOE, the EPA, and the Tennessee Department of Environment and Conservation signed a Federal Facilities Agreement focusing on resolving compliance issues associated with its environmental restoration. A few waste sites have been fully remediated but the bulk of the work remains to be done.

The Rocky Flats Plant

The Rocky Flats plant, located on a 27-km^2 site near Denver, Colorado, had a primary mission of shaping nuclear weapons components from plutonium and other metals (DOE, 1994d). The plant is no longer in production and the mission has changed to cleanup and waste management. Environmental contamination at more than 200 locations occurred as a result of past waste storage and disposal practices. The largest amount of contamination exits on-site, but off-site contamination at low levels exists in three reservoirs and one land area. Isotopes of plutonium represent much of the radioactive contamination but actual doses to nearby residents are extremely small (see Chapter 14). Nevertheless, DOE has funded an alternative water supply for the nearby town of Broomfield and has constructed catch basins and diversion ditches to manage water that could potentially become contaminated.

Some interim measures have been instituted to limit the spread of contamination. Systems to collect and treat contaminated groundwater have been installed in former waste-disposal areas known as the "881 Hillside" and the "902 Pad, Mound and East Trenches Area." About 350 acres (140 ha) of off-site land with low-level surface contamination are being treated by plowing and revegetation to limit migration. The situation at Rocky Flats could have been worse except that most of the plutonium and chemical hazardous wastes were transported to more isolated locations for disposal. Rocky Flats wastes disposed of in a landfill at the Idaho National Engineering Laboratory are the subject of a major remediation project discussed in the next section.

The Idaho Engineering Laboratory and the Pit 9 Project

The Idaho National Engineering Laboratory (INEL) is a 2300-km^2 multi-purpose DOE laboratory in southeastern Idaho that is best known for the development of nuclear reactors. It is part of the weapons complex because uranium recovered from spent nuclear fuel from the Naval Nuclear Propulsion Program was used in the Savannah River production reactors and because wastes from other weapons sites were shipped to INEL for disposal or management. Early waste practices resulted in environmental contamination at 350 sites that have been consolidated for management purposes into 10 waste area groups (DOE, 1994a).

Most of the mixed transuranic wastes (i.e., containing both radionuclides and other waste such as hazardous chemicals) resulting from the manufacture of nuclear weapons components at the Rocky Flats plant were shipped to INEL, where up until 1971 they were buried in drums and boxes without intention of further treatment. The containers have since leaked and contaminated soil in the disposal pits. The "Pit 9" project is a demonstration project to retrieve and treat the waste and contaminated soil from one of the INEL burial sites (Schwartz and Strider, 1995). The project is being conducted by the DOE as an interim action under CERCLA and is regulated by the EPA and the state of Idaho under a Federal Facility Agreement. The goals are to determine capabilities to cost-effectively retrieve and treat buried mixed waste and to obtain data on container integrity, contaminant migration, soil moisture, and waste characterization that will reduce uncertainties in remedial actions for the remaining waste buried at INEL. Although the Pit 9 interim action is not a "risk-driven" cleanup, buried transuranic contaminants are of concern at INEL and other DOE locations and are the focus of the demonstration.

Pit 9 covers an area of about 120 by 40 m and contains approximately 10,000 m^3 of soil beneath and between the buried waste and about 7000 m^3 of overburden soil. The average depth of the pit from soil surface to bedrock is approximately 5 m. Approximately 3000 m^3 of mixed transuranic wastes from Rocky Flats and approximately 1100 m^3 of low-level and mixed wastes from INEL were buried in Pit 9 from November 1967 to June 1969. Pit 9 is estimated to contain over 110 m^3 of organics and approximately 30 kg of transuranic radionuclides.

Pit 9 was selected as a demonstration site because it was one of the last disposal pits at INEL to receive Rocky Flats waste, disposal records are better for Pit 9 than for earlier disposal activities, and its wastes are representative of the wastes disposed of at INEL. Pit 9 is also a relatively small area located away from other disposal pits and trenches.

A treatment level of 10 nCi g^{-1} (0.37 MBq kg^{-1}) transuranic content has been established as the only radioactive waste treatment standard for material to be returned to Pit 9. Modeling demonstrated that the concentration limit would be protective of groundwater and other pathways, provided pit closure requirements were followed. Standards for residual amounts of chemical wastes to be returned to the pit have also been established.

Wastes at Pit 9 are to be retrieved and segregated using remote excavation and sorting technology within a mobile containment building. Excavated soil will be treated to reduce chemical and radionuclide contamination levels. Materials that do not meet the standards for reburial are to be stored near the site until disposal decisions can be made. Plasma arc hearth furnace technology, discussed earlier in this chapter, is expected to be used to decompose organic chemicals and stabilize the radioactive residue. Although removal and treatment facilities are being developed and constructed at Pit 9, no waste material has been actually removed or treated as of this writing. The remote retrieval and treatment technologies employed on the project are potentially applicable to the many tons of radioactive, hazardous, and mixed wastes buried and stored at INEL and at sites throughout the DOE complex.

Project Chariot: Remediation Carried to Excess

In 1958 the AEC identified a site in the Cape Thompson region of northwestern Alaska to test nuclear explosives for experimental harbor excavation as part of the "Plowshare" project (Environmental Restoration Division, 1994). The site, 200 miles north of the Arctic Circle and 32 miles (50 km) from the nearest populated area, is inaccessible for 10 months of the year but used for subsistence hunting by indigenous peoples. Although the experimental nuclear excavation, dubbed Project Chariot, was never conducted, radioactive soil containing about 25 mCi (1 GBq) of mixed fission products from the Sedan Crater at the Nevada test site was used to conduct some simple tracer experiments in 1962. This test by the U.S. Geological Survey was to determine how radionuclides would behave in the Alaskan tundra environment. After the tests were conducted over a 5-day period, the contaminated material was recovered, buried on site, and covered with 1.2 m of native soil, creating a small mound. There was no measurable radiation from the mound.

Thirty years later a researcher at the University of Alaska–Fairbanks became aware of the tracer study and informed the U.S. Army Corps of Engineers and the media of the presence of radionuclides on the site. This information triggered a series of events that culminated in a decision by DOE to remediate the site, despite the fact that there was no risk from the remaining 3 mCi (100 MBq) of ^{137}Cs activity. Approximately 125 m^3 of soil containing the activity was excavated, packaged, and transported to the

Nevada test site for disposal at a cost of $7 million. It is interesting to note that the average concentration in the waste, about 25 pCi g^{-1} (1 Bq g^{-1}), is similar to that measured in some fireplace wood ash in New England (see Chapter 5).

SOME CONCLUDING REMARKS

At this writing the subject of site remediation is in a state of flux owing to its relative immaturity. In the past few years, important policy decisions have committed the United States to remediation programs that will involve huge expenditures well into the twenty-first century. It is important to emphasize that the basic policy decisions were made without the benefit of formal risk assessments and cost–benefit analyses. The public was exposed to a barrage of news reports about leaking tanks, quantities of atmospheric and liquid releases to the environment, and nonquantitative statements about contamination. Most reports from the government agencies that describe the remedial programs and that were prepared for consumption by the public have not been risk assessments but rather statements of U.S. government action.

Most of the approximately $250 billion projected to be spent for cleanup of the weapons complex is being justified by the assumed danger to public health. That is a huge amount of money that exceeds by far the expenditures made to solve other major public health problems. Elimination of smallpox from the entire world cost the World Health Organization about $100 million, a feat that saved about two million lives per year. Public health officials expect big returns on investments made in their behalf.

The decision to proceed with remediation at this scale has now been made and it is perhaps unrealistic to expect that any change in policy is possible. However, one cannot help but wish that the remediation projects, except for those that seem especially urgent, had been delayed for a decade or so to permit the completion of much needed risk assessments as well as the research needed to develop efficient and cost-effective technology. Instead, political decisions are being driven by the fears of the general public concerning the effects of traces of radioactive or chemical substances in the environment. This is a subject that will occupy most of the next, and final, chapter of this book.

CHAPTER 16

Radiation Exposure and Risks: Some Contemporary Social Aspects

At the start of the World War II nuclear bomb development program, there was understandable apprehension about the hazards for both the workers and the general public. It had been nearly half a century since the discovery of radioactivity, but most scientists still regarded it as a dangerous novelty. Only about 1400 g of radium had been extracted from the earth's crust since 1895, yet radiation had been responsible for more than 100 deaths, mostly from the uses and misuses of radium but also among physicians and physicists who had worked with X rays. Because it was anticipated that thousands of tons of radioactive material would be handled in the course of World War II and afterward, the nervousness over radioactivity was understandable. The program could not proceed unless safe methods of conducting the project could be devised.

Another half-century has now passed since the end of World War II and the nuclear energy industries in many countries of the world have developed to maturity. In the United States, nearly 200,000 employees work in the nuclear production and research facilities of the U.S. government (DOE, 1993b), and about one million people use radioactive materials in power production, medical practice, and research. The safety record of the industry has been exemplary and compares favorably with the safety record of industry in general. After such a bad start during the first few decades of the twentieth century, how did the nuclear industry compile such an excellent safety record? And if the record of safety has been exemplary (though by no means perfect), how can one account for the widespread public fear of even the most benign applications of nuclear energy in the United States and many other countries? In this chapter we will first explore the evidence that the industry has in fact experienced an excellent safety record. This conclusion may be difficult for the general reader to accept because the electronic and printed media have for years been publishing reports to the contrary and greatly exaggerating minor mishaps.

The tragic experiences during the first few decades following the discoveries of X rays and radium made it obvious that useful applications of nuclear radiations and radioactivity would not be possible unless effective safety procedures could be developed. This resulted in formation of the International Commission on Radiological Protection in 1928 and establishment of its U.S. counterpart, the National Council on Radiation Protection and Measurements, in 1929. By the time World War II began in 1941, three basic standards for internal and external radiation were established that, unknown to anyone at the time, would soon prove to be of critical importance to safe practices when the wartime atomic energy program began.[1]

Another important influence was that the organizers of the wartime program and their postwar successors understood the need for both research and training centers. Major government-owned laboratories were established during World War II and have earned worldwide reputations for their research on the ionizing radiations and their effects. Additional funds have been supplied to support scholarly research in university laboratories, and fellowship programs were established to provide opportunities for graduate education in the fields of health physics and environmental hygiene. Finally, a system of strict federal regulation over the use of radioactive materials was put in place (Stannard, 1988; Taylor, 1981). All of these factors have contributed to an excellent safety record. Moreover, these activities generated a better understanding about the effects of ionizing radiations on human health than for any other of the many physical and chemical agents of disease that are known to be present in the environment.

Ironically, the tragic experience of the early decades must be regarded as a fortunate historical accident. Had the injuries caused by radium and

---

[1]See Chapter 3. A recommendation that the body burden of ^{226}Ra be limited to 0.1 Ci was made by the NCRP in 1941. (At that time the NCRP was known as the Advisory Committee on X-rays and Radium). A standard was also recommended to limit radon exposure. The radium standard proved useful for establishing permissible body burdens for other radionuclides. The NCRP had also recommended a limit for exposure to external gamma and X radiation by the mid-1930s.

X rays escaped notice, or had they not been investigated by such diligent and perceptive scientists and physicians, their potential dangers might not have been understood in time to permit development of the standards and procedures used during and after World War II.

THE OCCUPATIONAL SAFETY RECORD OF THE U.S. NUCLEAR ENERGY PROGRAM

Although this book has been concerned until now with radioactivity in the off-site environment, discussion of the workplace is important in this concluding chapter because the record of occupational health and safety provides one of the basic tests of whether an industry has been conducted in a responsible manner. It is axiomatic that the workers in an industry are at greater risk than the general public. Of necessity, an effective health and safety program must begin in the workplace. An analysis of the occupational safety record should serve as a useful indicator of the prudence with which responsibilities for protection of health and safety more generally have been discharged.

The DOE and its predecessor agencies have regularly summarized the radiation exposure of their employees in reports that also include statistics on the agency's experience with industrial accidents of all kinds, both those that involve radiation exposure and those that do not. The accident frequency among DOE employees has been consistently less than one-half of the average for all U.S. industry. Among eight types of industries for which the U.S. Department of Labor maintains records, only one, "Finance, Insurance, and Real Estate," had a lower accident frequency during 1991 than the DOE and its contractors. This is all the more an outstanding accomplishment when one considers that the DOE program involves heavy construction, chemical processing, and other types of industrial activities that are potentially more hazardous than those for most industries (DOE, 1993a; National Safety Council, 1994).

Despite the relative excellence of the 52-year-old program, there have been more than 400 work-related fatal accidents among government nuclear energy workers. All but six of these deaths resulted from the traditional causes of industrial accidents such as falls or falling objects, motor vehicle and aircraft mishaps, electric shock, burns, and explosions. The six exceptions were deaths that resulted from accidental massive radiation exposure. Three were caused by the explosion of the SL-1 experimental reactor (which was described in Chapter 12), and three were the result of criticality accidents during research in a weapons laboratory early in the program. The only other death that resulted from a radiation accident in the United States, and the only one that occurred in a privately operated industrial plant, took place in 1964 (Karas and Stanbury, 1965).

It is important to note that the six fatal accidents in the government facilities were all associated with experimental programs and that, as of this writing, the last such accident occurred more than 35 years ago. Moreover, 32 years have elapsed since the only fatal radiation accident in the U.S. private sector.

In addition to the seven fatalities, radiation accidents in the U.S. governmental and privately operated nuclear energy programs have also caused 25 cases of clinically observable radiation injury. These included beta-radiation burns of the skin (7 cases), excision of tissue in which plutonium or americium had lodged mechanically (3 cases), and symptoms of whole-body or partial-body overexposure (14 cases). Four of the latter cases were from radiation sources such as X-ray spectrometers, which are used as research tools in many industries. (Ricks and Fry, 1990).

Of the nearly 200,000 DOE and contractor employees, only six received doses in excess of the permissible annual dose of 0.05 Sv (5 rem) during 1990. As a result of the method of recording radiation exposure, the annual dose must be considered in two parts: the external radiation component that is incurred entirely in the year in which the dose is received, and the internal dose from long-lived radionuclides deposited in the body during past years. The doses received by most of the six workers who received in excess of 0.05 Sv (5 rem) was the result of past internal emitter accumulations. No employee received a dose greater than 0.03 Sv (3 rem) from the work performed during 1990 (DOE, 1993b).

The non-DOE nuclear industry is regulated by the states and by the NRC. The NRC publishes annual summaries of the radiation exposures experienced by certain categories of licensees (NRC, 1995). In 1994, reports were received from 303 licensees, of which 109 were operators of nuclear power reactors. In the selected categories, a total of 152,000 employees were monitored for radiation exposure. Of the 141,000 workers monitored at the nuclear power plant sites, none received doses in excess of the regulatory limit of 0.05 Sv (5 rem).

Only in uranium mining is there clear-cut evidence of delayed radiation effects among nuclear energy workers in the United States (UNSCEAR, 1994). Ironically, uranium mining is the one part of the nuclear energy program that had reason to take meticulous precautions, based on experience in the European mines, as was discussed in Chapters 2 and 6. Had the uranium mine atmospheres been controlled to meet the standard established in 1941 to control the radon hazard in another industry, the tragic epidemic of lung cancer among the uranium miners in the southwestern United States would have been avoided. Regrettably, regulation of the mines was not preempted by the federal government, as was the rest of the nuclear energy program, but was left to the individual states, which lacked either the means or the will to deal with the problem in an effective manner.

In 1995, when the nuclear industry was more than 50 years old, the mortality records of the nuclear workforce were studied in a combined analysis of seven major facilities in three countries—the United States, the United Kingdom, and Canada (Cardis *et al.,* 1995). The cohort included more than 95,000 radiation workers who had accumulated more than 2 million person-years of exposure with 15,825 deaths, of which 3976 were due to cancer. The study found a slightly elevated relative risk of leukemia (1.22). Cases of chronic lymphocytic leukemia, which is a disease not associated with radiation exposure, were excluded from the analysis. A slight increase of deaths from multiple myeloma was also found. There was no evidence of excess risk for other cancers. Since the industry is more than 50 years old, cases should be evident by now if they were caused in significant numbers from the practices adopted early in the program.

About 25% of all deaths in the United States are attributable to cancer. The aging nuclear energy workers are entering the cancer-prone years of life at a time of unprecedented litigious activity as a result of which numerous lawsuits are being brought against the U.S. government and private industry to seek compensation for alleged radiation injuries. Many people who lived near nuclear facilities are also bringing suit against the government or its contractors. By 1985 there were about 4000 claims for radiation injury pending against the U.S. government (Jose, 1985). The total has swelled considerably during the past decade. The legal dilemma is posed by the fact that the linear no-threshold hypothesis makes it impossible to prove that even the smallest dose did not cause any given case of cancer. As discussed in Chapter 2, it had been hoped that the question of causality could be judged in probabilistic terms, with the use of tables that take into consideration the many factors that must be considered (NAS/NRC, 1984; Department of Health and Human Services, 1985; Bond, 1982). However, the probabilistic approach has not found ready acceptance in the courts.

As mentioned earlier, the NRC publishes annual summaries of the radiation exposure of reactor workers. In 1994 there were 109 civilian nuclear power plants in operation in the United States, among which 141,900 workers were monitored for radiation exposure. The annual average dose for the 73,000 workers who received a measurable dose was 2.9 mSv (290 mrem), which represented a collective dose of about 215 person-Sv (NRC, 1995). If one assumes a fatal cancer risk coefficient for workers of 4×10^{-2} cases per person-Sv (ICRP, 1991), it can be estimated that the collective dose from a single year of exposure may result in about 9 radiation-induced fatal cancers among the 73,000 reactor workers with measurable dose. If the average worker continues to be so exposed for 20 years, which is probably an overestimate for the average worker, the lifetime expectation of radiation-induced cancer among the entire population of workers will be $20 \times 9 = 180$ cases, compared to an expected 18,000 or so cancer deaths that would occur in the absence of radiation exposure at the reactors. Put another way, the probability that one of the cancers was radiation-induced would be $180/18,000 = 0.01$, which serves to illustrate the basis of the existing legal dilemma. At present, epidemiological methods are not sufficiently sensitive to permit resolution of 1% excess risk and if they did it would not be possible to identify the one case per hundred that is radiation induced from the 99 cases per hundred that would occur in the absence of radiation exposure.

It is clear from the above how important it is that the nuclear utility industry maintain complete and accessible medical records and radiation data on all employees so that an actuarial base can be developed that may eventually help to answer the question of whether the incidence of cancer among reactor workers is greater than expected.

SOURCES OF EXPOSURE OF THE GENERAL POPULATION

The various sources of environmental radioactivity to which the public is exposed have been discussed in the preceding chapters. These included natural radioactivity, the nuclear power industry (from mining to waste disposal), fallout from nuclear weapons testing, and a number of miscellaneous sources, including the use of radioactive pharmaceuticals in the practice of medicine. However, one important source was not included—the diagnostic uses of radiations. It was estimated in 1981 that more than 300,000 X-ray units were being used for medical and dental diagnoses in the United States (NCRP, 1989b). Nearly 8 million doses of radionuclides were administered for diagnostic purposes. In 1989 the NCRP reviewed the doses received from medical sources in the United States, and the subject was reviewed worldwide by UNSCEAR (1993).

The average dose received from radiation administered for diagnostic purposes in the United States has been increasing from decade to decade. The X-ray equipment has been greatly improved and faster film has been developed, but these advantages have been moderately offset by increased utilization. The last major estimate in the United States was projected for 1980 based on data from Shleien *et al.* (1978) and Rosenstein (1976). At that time it was estimated that the mean dose to bone marrow was between

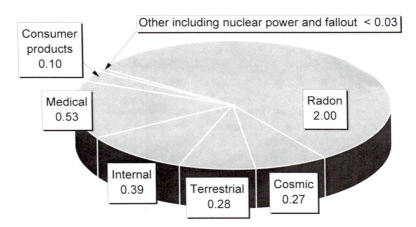

Average Effective Dose Equivalent in U.S.
(mSv per year)

Consumer products 0.10

Other including nuclear power and fallout < 0.03

Medical 0.53

Radon 2.00

Internal 0.39

Terrestrial 0.28

Cosmic 0.27

FIGURE 16-1   Estimates of average radiation dose to the U.S. population from all sources of exposure. The sources identified as internal, terrestrial, cosmic, and radon are natural. [Adapted from NCRP (1987b).]

0.75 mSv (75 mrem) and 1.15 mSv (115 mrem). This was a 17% increase since 1970 and a 38% increase since 1964 (NCRP, 1989b).

From all of the foregoing data, the sources of exposure of the general public can be approximated as shown in Fig. 16-1. The doses are taken from various sources and should be considered as rough approximations of how exposure of the general public is apportioned. Ninety-six percent of the total is contributed by two sources, natural radiation[2] and exposure from medical and dental uses. The estimates for natural radiation are probably more reliable than the estimate for medical and dental uses.

As discussed in Chapters 2 and 6, the dose received by the bronchial epithelium from inhalation of the naturally occurring decay products of radon is estimated to be about 20 mSv (2 rem) per year, very much higher than the effective dose equivalent shown in Fig. 16-1.

The impact of the nuclear energy industry on human health was examined in an EPA report (EPA, 1979), the purpose of which was to estimate the number of cancers caused by the emission of radionuclides into the atmosphere. The study included radium mining, milling, the radiopharmaceutical industry, test reactors, plutonium-fabrication facilities, and the ma-

[2]Natural radiation includes internal, terrestrial, cosmic and radon sources.

jor DOE research and production facilities. The number of cases of radiation-induced fatal cancers in the U.S. population from all of these facilities was estimated by EPA to be less than one cancer per year. The DOE routinely reports on the emissions to the atmosphere from its facilities. In 1992, which can be taken as a typical year, all sites continue to be within the limits set by the EPA (DOE, 1994f).

The civilian nuclear power plants were not included in the EPA report because the experience of that industry is covered elsewhere in periodic reports from the NRC. The NRC estimated that in 1988, when 95 commercial nuclear power plants were in operation in the United States, the collective dose received by 150 million people living within 80 km of the 71 different reactor sites was 0.75 person-Sv (75 person-rem) (Baker, 1992). The average dose commitments to individuals who lived near the power plants were all less than 0.2 μSv (0.02 mrem). This is somewhat less than the estimated radiation dose received on average from coal-fired power plants in the United States (NCRP, 1987d). In short, if the current models for estimating risk from exposure to airborne radioactivity are applied to the large complex of U.S. nuclear energy facilities, the total potential impact of the nuclear power plants on the incidence of cancer is less than that of the radioactive emissions from coal-fired power stations and very much less than the incremental doses received during high-altitude aircraft flights, and even lower than the differences due to the type of building construction or place of residence.

The risks from the routine releases of radioactive materials from the nuclear industry in general and the nuclear power plants in particular are trivial. However, the public is understandably concerned about catastrophic events that have a low probability of occurrence but can involve severe consequences. The Rasmussen Report (NRC, 1975) stated that the probability of reactor accidents with fatal consequences is orders of magnitude lower than that of other catastrophes caused by human activities, such as dam failures, explosions, or air crashes (see Chapter 8). Fatalities from reactor accidents are also far less probable than from most natural causes of disaster. The Three Mile Island accident, which was by far the most severe accident to a U.S. civilian nuclear power plant, involved no overexposure of either the employees or members of the public, but enormous economic consequences did result from destruction of the nuclear power plant. The Chernobyl accident has certainly increased the fears of the public about the dangers of catastrophic reactor accidents, but the designs of the Soviet graphite-moderated reactors did not provide basic safety features that have been incorporated routinely into the construction of civilian power plants in the United States and other Western countries (see Chapters 8 and 12).

THE DISPARITY BETWEEN ACTUAL AND PERCEIVED RISK

If the nuclear industry has had as safe a record as indicated by the facts, why is there such widespread public opposition to the development of nuclear energy? To some extent this certainly results from the association of radioactivity and nuclear weapons. Everyone knows that nuclear bombs can cause great destruction and produce lethal amounts of radioactivity, and this knowledge dominates the thinking of many members of the public when it is confronted with decisions involving even the most benign applications of nuclear energy. A more subtle contributing factor may be the enormous, and therefore confusing, range in the quantities of radioactive substances with which the public must deal. In this book, for example, the quantities of radioactivity discussed have been as small as attocuries ($\sim 4 \times 10^{-8}$ Bq) and as large as hundreds of megacuries ($\sim 4 \times 10^{18}$ Bq). This is a range of 26 orders of magnitude, a spread of values that is without precedent insofar as the public and most scientists are concerned. It is not a simple matter to place risk in perspective when the magnitude of risk must be related to such a wide range of values.

The fears of the public are greatly affected by knowledge that the ionizing radiations are capable of producing both cancer and genetic effects, but very few people are able to place the probabilities of occurrences into a frame of reference. Few people have a "feel" for probabilities. The fact that radioactivity produces cancer is easily comprehended, but for many people the perception of risk will not be influenced by an added statement that the additional risk is "only" one in a million. To many people, the qualitative fact that radiation can produce cancer is much more important than whether the probability of occurrence is 10^{-2} or 10^{-8}.

There is also insufficient awareness of the enormous amount of scientific knowledge on which the safety standards and operating guidelines are based. The U.S. government has spent several billion dollars on research concerned with the health effects of radiation (GAO, 1981). This scale of research effort far overshadows similar efforts in other areas of environmental protection. To the contrary, the attitude of the public is influenced by unfounded statements such as "nothing is known about the effects of low levels of radiation." Strictly speaking, we do know very little about the effects of doses below a few hundredths of a gray (a few rem) per year. Yet this is because the effects, if they occur at all, occur so infrequently that they cannot be measured, even with the best of clinical and epidemiological techniques. Estimates of the effects at such low doses are made by extrapolation of information obtained at higher doses and high rates. Most scientists believe that extrapolations made in this way tend to overestimate risks. Caveats in this regard are usually given in scientific reports, but they are regularly ignored by the media and are not familiar to the general public.

Why is there such a great gap between the actual and perceived risks? This can best be understood in relation to contemporary attitudes toward environmental risks more generally. Disparity between the actual and perceived risks is by no means limited to the nuclear industry. Public nervousness about relatively minor risks is one of the unfortunate results of the way the environmental movement developed.

EVOLUTION OF THE MODERN ENVIRONMENTAL MOVEMENT

The modern environmental movement started during the 1960s with a passion characteristic of that period. Those years were marked by vociferous opposition to the war in Vietnam, widespread and sometimes violent racial unrest, and the struggle of women for equal rights. The popular movements of the 1960s and early 1970s stirred intense emotions and the new environmentalism was no exception. Environmentalism has in fact evolved into one of the most important social developments of this century. Environmental protection has become an accepted responsibility of both government and the private sector. The need to analyze the environmental impact of technological innovations is now well accepted, and the legislative framework required for environmental protection has been established in the United States and other countries. In the United States, environmental protection initiatives have developed and attained enormous influence. The EPA (1990a) estimated that it costs taxpayers about $120 billion per year to implement the regulations that have been promulgated.

Most important of all, the public has developed an awareness of the fragility of our environment and the need to protect it. We all benefit from the efforts of the populist environmental movement.

The modern environmental movement[3], which has by now permeated many parts of our society, arose out of popular opposition to nuclear weapons production and testing in the mid-1950s. Disagreements about the health effects of radioactive fallout developed among scientists, and the subject even became a major issue in the 1956 presidential campaign. Scien-

---

[3]Here we refer to the "modern" environmental movement to differentiate it from previous environmental populism that took place beginning in the early nineteenth century and at the start of the twentieth century. However, those popular movements were mainly concerned with the deplorable living and working conditions that arose out of the industrial revolution (Eisenbud, 1978).

tists on both sides of the debate had access to the same data and their scientific conclusions were in essential agreement. What differed was the way their conclusions were presented. The two principal protagonists of that period were men of great scientific stature. Linus Pauling had received the Nobel Prize for chemistry and would later also receive the Nobel Peace Prize for his efforts to ban the testing of nuclear weapons. Willard Libby, then a member of the AEC, was also a distinguished chemist who was soon to receive the Nobel Prize for applying the radioactive properties of ^{14}C to the determination of the age of archaeological specimens. Using the best data available at that time, Libby calculated that the risk of developing cancer from exposure to the levels of radiation from worldwide fallout of the dust from nuclear and thermonuclear tests was on the order of one in a million. This is not a level of risk that would cause many people to be concerned. Pauling multiplied this probability of individual harm by the large numbers of people who were being exposed and predicted that a substantial number of cancers would develop. For the world population of 5 billion people, a risk of one in a million of developing leukemia would result in 5000 cases worldwide, surely something about which governments should be concerned.

In August 1963 a treaty was signed that banned the testing of nuclear weapons in the atmosphere. By coincidence, it was also the year in which the first order for a privately owned economically viable nuclear power plant was placed by a United States utility. The antinuclear movement, which was formed to protest the manufacturing and testing of weapons, soon shifted its focus to nuclear power.

Another event in 1963 was destined to be of even greater importance in the history of the environmental movement. It was the year in which "Silent Spring" was published, a book written by Rachel Carson that explained the subtle ways in which organic pesticides can move through the environment and endanger many forms of life. It was the Carson thesis, presented in eloquent prose, that there would someday be a "silent spring" because the pesticides were poisoning the food being eaten by the birds. The book had an explosive effect on public interest in the environment. The antinuclear activists whose interests until then had been limited to the dangers of radioactive fallout became aware that many of their concerns were applicable to chemicals as well. Like radioactive fallout, traces of man-made chemicals were found to be present around the globe. Like radionuclides, many of the chemicals were also possible human carcinogens that could pass insidiously along ecological pathways. Moreover, there were reasons to believe that, as in the case of radiation, there was no threshold for the effects of carcinogenic chemicals.

Global contamination by the radioactive fallout from weapons testing could be studied readily in the decade before publication of "Silent Spring" because instruments were already available that permitted detection of exceedingly small amounts of radioactivity. Until about 1960 this was not possible for toxic chemicals, because the methods of detection either did not exist or were overly tedious or expensive. Development of the atomic absorption spectrophotometer suddenly simplified the analyses of trace metals such as lead, arsenic, and mercury. At about the same time, improvements in the techniques of gas chromatography and mass spectroscopy increased the ease of detection of traces of complicated organic molecules, such as DDT and the family of polychlorinated biphenyls (PCBs). With these new methods of analysis, it was found that not only were many toxic substances distributed more widely in the environment than had formerly been realized but that seemingly insignificant amounts of DDT in pond water were found to increase in concentration, step by step, from phytoplankton, to zooplankton, crustaceans, small fish, larger fish, and so on to human beings and the other mammals and birds at the top of the food chain. The most alarming finding was that the DDT was concentrating in the tissues of birds to such an extent that their ability to reproduce was being impaired.

"Silent Spring" was unquestionably the single most influential factor in the sudden development of environmental populism in the 1960s. The public became aroused, environmental societies were formed in hundreds of communities, and some of the larger organizations developed powerful voices in the media as well as in legislative and judicial arenas. The widespread public concern resulted in political action, and in the decade that followed the legislative basis for much needed programs of environmental protection was established at the state and federal levels. The requirements for environmental impact analysis were enunciated in the National Environmental Policy Act (NEPA) and the EPA was established. The Clean Air and Clean Water Acts were passed, and more funds became available for scientists who were doing environmental protection research. For the first time the federal government assumed major financial responsibility for environmental protection, and large sums of money were made available for air and water pollution control. In many countries, environmental awareness grew rapidly, as did appreciation that the human species is part of a global ecological system. The realization developed that people are only temporary stewards of the world in which they live and have the responsibility to protect it for the generations to follow. The environmental awareness that has developed, and the ways in which the government and industry have restructured themselves to address environmental problems, has constituted a major social revolution. This was a long overdue and desirable development.

THE NEW ENVIRONMENTALISTS

By the end of the 1960s, the words "ecology" and "environment" had entered the popular vocabulary, and much attention was given to these subjects in the electronic and printed media. The movement was driven by a rapidly growing group of activists who became known as "environmentalists." Many of them had considerable charisma. Some had scientific training, but most did not. Some of them were elected or appointed officials who began to see the new environmentalism as a way to advance their political agendas.

Many environmentalists tended to be apocalyptic in their writings. Wave after wave of excitement about environmental matters spread through the country. In quick succession, the public soon learned about mercury in fish, carbon monoxide, a long list of carcinogens in foods, and low-level radiation. At the beginning of the environmental movement, newspapers and magazines regularly included special sections on the subject.

By the early 1970s, there was a growing feeling of discomfort among some of the scientists, engineers, and physicians who had been at the forefront of environmental protection for many years. They believed that environmental priorities were becoming misaligned. These professionals had traditionally worked without fanfare in a field in which there was little popular interest, and funds had been scarce. Suddenly the situation changed and the subject of the environment attracted the interests of the general public, all branches of government at all levels, and the printed and electronic media. Organizations with large public memberships suddenly found that they had achieved considerable political influence. Some of these societies, which had been respected for many decades for their efforts to educate the public about the need to protect the natural environment, became influential as interveners who blocked power plant construction, pipelines, highways, and dams. The National Audubon Society and the Sierra Club are two such examples. This was a positive development in some respects, but resulted in reduced recognition of the need for a careful analytical approach to complicated issues of public policy.

Gradually, the subject of "environment" bacame ill-defined and reports that appeared with regularity in the media were causing the public to become confused and increasingly alarmed. There seemed to be a new crisis each week. These often dealt with problems of which professional environmental health specialists were well aware, but the statements made by the media were too often exaggerated. Scientists who had little or no training or experience with environmental matters were proclaiming that

Lake Erie was dead, and that we should generate electricity from the sun and wind. Most such statements contained an element of truth: Lake Erie was badly polluted and its fauna decimated (but was far from "dead"), and it was an admirable objective to obtain electric energy from renewable sources (but this was not then or yet practical on a large scale). Many of the new environmental spokesmen wanted simple and quick answers to complicated questions, and this was not possible.

Much, though not all, of the public anxiety about environmental pollution resulted from reports of the presence in food, air, and water of trace amounts of chemicals that were known to have caused cancer when given in high doses to laboratory rodents. There is obvious reason for the public to be fearful of cancer. It is a dreaded disease that is the cause of more than one out of five deaths in the United States. But the role of pollution as a cause of cancer has been misunderstood. This seems to have originated from a 1964 report from the World Health Organization (WHO, 1964) in which it was stated that 60 to 90% of all cancers might be the result of some environmental factor. That report was issued only one year after publication of "Silent Spring," which warned about the dangers of polluting the environment with the toxic and carcinogenic chemicals being used in industry and agriculture.

The WHO report was widely misinterpreted to mean that 60 to 90% of all cancers were the result of industrial pollution of air, water, and food. This was not intended by the writers of the report, which referred to the environment in a far more general sense. Cigarette smoking was identified as the most important environmental factor, but personal hygiene and dietary customs were also stated to be important. Environmental pollution was properly included as a possible factor, but there was no intent to imply that it was the major cause of cancer. Nevertheless, for many years after its publication, the WHO report was widely cited in support of the thesis that environmental contamination was causing the prevalence of cancer to increase in the general population.

It is true that the number of deaths from cancer has been increasing from year to year, but a major reason for this has been the larger size and average age of the population. Many people who in former years would have died at a young age from tuberculosis or other communicable diseases are now living to old age. This increases the probability that they will eventually die of cancer, the incidence of which increases rapidly with age. It is necessary to correct the statistics for the effect of aging, which is done annually by the American Cancer Society, using statistics gathered by the National Center for Health Statistics. Their reports show that during the past 50 years the only major increase in the age-adjusted rates of cancer mortality has been from lung cancer in both men and women (Fig. 16-2). The practice

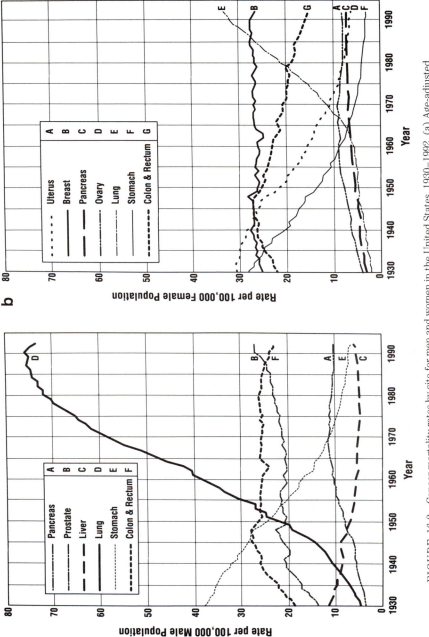

FIGURE 16-2 Cancer mortality rates by site for men and women in the United States, 1930–1992. (a) Age-adjusted cancer death rates for males per 100,000, adjusted to the 1970 U.S. standard population. (b) Age-adjusted cancer death rates for females per 100,000, adjusted to the 1970 U.S. standard population. Uterine rates are for cervix and corpus combined. [From American Cancer Society (1996).]

of smoking cigarettes increased in popularity among men during the World War I era, but because of the delay in the development of this disease, lung cancer did not begin to increase among males until the mid-1930s. By 1993, lung cancer accounted for 34% of the 263,000 deaths from all cancers among men in the United States (American Cancer Society, 1996).

The popularity of smoking among women lagged that of men by about the period between the two world wars. Lung cancer began to increase among women in the post-World War II period, and since 1985 the death rate from this self-induced disease has exceeded that from breast cancer, the disease that is most feared by women.

Except for lung cancer, there have been no dramatic increases in cancer mortality among the major cancer sites. To the contrary, there has been a notable reduction in the death rates from cancers of the uterus among women and stomach in both men and women. One must be cautious when interpreting these trends. The cancers shown are the major ones and are responsible for 72% of the reported deaths from cancer. There are many other sites that individually cause relatively few deaths compared to the data presented, but altogether they account for a considerable fraction of the total, about 28%. One major cancer that is not included is melanoma, a cancer of the skin that is estimated to have resulted in 6700 deaths in 1992. This cancer, most cases of which are believed to be caused by exposure to sunlight (Elder, 1995), has become more prevalent, possibly because of the increase in outdoor recreation.

There is widespread concern among the public that it is at increased risk of developing cancer because food, air, and water contain traces of artificially produced carcinogens. For the most part, this is unjustified by the facts. It has been shown that many of our foods contain naturally produced pesticides used by plants for self-protection, and that about half of these natural pesticides have been shown to be carcinogenic in rats. The food we consume contains naturally occurring pesticides in far greater amounts than man-made agricultural chemicals. It has been estimated that each day the average person consumes about 1500 mg of natural pesticides compared to less than 0.1 mg of synthetic pesticide residues (Ames and Gold, 1990). Why are people so concerned about the trace residues of agricultural chemicals in our food when similar chemicals are present in far greater amounts from naturally produced sources? An effect of this excessive concern is that regulations are promulgated that require the expenditure of huge amounts of money to reduce risks that are already infinitesimally small.

In the field of public health, tangible benefits can be achieved by relatively modest investments. The disproportionate amount of concern for chemical and radiation effects is seen in the unusually large expenditures to reduce the levels of risk (Tengs *et al.*, 1995; Cohen, 1980). The average cost to society of extending one life by one year has been estimated to be about $69 for the mandatory requirement for seat belts and $30,602 by reducing the density of utility poles from 40 per mile to 20 per mile. One life-year can be saved at a cost of $20,200 by screening for hypertension. In contrast, life-saving intervention by benzene control is estimated to cost more than $14 million per life-year and more than $33 billion per life-year saved by radionuclide emission control from uranium fuel-cycle facilities.

One can also make estimates of the cost of averting one premature death. If smoke detectors were made mandatory in all rooms used for sleeping, the cost would be $40,000 per life saved (Graham and Vaupel, 1981). In contrast, the NRC policy of justifying expenditures of $1000 per person-rem (0.01 person-Sv) avoided is equivalent to a cost of $2.5 million per hypothetical cancer prevented. One proposal by EPA to reduce the atmospheric emissions of ^{210}Po from five producers of elemental phosphorus would have cost about $200 million to avert one cancer (EPA, 1984b).

Although many people find it repugnant to place a dollar value on life, the fact remains that in the administration of public health programs it is customary and necessary to consider the costs of proposed programs in relation to the benefits that will result from them. In a well-operated community that is the way decisions are made concerning installation of traffic lights, addition of fire-fighting equipment, or the deployment of police and paramedical personnel.

Modern methods of environmental modeling make it possible to predict levels of exposure far into the future, and at levels of exposure that are beyond the measuring capabilities of even our most sophisticated instruments. In an environmental impact statement prepared by the DOE in connection with a proposed waste storage site, it was stated that a person living near the proposed site "would receive a 1.1×10^{-8} rem dose to bone over a 50-year period," and the expected health effects were accordingly estimated (DOE, 1979a). There is no longer any lower limit to the health effects that can be estimated by using risk coefficients based on the linear hypothesis in conjunction with computer programs that are capable of predicting (albeit with considerable uncertainty) the time and place of decay of the very last radioactive atoms discharged into a waste stream.

There is a legal principle that is relevant to this discussion—"de minimis non curat lex," or "the law does not concern itself with trifles." This maxim should be incorporated into regulatory procedures, not only for radioactive substances but for toxic materials as well. It would be much more logical if the regulatory agencies would define a level of de minimis dose—the lower limit of regulatory concern (Eisenbud, 1981b). A de minimis dose of 0.1 mSv (10 mrem) per year could be easily defended on the grounds that it is well within the limits of the variations in effective dose received from natural sources. Of course one could point out that a does of 0.1 mSv (10 mrem) per year received by 260 million people would result in a collective dose of 2.6×10^4 person-Sv (2.6×10^6 person-rem), which, with a risk coefficient for the public of about 0.05 cancers per Sv, would potentially result in about 1300 premature deaths per year. This would not be an inconsequential effect.

A fallacy in this argument, however, is that it is unlikely that the sum of the doses from the nuclear power industry and other sources of radioactivity (other than natural or medical sources) could produce an average dose of 10 mrem to 260 million people. Emissions from reactors, incinerators, or waste storage sites would result in measurable exposures to only a small fraction of the population. One exception would be with the military use of nuclear weapons, as was the case during the period of atmospheric testing. Another exception might be accumulation of long-lived nuclides such as ^{14}C, ^{129}I, and ^{85}Kr in the global atmosphere. However, these emissions would originate from a relatively few fuel-reprocessing plants for which controls could be installed at very low cost. Of course at the root of the matter is the question of whether the radiation effects can be extrapolated to such low doses. Reviews of low-dose radiation effects do not exclude the possibility that there may be no effects at doses on the order of natural background (NAS/NRC, 1990).

The vast majority of sources of radiation exposures could be excluded from regulatory concern if 0.1 Sv (10 mrem) per year were considered the de minimis dose. This value is well within the variations in natural radiation exposure that we encounter in normal living. In the early 1990s the NRC attempted to adopt a policy that would have defined 1 mrem (0.01 mSv) per year as the lower limit of regulatory concern, but it was soon withdrawn in the face of congressional and public opposition.

The United States has major public health problems, exemplified by the facts that neighborhood health centers have been closed in many communities for lack of funds, that many parents lack either the means or the motivation to have children inoculated against common communicable diseases, and that it is difficult for the poor to receive the family planning counseling they need. These are matters of environmental health. Society's failure to provide for these basic needs is reflected in the failure to eradicate measles when it was within our grasp to do so, and in the resurgence of tuberculosis and certain venereal diseases after they were nearly eliminated in this country.

Environmental organizations frequently dominate public hearings on matters such as proposed construction of incinerators, landfills, or changes in public laws that involve perceived relaxation of rules that govern the disposal of radioactive waste. In the typical case, opposition develops despite assurances from competent authorities that there will be no significant health effects. It is a great pity that similar activism has not been mobilized in support of programs to control more traditional problems in environmental health. Environmental activists should understand how much health officials would welcome their support at city council meetings where decisions often result in reduced budgetary provisions for matters such as rat control, housing inspection, street cleaning, and the hundreds of other traditional environmental problems with which the officials must contend. Rats and dirty streets, as well as drug peddlers and muggers, are also part of our environment and have far greater impact on human well-being than environmental matters of more popular current interest that often consume funds at a rate all out of proportion to their health hazards.

Why is there such anxiety about trace contaminants in the environment? These anxieties are reflected in the laws that are passed, many of which can be enforced only at great expense. Part of the answer is that some people are unwilling to agree that anyone should be required to accept any involuntary risk, however small, or however large the benefits. These are people with mind-sets of the kind that have blocked water fluoridation in many parts of the United States despite the great reduction in tooth decay that has been demonstrated for nearly 50 years in communities with fluoridated water, with no harmful effects. They are citizens who do not understand that trade-offs must be made between benefits and risks: they want only the one, not the other. That is never possible. Every technical innovation involves some level of risk, and it is up to society, using the methods of environmental impact analysis, to be certain that the benefits do outweigh the risks.

Public perception about radioactivity is also influenced by the widely publicized long lives of some of the waste products of the fission process. This is particularly true of the transuranic actinide elements such as plutonium. Yet this is a specious argument that overlooks the fact that the stable toxic elements such as lead, arsenic, and mercury do not decay spontaneously and could be said to have infinite half-lives. They are permanently present in the environment. Although it is true that gram for gram the long-lived nuclides are more hazardous than the stable elements, the latter exist in far greater quantities than the radionuclides. The hazardous characteristics of radioactive waste are not so much related to the fact that it has a long half-life, but rather to the manner in which it is being stored and the geochemical and hydrological mechanisms that may transport the material into the biosphere.

THE SPECIAL CASE OF PLUTONIUM

The case of plutonium illustrates how disparate the real and perceived risks can become. The public is accustomed to reading or hearing that plutonium causes cancer and "is the most dangerous known substance." Intense concerns are expressed by communities that learn that shipments containing plutonium will pass through their neighborhoods. We saw in Chapter 12 that the presence of traces of plutonium in soil near the Rocky Flats plant served as the basis for a major class-action suit against the company that operated the plant for DOE. However, after 50 years of experience with this artificially produced element, it can be said that, its reputation notwithstanding, the methods for its control have been effective in assuring safety to both those who work with it and the public potentially exposed to it as an environmental contaminant (Eisenbud, 1994b; Wheelright, 1995; Sutcliffe *et al.*, 1995).

Fears about the carcinogenic effects of plutonium developed on the atomic bomb project during World War II as a consequence of the tragic experience with radium during the preceding 40 years. When the wartime planners realized that ^{239}Pu was also long-lived, that it was also a bone-seeker and alpha emitter, and that it would be produced in metric ton quantities rather than in grams, understandable concerns emerged. What was not realized at the time was that there were major differences in the physical and chemical properties of the two elements that greatly diminished the hazard of Pu. Absorption of ingested radium is orders of magnitude greater than that of plutonium. Plutonium becomes immobile in soil because of the tenacity with which it is bound by soil minerals. There are also major differences in solubilities, with Pu being less soluble than silica in water. For all of these reasons and others, the methods used to provide for the safety of those exposed to plutonium have proved to be effective. Over about 50 years, the only known injury caused by Pu in the United States was a single case of bone cancer in a worker who was heavily exposed to Pu early in the program. Another persistent fear was that inhalation of plutonium particles would be especially carcinogenic because the alpha emissions would deliver high doses in the immediate vicinity of the inhaled particulates. As discussed in Chapter 2, this was later shown not to be the case.

PUBLIC PERCEPTION OF TECHNOLOGY

Though it is true that many new risks have resulted from technological developments, it is also true that modern technology has given us longer lives than in former times, with far fewer days of sickness during our productive years. Technology has also increased human productivity to such an extent that workers now produce greater wealth in fewer hours than in the past. The five-day, forty-hour week, annual vacations, travel, high-quality audio and video home entertainment, and many other advantages of modern living are the result of technological innovations that have taken place in this century. Yet many of these innovations have also brought new risks. Among the most obvious are the risks from production, distribution, and use of electrical power. From the mining of fuels to the use of electricity in the home, office, or factory, we have learned to accept these risks because of the associated benefits.

In the first half of this century, the acceptability of a risk being imposed on the public would be determined by the government in consultation with experts. The appropriate federal and state agencies, the National Academy of Sciences/National Research Council, the NCRP, committees of the professional societies, and individual experts were allowed to decide the course of action that best served the public interest. This is no longer the case. Citizens are now distrustful of authority and will no longer delegate to others the power to make decisions that affect their lives. People want to make their own decisions. Particularly in the environmental area, participatory democracy has replaced representative government and the word "expert" has developed a pejorative connotation.

The media also share the general public's distrust of scientific authority. This is less true of professional science writers because most of them are scientifically trained, and they know how to evaluate their sources of information. Unfortunately, many science writers are now less involved in reporting on environmental news than was formerly the case. They are devoting their attention to developments in other fields, such as molecular genetics, cosmology, materials science, and medicine. Some may avoid writing on environmental issues because they realize how politicized the subject has become. Perhaps they have trouble getting balanced reporting past their editors. For whatever reasons, one rarely sees the bylines of well-known science writers on environmental health subjects when they are discussed in the media.

Koren and Klein (1991) showed that media bias exists in reporting on radiation effects studies that are negative (show no effect), and that such reports are given less attention than those in which positive effects are shown. This source of bias in public perception is not limited to reporting on environmental matters, and is not easy to counter. Furthermore, sensational reporting is hard to refute. Letters sent to the newspapers by knowledgeable readers who want to correct even the most glaring errors are not often published. The mechanisms for correcting errors in television or radio reporting are almost nonexistent.

Sometimes it is said that the public is poorly informed about environmental matters because scientists do not communicate well with the public. The problem is not that scientists do not communicate well, but rather that they hardly communicate at all. They are rarely given an opportunity to communicate with the public on terms they think are fair. Scientists often decline to be interviewed by the media because they have had the experience of being quoted out of context or are rebutted by others who are more sensational, but who have less regard for the facts. All too often reporters with no background in science are given unreasonable deadlines by their editors, which may result in reporting that is less accurate and less thorough than desired.

The average person has no personal experience on which to form a judgment on risks associated with advanced technology, and thus his or her perspective must be influenced by external sources of information. Few people are familiar with the methods by which primary sources of information can be studied in libraries. Moreover, most people do not have either the time or the technical training to undertake meaningful literature research. Hence the printed and electronic media are the primary sources of information. Secondary sources, which undoubtedly exert some influence on the media, are the publications of public interest groups, many of which have strong antinuclear orientations.

A series of illuminating surveys of the attitudes of various leadership groups in the United States has been published by Rothman and Lichter (1985). In response to the question "Are nuclear plants safe?", only 6.4% of the leadership of public interest organizations answered in the affirmative. This compares with 60.2% for a random sample of 929 scientists selected from "American Men and Women of Science." Among a sample of 279 scientists selected from energy-related fields, 76% answered "yes," as did 98.7% of the nuclear energy experts. In contrast, only 14% of the directors and producers of motion pictures answered affirmatively, as did only 13% of the producers and directors of prime time television programs. The surveys further showed that 50% of journalists underestimated the extent of scientific support for nuclear energy, as did 85% of the directors, writers, and producers of television programs. This may explain why Ralph Nader, a nonscientist with well-known antinuclear views, was reported to be the most frequently interviewed "expert" on the subject of nuclear energy on prime time TV during the 1970s (Media Institute, 1979).

We must conclude that the overly cautious attitude of the public regarding nuclear matters is part of a much larger phenomenon that impacts not only on attitudes toward environmental radioactivity but toward other trace pollutants as well. How long it will last can only be a subject for conjecture.

At this writing, it has been with us for about 30 years and shows no signs of subsiding.

THE POSITIVE SIDE OF POPULAR ENVIRONMENTALISM

We have discussed some of the negative aspects of the modern environmental movement, among them the following. The media have been guilty of exaggerations and distortions, priorities have become badly misaligned, the subject of environmental protection has become deprofessionalized to a disturbing extent, and the rigidity of the positions taken by some environmental activists has many of the characteristics of religious dogma. Yet, despite these shortcomings, considerable good has come from environmentalism.

Toleration of a measure of extremism is the price we must pay for the benefits of our pluralistic form of society. In the curious, mysterious, and exasperating ways in which democracy works, many of the good things that have occurred would not have been possible without the populist shouting that often drowns out the more moderate tones of others.

Some may wish that the arguments of the populists could be expressed less stridently and that the arguments presented would be based more on hard facts and less on emotion, but in the long run the environmental extremists will be shown to have served society well. In a democracy it is sometimes necessary to achieve goals by language that is so strident as to be irritating to anyone trained to reason with disciplined caution. However, the caution inherent in the language of scientists involved in matters of public policy is no doubt even more irritating to the less patient activists and even to a concerned public.

However, by now the environmental populists have made their point. Their message has been delivered and has been understood clearly. What we now need is a period of collective introspection to permit us to reorder our environmental health priorities, while at the same time preserving the important public support that has developed for environmental protection during the past three decades. That our environmental priorities must be reevaluated is beyond doubt. We must ask whether the vast sums of money and effort being spent for successive reductions in emissions of trace radioactive substances from the nuclear industry, the many expensive and redundant procedures that are being designed into nuclear facilities, and many of the requirements mandated in the name of environmental protection in other areas of the economy could be better spent to solve more traditional environmental health problems that at the present time are receiving insufficient attention.

Above all, we must promote informed discussion of the pros and cons of proposals that involve environmental protection. If a person does not believe a particular proposal is worthy of financial support, he or she should have the right to argue against it without being branded as "antienvironmental"—which has become a modern form of heresy. We must recognize that many environmental decisions involve considerable economic and scientific complexity. Politicians should not be accused of being "antienvironmental" because they do not agree with the popular mood on any given issue.

Yet we must not return to the unrestricted industrial practices of the 1950s and 1960s, which were responsible for so much environmental contamination, but rather must ask whether we have passed the point of diminishing returns in some of our programs of pollution control, while at the same time neglecting more basic problems of environmental health protection. Environmental populism has reversed the swing of a pendulum that was moving in the wrong direction. History will record that the reversal was a major social and health development, for which we should all be grateful. However, we have gone too far in some directions and not far enough in others. Our society must take steps to assure that billions of dollars are not wasted on trivial environmental health risks.

WHAT ABOUT THE FUTURE OF NUCLEAR ENERGY?

The demand for electricity was growing at an annual rate of about 6% for several decades, but this was reduced markedly during the mid-1970s as a result of unforeseen increases that took place in the price of oil, the worldwide economic depression that followed, and subsequent, successful conservation efforts. At this writing, recovery from the global recession is occurring, but the demand for electricity is not likely to return to the former 6% annual growth. Many economists now predict that future growth will be about 2–2.5% per year. Though this is much less than earlier growth, even a 2% annual growth will require that the nation's installed generating capacity be doubled every 35 years. New power plants will be required in large numbers in the years ahead.

Solar and other renewable sources of energy are not likely to have a significant impact until at least the middle years of the twenty-first century, if even then. The same can be said about energy from nuclear fusion. Thus, for the foreseeable future, the nation will be required to depend on coal, oil, and nuclear power, but the cost of oil and our dependence on foreign sources make it a highly questionable choice of fuel for electrical generation.

Coal is a logical candidate to meet future needs, but there are obvious occupational safety and environmental problems associated with its use, of which the most intractable is likely to be atmospheric effects of carbon dioxide produced by the burning of coal (NAS/NRC, 1991). The likelihood that the increase in atmospheric carbon dioxide will eventually result in a global warming trend seems more plausible with each passing decade, and the social and economic consequences of this "greenhouse effect" are bound to receive increasing attention from energy policy makers. It could be a major reason for a resumption in the demand for nuclear power in the decades ahead. The requirement that automobile tailpipe emissions be reduced may also create a demand for electric-powered automobiles, as well as a greater dependence on electricity for heating homes and other buildings. These new markets for electric energy should increase the demand for nuclear power.

Whether nuclear power will achieve public acceptance in the years ahead will depend to a considerable extent on the reception given to the advanced reactor designs now being introduced (see Chapter 8). Society must also achieve consensus on the controversial question of how to dispose of high-level radioactive wastes.

Appendix

The Properties of Certain Radionuclides

This appendix summarizes some of the important properties of the nuclides that are of principal environmental interest. The biological and physical constants are subject to revision from time to time, and a reader requiring the most recent data should consult the primary sources of such information.

Explanations of the subheadings used and the principal citations are as follows:

Isotope: Name ($^A X$), where A is the atomic number and X is the element abbreviation.
Physical Half-life: Obtained from the NUDAT Data Base.[1]
Biological Half-life: ICRP Publication 30, Parts 1–4 (1979–1988), except for the alkaline earth elements, which are obtained from ICRP Publication 20 (1973a).
Natural Levels: See Chapter 6.
Specific Activity (elemental): For naturally occurring radionuclides, the elemental specific activity is given in units of activity per unit mass of the natural element.

[1]The NUclear DATa (NUDAT) data base has been derived from the radioactive decay data sets in the Evaluated Nuclear Structure Data File (ENSDF) data base. ENSDF is a computer file of evaluated experimental nuclear structure data maintained by the National Nuclear Data Center at Brookhaven National Laboratory. The data are published on a cyclic basis in which each nuclide is evaluated every 6 years. The data used are current as of December 1995. These data are also published periodically in: (1) Nuclear Data Sheets, (Academic Press, San Diego), for values of atomic weights greater than 45; (2) Nuclear Physics, (North Holland Publishing Co., Amsterdam), for atomic weights less than 45; and (3) Table of Isotopes, (John Wiley & Sons, Inc. New York), for all available isotopes.

Sources of Natural Radionuclides: See Chapter 6.

Principal Modes and Energies of Decay: Modes of decay and significant radiations with percent per transformation. The average and maximum energies are listed for β emitters; when a radionuclide emits betas from more than one transition, subscripts are used to distinguish them (From ENSDF Data Base. See *Physical Half-life* for reference.)

Principal Human Metabolic and Dosimetric Parameters, Daily Intake in Food and Fluids, and Body Burden: Element ingestion and inventory as defined for adult radiation workers (ICRP 1979–1988; ICRP, 1975).

Values for fraction transferred from gut to blood (f_1), annual limit of intake (ALI), and derived air concentration (DAC) are taken from Federal Guidance Report No. 11 (Eckerman *et al.*, 1988). These ALIs and DACs are based on not exceeding a 50-year committed effective dose equivalent of 50 mSv (5 rem) for stochastic effects and 500 mSv (50 rem) for nonstochastic effects from inhalation or ingestion of the radionuclide. The DAC is that concentration of radionuclide aerosol (1 μm AMAD) in air that if breathed by reference man for a working year of 2000 hours would result in one ALI by inhalation. All ALIs listed are for stochastic effects unless otherwise noted. The new ICRP recommendations of ALI corresponding to 20 mSv (2 rem) stochastic are available but are not included here (ICRP Publication 60, 1991). All dosimetric data presented here are for an occupationally exposed working adult male, based on "Reference Man" (ICRP Publication 23, 1975). ICRP age-dependent values are available for some radionuclides (ICRP, 1990, 1994a, 1995).

Inhalation Class (IC) describes the clearance rate of radioactive materials from the lungs. Classifications are: D (pulmonary half-times <10 days), W (pulmonary half-times 10–100 days), and Y (pulmonary half-times >100 days).

Additional Dosimetric Considerations: Exposure-to-dose conversion factors are also from Federal Guidance Report No. 11 (Eckerman *et al.*, 1988). Unless otherwise noted, the limiting coefficients are for the effective dose equivalent.

Whole-Body Retention Function: Retention of the stable form of the element in whole body (ICRP Publication 30, 1980; ICRP Publication 54, 1989).

Systemic Transfer Fractions: Fractional transfer coefficient from the transfer compartment to the specific organs and whole-body tissues (ICRP Publication 30, 1980).

Isotope: Tritium (^{3}H)

Half-life:

Physical: 12.3 years.

Biological: ~10 days (range 4–18 days).

Natural Levels: 190–930 Bq m^{-3} (5–25 pCi L^{-1}) of surface water, prior to the advent of nuclear weapons testing.

Sources: Cosmic-ray interactions with N and O; ternary fission, spallation from cosmic rays, ^{6}Li(n,α)^{3}H. World inventory of naturally produced tritium is 960 PBq (26 MCi). Production rates: 240 PBq per megaton (6.6 MCi per megaton) yield of thermonuclear weapon; in light-water reactors, produced primarily by ternary fission, 560–740 PBq MWe y (15–20 Ci MWe y), and secondarily by neutron interaction by light elements, 190–380 GBq MWe y for PWR and 11–33 GBq MWe y for BWR (5–10 Ci MWe y and 0.3–0.9 Ci MWe y, respectively).

Principal Modes and Energies of Decay (MeV): β^- ave., 0.0057; max., 0.0186 (100%).

Special Chemical and Biological Characteristics: Metabolized as H_2O. HT gas is inhaled and exhaled with only 0.005% of the activity being deposited in the lungs. The uptake of HTO vapor is near 100% for inhalation and ingestion. The DAC is adjusted to include skin absorption for water vapor.

Principal Organ: Whole body.

Amount of Element in Body: 7 kg.

Water Content of Body: 42 kg.

Daily Intake of Hydrogen: 350 g.

Principal Human Metabolic and Dosimetric Parameters:

f_1 = 1 (all forms).

ALI = 3000 MBq (8×10^4 μCi), inhalation and ingestion of water vapor.

DAC = 0.8 MBq m^{-3} (2×10^{-5} μCi cm^{-3}) water vapor,

= 2×10^4 MBq m^{-3} (0.5 μCi cm^{-3}), elemental.

Additional Dosimetric Considerations: Ingestion and inhalation, 1.73×10^{-11} Sv Bq^{-1} (6.4×10^{-2} mrem μCi^{-1}) for water vapor. The dose to humans peaked in the mid-1960s at about 2 μSv y^{-1} (0.2 mrem y^{-1}) with 10 nSv y^{-1} (0.001 mrem y^{-1}) from natural sources of tritium (NCRP 62, 1979).

Whole-Body Retention Function: $R(t) = e^{-0.693t/10}$, with t as elapsed time in days.

Isotope: Carbon-14 (^{14}C)

Half-life:

Physical: 5730 years.

Biological: 40 days.

Specific Activity in Living Matter and Atmospheric CO_2 Prior to Era of Nuclear Testing: 230 Bq kg^{-1} (6.1 pCi g^{-1}). Lower values in urban areas owing to fossil fuel combustion.

Sources: Cosmic-ray neutron activation, ^{14}N(n,p)^{14}C; nuclear weapons testing. Production rates at earth's surface: natural, 1.4 PBq y^{-1} (38 kCi y^{-1}); from thermonuclear weapons testing, 1.3 PBq per megaton (34 kCi per megaton).

Principal Modes and Energies of Decay (MeV): β^- ave., 0.0495; max., 0.156 (100%).

Special Chemical and Biological Characteristics: Metabolism varies with chemical compound of which it is incorporated.

Principal Organ: Whole body.

Amount of Element in Body: 16 kg.

Daily Intake of Element in Food and Fluids: 0.3 kg.

Principal Human Metabolic and Dosimetric Parameters:

f_1 = 1.

ALI   = 90 MBq (2000 μCi), inhalation or ingestion of labeled organic compounds,

       = 6 × 10^4 MBq (2 × 10^6 μCi), inhalation of CO,

       = 8000 MBq (2 × 10^5 μCi), inhalation of CO_2.

DAC = 0.04 MBq m^{-3} (1 × 10^{-6} μCi cm^{-3}), labeled organic compounds,

       = 30 MBq m^{-3} (7 × 10^{-4} μCi cm^{-3}), inhalation of CO,

       = 3 MBq m^{-3} (9 × 10^{-5} μCi cm^{-3}), inhalation of CO_2.

Additional Dosimetric Considerations: Ingestion, 5.64 × 10^{-10} Sv Bq^{-1} (2.09 mrem μCi^{-1}). Inhalation, 6.36 × 10^{-12} Sv Bq^{-1} (2.35 × 10^{-2} mrem μCi^{-1}) for CO_2, 7.83 × 10^{-13} Sv Bq^{-1} (2.9 × 10^{-3} mrem μCi^{-1}) for CO, and 5.64 × 10^{-10} Sv Bq^{-1} (2.09 mrem μCi^{-1}) for labeled organic compounds. The doses from natural sources are 24 mSv y^{-1} (2.4 mrem y^{-1}) to the skeletal tissues and 5 mSv y^{-1} (0.5 mrem y^{-1}) to the gonads (UNSCEAR, 1982).

Whole-Body Retention Function: $R(t) = 0.18e^{-0.693t/5} + 0.81e^{-0.693t/60} + 0.01e^{-0.693t/60000}$, with t as elapsed time in minutes.

Isotope: Potassium-40 (^{40}K)

Half-life:

   Physical: 1.277 × 10^9 years.

   Biological: 30 days.

Natural Levels: 0.0117% of total K.

Specific Activity: 31 kBq kg^{-1} (836 pCi g^{-1}).

Sources: Naturally occurring primordial nuclide.

Principal Modes and Energies of Decay (MeV): EC; β^- ave., 0.561; max., 1.31 (89.3%); γ, 1.46 (10.7%).

Special Chemical and Biological Characteristics: Element distributes throughout the body. Body burden decreases with age and muscle-wasting diseases.

Principal Organ: Whole body.

Amount of Element in Body: 140 g.

Daily Intake of Element in Food and Fluids: 3.3 g.

Amount of Radionuclide in Body: 0.1 μCi (4 kBq).

Principal Human Metabolic and Dosimetric Parameters:

   f_1 = 1 (all forms).

   ALI   = 10 MBq (400 μCi), IC-D (all forms),

          = 10 MBq (300 μCi), ingestion (all forms).

   DAC = 0.006 MBq m^{-3} (2 × 10^{-7} μCi cm^{-3}), IC-D (all forms).

Additional Dosimetric Considerations: Ingestion, 5.02 × 10^{-9} Sv Bq^{-1} (18.6 mrem μCi^{-1}). Inhalation, 3.34 × 10^{-9} Sv Bq^{-1} (12.4 mrem μCi^{-1}). This isotope delivers a dose of 18 mrem y^{-1} (0.18 mSv y^{-1}) to the gonads and other soft tissues and 14 mrem y^{-1} (0.14 mSv y^{-1}) to bone (NCRP, 1987a).

Whole-Body Retention Function: $R(t) = e^{-0.693t/30}$, with t as elapsed time in days.

Special Ecological Aspects: Stable K constitutes 2.59% of the earth's crust with 380 ppm in seawater; ^{40}K levels in soil are 0.037–1.1 Bq g^{-1} (1–30 pCi g^{-1}). ^{40}K is the predominant radioactive component in normal food and human tissue.

Isotope: Manganese-54 (^{54}Mn)

Half-life:

   Physical: 312.12 days.

   Biological: Multiple compartments with half-life for the transfer fractions: 4 days (0.1 in liver, 0.2 in other tissues), 40 days (bone, 0.15 in liver, 0.2 in other tissues).

Sources: Activation product produced by nuclear weapons testing and reactors—^{54}Cr(p,n)^{54}Mn and ^{54}Fe(n,p)^{54}Mn.

Principal Modes and Energies of Decay (MeV): EC; γ, 0.834 (100%).

Special Chemical and Biological Characteristics: Stable manganese is an essential trace element for both plants and animals. In animals, bone and brain retain Mn more avidly than do most other organs.

Principal Organ: Bone surface.

Amount of Element in Body: 12 mg.

Daily Intake of Element in Food and Fluids: 3.7 mg.

Principal Human Metabolic and Dosimetric Parameters:

   f_1 = 0.1 (all forms).

   ALI   = 30 MBq (800 μCi), IC-W (oxides, hydroxides, halides, and nitrates),

        = 30 MBq (900 μCi), IC-D (all other forms).

        = 70 MBq (2000 μCi), ingestion (all forms).

   DAC = 0.01 MBq m^{-3} (3 $\times$ 10^{-7} μCi cm^{-3}), IC-W (oxides, hydroxides, halides, and nitrates),

        = 0.01 MBq m^{-3} (4 $\times$ 10^{-7} μCi cm^{-3}), IC-D (all other forms).

Additional Dosimetric Considerations: Ingestion, 7.48 $\times$ 10^{-10} Sv Bq^{-1} (2.77 mrem μCi^{-1}). Inhalation, Class D is 1.42 $\times$ 10^{-9} Sv Bq^{-1} (5.25 mrem μCi^{-1}) and Class W is 1.81 $\times$ 10^{-9} Sv Bq^{-1} (6.7 mrem μCi^{-1}).

Whole-Body Retention Function: $R(t) = 0.3e^{-0.693t/4} + 0.7e^{-0.693t/40}$, with t as elapsed time in days.

Systemic Transfer Fractions (longer half-life listed second): Skeleton, 0.35; liver, 0.1 and 0.15; other tissues, 0.2 and 0.2.

Isotope: Iron-55 (^{55}Fe)

Half-life:

   Physical: 2.73 years.

   Biological: 2000 days.

Sources: Activation product produced by nuclear weapons testing and reactors—^{54}Fe(n,γ)^{55}Fe and ^{56}Fe(n,2n)^{55}Fe. Fission yield from weapons testing is 630 PBq per megaton (1.7 $\times$ 10^7 Ci per megaton).

Principal Modes and Energies of Decay (MeV): EC; Auger K electron, 0.00519 (60.7%); K$_{\alpha 1}$ X ray, 0.00590 (16.3%).

Special Chemical and Biological Characteristics: 70% of total body iron is bound in hemoglobin with the remainder stored in other constituents of blood.

Principal Organ: Whole body.

Amount of Element in Body: 4.2 g.

Daily Intake of Element in Food and Fluids: 16 mg.

Principal Human Metabolic and Dosimetric Parameters:

f_1 = 0.1.

ALI = 200 MBq (4000 μCi), IC-W (oxides, hydroxides, halides, and nitrates),
 = 70 MBq (2000 μCi), IC-D (all other forms),
 = 300 MBq (9000 μCi), ingestion (all forms).

DAC = 0.06 MBq m^{-3} (2 × 10^{-6} μCi cm^{-3}), IC-W (oxides, hydroxides, halides, and nitrates),
 = 0.03 MBq m^{-3} (8 × 10^{-7} μCi cm^{-3}), IC-D (all other forms).

Additional Dosimetric Considerations: Ingestion, 1.64 × 10^{-10} Sv Bq^{-1} (0.61 mrem μCi^{-1}). Inhalation, Class D is 7.26 × 10^{-10} Sv Bq^{-1} (2.69 mrem μCi^{-1}) and Class W is 3.61 × 10^{-10} Sv Bq^{-1} (1.34 mrem μCi^{-1}).

Whole-Body Retention Function: $R(t) = e^{-0.693t/2000}$, with t as elapsed time in days.

Systemic Transfer Fractions: Liver, 0.08; spleen, 0.013.

Special Ecological Aspects: Direct foliar deposition is the most important pathway for terrestrial plant contamination. Readily taken up into marine food chains owing to the low stable iron content of ocean water. Concentrations in the blood of Lapps and Eskimos are high owing to the air → lichen → reindeer (caribou) → human food chain. Also higher in consumers of fish.

Isotope: Iron-59 (^{59}Fe)

Half-life:

Physical: 44.503 days.

Biological: 2000 days.

Sources: Activation product produced by nuclear weapons testing and reactors— ^{58}Fe(n,γ)^{59}Fe. Fission yield from weapons testing is 81 PBq per megaton (2.2 × 10^6 Ci per megaton).

Principal Modes and Energies of Decay (MeV): β_1^- ave., 0.0808, 0.271 (45.3%); β_2^- ave., 0.149; max., 0.466 (53.1%); γ, 0.192 (3.08%), 1.10 (56.5%), 1.29 (43.2%).

Special Chemical and Biological Characteristics: 70% of total body iron is bound in hemoglobin with the remainder stored in other constituents of blood.

Principal Organ: Whole body.

Amount of Element in Body: 4.2 g.

Daily Intake of Element in Food and Fluids: 16 mg.

Principal Human Metabolic and Dosimetric Parameters:

f_1 = 0.1.

ALI = 20 MBq (500 μCi), IC-W (oxides, hydroxides, halides, and nitrates),
 = 10 MBq (300 μCi), IC-D (all other forms),
 = 30 MBq (800 μCi), ingestion (all forms).

DAC = 0.008 MBq m^{-3} (2 × 10^{-7} μCi cm^{-3}), IC-W (oxides, hydroxides, halides, and nitrates),
 = 0.005 MBq m^{-3} (1 × 10^{-7} μCi cm^{-3}), IC-D (all other forms).

Additional Dosimetric Considerations: Ingestion, 1.81 × 10^{-9} Sv Bq^{-1} (6.7 mrem μCi^{-1}). Inhalation, Class D is 4.00 × 10^{-9} Sv Bq^{-1} (14.8 mrem μCi^{-1}) and Class W is 3.30 × 10^{-9} Sv Bq^{-1} (12.2 mrem μCi^{-1}).

APPENDIX

Whole-Body Retention Function: See ^{55}Fe.
Systemic Transfer Fractions: Liver, 0.08; spleen, 0.013.

Isotope: Cobalt-60 (^{60}Co)

Half-life:
  Physical: 5.2714 years.
  Biological: 0.5 day (transfer compartment), 6 days (0.6 in all tissues), 60 days (0.2 in all tissues), 800 days (0.2 in other tissues).
Sources: Activation product produced by nuclear weapons testing and reactors— ^{59}Co(n,γ)^{60}Co and ^{60}Ni(n,p)^{60}Co.
Principal Modes and Energies of Decay (MeV): β^- ave., 0.0958; max., 0.318 (99%); γ, 1.33 (100%), 1.17 (100%).
Principal Organ: Liver and whole body.
Amount of Element in Body: 1.5 mg.
Daily Intake of Element in Food and Fluids: 300 μg.
Principal Human Metabolic and Dosimetric Parameters:
  ALI = 6 MBq (200 μCi), IC-W (oxides, hydroxides, halides, and nitrates), f_1 = 0.05,
      = 1 MBq (30 μCi), IC-Y (all other forms), f_1 = 0.05,
      = 20 MBq (500 μCi), ingestion of oxides, hydroxides, and trace inorganics, f_1 = 0.05,
      = 7 MBq (200 μCi), ingestion of organic complexed and other inorganics, f_1 = 0.3.
  DAC = 0.003 MBq m^{-3} (7 × 10^{-8} μCi cm^{-3}), IC-W (oxides, hydroxides, halides, and nitrates), f_1 = 0.05,
      = 5 × 10^{-4} MBq m^{-3} (1 × 10^{-8} μCi cm^{-3}), IC-Y (all other forms), f_1 = 0.05.
Additional Dosimetric Considerations: Ingestion, 2.77 × 10^{-9} Sv Bq^{-1} (10.2 mrem μCi^{-1}) for f_1 = 0.05 and 7.28 × 10^{-9} Sv Bq^{-1} (26.9 mrem μCi^{-1}) for f_1 = 0.3. Inhalation, Class W is 8.94 × 10^{-9} Sv Bq^{-1} (33.1 mrem μCi^{-1}) and Class Y is 5.91 × 10^{-8} Sv Bq^{-1} (219 mrem μCi^{-1}).
Whole-Body Retention Function: $R(t) = 0.5e^{-0.693t/0.5} + 0.3e^{-0.693t/6} + 0.1e^{-0.693t/60} + 0.1e^{-0.693t/800}$, with t as elapsed time in days.
Systemic Transfer Fractions: Liver, 0.05; other tissues, 0.45; directly to excretion, 0.5.

Isotope: Zinc-65 (^{65}Zn)

Half-life:
  Physical: 244.26 days.
  Biological: Skeleton, 400 days; other tissues will have two compartments (0.3 and 0.7) with 20 days and 400 days.
Sources: Activation product produced by nuclear weapons testing and reactors— ^{64}Zn(n,γ)^{65}Zn and ^{63}Cu(d,γ)^{65}Zn
Principal Modes and Energies of Decay (MeV): EC; γ, 1.12 (50.6%).

Special Chemical and Biological Characteristics: Uptake from gastrointestinal
   tract (f_1) dependent to some extent on fasting state and possibly dietary Zn levels.
Principal Organ: Whole body.
Amount of Element in Body: 2.3 g.
Daily Intake of Element in Food and Fluids: 13 mg.
Principal Human Metabolic and Dosimetric Parameters:
   f_1 = 0.5 (all forms).
   ALI   = 10 MBq (300 μCi), IC-Y (all forms),
         = 10 MBq (400 μCi), ingestion (all forms),
   DAC = 0.004 MBq m^{-3} (1 × 10^{-7} μCi cm^{-3}), IC-Y (all forms).
Additional Dosimetric Considerations: Ingestion, 3.90 × 10^{-9} Sv Bq^{-1} (14.4 mrem
   μCi^{-1}). Inhalation, 5.51 × 10^{-9} Sv Bq^{-1} (20.4 mrem μCi^{-1}).
Whole-Body Retention Function: $R(t) = 0.24e^{-0.693t/20} + 0.76e^{-0.693t/400}$, with t as
   elapsed time in days.
Systemic Transfer Fractions: Skeleton, 0.20; other tissues, 0.8.

Isotope: Krypton-85 (^{85}Kr)

Half-life:
   Physical: 10.756 years.
Sources: Fission product produced from weapons testing and fission reactors, yield
   ~15 TBq Mwe^{-1} y^{-1} (~400 Ci MWe^{-1} y^{-1}). Atmospheric burden mainly attributable
   to fuel reprocessing. Minor product of comic-ray interactions.
Principal Modes and Energies of Decay (MeV): β^- ave., 0.252; max., 0.687 (99.6%);
   γ, 0.514 (0.434%).
Special Chemical and Biological Characteristics: Noble gas; inhalation and inges-
   tion doses are small compared to doses from submersion. Diffusion is sole mechanism of absorption into body.
Principal Organ: Skin and whole-body external irradiation.
Principal Human Metabolic and Dosimetric Parameters:
   DAC = 5 MBq m^{-3} (1 × 10^{-4} μCi cm^{-3}), submersion limited.
Additional Dosimetric Considerations: Exposure to dose conversion factors for
   submersion—skin, 4.66 × 10^{-11} Sv hr^{-1} Bq^{-1} m^{-3} (1.7 × 10^5 mrem hr^{-1} μCi^{-1}
   cm^{-3}); effective, 4.70 × 10^{-13} Sv hr^{-1} Bq^{-1} m^{-3} (1.7 × 10^3 mrem hr^{-1} μCi^{-1} cm^{-1}).

Isotope: Strontium-89 (^{89}Sr)

Half-life:
   Physical: 50.53 days.
   Biological: Retention is described by a complicated power function equation
      (ICRP Publication 20, 1973).
Sources: Fission product produced from weapons testing and fission reactors.
Principal Modes and Energies of Decay (MeV): β^- ave., 0.583; max., 1.50 (100%).
Special Chemical and Biological Characteristics: Alkaline earth element, with
   tendency to concentrate throughout the mineralized bone volume.

Principal Organs: Mineralized bone volume and lower large intestine.

Amount of Element in Body: 320 mg.

Daily Intake of Element in Food and Fluids: 1.9 mg.

Principal Human Metabolic and Dosimetric Parameters:

ALI $\,= 5$ MBq (100 μCi), IC-Y (SrTiO$_3$), $f_1 = 0.01$,

$\qquad = 30$ MBq (800 μCi), IC-D (all other forms), $f_1 = 0.3$,

$\qquad = 20$ MBq (600 μCi), ingestion of soluble salts, $f_1 = 0.3$,

$\qquad = 20$ MBq (500 μCi), ingestion of SrTiO$_3$, $f_1 = 0.01$.

DAC $= 0.002$ MBq m^{-3} (6 $\times$ 10^{-8} μCi cm^{-3}), IC-Y (SrTiO$_3$), $f_1 = 0.01$,

$\qquad = 0.01$ MBq m^{-3} (4 $\times$ 10^{-7} μCi cm^{-3}), IC-D (all other forms), $f_1 = 0.3$.

Additional Dosimetric Considerations: Ingestion, 2.5×10^{-9} Sv Bq^{-1} (9.25 mrem μCi^{-1}) for $f_1 = 0.3$ and 2.89×10^{-8} Sv Bq^{-1} (107 mrem μCi^{-1}) to the lower large intestine for $f_1 = 0.01$. Inhalation, Class D is 1.79×10^{-9} Sv Bq^{-1} (6.62 mrem μCi^{-1}) and Class Y is 1.12×10^{-8} Sv Bq^{-1} (41.4 mrem μCi^{-1}).

Isotope: Strontium-90 (^{90}Sr)

Half-life:

Physical: 28.78 years.

Biological: Retention is described by a complicated power function equation (ICRP Publication 20, 1973).

Sources: Fission product produced from weapons testing and fission reactors.

Principal Modes and Energies of Decay (MeV): β^- ave., 0.196; max., 0.546 (100%).

Special Chemical and Biological Characteristics: Alkaline earth element, with tendency to concentrate throughout the mineralized bone volume.

Principal Organ: Mineralized bone volume.

Amount of Element in Body: 320 mg.

Daily Intake of Element in Food and Fluids: 1.9 mg.

Principal Human Metabolic and Dosimetric Parameters:

ALI $\,= 0.1$ MBq (4 μCi), IC-Y (SrTiO$_3$), $f_1 = 0.01$,

$\qquad = 0.7$ MBq (20 μCi), IC-D (all other forms), $f_1 = 0.3$,

$\qquad = 1$ MBq (30 μCi), ingestion of soluble salts, $f_1 = 0.3$,

$\qquad = 20$ MBq (400 μCi), ingestion of SrTiO$_3$, $f_1 = 0.01$.

DAC $= 6 \times 10^{-5}$ MBq m^{-3} (2 $\times$ 10^{-9} μCi cm^{-3}), IC-Y (SrTiO$_3$), $f_1 = 0.01$,

$\qquad = 3 \times 10^{-4}$ MBq m^{-3} (8 $\times$ 10^{-9} μCi cm^{-3}), IC-D (all other forms), $f_1 = 0.3$.

Additional Dosimetric Considerations: Ingestion, 4.19×10^{-7} Sv Bq^{-1} (1.55 $\times$ 10^3 mrem μCi^{-1}) to the bone surfaces for $f_1 = 0.3$ and 3.23×10^{-9} Sv Bq^{-1} (12 mrem μCi^{-1}) for $f_1 = 0.01$. Inhalation, Class D is 7.27×10^{-7} Sv Bq^{-1} (2.69 $\times$ 10^3 mrem μCi^{-1}) to the bone surfaces and Class Y is 3.51×10^{-7} Sv Bq^{-1} (1.3 $\times$ 10^3 mrem μCi^{-1}).

Special Ecological Aspects: Incorporates into the calcium pool of the biosphere. Principal ecological pathway is grass $\rightarrow$ cow's milk $\rightarrow$ human. The short-lived decay product ^{90}Y, β^- decays with an average energy of 0.93 MeV (2.28 MeV max.) that contributes to the internal dose of ^{90}Sr.

Isotope: Technetium-99 (^{99}Tc)

Half-life:
   Physical: 2.113×10^5 years.
   Biological: 0.02 day in transfer compartment; 0.5 day for thyroid; other tissues
      of the body described with three compartments with 1.6, 3.7, and 22 days.
Sources: Fission product produced from weapons testing and fission reactors, re-
   leased to the environment mainly from weapons testing and fuel reprocessing.
Principal Modes and Energies of Decay (MeV): β^- ave., 0.0846; max., 0.294
   (100%).
Special Chemical and Biological Characteristics: Most stable as the very soluble
   pertechnetate anion (TcO_4^-).
Principal Organ: Stomach wall and whole body.
Principal Human Metabolic and Dosimetric Parameters:
   f_1 = 0.8 (all forms).
   ALI = 20 MBq (700 μCi), IC-W (oxides, hydroxides, halides, and nitrates),
         = 200 MBq (5000 μCi), IC-D (all other forms),
         = 100 MBq (4000 μCi), ingestion (all forms).
   DAC = 0.01 MBq m^{-3} (3×10^{-7} μCi cm^{-3}), IC-W (oxides, hydroxides, halides,
         and nitrates),
         = 0.08 MBq m^{-3} (2×10^{-6} μCi cm^{-3}), IC-D (all other forms).
Additional Dosimetric Considerations: Ingestion, 3.95×10^{-10} Sv Bq^{-1} (1.46 mrem
   μCi^{-1}). Inhalation, Class D is 2.47×10^{-9} Sv Bq^{-1} (9.14 mrem μCi^{-1}) to the
   stomach wall and Class W is 2.25×10^{-9} Sv Bq^{-1} (8.33 mrem μCi^{-1}).
Whole-Body Retention Function: $R(t) = 0.76e^{-0.693t/1.6} + 0.19e^{-0.693t/3.7} + 0.043e^{-0.693t/22}$, with t as elapsed time in days.
Systemic Transfer Fractions: Thyroid, 0.04; liver, 0.1; stomach wall, 0.03; other
   tissues, 0.83.
Special Ecological Aspects: TcO_4^- readily available for plant root uptake (plant/
   soil concentration ratio of ~20). High mobility and apparent short half-times in
   vegetation may limit dosimetric consequences from plant ingestion (Hoffman *et
   al.*, 1982).

Isotope: Iodine-129 (^{129}I)

Half-life:
   Physical: 1.57×10^7 years.
   Biological: Thyroid, 120 days; rest of tissues, 12 days.
Sources: Naturally produced in the upper atmosphere from interactions of high-
   energy particles with xenon and, to a lesser extent, in the lithosphere from the
   spontaneous fission of ^{238}U. Also produced by weapons testing at a rate of
   1.9 GBq per megaton (50 mCi per megaton) and from light-water reactors at
   4.8 kBq per megawatt-day (1.3 μCi per megawatt-day).
Principal Modes and Energies of Decay (MeV): β^- ave., 0.0490; max., 0.154
   (100%); γ, 0.0376 (7.5%).

Special Chemical and Biological Characteristics: Assume 0.3 for compartment transfer fraction into a 20-g thyroid.

Principal Organ: Thyroid.

Amount of Element in Body: 11 mg.

Daily Intake of Element in Food and Fluids: 200 μg.

Principal Human Metabolic and Dosimetric Parameters:
   f_1 = 1 (all forms).
   ALI = 0.3 MBq (9 μCi), IC-D (all forms),
       = 0.2 MBq (5 μCi), ingestion (all forms).
   DAC = 1 $\times$ 10^{-4} MBq m^{-3} (4 $\times$ 10^{-9} μCi cm^{-3}), IC-D (all forms).

Additional Dosimetric Considerations (all factors are for dose to the thyroid): Ingestion, 2.48 $\times$ 10^{-6} Sv Bq^{-1} (9.18 $\times$ 10^3 mrem μCi^{-1}). Inhalation, 1.56 $\times$ 10^{-6} Sv Bq^{-1} (5.77 $\times$ 10^3 mrem μCi^{-1}).

Special Ecological Aspects: Prior to weapons testing, ∼30 Ci (1.1. TBq) resided in the hydrosphere and 10 Ci (370 GBq) in the lithosphere from natural sources. Weapons testing approximately doubled the natural inventory (NCRP, 1983).

Isotope: Iodine-131 (^{131}I)

Half-life:
   Physical: 8.04 days.
   Biological: Thyroid, 120 days; rest of tissues, 12 days.

Sources: Fission product produced from weapons testing at a rate of 4.6 EBq per megaton (125 MCi per megaton). Attains an equilibrium of inventory at steady-state condition in a reactor of 960 TBq per MW$_t$ (26 kCi per MW$_t$).

Principal Modes and Energies of Decay (MeV): β_1^- ave., 0.0694; max., 0.248 (2.1%); β_2^- ave., 0.0966; max., 0.334 (7.27%); β_3^- ave., 0.192; max., 0.606 (89.9%); γ, 0.0802 (2.62%), 0.284 (6.14%), 0.365 (81.7%), 0.637 (7.17%).

Special Chemical and Biological Characteristics: Assume 0.3 for transfer into thyroid.

Principal Organ: Thyroid.

Amount of Element in Body: See ^{129}I.

Daily Intake of Element in Food and Fluids: See ^{129}I.

Principal Human Metabolic and Dosimetric Parameters:
   f_1 = 1 (all forms).
   ALI = 2 MBq (50 μCi), IC-D (all forms),
   DAC = 7 $\times$ 10^{-4} MBq m^{-3} (2 $\times$ 10^{-8} μCi cm^{-3}), IC-D (all forms).

Additional Dosimetric Considerations (all factors are for dose to the thyroid): Ingestion, 4.76 $\times$ 10^{-7} Sv Bq^{-1} (1.76 $\times$ 10^3 mrem μCi^{-1}). Inhalation, 2.92 $\times$ 10^{-7} Sv Bq^{-1} (1.08 $\times$ 10^3 mrem μCi^{-1}). Dose commitments are ∼10 times greater for a 1-year-old child.

Special Ecological Aspects: Principal route of human absorption is via fresh milk by the grass → cow → milk → human food chain, with rapid distribution of milk being a primary consideration. Milk content of radioiodine reaches a peak 3 days after deposition on forage. Effective half-times of removal from grass is about 5 days. 9.1 $\times$ 10^{-2} μCi L^{-1} (3.4 kBq L^{-1}) of milk can be expected per μCi (37 kBq) of ^{131}I deposited on 1 m^2 of grass. The ratio pCi kg^{-1} grass : pCi m^{-3} air (or kBq kg^{-1} grass : kBq m^{-3} air) is 4200. The ratio pCi L^{-1} milk : pCi kg^{-1} grass (or kBq L^{-1} milk : kBq kg^{-1} grass) is 0.15 (Soldat, 1965).

Isotope: Cesium-134 (^{134}Cs)

Half-life:
   Physical: 2.065 years.
   Biological: 110 days.
Sources: Activation product ^{133}Cs(n,γ)^{134}Cs from fission reactors at a rate of 110 GBq per megawatt-day (3 Ci per megawatt-day), ^{134}Cs/^{137}Cs = 0.4–0.6. A normal constituent in low-level liquid effluent from commercial reactors.
Principal Modes and Energies of Decay (MeV): EC; β_1^- ave., 0.023; max., 0.886 (27.3%); β_2^- ave., 0.123; max., 0.415 (2.51%); β_3^- ave., 0.210; max., 0.658 (70.2%); γ, 0.605 (97.6%), 0.563 (8.35%), 0.796 (85.5%), 0.802 (8.69%), 1.37 (3.01%).
Special Chemical and Biological Characteristics: Metabolism resembles potassium and distributes uniformly throughout the body. Convenient to express concentration in biological material relative to the K content.
Principal Organ: Whole body.
Amount of Element in Body: 1.5 g.
Daily Intake of Element in Food and Fluids: 10 μg.
Principal Human Metabolic and Dosimetric Parameters:
   f_1 = 1 (all forms).
   ALI = 4 MBq (100 μCi), IC-D and IC-W (all forms),
       = 3 MBq (70 μCi), ingestion (all forms).
   DAC = 0.002 MBq m^{-3} (4 × 10^{-8} μCi cm^{-3}), IC-D (all forms).
Additional Dosimetric Considerations: Ingestion, 1.98 × 10^{-8} Sv Bq^{-1} (73.3 mrem μCi^{-1}). Inhalation, 1.25 × 10^{-8} Sv Bq^{-1} (46.3 mrem μCi^{-1}).
Whole-Body Retention Function: $R(t) = 0.1e^{-0.693t/2} + 0.9e^{-0.693t/110}$, with t as elapsed time in days.

Isotope: Cesium-137 (^{137}Cs)

Half-life:
   Physical: 30.17 years.
   Biological: 110 days.
Sources: Fission product produced from weapons testing at a rate of 6.3 PBq per megaton (0.17 MCi per megaton). Also produced from fission reactors, a normal trace constituent in low-level liquid effluent from commercial reactors.
Principal Modes and Energies of Decay (MeV): β_1^- ave., 0.174; max., 0.514 (94.4%); β_2^- ave., 0.417; max., 1.18 (5.6%); γ, 0.662 (85.1%) from ^{137m}Ba.
Special Chemical and Biological Characteristics: See ^{134}Cs.
Principal Organ: Whole body.
Amount of Element in Body: 1.5 mg.
Daily Intake of Element in Food and Fluids: 10 μg.
Principal Human Metabolic and Dosimetric Parameters:

f_1 = 1.
ALI = 6 MBq (200 μCi), IC-D (all forms),
     = 4 MBq (100 μCi), ingestion (all forms).
DAC = 0.002 MBq m^{-3} (6 × 10^{-8} μCi cm^{-3}), IC-D (all forms).
Additional Dosimetric Considerations: Ingestion, 1.35 × 10^{-8} Sv Bq^{-1} (50 mrem μCi^{-1}). Inhalation, 8.63 × 10^{-9} Sv Bq^{-1} (31.9 mrem μCi^{-1}).
Special Ecological Aspects: Higher body burdens in Lapps and Eskimos due to the pathway air → lichen → reindeer (caribou) → human food chain; strong retention to clay soils limits plant uptake; solubility much greater in marine than fresh water; uptake by fish controlled by potassium ion content of water.

Isotope: Lead-210 (^{210}Pb)—Ra D

Half-life:
  Physical: 22.3 years.
  Biological: Three compartments with 12, 180, and 10,000 days are seen for skeleton and other tissues.
Natural Levels: Soil range similar to ^{238}U; annual average air concentration range from 0.2 to 1.5 mBq m^{-3} (5–40 × 10^{-3} pCi m^{-3}).
Sources: Naturally occurring from uranium decay chain.
Principal Modes and Energies of Decay (MeV): β_1^- ave., 0.00416; max., 0.0165 (84%); β_2^- ave., 0.0162; max, 0.064 (16%); γ, 0.0465 (4.25%); ^{210}Bi X ray, 0.013.
Special Chemical and Biological Characteristics: Lead can be substituted for calcium in apatite (mineral bone) and, accordingly, is uniformly distributed throughout the bone volume.
Principal Organ: Bone surface.
Amount of Element in Body: 120 mg.
Amount of Element in Skeleton: 110 mg.
Daily Intake in Food and Fluids: 440 μg of element with 3 pCi (0.11 Bq) ^{210}Pb (UNSCEAR 1982).
Amount of Radionuclide in Body: 160–860 pCi (6–32 Bq) in skeleton, mean of 380 pCi (14 Bq) (NCRP, 1984b).
Principal Human Metabolic and Dosimetric Parameters:
  f_1 = 0.2.
  ALI = 0.009 MBq (0.2 μCi), IC-D (all forms),
       = 0.02 MBq (0.6 μCi), ingestion (all forms).
  DAC = 4 × 10^{-6} MBq m^{-3} (1 × 10^{-10} μCi cm^{-3}), IC-D (all forms).
Additional Dosimetric Considerations (all factors are for dose to the bone surface): Ingestion, 2.16 × 10^{-5} Sv Bq^{-1} (7.99 × 10^4 mrem μCi^{-1}). Inhalation, 5.47 × 10^{-5} Sv Bq^{-1} (2.02 × 10^5 mrem μCi^{-1}). Naturally present ^{210}Pb and its decay products are estimated to contribute annually about 0.14 mSV (14 mrem) to soft tissue and bone marrow and 0.7 mSv (70 mrem) to bone surface (NCRP, 1987a).
Systemic Transfer Fractions: Skeleton, 0.55; liver, 0.25; kidneys, 0.02; other tissues, 0.18.
Skeleton Transfer Fractions: 0.6 (12 days), 0.2 (180 days), 0.2 (10,000 days).

All Other Tissue and Organ Transfer Fractions: 0.8 (12 days), 0.18 (180 days), 0.02 (10,000 days).

Isotope: Polonium-210 (^{210}Po)

Half-life:
   Physical: 138.38 days,
   Biological: 50 days,
Natural Levels: Soil range similar to ^{238}U; annual average air concentration range from 0.2 to 1.5 mBq m^{-3} (5–40 × 10^{-3} pCi m^{-3}).
Sources: Naturally occurring from uranium decay chain and ^{209}Bi(n,γ)^{210}Bi β^- → ^{210}Po.
Principal Modes and Energies of Decay (MeV): α, 5.31 (100%).
Special Chemical and Biological Characteristics: Tendency for radiocolloid formation.
Principal Organs: Liver, kidney, and spleen.
Amount of Element in Body: ~40 Bq (1 nCi), with about 60% in the mineralized bone.
Daily Intake in Food and Fluids: ~100 mBq.
Principal Human Metabolic and Dosimetric Parameters:
   f_1 = 0.1 (all forms).
   ALI   = 0.02 MBq (0.6 μCi), IC-W (oxides, hydroxides, and nitrates),
         = 0.02 MBq (0.6 μCi), IC-D (all others),
         = 0.1 MBq (3 μCi), ingestion (all forms).
   DAC = 1 × 10^{-5} MBq m^{-3} (3 × 10^{-10} μCi cm^{-3}), IC-W (oxides, hydroxides, and nitrates),
         = 1 × 10^{-5} MBq m^{-3} (3 × 10^{-10} μCi cm^{-3}), IC-D (all others).
Additional Dosimetric Considerations: Ingestion, 5.14 × 10^{-7} Sv Bq^{-1} (1.9 × 10^3 mrem μCi^{-1}). Inhalation, Class D is 2.54 × 10^{-6} Sv Bq^{-1} (9.4 × 10^3 mrem μCi^{-1}) and Class W is 2.32 × 10^{-6} Sv Bq^{-1} (8.58 × 10^3 mrem μCi^{-1}).
Whole-Body Retention Function: $R(t) = e^{-0.693t/50}$, with t as elapsed time in days.
Systemic Transfer Fractions: Liver, 0.1; kidneys, 0.1; spleen, 0.1; other tissues, 0.70.
Special Ecological Aspects: Higher body burdens in Lapps and Eskimos due to the air → lichen → reindeer (caribou) → human food chain; high burdens from fish and shellfish diets; also present in tobacco smoke.

Isotope: Radon-222 (^{222}Rn)

Half-life:
   Physical: 3.8235 days.
Natural Levels: Soil gas range from 4 to 40 kBq m^{-3} (10^2–10^3 pCi L^{-1}); annual average air concentration range in the United States from 0.016 pCi L^{-1} (0.6 Bq m^{-3}) to 0.75 pCi L^{-1} (28 Bq m^{-3}). Indoor concentrations variable owing to location, ventilation, and source strength.
Sources: Naturally occurring from uranium decay chain. Equilibrium global inventory ~40 MCi (1.5 EBq).

Principal Modes and Energies of Decay (MeV): α, 5.49 (99.9%).

Special Chemical and Biological Characteristics: Noble gas; inhalation and ingestion doses are small compared to doses from radioactive progeny. Doses primarily from short-lived progeny (^{218}Po, ^{214}Pb, ^{214}Bi, ^{214}Po) attached to inhaled dust particles.

Principal Organ: Lungs (from progeny).

Principal Human Metabolic and Dosimetric Parameters: Estimates of the average annual dose to the basal cells of the bronchial epithelium of the lung of U.S. and Canadian residents from radon decay products is 24 mSv (2400 mrem). This calculation assumed an average radon concentration of 30 Bq m^{-3} (0.8 pCi L^{-1}) and an equilibrium fraction of 0.5 (NCRP, 1987a). The annual effective dose equivalent received under these conditions was estimated to be 2 mSv (200 mrem).

Additional Dosimetric Considerations: One Working Level Month (WLM) is approximately equal to annual exposure to an average of 150 Bq m^{-3} (4 pCi L^{-1}) of ^{222}Rn if the decay products are in 50% equilibrium with the ^{222}Rn. One WLM exposure would occur from being exposed to 1 Working Level (WL) for a period of one working month, that is, 170 hours. One WL is defined as that concentration of radon decay products that has a potential alpha energy release of 1.3 × 10^5 MeV L^{-1} (2 × 10^{-5} J m^{-3}) of air. This is equivalent to a ^{222}Rn concentration of 100 pCi L^{-1} (3.7 Bq L^{-1}), in equilibrium with its short-lived progeny. EPA recommends that residential levels be less than 150 Bq m^{-3} (4 pCi L^{-1}), with reduction of levels below 75 Bq m^{-3} (2 pCi L^{-1}) if possible. See Chapter 6 for more detailed information. Primary Guide is 4 WLM.

Isotope: Radium-226 (^{226}Ra)

Half-life:
  Physical: 1.60 × 10^3 years.
  Biological: Retention is described by a complicated power function equation (ICRP Publication 20, 1973).

Natural Levels: Soil range similar to ^{238}U (see Table 6-8).

Sources: Naturally occurring from ^{238}U decay chain.

Principal Modes and Energies of Decay (MeV): α, 4.78 (94.5%), 4.61 (5.55%); γ, 0.186 (3.5%).

Special Chemical and Biological Characteristics: Deposits in bone with nonuniform distribution, following decay of ^{226}Ra in bone; ~70% of ^{222}Rn diffuses into the blood and is exhaled.

Principal Organ: Mineralized bone volume.

Amount in Body: 1.1 Bq (31 pCi) with 1 Bq (27 pCi) found in the skeleton.

Daily Intake in Food and Water: 0.09 Bq (2.3 pCi).

Principal Human Metabolic and Dosimetric Parameters:
  f_1 = 0.2 (all forms).
  ALI = 0.02 MBq (0.6 μCi), IC-W (all forms),
      = 0.07 MBq (2 μCi), ingestion (all forms).
  DAC = 1 × 10^{-5} MBq m^{-3} (3 × 10^{-10} μCi cm^{-3}, IC-W (all forms).

Additional Dosimetric Considerations: Ingestion, 6.83×10^{-6} Sv Bq^{-1} (2.53×10^4 mrem μCi^{-1}) to the bone surface. Inhalation, 2.32×10^{-6} Sv Bq^{-1} (8.58×10^3 mrem μCi^{-1}). Dose commitments based on ^{226}Ra plus daughters assuming $\sim$30% ^{222}Rn retention.

Special Ecological Aspects: Radium is chemically similar to calcium and is absorbed from the soil by plants and passed up the food chain to humans. Penna Franca found the radium content of Brazil nuts to range between 273 and 7100 pCi kg^{-1} (10–260 Bq kg^{-1}), with only 3 out of 15 samples assaying less than 1000 pCi kg^{-1} (37 Bq kg^{-1}). The radioactivity was about equally divided between ^{226}Ra and ^{228}Ra and was not related to the radium or barium content of the soil in which the tree is grown. The radium concentration of Brazil nuts is on the order of 1000 times greater than that in the foods that comprise the average diet in the United States. However, Gabay and Sax (1969) showed that most of the radium from ingested Brazil nuts is not retained.

Isotope: Radium-228 (^{228}Ra)

Half-life:
  Physical: 5.75 years.
  Biological: Retention is described by a complicated power function equation (ICRP Publication 20, 1973).
Natural Levels: Soil range similar to ^{232}Th.
Sources: Naturally occurring decay product of ^{232}Th.
Principal Modes and Energies of Decay (MeV): β_1^- ave., 0.00374; max., 0.0149 (40%); β_1^- ave., 0.00994; max., 0.039 (60%).
Special Chemical and Biological Characteristics: See ^{226}Ra. No loss of progeny (^{220}Rn) owing to short half-life of 56 seconds.
Principal Organ: Mineralized bone volume.
Amount of Element in Body: Normal body burden of ^{228}Ra believed to be 0.25–0.5 that of ^{226}Ra.
Daily Intake in Food and Water: Believed to be 0.25 to 0.5 that of ^{226}Ra.
Principal Human Metabolic and Dosimetric Parameters:
  f_1 = 0.2 (all forms).
  ALI = 0.04 MBq (1 μCi), IC-W (all forms),
       = 0.09 MBq (2 μCi), ingestion (all forms).
  DAC = 2×10^{-5} MBq m^{-3} (5×10^{-10} μCi cm^{-3}), IC-W (all forms).
Additional Dosimetric Considerations: Ingestion, 5.82×10^{-6} Sv Bq^{-1} (2.15×10^4 mrem μCi^{-1}) to the bone surface. Inhalation, 1.29×10^{-6} Sv Bq^{-1} (4.77×10^3 mrem μCi^{-1}).

Isotope: Thorium-230 (^{230}Th)

Half-life:
  Physical: 7.538×10^4 years.
  Biological: Bone, 8000 days; liver and other tissues, 700 days; transfer compartment, 0.5 days.

Natural Levels: Soil levels similar to ^{238}U.

Sources: Naturally occurring decay product of ^{238}U chain.

Principal Modes and Energies of Decay (MeV): α, 4.69 (76.3%), 4.62 (23.4%).

Special Chemical and Biological Characteristics: See ^{232}Th.

Principal Organ: Bone surface.

Amount of Element in Body: 30 μg found in mineralized bone.

Daily Intake in Food and Fluids: 3 μg (total thorium).

Amount of Radionuclide in Body: ~3 pCi (110 mBq) (Wrenn *et al.*, 1981).

Principal Human Metabolic and Dosimetric Parameters:

f_1 = 2×10^{-4} (all forms).

ALI = 6×10^{-4} MBq (0.02 μCi), IC-Y (oxides),

     = 2×10^{-4} MBq (0.006 μCi), IC-W (all other forms),

     = 0.1 MBq (4 μCi), ingestion (all forms).

DAC = 2×10^{-7} MBq m^{-3} (6×10^{-12} μCi cm^{-3}), IC-Y (oxides),

     = 1×10^{-7} MBq m^{-3} (3×10^{-12} μCi cm^{-3}), IC-W (all other forms).

Additional Dosimetric Considerations (all factors are for dose to the bone surface): Ingestion, 3.60×10^{-6} Sv Bq^{-1} (1.33×10^4 mrem μCi^{-1}). Inhalation, Class W is 2.16×10^{-3} Sv Bq^{-1} (7.99×10^6 mrem μCi^{-1}) and Class Y is 8.71×10^{-4} Sv Bq^{-1} (3.22×10^6 mrem μCi^{-1}).

Whole-Body Retention Function: $R(t) = e^{-0.693t/50}$, with t as elapsed time in days.

Systemic Transfer Fractions: Bone, 0.7; liver, 0.04; other tissues, 0.16.

Special Ecological Aspects: ^{230}Th is found in elevated concentrations in uranium mill tailings and certain phosphate fertilizers.

Isotope: Thorium-232 (^{232}Th)

Half-life:

Physical: 1.405×10^{10} years.

Biological: Bone, 8000 days; liver and other tissues, 700 days; transfer compartment, 0.5 days.

Natural Levels: Range from 1.6 to 20 ppm in the common rock types with a crustal average of 10.7 ppm (also see Table 6-6).

Sources: Naturally occurring primordial nuclide.

Principal Modes and Energies of Decay (MeV): α, 4.01 (77.8%), 3.95 (22.1%); γ, 0.0638 (0.267%).

Special Chemical and Biological Characteristics: Hydroxides and oxides are insoluble; nitrates, sulfates, chlorides, and perchloride salts are readily soluble. Tendency to accumulate on bone surfaces.

Principal Organ: Bone surface.

Amount of Element in Body: 30 μg.

Daily Intake in Food and Fluids: 3 μg (total thorium).

Amount of Radionuclide in Body: ~1.3 pCi (48 mBq) (Wrenn *et al.*, 1981).

Principal Human Metabolic and Dosimetric Parameters:

$f_1 = 2 \times 10^{-4}$ (all forms).

ALI = 1×10^{-4} MBq (0.003 μCi), IC-Y (oxides),

     = 4×10^{-5} MBq (0.001 μCi), IC-W (all other forms),

$= 0.03$ MBq (0.7 μCi), ingestion (all forms).

DAC $= 4 \times 10^{-8}$ MBq m^{-3} (1×10^{-12} μCi cm^{-3}), IC-Y (oxides),

$= 2 \times 10^{-8}$ MBq m^{-3} (5×10^{-13} μCi Cm^{-3}), IC-W (all other forms).

Additional Dosimetric Considerations (all factors are for dose to the bone surface): Ingestion, 1.85×10^{-5} Sv Bq^{-1} (6.85×10^{4} mrem μCi^{-1}). Inhalation, Class W is 1.11×10^{-2} Sv Bq^{-1} (4.11×10^{7} mrem μCi^{-1}) and Class Y is 4.99×10^{-3} Sv Bq^{-1} (1.85×10^{7} mrem μCi^{-1}).

Systemic Transfer Fractions: Bone, 0.7; liver, 0.04; other tissues, 0.16.

Special Ecological Aspects: Thorium-bearing minerals result in anomalously high natural levels in certain areas in Brazil, India, and China. Depending on type of rock, the concentration of ^{232}Tn in the earth's crust is 2–20 μg g^{-1} and about 10 μg g^{-1} for normal soil.

Isotope: Uranium-235 (^{235} U)

Half-life:

Physical: 7.038×10^{8} years.

Biological: Two compartments for each of the following: mineralized bone, 20 and 5000 days; kidney and other tissues, 6 and 1500 days

Natural Levels: 0.72% of ^{238}U by mass.

Sources: Naturally occurring primordial nuclide, 0.720% of natural uranium in the earth's crust. Enriched with respect to ^{238}U in fissionable material for reactors and weapons.

Specific Activity: 15.8 nCi g^{-1} (582 kBq kg^{-1}) or ~64 kg mCi^{-1} (per 37 MBq).

Principal Modes and Energies of Decay (MeV): α, 4.22 (5.7%), 4.32 (4.4%), 4.40 (55%), 4.56 (4.2%), 4.37 (17%); γ, 0.196 (61%).

Special Chemical and Biological Characteristics: See ^{238}U.

Principal Organ: Mineralized bone volume.

Amount of Element in Body: 90 μg with 59 μg found in the skeleton.

Daily Intake in Food and Fluids: 1.9 μg.

Principal Human Metabolic and Dosimetric Parameters:

ALI $= 0.05$ MBq (1 μCi), IC-D (UO$_2$, U$_3$O$_8$), $f_1 = 0.05$,

$= 0.03$ MBq (0.8 μCi), IC-W (UO$_3$, UF$_4$, UCl$_4$) $f_1 = 0.05$,

$= 0.002$ MBq (0.04 μCi), [UF$_6$, UO$_2$F$_2$, UO$_2$(NO$_3$)$_2$], $f_1 = 0.002$,

$= 0.5$ MBq (10μCi), ingestion of hexavalent forms, $f_1 = 0.05$,

$= 7$ MBq (200 μCi), ingestion of insoluble forms, $f_1 = 0.002$.

DAC $= 2 \times 10^{-5}$ MBq m^{-3} (6×10^{-10} μCi cm^{-3}), IC-D (UO$_2$,U$_3$O$_8$), $f_1 = 0.05$,

$= 1 \times 10^{-5}$ MBq m^{-3} (3×10^{-10} μCi cm^{-3}), IC-W (UO$_3$, UF$_4$, UCl$_4$),

$f_1 = 0.05$,

$= 6 \times 10^{-7}$ MBq m^{-3} (2×10^{-11} μCi cm^{-3}), IC-Y [UF$_6$, UO$_2$,F$_2$, UO$_2$(NO$_3$)$_2$], $f_1 = 0.002$.

Additional Dosimetric Considerations: Ingestion, 1.05×10^{-6} Sv Bq^{-1} (3.89×10^{3} mrem μCi^{-1}) to the bone surface for $f_1 = 0.05$ and 7.22×10^{-9} Sv Bq^{-1} (26.7 mrem μCi^{-1}) for $f_1 = 0.002$. Inhalation, Class D is 1.01×10^{-5} Sv Bq^{-1} (3.74×10^{4} mrem μCi^{-1}) to the bone surface, Class W is 1.97×10^{-6} Sv Bq^{-1}

$(7.29 \times 10^{-3}$ mrem $\mu\text{Ci}^{-1})$, and Class Y is 3.32×10^{-5} Sv Bq^{-1} $(1.23 \times 10^5$ mrem $\mu\text{Ci}^{-1})$.

Bone Retention Function: $R(t) = 0.2e^{-0.693t/20} + 0.023e^{-0.693t/5000}$, with t as elapsed time in days.

Kidney and Other Tissues: $R(t) = 0.12e^{-0.693t/6} + 0.00052e^{-0.693t/1500}$, with t as elapsed time in days.

Systemic Transfer Fractions (longer half-life listed second): Bone, 0.2 and 0.023; kidney and other tissues, 0.12 and 0.00052.

Isotope: Uranium-238 (^{238}U)

Half-life:

Physical: 4.468×10^9 years.

Biological: Two compartments for each of the following: mineralized bone, 20 and 5000 days; kidney and other tissues, 6 and 1500 days.

Natural Levels: 0.5 to 4.7 ppm uranium in the common rock types (also see Table 6-6).

Sources: Naturally occurring primordial nuclide, 99.2745% by weight in natural uranium.

Specific Activity: 330 nCi g^{-1} (12 kBq g^{-1}) or 3 g μCi^{-1} (per 37 kBq).

Principal Modes and Energies of Decay (MeV): α, 4.21 (77%), 4.15 (23%).

Special Chemical and Biological Characteristics: Since ^{Nat}U has a low specific activity, chemical damage to the kidneys is likely to be more limiting than radiation damage. However, radiation damage to the lung and kidney must be considered if exposure is to enriched uranium.

Principal Organ: Mineralized bone volume.

Amount of Element in Body: 90 μg with 59 μg found in the skeleton.

Daily Intake in Food and Fluids: 1.9 μg.

Principal Human Metabolic and Dosimetric Parameters:

ALI = 0.05 MBq (1 μCi), IC-D (UO$_2$, U$_3$O$_8$), f_1 = 0.05,
   = x0.03 MBq (0.8 μCi) IC-W (UO$_3$, UF$_4$, UCl^4), f_1 = 0.05,
   = 0.002 MBq (0.04 μCi), IC-Y [UF$_6$, UO$_2$F$_2$, UO$_2$(NO$_3$)$_2$], f_1 = 0.002,
   = 0.5 MBq (10 μCi), ingestion of hexavalent forms, f_1 = 0.05,
   = 8 MBq (200 μCi), ingestion of insoluble forms, f_1 = 0.002.

DAC = 2 $\times$ 10^{-5} MBq m^{-3} (6 $\times$ 10^{-10} μCi cm^{-3}), IC-D (UO$_2$, U$_3$O$_8$), f_1 = 0.05,
   = 1 $\times$ 10^{-5} MBq m^{-3} (3 $\times$ 10^{-10} μCi cm^{-3}), IC-W (UO$_3$, UF$_4$, UCl$_4$),
      f_1 = 0.05,
   = 7 $\times$ 10^{-7} MBq m^{-3} (2 $\times$ 10^{-11} μCi cm^{-3}), IC-Y [UF$_6$,
      UO$_2$F$_2$, UO$_2$(NO$_3$)$_2$], f_1 = 0.002,

Additional Dosimetric Considerations: Ingestion, 1.01×10^{-6} Sv Bq^{-1} $(3.74 \times 10^{-3}$ mrem $\mu\text{Ci}^{-1})$ to the bone surface for f_1 = 0.05 and 6.42×10^{-9} Sv Bq^{-1} (23.8 mrem μCi^{-1}) for f_1 = 0.002. Inhalation, Class D is 9.78×10^{-6} Sv Bq^{-1} $(3.62 \times 10^{-4}$ mrem $\mu\text{Ci}^{-1})$ to the bone surface, and Class W is 1.90×10^{-6} Sv Bq^{-1} $(7.03 \times 10^{-3}$ mrem $\mu\text{Ci}^{-1})$, and Class Y is 3.20×10^{-5} Sv Bq^{-1} $(1.18 \times 10^{-5}$ mrem $\mu\text{Ci}^{-1})$.

Bone Retention Function: $R(t) = 0.2e^{-0.693t/20} + 0.023e^{-0.693t/5000}$, with t as elapsed time in days.

Kidney and Other Tissues: $R(t) = 0.12e^{-0.693t/6} + 0.00052e^{-0.693t/1500}$, with t as elapsed
time in days.
Systemic Transfer Fractions (longer half-life listed second): Bone, 0.2 and 0.023;
kidney and other tissues, 0.12 and 0.00052.

Isotope: Neptunium-237 (^{237}Np)

Half-life:
    Physical: 2.14×10^6 years.
    Biological: Bone, 50 years; liver, 20 years; gonads, considered permanent.
Sources: Produced from neutron capture reaction of ^{238}U and decay of ^{241}Am.
Principal Modes and Energies of Decay (MeV): α, 4.64 (6.18%), 4.65 (3.32%),
    4.77 (8%), 4.79 (47%), 4.77 (25%), 4.82 (2.5%); γ, 0.0294 (15%), 0.00822 (9%),
    0.0865 (12.4%).
Special Chemical and Biological Characteristics: No data on distribution or reten-
    tion in humans; based solely on animal data. The f_1 value below from ICRP
    Publication 30 (1980) recommendations is likely to overestimate true gut absorp-
    tion by a factor of 10. Similarly, systemic transfer functions to the bone and liver
    may be better represented by 0.60 and 0.15, respectively (Thompson, 1982).
    Uncertainty exists with respect to the uniformity of bone deposition.
Principal Organ: Bone surface.
Principal Human Metabolic and Dosimetric Parameters:
    $f_1 = 0.001$ (all forms).
    ALI $= 2 \times 10^{-4}$ MBq (0.004 μCi), IC-W (all forms),
        $= 0.02$ MBq (0.5 μCi), ingestion (all forms).
    DAC $= 6 \times 10^{-8}$ MBq m^{-3} (2×10^{-12} μCi cm^{-3}), IC-W (all forms).
*Additional Dosimetric Considerations (all factors are for dose to the bone
    surface):* Ingestion, 2.72×10^{-5} Sv Bq^{-1} (1.01×10^{-5} mrem μCi^{-1}). Inhalation,
    3.27×10^{-3} Sv Bq^{-1} (1.21×10^{-7} mrem μCi^{-1}).
Systemic Transfer Fractions: Mineralized bone, 0.75; liver, 0.15; gonads, 0.00035
    (testes) and 0.00011 (ovaries).

Isotope: Plutonium-238 (^{238}Pu)

Half-life
    Physical: 87.7 years.
    Biological: Bone, 50 years; liver, 20 years; gonads, considered permanent.
Sources: Minor constituent of fallout ($\sim$3.5% ^{239,240}Pu by activity). Produced for
    radioisotope thermal generators (RTGs) by ^{237}Np(n,β^-)^{238}Pu.
Principal Modes and Energies of Decay (MeV): α, 5.50 (70.9%), 5.46 (29%); L
    X ray 0.0136 (11.7%).
Special Chemical and Biological Characteristics: General forms are insoluble
    fluorides, hydroxides, and oxides. Solubility in water is dependent on redox, pH,
    and organic ligands present. Metabolism is described in detail in ICRP Publication
    48 (1986).
Principal Organs: Bone surfaces, liver.

Amount of Element in Body: Trace.

Principal Human Metabolic and Dosimetric Parameters:

ALI = 7×10^{-4} MBq (0.02 μCi), IC-Y (oxides), $f_1 = 1 \times 10^{-5}$,

= 3×10^{-4} MBq (0.007 μCi), IC-W (all other forms), $f_1 = 0.001$,

= 3 MBq (90 μCi), ingestion of oxides, $f_1 = 1 \times 10^{-5}$,

= 0.3 MBq (9 μCi), ingestion of nitrates, $f_1 = 1 \times 10^{-4}$,

= 0.03 MBq (0.9 μCi), ingestion of all other forms, $f_1 = 0.001$.

DAC = 3×10^{-7} MBq m^{-3} (8×10^{-12} μCi cm^{-3}), IC-Y (oxides), $f_1 = 1 \times 10^{-5}$,

= 1×10^{-7} MBq m^{-3} (3×10^{-12} μCi cm^{-3}), IC-W (all other forms),

$f_1 = 0.001$.

Additional Dosimetric Considerations: Ingestion, 1.58×10^{-5} Sv Bq^{-1} (5.85×10^4 mrem μCi^{-1}) to the bone surface for $f_1 = 1 \times 10^{-3}$, 1.58×10^{-6} Sv Bq^{-1} (5.85×10^3 mrem μCi^{-1}) for $f_1 = 1 \times 10^{-4}$, and 1.58×10^{-7} Sv Bq^{-1} (585 mrem μCi^{-1}) for $f_1 = 1 \times 10^{-5}$. Inhalation, Class W is 1.90×10^{-3} Sv Bq^{-1} (7.03×10^6 mrem μCi^{-1}) to the bone surface and Class Y is 7.79×10^{-5} Sv Bq^{-1} (2.88×10^5 mrem μCi^{-1}).

Systemic Transfer Fractions: Skeleton, 0.45; liver, 0.45; gonads, 0.00035 (testes) and 0.00011 (ovaries).

Isotope: Plutonium-239 (^{239}Pu)

Half-life:

Physical: 2.41×10^4 years.

Biological: Bone, 50 years; liver, 20 years; gonads, considered permanent.

Sources: Produced in fission reactors by neutron irradiation of ^{238}U. Used in nuclear weapons and as fuel for fission reactors. Also fallout product from nuclear weapons testing.

Principal Modes and Energies of Decay (MeV): α, 5.16 (73.3%).

Special Chemical and Biological Characteristics: See ^{238}Pu.

Principal Organ: Bone surfaces.

Amount of Element in Body: Trace.

Principal Human Metabolic and Dosimetric Parameters:

ALI = 6×10^{-4} MBq (0.02 μCi), IC-Y (oxides) $f_1 = 1 \times 10^{-5}$,

= 2×10^{-4} MBq (0.006 μCi), IC-W (all other forms) $f_1 = 0.001$,

= 3 MBq (80 μCi), ingestion of oxides, $f_1 = 1 \times 10^{-5}$,

= 0.3 MBq (8 μCi), ingestion of nitrates, $f_1 = 1 \times 10^{-4}$,

= 0.03 MBq (08 μCi), ingestion of all other forms, $f_1 = 0.001$.

DAC = 3×10^{-7} MBq m^{-3} (7×10^{-12} μCi cm^{-3}), IC-Y (oxides) $f_1 = 1 \times 10^{-5}$,

= 1×10^{-7} MBq m^{-3} (3×10^{-12} μCi cm^{-3}), IC-W (all other forms),

$f_1 = 0.001$

Additional Dosimetric Considerations (all factors are for dose to the bone surface): Ingestion, 1.76×10^{-5} Sv Bq^{-1} (6.51×10^4 mrem μCi^{-1}) for $f_1 = 1 \times 10^{-3}$, 1.76×10^{-6} Sv Bq^{-1} (6.51×10^3 mrem μCi^{-1}) for $f_1 = 1 \times 10^{-4}$, and $1.76 \times$ Sv Bq^{-1} (651 mrem μCi^{-1}) for $f_1 = 1 \times 10^{-5}$. Inhalation, Class W is 2.11×10^{-3} Sv Bq^{-1} (7.81×10^6 mrem μCi^{-1}) and Class Y is 8.21×10^{-4} Sv Bq^{-1} (3.04×10^{-6} mrem μCi^{-1}).

Systemic Transfer Fractions: Skeleton, 0.45; liver, 0.45; gonads, 0.00035 (testes) and 0.00011 (ovaries).

Isotope: Americium-241 (^{241}Am)

Half-life:

    Physical: 432.2 years.

    Biological: Bone, 50 years; liver, 20 years; gonads, considered permanent.

Sources: Decay product of ^{241}Pu.

Principal Modes and Energies of Decay (MeV): α, 5.49 (84.5%), 5.44 (13%); γ, 0.0595 (35.9%), 0.0263 (2.4%).

Special Chemical and Biological Characteristics: Oxides and hydroxides of Am(III) are relatively insoluble.

Principal Organs: Bone surfaces, liver.

Amount of Element in Body: Trace.

Principal Human Metabolic and Dosimetric Parameters:

    $f_1 = 0.001$ (all forms).

    ALI $= 2 \times 10^{-4}$ MBq (0.006 μCi), IC-W (all forms),

        $= 0.03$ MBq (0.8 μCi), ingestion (all forms).

    DAC $= 1 \times 10^{-7}$ MBq m^{-3} (3×10^{-12} μCi cm^{-3}, IC-W (all forms).

Additional Dosimetric Considerations (all factors are for dose to the bone surface): Ingestion, 1.81×10^{-5} Sv Bq^{-1} (6.7×10^4 mrem μCi^{-1}). Inhalation, 2.17×10^{-3} Sv Bq^{-1} (8.03×10^6 mrem μCi^{-1}).

Systemic Transfer Fractions: Skeleton, 0.45; liver, 0.45; gonads, 0.00035 (testes) and 0.00011 (ovaries).

References

Aarkrog, A. (1971a). Radioecological investigations of plutonium in an arctic marine environment. *Health Phys.* **20,** 31–47.

Aarkrog, A. (1971b). Prediction models for ^{90}Sr and ^{137}Cs levels in the human food chain. *Health Phys.* **20,** 297–312.

Aarkrog, A., Dahlgaard, H., and Nilsson, K. (1984). Further studies of plutonium and americium at Thule, Greenland. *Health Phys.* **46,** 29–44.

Abbott, M. L., Fraley, L., Jr., and Reynolds, T. D. (1991). Root profiles of selected cold desert shrubs and grasses in disturbed and undisturbed soils. *Environ. Exp. Bot.* **31,** 165–178.

Adler, H. L., and Weinberg, A. M. (1978). An approach to setting radiation standards. *Health Phys.* **34,** 719–720.

Adriano, D. C., Pinder III, J. E., McLeod, K. W., Corey, J. C., and Boni, A. L. (1982). Plutonium contents and fluxes in a soybean crop ecosystem near a nuclear fuel chemical separation plant. *J. Environ. Qual.* **11,** 506–511.

AEC (1948). "April 2, 1948, Report of the Safety and Industrial Health Advisory Board." p. 67. U.S. Atomic Energy Commission, Washington, D.C.

AEC (1957a). Theoretical possibilities and consequences of major accidents in large nuclear power plants. Report WASH-740. U.S. Atomic Energy Commission, Washington, D.C.

AEC (1957b). "Atomic Energy Facts." U.S. Atomic Energy Commission, Washington, D.C.

AEC (1960a). "Hazards Summary Report on Consolidated Edison Thorium Reactor." Docket No. 50-3, Exhibit K-5 (Revision 1). U.S. Atomic Energy Commission, Washington, D.C.

AEC (1960b). Summary of available data on the strontium-90 content of foods and of total diets in the United States. Report HASL-90. U.S. Atomic Energy Commission, New York.

AEC (1961). Investigation board report on the SL-1 accident. U.S. Atomic Energy Commission, Washington, D.C.

AEC (1969). Press release No. M-257. U.S. Atomic Energy Commission, Washington, D.C.

AEC (1970a) "The Nuclear Industry—1970." U.S. Atomic Energy Commission, Washington, D.C.

AEC (1970b). Survey report on structural design of piping systems and components, TID-25553. U.S. Atomic Energy Commission, Washington, D.C.

AEC (1972a). "Biomedical Implications of Radiostrontium Exposure." AEC Symposium Series. U.S. Atomic Energy Commission, Washington, D.C.

AEC (1972b). Grand Junction remedial action criteria. *Fed. Regist.* **12/6/72,** 25918–25919.

AEC (1973). Report on the investigation of the 106 T tank leak at the Hanford Reservation, Richland, Washington. Richland Operations Office, Richland, Washington.

Agricola, G. (1956). "De Re Metallica" (translated by H. Hoover and L. Hoover). Reprinted by Dover, New York (1950).

Ahearne, J. F. (1993). The future of nuclear power. *Am. Sci.* **81,** 24–35.

Ainsworth, C. C., Hill, R. L., Cantrell, K. J., Kaplan, D. I., Norton, R. L., Aaberg, R. L., and Stetar, E. A. (1994). Reconcentration of radioactive material released to sanitary sewer systems, Report PNL-10193 and NUREG/CR-6289. Pacific Northwest Laboratory, Richland, Washington.

Akopova, A. B., Dudkin, V. E., Kovalev, E. E., Magradze, N. V., and Potapov, Y. V. (1987). Linear energy transfer spectra of cosmic radiation aboard the Cosmos-1129 satellite. *Radiat. Protect. Dosim.* **18,** 153–156.

Alavanja, M. C. R., Brownson, R. C., Lubin, J. L., Berger, E., Chiang, J., and Boice, J. D. (1994). Residential radon exposure and lung cancer among non-smoking women. *J. Nat. Cancer Inst.* **86,** 1829–1837.

Albenesius, E. L. (1959). Tritium as a product of fission. *Phys. Rev. Lett.* **3,** 274.

Albert, R. E. (1966). "Thorium—Its Industrial Hygiene Aspects." Academic Press, Orlando, Florida.

Albert, R. E., and Altshuler, B. (1973). Considerations relating to the formulation of limits for unavoidable population exposures to environment carcinogens. *In* "Radionuclide Carcinogenesis" (C. L. Sanders, R. H. Busch, J. E. Ballou, and D. D. Mahlum, eds.), AEC Symp. Ser., CONF-72050, pp. 233–253. NTIS, Springfield, Virginia.

Alberts, J. J., Bobula, C. M., and Farras, D. T. (1980). A comparison of the distribution of industrially released ^{238}Pu and fallout ^{239,240}Pu in temperate, northern United States soils. *J. Environ. Qual.* **9,** 592–596.

Alexakhin, R. M. (1990). ^{137}Cs migration in agroecosystems of Byelorussian woodlands. *In* "Proceedings of First International Conference on the Biological and Radioecological Aspects of the Chernobyl Accident." USSR Academy of Sciences, Moscow.

Alexander, L. T. (1967). Depth of penetration of the radioisotopes strontium-90 and cesium-137, Fallout Program Quarterly Summary, Report HASL-183. U.S. Atomic Energy Commission, New York.

Alexander, L. T., Hardy, E. P., and Hollister, H. L. (1960). Radioisotopes in soils: Particularly with reference to strontium-90. *In* "Radioisotopes in the Biosphere" (R. S. Caldecott and L. A. Snyder, eds.), pp. 3–22. Univ. of Minnesota Press, Minneapolis.

Alexiades, V., Jacobs, G. K., and Dunbar, N. W. (1994). Constraints on mass balance of soil moisture during *in situ* vitrification. *Environ. Geol.* **23,** 83–88.

Alter, H. W., and Oswald, R. A. (1987). Nationwide distribution of indoor radon measurements: A preliminary data base. *J. Air Pollut. Control Assoc.* **37,** 227–231.

Altshuler, B., Nelson, N., and Kuschner, M. (1964). Estimation of the lung tissue dose from the inhalation of radon and daughters. *Health Phys.* **10,** 1137–1162.

Alvarez, M. C., and Garzon, L. (1989). Assessment of radiological emissions from Spanish coal power plants: Radioactive releases and associated risks. *Health Phys.* **57,** 765–769.

American Cancer Society (1996). "Cancer Statistics." Published annually in *Ca, A Cancer Journal for Clinicians,* American Cancer Society.

American Conference of Governmental Industrial Hygienists (1974). "Air Sampling Instruments for Evaluation of Atmospheric Contaminants," 4th Ed. American Conference of Governmental Industrial Hygienists, Washington, D.C.

American National Standards Institute (1975). "Performance, Testing and Procedural Specifications for Thermoluminescence Dosimetry (Environmental Applications)." ANSI N545-1975. American National Standards Institute, New York.

American Nuclear Society (1971). Symposium for reactor containment spray system technology. *Nuclear Technol.* **10,** 400.

American Nuclear Society (1984). "Report of Special Committee on Source Terms." ANS, La Grange Park, Illinois.

American Nuclear Society (1986). The Chernobyl accident. *Nucl. News (La Grange Park, Ill.)* **June.**

American Nuclear Society (1992). The new reactors. *Nucl. News (La Grange Park, Ill.)* **35**(12), 65–90.

American Public Health Association (1985). "Standard Methods for the Examination of Water and Wastewater," 16 Ed. American Public Health Association, New York.

Ames, B. N., and Gold, L. S. **(19**90). Chemical carcinogenesis: Too many rodents. *Proc. Nat. Acad. Sci. U.S.A.* **87,** 7772–7776.

Anantatmula, R. P., Schwenk, E. B., and Danielson, M. J. (1994). Characterization of the corrosion behavior of the carbon steel linear in Hanford Site single shell tanks. Report WHC-EP-0772, June. Westinghouse Hanford Company, Richland, Washington.

Anderson, E. C. (1953). The production and distribution of natural radiocarbon. *Annu. Rev. Nucl. Sci.* **2,** 63.

Anderson, M. P., and Woessner, W. W. (1992). "Applied Groundwater Modeling." Academic Press, San Diego.

Anderson, W., and Turner, R. C. (1956). Radon content of the atmosphere. *Nature (London)* **178,** 203.

Anderson, W., Mayneord, W. V., and Turner, R. C. (1954). The radon content of the atmosphere. *Nature (London)* **74,** 424.

Angelovic, J. W., White, J. C., and Davis, E. M. (1969). Interaction of ionizing radiation, salinity, and temperature on the estuarine fish *Fundulus heteroclitus. In* Proceedings of the Second National Symposium on Radioecology, 1969," CONF-670503, pp. 131–141. U.S. Atomic Energy Commission, Washington, D.C.

Ansoborlo, E., Henge-Napoli, M. H., Donnadieu-Claraz, M., Roy, M., and Pihet, P. (1994). Industrial exposure to uranium aerosols at laser enrichment processing facilities. *Radiat. Protect. Dosim.* **53**(1–4), 163–167.

Anspaugh, L. R., Shinn, J. H., Phelps, P. L., and Kennedy, N. C. (1975). Resuspension and redistribution of plutonium in soils. *Health Phys.* **29,** 571–582.

Anspaugh, L. R., Catlin, R. J., and Goldman, M. (1988). The global impact of the Chernobyl reactor accident. *Science* **242,** 1513–1519.

Archer, V. E. (1981). Health concerns in uranium mining and milling. *J. Occup. Med.* **23,** 502–505.

Arnold, L. (1992). "Windscale 1957." Macmillan, London.

Arthur, III, W. J., Markham, O. D. (1984). Polonium-210 in the environment around a radioactive waste disposal area and phosphate ore processing plant. *Health Phys.* **46,** 793–799.

Atomic Industrial Forum (1961). p. 27. Washington, D.C. (Obtainable from Nuclear Energy Institute.)

Auerbach, S. I., Nelson, D. J., Kaye, S. V., Reichle, D. E., and Coutant, C. C. (1971). Ecological considerations in reactor power plant siting. *In* "Proceedings on the Environmental Aspects of Nuclear Power Stations, 1971." International Atomic Energy Agency, Vienna.

Aulenbach, D. B., and Davis, R. E. (1976). Long-term consumption of mineral water containing natural Ra-226. *In* "Proceedings of the Tenth Midyear Topical Symposium." Health Physics Society. Rensselaer Polytechnic Institute, Troy, New York.

Auxier, J. A., and Dickson, H. W. (1983). Concern over recent use of the ALARA philosophy. *Health Phys.* **44,** 595–600.

Auxier, J. A., Cheka, J. S., Haywood, F. F., Jones, T. D., and Thorngate, J. H. (1966). Free-field radiation-dose distributions from the Hiroshima and Nagasaki bombings. *Health Phys.* **12,** 425–429.

Axelson, O., and Sundell, L. (1978). Mining, lung cancer and smoking. *Scand. J. Work Environ. Health* **4,** 46.

Babad, H., Eberlein, S. J., Johnson, G. D., Meacham, J. E., Osborne, J. W., Payne, M. A., and Turner, D. A. (1995). Progress in resolving Hanford Site high level waste safety issues. Report WHC-SA-2639-FP. Westinghouse Hanford Company, Richland, Washington.

Baker, D. A. (1992). Population dose commitments due to radioactive releases from nuclear power plant sites in 1988. Report NUREG/CR-2850 (PNL-4221), Vol. 10. Pacific Northwest Laboratory, Richland, Washington.

Bakulin, V. N., Senko, E. E., and Starov, B. G. (1970). Investigation of turbulent exchange and wash-out by measurement of natural radioactivity in surface air. *J. Geophys. Res.* **75,** 3669–3674.

Balanov, M. I. (1993). Overview of doses to the Soviet population from the Chernobyl accident and the protective actions applied. *In* "The Chernobyl Papers 1" (S. E. Merwin and M. I. Balanov, eds.), pp. 23–45. Research Enterprises, Richland, Washington.

Balter, M. (1995). Filtering a river of cancer data. *Science* **267,** 1064–1068.

Baranov, A. E., and Guskova, A. K. (1990). Acute radiation disease in the Chernobyl accident victims. *In* "The Medical Basis for Radiation Accident Preparedness II" (R. C. Ricks and S. A. Fry, eds.), pp. 79–87. Elsevier, New York.

Barcinski, M. A., Abreu, M. C. A., de Almeida, J. C. C., Naya, J. M., Fonseca, L. G., and Castro, L. E. (1975). Cytogenetic investigations in a Brazilian population living in a high natural radioactivity. *Am. J. Hum. Genet.* **27,** 802–806.

Barish, R. J. (1995). Health physics and aviation: 1990–1994. *Health Phys.* **69,** 538–542.

Barnaby, F. (1982). The effect of a global war: The arsenals. *Ambio* **11,** 76.

Barnard, R. W., Wilson, M. L., Dockery, H. A., Gauthier, J. H., Kaplan, P. G., Eaton, R. R., Bingham, F. W., and Robey, T. H. (1992). An initial total-system performance assessment for Yucca Mountain. Report SAND 91–2795. Sandia National Laboratories, Albuquerque, New Mexico.

Barreira, F. (1961). Concentration of atmospheric radon and wind directions. *Nature (London)* **190,** 1092.

Bartholomay, R. C., Orr, B. R., Liszewski, M. J., and Jensen, R. G. (1995). Hydrologic conditions and distribution of selected chemical constituents in water, Snake River Plain Aquifer, Idaho National Engineering Laboratory, Idaho, 1989 to 1991. Water Resources Investigations Report 95–4175. U.S. Geological Survey, Denver, Colorado.

Bartlett, B. O., and Russell, R. S. (1966). Prediction of future levels of long-lived fission products in milk. *Nature (London)* **209,** 1062–1065.

Battist, L., Buchanan, J., Congel, F., Nelson, C., Nelson, M., Peterson, H., and Rosenstein, M. (1979). "Population Dose and Health Impact of the Accident at the Three Mile Island Nuclear Station, a Preliminary Assessment for the Period March 28 through April 7, 1979." U.S. Government Printing Office, Washington, D.C.

Baverstock, K. F., Frank, B., and Simon, S. L. (1994). Summary report of first phase: Determining compliance with agreed limits for total annual dose-rate on Rongelap Island and actinide contamination of soils on Rongelap Island and neighboring islands. Rongelap Resettlement Project, Majuro, Marshall Islands.

Beasley, T. M., and Palmer, H. E. (1966). Lead-210 and polonium-210 in biological samples from Alaska. *Science* **152,** 1062–1063.

Beck, H. L. (1966). Environmental gamma radiation from deposited fission products, 1960–1964. *Health Phys.* **12,** 313–322.

Beck, H. L. (1972). The physics of environmental radiation fields. *In* "The Natural Radiation Environment II" (J. A. S. Adams, W. M. Lowder, and T. F. Gesell, eds.), U.S. Department of Energy CONF-720805-P1, pp. 101–133. NTIS, Spring-field, Virginia.

Beck, H. L. (1980). Factors for radionuclides deposited on the ground. EML Report 378. U.S. Department of Energy, New York.

Beck, H. L., and de Planque, G. (1968). The radiation field in air due to distributed gamma-ray sources in the ground. Report HASL-195. U.S. Atomic Energy Commission, New York.

Beck, H. L., Lowder, W. M., Bennett, B. G., and Condon, W. J. (1966). Further studies of external environmental radiation. Report HASL-170. U.S. Atomic Energy Commission, New York.

Beck, H. L., Lowder, W. M., Bennett, B. G., and Condon, W. G. (1971). In situ external environmental gamma ray measurements using Ge (Li) and NaI (T1) spectrometry and pressurized ion chambers. Report SM/148-2. International Atomic Energy, Vienna.

Beck, H. L., Gogolak, C. V., Miller, K. M., and Lowder, W. M. (1980). Perturbations on the natural radiation environment due to the utilization of coal is an energy source. *In* "The Natural Radiation Environment III" (T. F. Gesell and W. M. Lowder, eds.), CONF-89-42, Vol. 2, pp. 1521–1558. U.S. Department of Energy, Washington, D.C.

Beck, H. L., Helfer, I. K., Bouville, A., and Dreicer, M. (1990). Estimates of fallout in the continental U.S. from Nevada weapons testing based on gummed-film monitoring data. *Health Phys.* **59,** 565–576.

Becker, K. (1973). "Solid State Dosimetry." CRC Press, Cleveland, Ohio.

Becquerel, H., and Curie, P. (1901). Action physiologique des rayons du radium. *C. R. Acad. Sci.* **132,** 1289–1291.

Bedrosian, P. H., Easterly, D. G., and Cummings, S. L. (1970). Radiological survey around power plants using fossil fuel. Report EERL-71-3. U.S. Environmental Protection Agency, Rockville, Maryland.

Behounek, F. (1970). History of exposure of miners to radon. *Health Phys.* **19,** 56–57.

Bennett, B. G. (1972). Estimation of ^{90}Sr levels in the diet. HASL Report 246. U.S. Atomic Energy Commission, New York.

Bennett, B. G. (1978). Environmental aspects of americium. EML Report 348. U.S. Department of Energy, New York.

Bennett, B. G., and Bouville, A. (1988). Radiation doses in countries of the northern hemisphere from the Chernobyl nuclear reactor accident. *Environ. Int.* **14,** 75–82.

Bennett, G. L. (1981). Overview of the U.S. flight safety process for space nuclear power. *Nuclear Saf.* **22,** 423–434.

Bennett, G. L., Lombardo, J. J., and Rock, B. J. (1984). U.S. radioisotope thermoelectric generator space operating experience (June 1961–December 1982). *Nucl. Eng. (Inst. Nucl. Eng.)* **25,** 49–58.

Beral, V. (1996). Thyroid cancer around Chernobyl. Paper presented at the Thirty-Second Annual Meeting of the National Council on Radiation Protection and Measurements, April 3–4, 1996, Bethesda, Maryland (proceedings in preparation).

Berlin, R. E., and Stanton, C. C. (1988). "Radioactive Waste Management." Wiley, New York.

Bernhardt, D. E. (1992). Depleted uranium cleanup criteria and associated risks. *Health Phys.* **June 62**(Suppl. 6) S37.

Bernhardt, D. E. (1995). Personal communication.

Bernhardt, D. E., Owen, D. H., and Rogers, V. C. (1996). Assessments of NORM in pipe from oil and gas production. *In* "NORM/NARM: Regulation and Risk Assessment, Proceedings of the 29th Midyear Topical Meeting of the Health Physics Society," Health Physics Society, McLean, Virginia.

Bertini, H. W. (1980). Descriptions of selected accidents that have occurred at nuclear reactor facilities. Report ORNL/NSIC-176. Oak Ridge National Laboratories, Oak Ridge, Tennessee.

Bjornerstedt, R., and Engstrom, A. (1960). Radioisotopes in the skeleton: Dose implications based on microscopic distribution. *In* "Radioisotopes in the Biosphere," Chap. 27. Univ. of Minnesota Press, Minneapolis.

Black, D. (1984). "Investigation of the Possible Increased Incidence of Cancer in West Cumbria." Report of the Independent Advisory Group. Her Majesty's Stationery Office, London.

Blanchard, R. L., and Holaday, D. A. (1960). Evaluation of radiation hazards created by thoron and thoron daughters. *Am. Ind. Hyg. Assoc. Q.* **21,** 201–206.

Blanchard, R. L., Fowler, T. W., Horton, T. R., and Smith, J. M. (1982). Potential health effects of radioactive emissions from active surface and underground uranium mines. *Nucl. Saf.* **23,** 439–450.

Blaylock, B. G. (1982). Radionuclide databases available for bioaccumulation factors for freshwater fish. *Nucl. Saf.* **23,** 427–438.

Blaylock, B. G., and Trabalka, J. R. (1978). Evaluating the effects of ionizing radiation on aquatic organisms. *Adv. Radiat. Biol.* **7,** 103–151.

Blifford, I. H., Lockhart, L. B., *et al.* (1952). On the natural radioactivity of the air. Report 4036. Naval Research Laboratory, Washington, D.C.

Blomeke, J. O., and Harrington, F. E. (1968). Waste management at nuclear power stations. *Nucl. Saf.* **9,** 239–248.

Blum, M., and Eisenbud, M. (1967). Reduction of thryoid irradiation from ^{131}I by potassium iodide. *JAMA, Am. Med. Assoc.* **200,** 1036–1040.

Blush, S. M., and Heitman, T. H. (1995). Train wreck along the river of money—An evaluation of the Hanford cleanup. U.S. Senate Committee on Energy and Natural Resources, Washington, D.C.

Bodnar, L. Z., and Percival, D. R. (1982). RESL Analytical Chemistry Branch procedures manual. Report IDO-12096. U.S. Department of Energy, Idaho Falls, Idaho.

Bond, V. P. (1982). The conceptual basis for evaluating risk from low-level radiation exposure. Issues in setting radiation standards. *In* Proceedings of the Seventeenth Annual Meeting of NCRP," pp. 25–26. National Council on Radiation Protection and Measurements, Bethesda, Maryland.

Bondietti, E. A., and Francis, C. W. (1979). Geologic migration potentials of technetium-99 and neptunium-237. *Science* **203,** 1337–1340.

Bondietti, E. A., Trabalka J. R., Garten, C. T., and Killough, G. G. (1979). Biogeochemistry of actinides: A nuclear fuel cycle perspective. Radioactive waste in geological storage. *Am. Chem. Soc. Symp. Ser.* **100,** 241–266.

Boone, F. W., Kantelo, M. V., Mayer, P. G., and Palms, J. M. (1985). Residence half-times of ^{129}I in undisturbed surface soils based on measured soil concentration profiles. *Health Phys.* **48,** 401–413.

Bowen, V. T., Noshkin, V. E., Livingston, H. D., and Volchok, H. L. (1980). Fallout radionuclides in the Pacific Ocean: Vertical and horizontal distributions, largely from Geosec stations. *Earth Planet. Sci. Lett.* **49,** 411–443.

Brazilian Academy of Sciences (Academia Brasileira de Ciencias). (1977). "Proceedings of the International Symposium on Areas of High Natural Radioactivity." Brazilian Academy of Sciences, Rio de Janeiro, Brazil.

Brecker, R., and Brecker, E. (1969). "The Rays: A History of Radiology in the United States and Canada." Williams & Wilkins, Baltimore, Maryland.

Breshears, D. D., Kirchner, T. B., and Whicker, F. W. (1992). Contaminant transport through agroecosystems: Assessing relative importance of environmental, physiological, and management factors. *Ecol. Appl.* **2,** 285–297.

Breslin, A. J., and Glauberman, H. (1970). Investigation of radioactive dust dispersed from uranium tailings piles. *In* "Environmental Surveillance in the Vicinity of Nuclear Facilities" (W. C. Reinig, ed.). Thomas, Springfield, Illinois.

Breslin, A. J., and Harris, W. B. (1952). "Use of Thoriated Tungsten Electrodes in Inert Gas Shielded Arc Welding—Investigation of Potential Hazard." Industrial Hygiene Quarterly, December, 1952. U.S. Atomic Energy Commission, New York Operations Office.

Brewer, A. W. (1949). Water vapour distribution in the stratosphere. *Q. J. R. Meterorol. Soc.* **75,** 351–363.

Briggs, G. A. (1984). Plume rise and buoyancy effects. *In* "Atmospheric Science and Power Production" (D. Randerson, ed.), Report DOE/TIC-27601, pp. 327–366. U.S. Department of Energy, Washington, D.C.

Brinck, W. L., Schliekelman, R. J., Bennet, D. L., Bell, C., and Markwood, I. M. (1978). Radium removal efficiencies in water treatment processes. *J. Am. Water Works Assoc.* **70,** 31–43.

British Nuclear Fuels Limited (1984). Annual report on radioactive discharges and monitoring of the environment, 1978. Health and Safety Directorate, Risley, Warington, Cheshire, U.K.

Brittan, R. O., and Heap, J. C. (1958). Reactor containment. *In* "Proceedings of the Second UN Conference on Peaceful Uses of Atomic Energy, Geneva," p. 437. United Nations, New York.

Broecker, W. S., Goddard, J., and Sarmiento, J. L. (1976). The distribution of ^{226}Ra in the Atlantic Ocean. *Earth Planet. Sci. Lett.* **32,** 220–235.

Broseus, R. W. (1970). Cesium-137/strontium-90 ratios in milk and grass from Jamaica. M.S. thesis, New York University, New York.

Brown, S. H., and Smith, R. C. (1980). A model for determining the overall release rate and annual source term for a commercial *in situ* leach uranium facility. *In* "Uranium Resources Technology," pp. 794–800. Seminar III. Colorado School of Mines, Boulder, Colorado.

Buchanan, J. R. (1963). SL-1 final report. *Nucl. Saf.* **4,** 83–86.

Bureau of Radiological Health (1970). "Radiological Health Handbook," Superintendent of Documents, Washington, D.C.

Burnett, T. J. (1970). A derivation of the factor of 700 for [131]I. *Health Phys.* **18,** 73–75.

Burns, P. A., Cooper, M., Johnston, P., Martin, L., and Williams, G. (1994). Determination of the ratios of [239]Pu and [240]Pu, to [241]Am for nuclear weapons test sites in Australia. *Health Phys.* **67,** 226–232.

Burson, Z., and Lushbaugh, C. C. (1990). The 1983–1984 Ciudad Juarez ^{60}Co accident. *In* "The Medical Basis for Radiation Accident Preparedness II (R. C. Ricks and S. A. Fry, eds.), pp. 13–23. Elsevier, New York.

Burton, E. F. (1904). A radioactive gas from crude petroleum. *Philos. Mag. Ser. 6,* **8,** 498–508.

Butherus, M. C., Brodeur, J., Varhus, D., and Ruud, C. (1995). Hanford tank farms vadose zone monitoring and characterization project. *In* "Proceedings of the Symposium on Environmental Restoration 95," August 13–17 in Denver, Colorado, U.S. Department of Energy Office of Environmental Restoration, Washington, D.C.

Cain, J. A., and Boothroyd, J. C. (1983). "Environmental Geology," Burgess, Minneapolis, Minnesota.

Caldwell, G. G., Kelley, D. B., Zach, M., Falk, H., and Health, C. (1983). Mortality and cancer frequency among military nuclear test (SMOKY) participants, 1957 through 1979. *JAMA, J. Am. Med. Assoc.* **250,** 620–624.

Cameron, J. R., Suntharalingam, N., and Kennedy, G. N. (1968). "Thermoluminescent Dosimetry." Univ. of Wisconsin Press, Madison.

Campos, M. J., Penna-Franca, E., Labao, N., Trinidade, H., and Sachett, J. (1986). Migration of radium from the ore deposit of the Moro de Ferro, Pocos de Caldas, Brazil. *J. Environ. Radioact.* **3,** 145–161.

Cantril, S. T., and Parker, H. M. (1945). The tolerance dose. MDDC-110. U.S. Atomic Energy Commission, Washington, D.C.

Cardis, E., Gilbert, E., Carpenter, L., Howe, G., Kato, I., Armstrong, B. K., Beral, V., Cowper, G., Douglas, A., Fix, J., Fry, S. A., Kaldor, J., Lave, C., Salmon, L., Smith, P. G., Voelz, G. L., and Wiggs, L. D. (1995). Effects of low doses and dose rates of external ionizing radiation: Cancer mortality among nuclear industry workers in three countries. *Radiat. Res.* **142,** 117–132.

Carhart, R. A., Policasro, A. J., Wastag, M., and Coke, L. (1989). Evaluation of eight short-term long-range transport models using field data. *Atmos. Environ.* **23,** 85–105.

Catlin, R. J. (1980). Assessment of the surveillance program of the high-level waste storage tanks at Hanford. Report to the U.S. Department of Energy, Assistant Secretary for the Environment, Washington, D.C.

Chamberlain, A. C. (1955). Aspects of travel and deposition of aerosol and vapor clouds. Report HP/R1261. U.K. Atomic Energy Authority, London, UK.

Chamberlain, A. C. (1960). Aspects of the deposition of radioactive and other gases and particles. *Int. J. Air Pollut.* **3,** 63–88.

Chamberlain, A. C. (1970). Interception and retention of radioactive aerosols by vegetation. *Atmos. Environ.* **4,** 57–78.

Chamberlain, A. C., and Chadwick, R. C. (1966). Transport of iodine from atmosphere to ground. *Tellus* **18,** 226–237.

Chang, T. Y., Cheng, W. L., and Weng, P. S. (1974). Potassium, uranium and thorium content of building materials of Taiwan. *Health Phys.* **27,** 385–387.

Chang, W. P., and Kau, J. (1993). Exposure to high doses of radiation. *Lancet* **341,** 750.

ChemRisk (1994). Dose assessment for historical contaminant releases from Rocky Flats. Project Task 8. Colorado Department of Health and Environment, Denver, Colorado.

Chepil, W. A. (1957). Erosion of soil by wind. *In* "The Yearbook of Agriculture," pp. 308–314. U.S. Department of Agriculture, Washington, D.C.

Chieco, N. A., Bogen, D. C., and Knutson, E. O., eds. (1990). EML procedures manual. Report HASL-300. DOE Environmental Measurements Laboratory, New York.

Clapp, C. A. (1934). "Cataract." Lea & Febiger, Philadelphia, Pennsylvania.

Clark, H. M. (1954). The occurrence of an unusually high-level radioactive rainout in the area of Troy, N.Y. *Science* **119,** 619–622.

Clarke, R. H. (1974). An analysis of the Windscale accident using the WEERIE code. *Am. Nucl. Sci. Eng.* **1,** 73.

Clarke, R. H. (1990). The 1957 Windscale accident revisited. *In* "The Medical Basis for Radiation Accident Preparedness II" (R. C. Ricks and S. A. Fry, eds.), pp. 281–289. Elsevier, New York.

Claus, W. D. (1958). "Radiation Biology and Medicine." Addison–Wesley, Reading, Massachusetts.

Clayton, G. D., Arnold, J. R., and Patty, F. A. (1955). Determination of sources of particulate atmospheric carbon. *Science* **122,** 751–753.

Clegg, J. W., and Foley, D. D. (1958). "Uranium Ore Processing." Addison–Wesley, Reading, Massachusetts.

Clemente, G. F., Renzetti, A., Santori, G., Steinhäusler, F., and Pohl-Rüling, J. (1984). Relationship between the ^{210}Pb content of teeth and exposure to Rn and Rn daughters. *Health Phys.* **47,** 253–262.

Clements, W., and Wilkening, M. (1974). Atmospheric pressure effects on radon transport across the earth–air interface. *J. Geophys. Res.* **79,** 5025–5029.

Cobb, J. C., Eversole, C., Archer, P. G., Taggart, R., and Efurd, D. (1983). Plutonium burdens in people living around the Rocky Flats Plant. Report PB 83-137372. NTIS, Springfield, Virginia.

Cochran, T. C., Arkin, W. M., Norris, R. S., and Sands, J. J. (1989). "Nuclear Weapons Data Book V. Soviet Nuclear Weapons." Harper & Row, New York.

Codman, E. A. (1902). A study of the cases of accidental X-ray burns hitherto recorded. *Philadelphia Medical Journal,* 438–442.

COGEMA (1992). Annual report for 1992, Compagnie Generale des Matieres Nucleaires. Velizy-Villacoublay, France.

Cohen, B. L. (1980). Society's valuation of life saving in radiation protection and other contexts. *Health Phys.* **38,** 33–51.

Cohen, B. L. (1982). Effects of ICRP 30 and the 1980 BEIR report on hazard assessment of high level wastes. *Health Phys.* **42,** 133–143.

Cohen, B. L. (1986). A national survey of Rn-222 in U.S. homes and correlating factors. *Health Phys.* **51,** 175–183.

Cohen, B. L. (1989). Measured radon levels in U.S. homes. *In* "Radon, Proceedings of the Twenty-Fourth Annual Meeting of the National Council on Radiation Protection and Measurements" (N. H. Harley, ed.), NCRP Proc. 10, pp. 170–181. National Council on Radiation Protection and Measurements, Bethesda, Maryland.

Cohen, B. L. (1991). Variation of radon levels in U.S. homes correlated with house characteristics, location, and socioeconomic factors. *Health Phys.* **60,** 631–642.

Cohen, B. L., Kulwicki, D. R., Warner, K. R., Jr., and Grassi, C. L. (1984). Radon concentrations inside public and commercial buildings in the Pittsburgh area. *Health Phys.* **47,** 399–405.

Cohen, B. S., Eisenbud, M., and Harley, N. H. (1980). Measurement of the alpha-radioactivity on the mucosal surface of the human bronchial tree. *Health Phys.* **39,** 619–632.

Colby, L. J. (1976). Fuel reprocessing in the U.S.: A review of problems and some solutions. *Nucl. News* **January.**

Cole, L. A. (1989). Much ado about radon. *The Sciences* **January.**

Cole, L. A. (1993). "Element of Risk: The Politics of Radon." AAAS Press, Washington, D.C.

Colli, A. (1990). The COMPLY program for demonstrating compliance with national radionuclide air emission standards. *Health Phys.* **58,** 411–416.

Collins, J. T., Bell, M. J., and Hewitt, W. M. (1978). Radioactive waste source terms. *In* "Nuclear Power Waste Technology" (A. A. Moghissi, H. W. Godbee, M. S. Ozker, and M. W. Carter, eds.), pp. 167–199. American Society of Mechanical Engineering, New York.

Colorado Committee for Environmental Information (1970). "Report on the Dow Rocky Flats Fire: Implications of Plutonium Releases to the Public Health and Safety." Subcommittee on Rocky Flats, Boulder Colorado, Jan 13, 1970, HASL-235 (Ref. 2, p. 38). U.S. Atomic Energy Commission, New York.

Comar, C. (1979). Risk: A pragmatic *de minimis* approach. *Science* **203,** 133–143.

Comar, C. L., and Wasserman, R. H. (1960). Radioisotope absorption and methods of elimination: Differential behavior of substances in metabolic pathways. *In* "Radioisotopes in the Biosphere" (R. S. Caldecott and L. A. Snyder, eds.), pp. 526–540. Univ. of Minnesota Press, Minneapolis, Minnesota.

Comar, C. L., Wasserman, R. H., and Nold, M. M. (1956). Strontium calcium discrimination factors in the rat. *Proc. Soc. Exp. Biol. Med.* **92,** 859–863.

Committee on Armed Services (1981). "Hearings on Uranium Ore Residues: Potential Hazards and Disposition, June 24–25, 1981." U.S. Government Printing Office, Washington, D.C.

Committee on Technical Bases for Yucca Mountain Standards (1995). "Technical Bases for Yucca Mountain Standards." National Academy Press, Washington, D.C.

Conard, R. A. (1984). Late radiation effects in Marshall Islanders exposed to fallout 28 years ago. *In* "Radiation Carcinogenesis: Epidemiology and Biological Significance" (J. D. Boice and J. F. Fraumeni, eds.), pp. 57–71. Raven, New York.

Conard, R. A. (1992). Fallout—The experiences of a medical team in the care of a Marshallese population accidentally exposed to fallout. Report BNL 46444. Brookhaven National Laboratory, Upton, Long Island, New York.

Conard, R. A., *et al.* (1970). Medical survey of the people of Rongelap and Utirik Islands thirteen, fourteen, and fifteen years after exposure to fallout radiation (March 1967, March, 1968 and March, 1969). Report BNL 50220 (T-562). Brookhaven National Laboratory, Upton, Long Island, New York.

Conard, R. A., Paglia, D. E., Larsen, P. R., Sutow, W. W., Dobyns, B. M., Robbins, J., Krotosky, W. A., Field, J. B., Rall, J. E., and Wolff, J. (1980). Review of medical findings in a Marshallese population twenty-six years after accidental exposure to radioactive fallout. Report BNL 52161. Brookhaven National Laboratory, Upton, Long Island, New York.

Coplan, B. V., and Baron, J. S. (1978). Treatment of liquid radwastes. *In* "Nuclear Power Waste Technology" (A. A. Moghissi, H. W. Godbee, M. S. Ozker, and M. W. Carter, eds.), pp. 233–272. American Society of Mechanical Engineering, New York.

Corbett, J. O. (1983). The radiation dose from coal burning: A review of pathways and data. *Radiat. Protect. Dosim.* **4,** 5–19.

Cothern, C. R., and Lappenbusch, W., eds. (1985). Special issue. *Health Phys.* **48**(No. 5), 529–712.

Cotten, G. B. (1995). Design considerations for pump-and-treat remediation based on characterization of industrial injection wells. Proceedings of the symposium. *In* "Environmental Restoration 95, August 13–17, Denver, Colorado. U.S. Department of Energy Office of Environmental Restoration, Washington, D.C.

Cottrell, W. B. (1974). Control of radioactive wastes in operating nuclear facilities. *In* "Human and Ecologic Effects of Nuclear Power Plants" (L. A. Sagan, ed.), pp. 72–131. Thomas, Springfield, Illinois.

Cowan, G. A. (1976). A natural fission reactor. *Sci. Am.* **235,** 36–47.

Cowser, K. E. (1964). Current practices in the release and monitoring of ^{131}I at NRTS, Hanford, Savannah River and ORNL. ORNL-NSIC3. Oak Ridge National Laboratory, Oak Ridge, Tennessee.

Crabtree, J. (1959). The travel and diffusion of the radioactive material emitted during the Windscale accident. *Q. J. R. Meteorol. Soc.* **85,** 362–370.

Crawford, T. V. (1978). Atmospheric transport of radionuclides: Report of working group on atmospheric dispersion, deposition and resuspension. *In* Proceedings of the Workshop on the Evaluation of Models Used for the Environmental Assessment of Radionuclides Releases," ORNL Report No. CONF-770901, p. 5. Oak Ridge National Laboratory, Oak Ridge, Tennessee.

Crick, M. J., and Lindsley, G. S. (1984). An assessment of the radiological impact of the Windscale reactor fire, October, 1957. *Int. J. Radiat. Biol.* **46,** 479–506.

Crim, E. M., and Bradley, T. D. (1995). Measurements of air concentrations of thorium during grinding and welding operations using thoriated tungsten electrodes. *Health Phys.* **68,** 719–722.

Crocker, G. R., O'Connor, J. D., and Freiling, E. C. (1966). Physical and radiochemical properties of fallout particles. *Health Phys.* **12,** 1099–1104.

Croft, A. G. (1980). ORIGEN2—A revised and updated version of the Oak Ridge isotope generation and depletion code. Report ORNL-5621. Oak Ridge National Laboratory, Oak Ridge, Tennessee.

Cronkite, E. P. (1961). Evidence for radiation and chemicals as leukemogenic agents. *Environ. Health* **3,** 297.

Cross, F. T., Harley, N. H., and Hofmann, W. (1985). Health effects and risks from ^{222}Rn in drinking water. *Health Phys.* **48,** 649–670.

Cuddihy, R. G. (1982). Risks of radiation-induced lung cancer. *In* "Critical Issues in Setting Radiation Dose Limits," pp. 133–152. Proceedings of the 17th Annual Meeting of NCRP. National Council on Radiation Protection and Measurements, Bethesda, Maryland.

Cuddihy, R. G., MacClellan, R. O., and Griffith, W. C. (1979). Variability in target organ deposition among individuals exposed to toxic substances. *Toxicol. Appl. Pharmacol.* **49,** 179–187.

Cunningham, K. I., and LaRock, E. J. (1991). Recognition of microclimate zones through radon mapping, Lechuguilla Cave, Carlsbad Caverns National Park, New Mexico. *Health Phys.* **61,** 493–500.

Curling, C. A., Rudnick, S. N., Ryan, P. B., and Moeller, D. W. (1990a). Optimization of filtration for reduction of lung dose from Rn decay products. Part I—Theoretical. *Health Phys.* **59,** 267–275.

Curling, C. A., Rudnick, S. N., Harrington, D. P., and Moeller, D. W. (1990b). Optimization of filtration for reduction of lung dose from Rn decay products: Part II—Experimental. *Health Phys.* **59,** 277–285.

Curtis, S. B., Dye, D. L., and Sheldon, W. R. (1966). Hazard from highly ionizing radiation in space. *Health Phys.* **12,** 1069–1075.

Cuthbert, F. L. (1958). "Thorium Production Technology." Addison–Wesley, Reading, Massachusetts.

Darby, S. C., Kendall, G. M., Fell, T. P., O'Hagen, J. A., Muirhead, C. R., Ennis, J. R., Ball, A. M., Dennis, J. A., and Doll, R. (1988). A summary of mortality and incidence of cancer in men from the United Kingdom who participated in the United Kingdom's atmospheric nuclear weapons tests and experimental programmes. *Br. Med. J.* **296,** 332–338.

da Silva, C. J., Delgado, J. U., Luiz, M. T. B., Cunha, P. G., and de Barros, P. D. (1991). Considerations related to the decontamination of houses in Goiania: Limitation and implications. *Health Phys.* **60,** 87–90.

Davis, J. J. (1963). Cesium and its relationship to potassium in ecology. *In* "Radioecology" (V. Schultz and A. W. Klement, Jr., eds.), p. 539 ff. Reinhold, New York.

Davis, J. J., Perkins, R. W., Palmer, R. F., Hanson, W. C., and Cline, J. F. (1958). Radioactive materials in aquatic and terrestrial organisms exposed to reactor effluent water. *In* Proceedings of the Second International Conference on Peaceful Uses of Atomic Energy." United Nations, New York.

de Bartoli, M. C., and Gaglione, P. (1969). SNAP plutonium-238 fallout at Ispra, Italy. *Health Phys.* **16,** 197–204.

Defense Nuclear Agency (1981). The radiological cleanup of Eniwetak Atoll. Defense Nuclear Agency, Washington, D.C.

Degteva, M. O. (1996). Environmental dose reconstruction for the Urals population. *In* "Proceedings of the 32nd Annual Meeting of the NCRP, April 3–4, 1996," in press. National Council on Radiation Protection and Measurements, Bethesda, Maryland.

Denver, D. (1994). High level nuclear waste—A Florida perspective. *Health Physics Newsletter* **22**(9), 5–7.

Department of Health and Human Services (1985). Report of the National Institutes of Health Ad Hoc Working Group to Develop Radioepidemiological Tables. Publ. No. 85-2748. NIH, Washington, D.C.

Department of Veterans Affairs (1995). VA fact sheet—VA programs for Persian Gulf veterans, June. Department of Veterans Affairs, Washington, D.C.

de Planque, G., and Gesell, T. F. (1986). Environmental measurements with thermoluminescence dosimeters—Trends and issues. *Radiat. Protect. Dosim.* **17**, 193–200.

de Villiers, A. J., and Windish, J. P. (1964). Lung cancer in fluorospar mining community. I: Radiation, dust and mortality experience. *Br. J. Ind. Med.* **21**, 94–109.

Dickerson, M. H., Foster, T., and Gudiksen, P. H. (1984). Experimental and model transport and diffusion studies in complex terrain with emphasis on tracer studies. *Boundary-Layer Meteorology* **30**, 333–350.

DiNunno, J. J., Anderson, F. D., Baker, R. E., and Waterfield, R. L. (1962). Calculation of distance factors for power and test reactor sites. Report TID 14844. U.S. Atomic Energy Commission, Washington, D.C.

Dirkes, R. L., and Hanf, R. W. (1995). Hanford site environmental report for calendar year 1994. Report PNL-10574. Pacific Northwest Laboratory, Richland, WA.

Dobson, G. M. B. (1956). Origin and distribution of the polyatomic molecules in the atmosphere. *Proc. R. Soc., Ser. A* **236**, 187–193.

DOE (1978). Long-term management of defense high-level radioactive wastes. Report DOE/EIS-0023-D. U.S. Department of Energy, Washington, D.C.

DOE (1979a). Draft environmental impact statement, Waste Isolation Pilot Plant (April, 1979), pp. 9–55. U.S. Department of Energy, Washington, D.C.

DOE (1979b). Final environmental impact statement: Management of commercially generated radioactive waste. DOE/EIS-0046G (3 vols.), p. 126. U.S. Department of Energy, Washington, D.C.

DOE (1980). Grand Junction remedial action program. DOE/EV/01621-T1. U.S. Department of Energy, Washington, D.C.

DOE (1981). Background report for the uranium mill tailings remedial action program. Report DOE/EP-001. U.S. Department of Energy, Washington, D.C.

DOE (1983). Spent fuel and radioactive waste inventories, projections, and characteristics. Report DOE/NE-0017/2. NTIS, Springfield, Virginia.

DOE (1984a). Commercial power 1984: Prospects for the United States and the world, Energy Information Administration Report DOE/EIA 0438(84). NTIS, Springfield, Virginia.

DOE (1984b). 10 CFR Part 960, Nuclear Waste Policy Act of 1982: General guidelines for the recommendation of sites for the nuclear waste repositories. *Fed. Regist.* **49**(236), 44714–47770.

DOE (1985a). Domestic uranium mining and milling industry—1984 viability assessment. Report DOE/EIA-0477. U.S. Department of Energy, Washington, D.C.

DOE (1985b). Energy systems acquisition project plan: Formerly used MED/AEC sites remedial action program (FUSRAP). Oak Ridge Operations Office (April, 1985), Oak Ridge, Tennessee.

DOE (1985c). Office of Civilian Radioactive Waste Management annual report to Congress. Report DOE/RW-0004/1. NTIS, Springfield, Virginia.

DOE (1985d). Announced U.S. nuclear tests. Report NVO 209 (January, 1985). U.S. Department of Energy, Washington, D.C.

DOE (1987). Health and environmental consequences of the Chernobyl nuclear power plant accident. Report DOE/ER-0332. U.S. Department of Energy, Washington, D.C.

DOE (1988). Site characterization plan: Yucca Mountain site, Nevada research and development area, Nevada. Report DOE/RW-0199. Office of Civilian Radioactive Waste Management, U.S. Department of Energy, Washington, D.C.

DOE (1989). Environmental restoration and waste management five-year plan. Report DOE/S-0070. U.S. Department of Energy, Washington, D.C.

DOE (1991a). Domestic uranium mining and milling industry—1990 viability assessment. Report DOE/EIA-0477 (90). U.S. Department of Energy, Washington, D.C.

DOE (1991b). Uranium mill tailings remedial action (UMTRA) program. Report DOE/UMTRA-400124-0167. U.S. Department of Energy, Albuquerque, New Mexico.

DOE (1991c). Environmental regulatory guide for radiological effluent monitoring and environmental surveillance. Report DOE/EH-0173T. U.S. Department of Energy, Washington, D.C.

DOE (1993a). Occupational injury and property damage summary, January–December 1993. Report DOE/EH/01570. U.S. Department of Energy, Washington, D.C.

DOE (1993b). Radiation exposures for DOE and DOE contractor employees—1990. Report DOE/EH-0287T. U.S. Department of Energy, Washington, D.C.

DOE (1994a). Environmental restoration activities at Idaho Operations Office. Report DOE/EM-0044P. U.S. Department of Energy Environmental Management Office, Washington, D.C.

DOE (1994b). Environmental restoration activities at Oak Ridge Operations Office. Report DOE/EM-0046P. U.S. Department of Energy Environmental Management Office, Washington, D.C.

DOE (1994c). Environmental restoration activities at Richland Operations Office. Report DOE/EM-0047P. U.S. Department of Energy Environmental Management Office, Washington, D.C.

DOE (1994d). Environmental restoration activities at Rocky Flats Office. Report DOE/EM-0048P. U.S. Department of Energy Environmental Management Office, Washington, D.C.

DOE (1994e). Environmental restoration activities at Savannah River Operations Office. Report DOE/EM-0050P. U.S. Department of Energy Environmental Management Office, Washington, D.C.

DOE (1994f). Summary of radionuclide air emissions from the Department of Energy Facilities for CY 1992. Report DOE/EH-0360. February. U.S. Department of Energy Environmental Management Office, Washington, D.C.

DOE (1994g). How can I find information? Report DOE/EM-0203. U.S. Department of Energy Environmental Management Office, Washington, D.C.

DOE (1994h). Department of Energy programmatic spent nuclear fuel management and Idaho National Engineering Laboratory environmental restoration and waste management programs draft environmental impact statement. Vol. 1, Report DOE/EIS-0203-D. U.S. Department of Energy, Idaho Falls, Idaho.

DOE (1995a). Closing the circle on the splitting of the atom—The environmental legacy of nuclear weapons production in the United States and what the Department of Energy is doing about it. U.S. Department of Energy Office of Environmental Management, Washington, D.C.

DOE (1995b). Draft programmatic environmental impact statement for the uranium mill tailings remedial action ground water project. Report DOE/EID-0198. National Technical Information Service, Springfield, Virginia.

DOE (1995c). Environmental management 1995: Progress and plans of the environmental management program. Report DOE/EM-0228. U.S. Department of Energy, Washington, D.C.

DOE (1995d). Estimating the cold war mortgage; The 1995 baseline environmental management report. Vol. I. U.S. Department of Energy, NTIS, Springfield, Virginia.

DOE (1995e). Site characterization progress report: Yucca Mountain, Nevada. Report DOE/RW-0463. U.S.Department of Energy Office of Civilian Radioactive Waste Management, Washington, D.C.

DOE (1995f). The formerly utilized sites remedial action program (FUSRAP). Report DOE/EM-0233. U.S. Department of Energy Office of Environmental Restoration, Washington, D.C.

DOE (1995g). UMTRA: Uranium mill tailings remedial action project—Annual report for FY 1995. U.S. Department of Energy Office of Environmental Management, Washington, D.C.

DOE (1996). The 1996 baseline environmental management report. Report DOE/EM-0290. U.S. Department of Energy, Washington, D.C.

DOE Carlsbad Area Office (1995). Title 30 CFR 191 compliance certification application for the Waste Isolation Pilot Plant (7/31/95 update). Report DRAFT-DOE/CAO-2056. U.S. DOE Carlsbad Area Office, Carlsbad, New Mexico.

DOE Environmental Measurements Laboratory (formerly U.S. AEC Health and Safety Laboratory). Data from ^{90}Sr analyses of foods obtained at market in New York and San Francisco have been published at intervals for many years.

DOE-ID (1987). 1986 environmental monitoring program report for the Idaho National Engineering Laboratory Site. Report DOE/ID-12082(86). U.S. Department of Energy Idaho Operations Office, Idaho Falls, Idaho.

DOE-ID (1988). 1987 environmental monitoring program report for the Idaho National Engineering Laboratory Site. Report DOE/ID-12082(87). U.S. Department of Energy Idaho Operations Office, Idaho Falls, Idaho.

DOE-ID (1989). The Idaho National Engineering Laboratory site environmental monitoring program report for calendar year 1988. Report DOE/ID-12082(88). U.S. Department of Energy Idaho Operations Office, Idaho Falls, Idaho.

DOE-ID (1991). Idaho National Engineering Laboratory historical dose evaluation. Report DOE/ID-12119, Vol. 1. U.S. Department of Energy Field Office Idaho, Idaho Falls, Idaho.

DOE Office of Environmental Management (1995). Draft waste management programmatic environmental impact statement for managing treatment, storage, and disposal of radioactive and hazardous waste. Report DOE/EIS-0200-D. U.S. Department of Energy, Washington, D.C.

Dohrenwend, B. P., Dohrenwend, B. S., Warheit, G. J., Bartlett, G. S., Goldsteen, R. L., Goldsteen, K., and Martin, J. L. (1981). Stress in the community: A report to the President's Commission on the Accident at Three Mile Island. The Three Mile Island nuclear accident: Lessons and implications. *Ann. N.Y. Acad. Sci.* **365,** 159–174.

Doi, M., and Kobayashi, S. (1994). Vertical distribution of outdoor radon and thoron in Japan using a new discriminative dosimeter. *Health Phys.* **67,** 385–392.

Dolphin, G. W. (1968). The risk of thyroid cancer following irradiation. *Health Phys.* **15,** 219–228.

Dolphin, G. W. (1971). Dietary intakes of iodine and thyroid dosimetry. *Health Phys.* **21,** 711–713.

Donahue, R. L., Miller, R. W., and Shickluna, J. C. (1977). "Soils: An Introduction to Soils and Plant Growth," 4th Ed. Prentice–Hall, Englewood Cliffs, New Jersey.

Dorian, D. (1991). Legal standards concerning health and ecological implications of radioactively contaminated environments. *In* "Proceedings of the Twenty-Sixth Annual Meeting." National Council on Radiation Protection and Measurements, Bethesda, Maryland.

Dragonette, K. S. (1995). Definition of the problem. *In* "Radioactive and Mixed Waste—Risk as a Basis for Waste Classification." NCRP Symposium Proceedings No. 2. National Council on Radiation Protection and Measurements, Bethesda, Maryland.

Dreicer, M., Hakonson, T. E., and White, G. C. (1984). Rain splash as a mechanism for soil contamination of plant surfaces. *Health Phys.* **46,** 177–187.

Drew, R. T., and Eisenbud, M. (1966). The natural radiation dose to indigenous rodents on the Morro do Ferro, Brazil. *Health Phys.* **12,** 1267–1274.

Drinker, P., and Hatch, T. (1954). "Industrial Dust," 2nd Ed. McGraw–Hill, New York.

DuFrain, R. J., Littlefield, L. G., Joiner, E. J., and Frome, E. L. (1980). *In vitro* human cytogenetic dose–response systems. *In* "The Medical Basis for Radiation Accident Preparedness," pp. 358–374. Elsevier/North-Holland, New York.

Dunlap, J. H. (1995a). REAC/TS. *Health Physics Society Newsletter* **23**(1), 24.

Dunlap, J. H. (1995b). Recent ^{226}Ra incident. *Health Physics Society Newsletter* **23**(12), 5–6.

Dunning, G. M. (1962). Fallout from USSR 1961 nuclear tests. Report TID-14377. U.S. Atomic Energy Commission, Washington, D.C.

Dunster, H. J. (1958). The disposal of radioactive liquid wastes into coastal waters. *In* "Proceedings of the Second UN Conference on Peaceful Uses of Atomic Energy." United Nations, New York.

Dunster, H. J. (1969). United Kingdom studies on radioactive releases in the marine environment. *In* "Biological Implications of the Nuclear Age." U.S. Atomic Energy Commission, Washington, D.C.

Dunster, H. J., Howells, H., and Templeton, W. L. (1958). District surveys following the Windscale incident, October, 1957. *In* "Proceedings of the Second UN Conference on Peaceful Uses of Atomic Energy." United Nations, New York.

Duursma, E. K., and Gross, M. C. (1971). Marine sediments and radioactivity. *In* "Radioactivity in the Marine Environment," pp. 147–160. National Academy of Sciences, Washington, D.C.

Eckerman, K. F., Wolbarst, A. B., and Richardson, A. C. B. (1988). Limiting values of radionuclide intake and air concentration and dose conversion factors for inhalation, submersion, and ingestion. Federal Guidance Report 11, EPA-520/1-88-020. U.S. Environmental Protection Agency, Washington, D.C.

EG&G Berthold Company (1992). Private communication.

Eisenbud, M. (1978). "Environment, Technology and Health: Human Ecology in Historic Perspective." New York Univ. Press, New York.

Eisenbud, M. (1980). Radioactive wastes from biomedical institutions (editorial). *Science* **207**, 1299.

Eisenbud, M. (1981a). The status of radioactive waste management: Needs for reassessment. *Health Phys.* **40**, 429–437.

Eisenbud, M. (1981b). The concept of de minimis dose. "Quantitative Risk in Standard Setting," pp. 64–75. National Council on Radiation Protection and Measurements, Bethesda, Maryland.

Eisenbud, M. (1982). Origins of the standards for control of beryllium disease (1947–1949). *Environ. Res.* **27**, 79–88.

Eisenbud, M. (1990). "An Environmental Odyssey." Univ. of Washington Press, Seattle.

Eisenbud, M. (1991). An overview of sites contaminated by radioactivity. *In* "Proceedings of the Twenty-Sixth Annual Meeting of the NCRP, Health and Ecological Implications of Radioactively Contaminated Sites" (C. R. Richmond, ed.), pp. 5–20. National Council on Radiation Protection and Measurements, Bethesda, Maryland.

Eisenbud, M. (1992). Exposure from thorium to the residents of high background areas. *Health Phys.* **62**(Suppl. 6), S39.

Eisenbud, M. (1994a). Testimony at Oversight Hearing of the Committee on Natural Resources, House of Representatives, 103rd Congress, February 24, 1994. Letters describing the AEC monitoring network are included. U.S. Government Printing Office, Washington, D.C.

Eisenbud, M. (1994b). The plutonium perception. *Health Physics Society Newsletter* **22**(12), 2.

Eisenbud, M., and Harley, J. H. (1953). Radioactive dust from nuclear detonations. *Science* **117**, 141–147.

Eisenbud, M., and Harley, J. H. (1956). Radioactive fallout through September, 1955. *Science* **124**, 251–255.

Eisenbud, M., and Petrow, H. (1964). Radioactivity in the atmospheric effluents of nuclear power plants that use fossil fuels. *Science* **144**, 288–289.

Eisenbud, M., and Quigley, J. (1956). Industrial hygiene of uranium processing. *AMA Arch. Ind. Health* **14**, 12–22.

Eisenbud, M., Mochizuki, Y., Goldin, A. S., and Laurer, G. R. (1962). Iodine-131 dose from Soviet nuclear tests. *Science* **136**, 370–374.

Eisenbud, M., Mochizuki, Y., and Laurer, G. L. (1963a). ^{131}I dose to human thyroids in New York City from nuclear tests in 1962. *Health Phys.* **9**, 1291–1298.

Eisenbud, M., Pasternack, B., Laurer, G. R., Mochizuki, Y., Wrenn, M. E., Block, L., and Mowafy, R. (1963b). Estimation of the distribution of thyroid doses in a population exposed to I-131 from weapons tests. *Health Phys.* **9**, 1281–1290.

Eisenbud, M., Petrow, H., Drew, R., Roser, F. X., Kegel, G., and Cullen T. L. (1964). Naturally occurring radionuclides in foods and waters from the Brazilian areas of high radioactivity. *In* "The Natural Radiation Environment" (J. A. S. Adams and W. M. Lowder, eds.), pp. 837–854. Univ. of Chicago Press, Chicago.

Eisenbud, M., Krauskopf, K., Penna Franca, E., Lei, W., Ballad, R., Linsalata, P., and Fujimori, K. (1984). Natural analogues for the transuranic actinide elements: An investigation in Minas Gerais, Brazil. *Environ. Geol. Water Sci.* **6**(1), 1–9.

Elder, D. E. (1995). Skin cancer. *Cancer Suppl.* **75**, 245–256.

Electric Power Research Institute (1982). Survey of plume models for atmospheric application. Report EA-2243. Energy Analysis and Environment Div., Palo Alto, California.

Electric Power Research Institute (1989). New interest in passive reactor designs. *EPRI Journal,* **April/May.**

Electric Power Research Institute (1990). The cleanup of Three Mile Island Unit 2. Report NP 6931. Electric Power Research Institute, Palo Alto, California.

Emery, R., Jeanette, M., and Sprau, D. (1992). Characterization of low-level radioactive waste generated by a large university/hospital complex. *Health Phys.* **62**, 183–185.

Environmental Protection Department (1994). Fernald 1993 site environmental report. FEMP-2342. NTIS, Springfield, Virginia.

Environmental Restoration Division (1994). Project Chariot site assessment and remedial action final report. U.S. Department of Energy Nevada Operations Office, Las Vegas, Nevada.

EPA (1972). Environmental radioactivity surveillance guide. Report of the Surveillance and Inspection Division. EPA Office of Radiation Programs, Washington, D.C.

EPA (1979). Radiological impact caused by emissions of radionuclides into air in the United States. Report EPA 520/7-79-006. U.S. Environmental Protection Agency, Washington, D.C.

EPA (1982). Final environmental impact statement for remedial action standards for inactive uranium processing sites. Report EPA 520/4-82-013-1. U.S. Environmental Protection Agency, Washington, D.C.

EPA (1983). Environmental standards for uranium and thorium mill tailings at licensed commercial processing sites. *Fed. Regist.* **48**(No. 196), Oct. 7.

EPA (1984a). Background information document (integrated risk assessment); Final rule for radionuclides. Report EPA 520/1-84-0002-2, Vol. 2. U.S. Environmental Protection Agency, Washington, D.C.

EPA (1984b). Report on the scientific basis of EPA's proposed national emission standards for hazardous air pollutants for radionuclides. Report of Sub-committee of the Science Advisory Board, Washington, D.C.

EPA (1985). Environmental standards for the management and disposal of spent nuclear fuel, high level and transuranic radioactive wastes. 40 CFR Part 191. U.S. Environmental Protection Agency, Washington, D.C.

EPA (1986a). A citizen's guide to radon. Report OPA-86-004. U.S. Environmental Protection Agency, Washington, D.C.

EPA (1986b). RCRA ground-water monitoring technical enforcement guidance document. EPA Report OSWER-9950.1. U.S. Environmental Protection Agency, Washington, D.C.

EPA (1989a). Radon reduction methods (3rd Ed.). Report RD-681. U.S. Environmental Protection Agency, Washington, D.C.

EPA (1989b). User's guide for the COMPLY code. EPA Report 520/1-89-003. U.S. Environmental Protection Agency, Washington, D.C.

EPA (1990a). Environmental investments: The costs of a clean environment. Report EPA-230-12-90-084. U.S. Environmental Protection Agency, Washington, D.C.

EPA (1990b). Idaho radionuclide study. EPA Report 520/6-90/008-R-92-011. U.S. Environmental Protection Agency, Las Vegas, Nevada.

EPA (1990c). National primary drinking water regulations. Title 40, Code of Federal Regulations, Part 141. U.S. Environmental Protection Agency, Washington, D.C.

EPA (1991a). National primary drinking water regulations, radionuclides; Proposed rule. *Fed. Regist.* **56**(No. 138), 33050–33123.

EPA (1991b). Technical support document to the citizen's guide to radon. Report 400/R-92/001. U.S. Environmental Protection Agency, Washington, D.C.

EPA (1992a). A citizen's guide to radon (2nd Ed.). Report 402-K92-0001. U.S. Environmental Protection Agency, Washington, D.C.

EPA (1992b). National residential radon survey. EPA Report 402-R-92-011. U.S. Environmental Protection Agency, Washington, D.C.

EPA (1993). Environmental radiation protection standards for management and disposal of spent nuclear fuel, high-level and transuranic radioactive wastes. Title 40 of the Code of Federal Regulations, Part 191. U.S. Government Printing Office, Washington, D.C.

EPA (1994). Title 40, Code of Federal Regulations, Part 61.122. U.S. Environmental Protection Agency, Washington, D.C.

EPA (1995). Groundwater standards for remedial actions at inactive uranium processing sites. *Fed. Regist.* **60**(7), 2854–2871 (January 11).

Etherington, H., ed. (1958). "Nuclear Engineering Handbook." McGraw–Hill, New York.

Evans, R. D. (1943). Protection of radium dial workers and radiologists from injury by radium. *J. Ind. Hyg. Toxicol.* **25**, 253–269.

Evans, R. D. (1966). The effect of skeletally deposited alpha-ray emitters in man. *Br. J. Radiol.* **39**, 881–895.

Evans, R. D. (1967). The radium standard for boneseekers—Evaluation of the data on radium patients and dial painters. *Health Phys.* **13**, 267–278.

Evans, R. D., Keane, A. T., Kolenkow, R. J., Neal, W. R., and Shanaham, M. M. (1969). Radiogenic tumors in the radium and mesothorium cases studied at M.I.T. *In* "Delayed Effects of Bone-Seeking Radionuclides" (C. W. Mays *et al.*, eds.). Univ. of Utah Press, Salt Lake City.

Failla, H. P. (1932). Radium protection. *Radiology* **19**, 12–21.

Faisca, M. C., Teixeira, M. M. G. R., and Bettencourt, A. O. (1992). Indoor radon concentrations in Portugal—A national survey. *Radiat. Protect. Dosim.* **45**(Suppl. 1–4), 465–467.

FAO (1960). Radioactive materials in food and agriculture. Food and Agriculture Organization of the United Nations, Rome.

FAO (1995). Guidelines for levels for radionuclides in food following accidental nuclear contamination for use in international trade. *In* "Joint FAO/WHO Food Standards Programme," Codex Alimentarius Vol. 1A, Section 6.2. Food and Agriculture Organization of the United Nations, Rome.

Farber, S. A., and Hodgdon, A. D. (1991). Cesium-137 in woodash—Results of a nationwide survey. *Health Phys.* **60**(Suppl. 2), S76.

Farris, W. T., Napier, B., Eslinger, P., Ikenberry, T., Shipler, D., and Simpson J. (1994a). Atmospheric pathway dosimetry report, 1944–1992. Report PNWD-2228-HEDR. Pacific Northwest Laboratories, Richland, Washington.

Farris, W. T., Napier, B., Simpson, J., Snyder, S., and Shipler, D. (1994b). Columbia River pathway dosimetry report, 1944–1992. Report PNWD-2227 HEDR. Pacific Northwest Laboratories, Richland, Washington.

Federal Radiation Council (1960). Background material for the development of radiation protection standards. Report No. 1. U.S. Government Printing Office, Washington, D.C.

Feely, H. W. (1960). Strontium-90 content of the stratosphere. *Science* **131**, 645–649.

Feely, H. W., Heyer, I. K., Juzdam, Z. R., Klusek, C. S., Larson, R. J., Leifer, R., and Sanderson, C. E. (1988). Fallout in the New York metropolitan area following the Chernobyl accident. *J. Environ. Radioact.* **7**, 177–191.

Ferlic, K. P. (1983). Fallout: Its characteristics and management. Report AFRI TR83-5. Armed Forces Radiobiology Research Institute, Bethesda, Maryland.

Fernald Citizen's Task Force (1995). Recommendations on remediation levels, waste disposition, priorities and future use. Fernald Citizen's Task Force, Ross, Ohio.

Fernald Environmental Management Project (1995). Fernald environmental remediation progress status report. U.S. Department of Energy, Washington, D.C.

Fesenko, S. V., Alexakin, R. M., Spiridinov, S. I., and Sansharova, N. I. (1995). Dynamics of ^{137}Cs concentration in argicultural products in areas of Russia contaminated as a result of the accident at the Chernobyl nuclear power plant. *Radiat. Protect. Dosim.* **60**, 155–166.

Fetter, C. W., Jr. (1980). "Applied Hydrology." Charles E. Merrill, Columbus, Ohio.

Field, R. W., Kross, B. C., Weih, L. M., Vust, L. J., and Nicholson, H. F. (1993). Factors associated with elevated Rn-222 levels in Iowa. *Health Phys.* **65**, 178–184.

Finck, R., ed. (1990). Emergency preparedness for nuclear-powered satellites. Office of Economic Cooperation and Development, Paris.

Fisenne, I. M. (1993). Initial study of Pb-210 in indoor air. *Health Phys.* **64**, 423–425.

Fisenne, I. M., and Keller, H. W. (1970). Radium-226 in the diet of two U.S. cities. Report HASL-224. U.S. Atomic Energy Commission, New York.

Fisenne, I. M., and Welford, G. A. (1986). Natural uranium concentrations in soft tissues and bone of New York City residents. *Health Phys.* **50**, 739–746.

Fisenne, I. M., Welford, G. A., Perry, P., Baird, R., and Keller, H. W. (1978). Distributional ^{234}U, ^{238}U, ^{226}Ra, ^{210}Pb and ^{210}Po in soil. *Environ. Int.* **1**, 245–246.

Fisenne, I. M., Cohen, N., Neton, J. W., and Perry, P. (1980). Fallout plutonium in human tissues from New York City. *Radiat. Res.* **83**, 162–168.

Fisenne, I. M., Keller, H. W., and Harley, N. H.(1981). Worldwide measurement of [226]Ra in human bone: Estimate of skeletal alpha dose. *Health Phys.* **40,** 163–171.

Fisenne, I. M., Perry, P. M., Decker, K. M., and Keller, H. W. (1987). The daily intake of [234]U, [235]U and [238]U, [228]Th, [230]Th, [232]Th, [226]Ra and [228]Ra by New York City residents. *Health Phys.* **53,** 357–363.

Fisher, H. B., List, E. J., Koh, R. C. Y., Imberger, I., and Brooks, N. H. (1979). "Mixing in Inland and Coastal Waters." Academic Press, Orlando, Florida.

Fitzsimmons, C. K., McNelis, D. N., and Wruble, D. T. (1972). Tritium activity in urine from luminous dial wrist watches. *Health Phys.* **22,** 514–516.

Florida (1994). Florida Administrative Code, Chap. 10D-91.1104.

Flynn, C. B. (1981). Local public opinion. The Three Mile Island nuclear accident: Lessons and implications. *Ann. N.Y. Acad. Science,* **365,** 146–158.

Folsom, T. R., and Vine, A. C. (1957). Tagged water masses for studying the oceans. Report 551. National Academy of Sciences, Washington, D.C.

Foster, G. R., and Hakonson, T. E. (1986). Erosional losses of fallout plutonium. *In* "Environmental Research for Actinide Elements" (J. E. Pinder III, ed.). NTIS, Springfield, Virginia.

Foster, R. F. (1959). Distribution of reactor effluent in the Columbia River. Industrial radioactive waste disposal. Hearings before Joint Committee on Atomic Energy.

Foster, R. F., Ophel, I. L., and Preston, A. (1971). Evaluation of human radiation exposure. *In* "Radioactivity in the Marine Environment," pp. 240–260. National Academy of Sciences, Washington, D.C.

Fowler, S. W., Baxter, M. S., Hamilton, T. F., Miquel, J.-C., Osvath, O., Povinec, P. P., and Scott, E. M. (1994). International assessment programs related to radioactive waste dumping in the Arctic seas. *Arctic Research of the United States* **8**(Spring), 92–100.

Frederickson, L., *et al.* (1958). Studies of soil–plant–animal interrelationship with respect to fission products. *In* "Proceedings of the Second UN International Conference on Peaceful Uses of Atomic Energy," p. 177. United Nations, New York.

Freiling, E. C., and Kay, M. A. (1965). Radionuclide fractionation in air burst debris. Report USNRDL-TR-933. U.S. Naval Radiological Defense Laboratory.

Fresco, J., Jetter, E., and Harley, J. (1952). Radiometric properties of the thorium series. *Nucleonics* **10,** 60.

Freudenthal, P. C. (1970a). Aerosol scavenging by ocean spray. Report HASL-232. U.S. Atomic Energy Commission, New York.

Freudenthal, P. C. (1970b). Strontium-90 concentrations in surface air: North America versus Atlantic Ocean from 1966 to 1969. *J. Geophys. Res.* **75,** 4089–4096.

Fry, R. J. M., Powers-Risius, P., Alpen, E. L., and Ainsworth, E. J. (1985). High-LET carcinogenesis. *Radiat. Res.* **104,** S180–S195.

Gabay, J. J., and Sax N. I. (1969). Retention of radium due to ingestion of Brazil nuts. *Health Phys.* **16,** 812–813.

Gahr, W. N. (1959). Uranium mill wastes. Industrial radioactive waste disposal. Hearings before Joint U.S. Congressional Committee on Atomic Energy.

Gallaghar, R. G., and Saenger, E. L. (1957). Radium capsules and their associated hazards. *Am. J. Roentgenol. Radium Ther. Nucl. Med.* **77,** 511–523.

GAO (1981). Problems in assessing the cancer risks of low-level ionizing radiation. U.S. General Accounting Office, Washington, D.C.

Garner, R. J. (1960). An assessment of the quantities of fission products likely to be found in milk in the event of fallout of aerial contamination of agricultural land. *Nature (London)* **186,** 1063.

Garten, C. T. (1978). A review of parameter values used to assess the transport of Pu, U and Th, in terrestrial food chains. *Environ. Res.* **17,** 437–452.

Geimer, R., Dwight, C., and McClellan, G. (1994). The plasma hearth demonstration project for mixed waste treatment. *In* "Proceedings of the Nuclear and Hazardous Waste Management International Topical Meeting Held in Atlanta, Georgia, August 14–18, 1994." American Nuclear Society, La Grange Park, Illinois.

General Electric Company (1985). Final safety analysis report for the *Galileo* mission and the *Ulysses* mission. Report GESP 7200. General Electric Company, New York.

George, A. C. (1996). State-of-the-art instrumentation for measuring radon/thoron and their progeny in dwellings—A review. *Health Phys.* **70,** 451–463.

Gephart, R. E., and Lundgren, R. E. (1995). Hanford tank cleanup: A guide to understanding the technical issues. Report PNL-10773. Pacific Northwest Laboratory, Richland, Washington.

German Federal Minister of Research and Technology (1979). The German risk study: Summary. Gesellschaft fur Reaktorsicherheit (GRS) (Reactor Safety Company), Cologne.

Gerusky, T. M. (1981). Three Mile Island: Assessment of radiation exposures and environmental contamination. The Three Mile Island nuclear accident: Lessons and implications. *Ann. N.Y. Acad. Sci.* **365,** 54–62.

Gesell, T. F. (1975). Some radiological health aspects of radon-222 in liquified petroleum gas. *In* "Noble Gases" (R. E. Stanley and A. A. Moghissi, eds.), ERDA TIC Report CONF. 730915, pp. 612–629. U.S. Energy Research and Development Agency, Washington, D.C.

Gesell, T. F. (1982). Environmental monitoring with thermoluminescence dosimetry. *IEEE Transactions on Nuclear Science,* **NS-29**(No. 3), 1225–1232.

Gesell, T. F. (1983). Background atmospheric ^{222}Rn concentrations outdoors and indoors: A review. *Health Phys.* **45,** 289–302.

Gesell, T. F., and Prichard, H. M. (1975). The technologically enhanced natural radiation environment. *Health Phys.* **28,** 361–366.

Gesell, T. F., Johnson, R. H., and Bernhardt, D. E. (1975). Some radiological health aspects of radon-22 in liquified petroleum gas. Report EPA-520 1-75-002. U.S. Environmental Protection Agency, Washington, D.C.

Giardina, P. A. *et al.* (1977). Summary report on the low-level radioactive burial site, West Valley, New York (1963–1975). U.S. Environmental Protection Agency, Washington, D.C.

Gifford, F. A. (1968). An outline of theories of diffusion in the lower layers of the atmosphere. *In* "Meteorology and Atomic Energy—1968" (D. H. Slade, ed.), Report TID-24190, pp. 65–116. U.S. Atomic Energy Commission, Washington, D.C.

Gifford, F. A. (1974). Power reactor siting: A summary of U.S. practice. *In* "Human and Ecologic Effects of Nuclear Power Plants" (L. A. Sagan, ed.), pp. 46–71. Thomas, Springfield, Illinois.

Gilinsky, V. (1978–1979). Plutonium, proliferation, and the price of reprocessing. *Foreign Affairs* **57,** 374–386.

Gilkeson, R. H., Perry, E. C., Cowart, J. B., and Holtzman, R. B. (1984). Isotope studies of the natural sources of radium in groundwater in Illinois. Report UILU-WRC-84-187. University of Illinois, Urbana.

Ginzburg, H. M. (1993). The psychological consequences of the Chernobyl accident—Findings from the International Atomic Energy Agency study. *Public Health Rep.* **108,** 184–192.

Glasstone, S. (1955). "Principles of Nuclear Reactor Engineering." Van Nostrand–Reinhold, New York.

Glasstone, S. (1962). "The Effects of Nuclear Weapons." U.S. Atomic Energy Commission, Washington, D.C.

Glasstone, S. (1994). "Elements of Nuclear Engineering." Chapman & Hall, New York.

Glasstone, S., and Dolan, P. J. (1977). "The Effects of Nuclear Weapons." U.S. Department of Defense, Washington, D.C.

Golay, M. W., and Todreas, N. E. (1990). Advanced light water reactors. *Sci. Am.* **April.**

Gold, S., Barkhau, H. W., Shleien, B., and Kahn, B. (1964). Measurement of naturally occurring radionuclides in air. *In* "The Natural Radiation Environment" (J. A. S. Adams and W. M. Lowder, eds.), pp. 369–382. Univ. of Chicago Press, Chicago.

Goldchert, N. W., Duffy, T. L., and Sedlet, J. (1985). Environmental monitoring at Argonne National Laboratory. Report ANL 85-17. Argonne National Laboratory, Argonne, Illinois.

Goldman, M., and Yaniv, S. S. (1978). Naturally occurring radioactivity in ophthalmic glass. *In* "Radioactivity in Consumer Products" (A. A. Moghissi, P. Paras, M. W. Carter, and R. F. Barker, eds.), Report NUREG/CP-0001, pp. 227–240. U.S. Nuclear Regulatory Commission, Washington, D.C.

Goldsmith, W. A. (1976). Radiological aspects of inactive uranium-milling sites: An overview. *Nucl. Saf.* **17,** 722–732.

Goodjohn, A. J., and Fortescue, P. (1971). "Environmental Aspects of High Temperature Gas-Cooled Reactors." Proceedings of the American Power Conference. Illinois Institute of Technology, Chicago.

Graham, J. D., and Vaupel, J. W. (1981). Value of a life: What difference does it make? *Risk Analysis* **1**(1), 89–95.

Grasty, R. L. (1994). Summer outdoor radon variations in Canada and their relationship to soil moisture. *Health Phys.* **66,** 185–93.

Green, B. M. R., Hughes, J. S., Lomas, P. R., and Janssens, A. (1992). Natural radiation atlas of Europe. *Radiat. Protect. Dosim.* **45,** 491–493.

Green, S. W. (1993). Progress update at Weldon Spring: Focus on building demolition. *Health Phys.* **64**(Suppl. 6), S52.

Grossman, R. F., and Holloway, R. W. (1985). Concentrations of krypton-85 near the Nevada Test Site. *Environ. Sci. Technol.* **19,** 1128–1131.

Groves, L. R. (1962). "Now It Can be Told: The Story of the Manhattan Project." Harper, New York.

Grübbe, E. H. (1933). Priority in the therapeutic use of X-rays. *Radiology* **21,** 156.

Guimond, R. J. (1978). The radiological aspects of fertilizer utilization. *In* "Radioactivity in Consumer Products" (A. A. Moghissi, P. Paras, M. W. Carter, and R. F. Barker, eds.), U.S. Nuclear Regulatory Commission Report NUREG/CPOOO3, pp. 31–393. NTIS, Springfield, Virginia.

Guimond, R. J., and Windham, S. T. (1980). Radiological evaluation of structures constructed on phosphate-related land. *In* "The Natural Radiation Environment III" (T. F. Gesell and W. M. Lowder, eds.), CONF-78042, Vol. 2, pp. 1457–1475. U.S. Department of Energy, Washington, D.C.

Guimond, R. J., and Windham, S. T. (1985). Radioactivity distribution in phosphate products, by-products, effluents and wastes. Report ORP/CSD-75-3. U.S. Environmental Protection Agency, Washington, D.C.

Guskova, A. K., Gusev, I. A., Moiseev, A. A., and Nugis, V. J. (1990). The estimate of the internal exposure importance in immediate early effects for Chernobyl nuclear power plant accident victims. *In* "The Medical Basis for Radiation Accident Preparedness II" (R. C. Ricks and S. A. Fry, eds.), pp. 231–241. Elsevier, New York.

Gustafson, P. F. (1969). Cesium-137 in freshwater fish during 1954–1965. *In* "Proceedings of The Second National Symposium on Radioecology, 1969," CONF-670503, pp. 249–257. U.S. Atomic Energy Commission, Washington, D.C.

Gustafson, P. F., Nelson, D. M., Brar, S. S., and Muniak, S. E. (1970). Recent trends in radioactive fallout. Report ANL-7760, Part III, p. 246. Argonne National Laboratory, Argonne, Illinois.

Gwaltney, R. C. (1969). Missile generation and protection in light-water-cooled reactors. *Nucl. Saf.* **10,** 300–307.

Hairr, L. (1974). An investigation of mechanisms of radiocesium cycling in estuarine sediments. Ph.D. dissertation, New York University, New York.

Halford, D. K., Millard, J. B., and Markham, O. D. (1981). Radionuclide concentrations in waterfowl using a liquid radioactive waste disposal area and the potential radiation dose to man. *Health Phys.* **40,** 173–181.

Halford, D. K., Markham, O. D., and Dickson, R. L. (1982). Radiation doses to waterfowl using a liquid radioactive waste disposal area. *J. Wildl. Manage.* **46,** 905–914.

Halitsky, J. (1968). Gas diffusion near buildings. *In* Meteorology and Atomic Energy" (D. H. Slade, ed.), Report TID-24190, pp. 221–255. U.S. Atomic Energy Commission, Washington, D.C.

Hallden, N. A., Fisenne, I. M., Ong, L. D. Y., and Harley, J. H. (1961). "Radioactive Decay of Weapons Debris." Report HASL-117. U.S. Atomic Energy Commission, New York.

Hamilton, E. I. (1971). Relative radioactivity in building materials. *Am. Ind. Hygiene Assoc. J.* **32,** 398–403.

Hanna, S. R., Briggs, C. A., and Hosker, R. P., Jr. (1982). Handbook on atmospheric diffusion. Report DOE/TIC-11223. NTIS, Springfield, Virginia.

Hansen, W. G., *et al.* (1964). Farming practices and concentrations of fission products in milk. Publ. 999-R-6. U.S. Public Health Service, Washington, D.C.

Hanslick, E., and Mansfield, A. (1990). Removal of radium from drinking water. *In* "The Environmental Behavior of Radium," IAEA Technical Report Series 310 (2 vols.). International Atomic Energy Agency, Vienna and UNIPUB, Lanham, Maryland.

Hanson, W. C., ed. (1980). Transuranic elements in the environment. Report DOE/TIC-22800. Technical Information Center, U.S. Department of Energy, Washington, D.C.

Hardy, E. P. (1974a). Depth distributions of global fallout of ^{90}Sr, ^{137}Cs and ^{239,240}Pu in sandy loam soil. Fallout Program Quarterly Summary Report HASL-286, pp. I-2 to I-10. U.S. Atomic Energy Commission, New York.

Hardy, E. P., Jr. (1974b). Worldwide distribution of plutonium. *In* "Plutonium and Other Transuranic Elements," Report WASH 1359, pp. 115–128. U.S. Atomic Energy Commission, Washington, D.C.

Hardy, E., and Alexander, L. T. (1962). Rainfall and deposition of strontium-90 in Challam County, Washington. *Science* **136**, 881–882.

Harley, J. H. (1952). A study of the airborne daughter products of radon and thoron. Ph.D. dissertation, Rensselaer Polytechnic Institute, Troy, New York.

Harley, J. H. (1956). Operation TROLL. Report NYOO 4656. U.S. Atomic Energy Commission, New York.

Harley, J. H. (1969). Radionuclides in food. *In* "Biological Implications of the Nuclear Age," AEC Symp. Ser. 16. U.S. Atomic Energy Commission, Washington, D.C.

Harley, N., and Cohen, B. (1980). Polonium-210 in tobacco. *In* "Radioactivity in Consumer Products" (A. A. Moghissi, P. Paras, M. W. Carter, and R. F. Barker, eds.), U.S. Nuclear Regulatory Commission Report NUREG/CPOOO3, pp. 199–216. NTIS, Springfield, Virginia.

Harley, N. H. (1984). Comparing radon daughter dose: Environmental vs. underground exposure. *Radiat. Protect. Dosim.* **7**, 371–375.

Harper, W. R. (1961). "Basic Principles of Fission Reactors." Wiley–Interscience, New York.

Harris, K. (1986). Personal communication. Pittway Corp., Northbrook, Illinois.

Hartung, F. H., and Hesse, W. (1879). Die Lungenkrebs, die Bergkrankheit, in den Schneeberger Gruben. *Vierteljahresschr. Gerichtl. Med. Oeff. Gesundheitwessen* **30**, 296.

Harvey, R. S. (1970). Temperature effects on the sorption of radionuclides by freshwater algae. *Health Phys.* **19**, 293–297.

Hatch, M. C., Wallenstein, S., Beyea, J., Nieves, J. W., and Susser, M. (1991). Cancer rates after the Three Mile Island nuclear accident and proximity of residence to the plant. *Am. J. Public Health* **81**, 719–724.

Hawley, C. A., Jr., Sill, C. W., Voelz, G. L., and Islitzer, N. F. (1964). Controlled environmental radioiodine tests at the National Reactor Testing Station, IDO-12035. U.S. Atomic Energy Commission, Idaho Operations Office, Idaho Falls, Idaho.

Health Physics (1983). Special issue. Radiobiology of radium and the actinides in man. *Health Phys.* **44**(Suppl.) 1–583.

Health Physics (1991). Special issue. The Goiania radiation accident. *Health Phys.* **50**, 1–113.

Healy, J. W. (1981). Statement before Procurement and Military Systems Subcommittee, Committee on Armed Services, House of Representatives, June 24–25, 1981. U.S. Government Printing Office, Washington, D.C.

Hebert, M. B., and Scott, L. M. (1993). A radiological evaluation of naturally occurring radioactive material (NORM) associated with six reclaimed tank battery sites. *Health Phys.* **64**(Suppl. 6), S18.

Heeb, C. M., and Bates, D. J. (1994). Radionuclide releases to the Columbia River from Hanford Operations, 1944–1971. Report PNWD-2223 HEDR. Pacific Northwest Laboratories, Richland, Washington.

Heinemann, K., and Vogt, K. J. (1979). Messungen zur ablagerungand biologischen Halbwertszeit von jod auf Vegetation. Norderney, Federal Republic of Germany (October 2–6, 1978).

Hemplemann, L. H., Hall, W. J., Philips, M., *et al.* (1975). Neoplasms in persons treated with X rays in infancy. *J. Nat. Cancer Inst.* **55**, 519–530.

Hering, S., ed. (1989). "Air Sampling Instruments for Evaluation of Atmospheric Contaminants." American Conference of Governmental Industrial Hygienists, Cincinnati, Ohio.

Hess, C. T., Casparius, R. E., Norton, S. A., and Brutsaert, W. F. (1980). Investigations of natural levels of radon-222 in groundwater in Maine for assessment of related health effects. *In* "The Natural Radiation Environment III" (T. F. Gesell and W. M. Lowder, eds.), U.S. Department of Energy CONF-708422, Vol. 1, pp. 529–546. NTIS, Springfield, Virginia.

Hess, C. T., Norton, S. A., Brutsaert, W. F., Lowry, J. F., Weiffenbach, C. V., Casparius, R. E., Coombs, E. G., and Brandow, J. E. (1981). Investigation of ^{222}Rn, ^{226}Ra, and U in air and groundwaters of Maine. Report B-017-Me. Land and Water Resources Center, University of Maine at Orono.

Hess, C. T., Weiffenbach, C. V., and Norton, S. A. (1983). Environmental radon and cancer correlations in Maine. *Health Phys.* **45**, 339–348.

Hess, C. T., Michel, J., Horton, T. R., Prichard, H. M., and Coniglio, W. A. (1985). The occurrence of radioactivity in public water supplies in the United States. *Health Phys.* **48**, 553–586.

Hewlett, R. C., and Anderson, O. E. (1962). "The New World" (Volume I of a history of the Atomic Energy Commission). Penn. State Univ. Press, University Park, Pennsylvania.

Hewlett, R. C., and Duncan, F. (1974). "Nuclear Navy, 1946–1972." Univ. of Chicago Press, Chicago.

Hill, C. R. (1966). Polonium-210 content of human tissues in relation to dietary habit. *Science* **152**, 1261–1262.

Hillel, D. (1971). "Soil and Water: Physical Principles and Processes." Academic Press, Orlando, Florida.

Hillman, D. J., Green, S. W., and French, T. (1994). ALARA design review for a chemical stabilization/solidification facility at the Weldon Spring Site remedial action project. *Health Phys.* **66**(Suppl. 6), S19.

Hilsmeier, W. F., and Gifford, F. A., Jr. (1962). Graphs for estimating atmospheric dispersion. Report ORO-545. U.S. Atomic Energy Commission, Washington, D.C.

Hinds, W. C. (1982). "Aerosol Technology—Properties, Behavior and Measurement of Airborne Particles." Wiley, New York.

Hisamatsu, S., and Sakanoue, M. (1978). Determination of transuranium elements in a so-called "Bikini Ash" sample and in marine sediment samples collected near Bikini Atoll. *Health Phys.* **35**, 301–307.

Hodges, P.C. (1964). "The Life and Times of Emil H. Grübbe." Univ. of Chicago Press, Chicago.

Hodgin, C. R. (1985). Terrain Responsive Atmospheric Code—Puff growth and shape. Rocky Flats Plant Report RFP-3683. Rocky Flats Plant, Golden, Colorado.

Hodgin, C. R. (1991). Terrain Responsive Atmospheric Code—Transport and dispersion. Rocky Flats Plant Report RFP-4516. Rocky Flats Plant, Golden, Colorado.

Hoffman, D. A. (1984). Late effects of I-131 therapy in the United States. *In* "Radiation Carcinogenesis: Epidemiology and Biological Significance" (J. D. Boice and J. F. Fraumeni, eds.), pp. 273–280. Raven, New York.

Hoffman, F. O., Garten, C. T., Lucas, D. M., and Huckabee, J. W. (1982). Environmental behavior of technetium in soil and vegetation: Implications for radiological assessments. *Environ. Sci. Technol.* **16**, 214–217.

Hoffman, F. O., Simon, S. L., and Thiessen, K. M. (1995). The role of uncertainty in dose reconstruction and risk assessment. Manuscript submitted for the Proceedings of the Thirty-First Annual Meeting of the National Council on Radiation Protection and Measurements, Arlington, Virginia, April 12–13, 1995. (To be published by the National Council on Radiation Protection and Measurements, Bethesda, Maryland.)

Holaday, D. A. (1959). The nature of wastes produced in the mining and milling of ores. Industrial Radioactive Waste Disposal. Hearings before Joint Committee on Atomic Energy, Washington, D.C.

Holaday, D. A. (1969). History of the exposure of miners to radon. *Health Phys.* **16**, 547–552.

Holaday, D. A., Rushing, D. E., Coleman, R. D., Woolrich, P. F., Kusnetz, H. L., and Bale, W. F. (1957). Control of radon and daughters in uranium mines and calculations on biologic effects, Public Health Service Publ. No. 494. U.S. Government Printing Office, Washington, D.C.

Holcomb, W. F. (1980). Inventory (1962–1978) and projections (to 2000) of shallow land burial of radioactive wastes at commercial sites: An update. *Nucl. Saf.* **21**(3), 380–388.

Holcomb, W. F. (1982). A history of ocean disposal of low level radioactive waste. *Nucl. Saf.* **23**, 183–197.

Holland, J. Z. (1953). A meteorological survey of the Oak Ridge area. Report ORO-99. U.S. Atomic Energy Commission, Washington, D.C.

Holland, J. Z. (1959). Stratospheric radioactivity data obtained by balloon sampling. Fallout from nuclear weapons tests. Hearings before Congressional Joint Committee on Atomic Energy. U.S. Government Printing Office, Washington, D.C.

Holleman, D. F., Luick, J. R., and Whicker, F. W. (1971). Transfer of radiocesium from lichen to reindeer. *Health Phys.* **21**, 657–666.

Holloway, D. (1994). "Stalin and the Bomb." Yale Univ. Press, New Haven, Connecticut.

Holm, L. E., Wiklund, K. E., Lundell, G. E., Bergman, N. A., Bjelkengren, G., Ceder-
quist, E. S., Ericson, U.-B., Larsson, L.-G., Lidberg, M. E., Lindberg, R. S.,
Wicklund, H. V., and Boice, J. C. (1988). Thyroid cancer after diagnostic doses of
iodine-131: A retrospective cohort study. *J. Natl. Cancer Inst.* **80,** 1132–1138.

Holm, W. M. (1978). Radium in consumer products: An historical perspective. *In*
"Radioactivity in Consumer Products" (A. A. Moghissi, P. Paras, M. W. Carter,
and R. F. Barker, eds.), Report NUREG/CP-0001, pp. 118–121. U.S. Nuclear
Regulatory Commission, Washington D.C.

Holtzman, R. B. (1964). Lead-210 (RaD) and polonium-210 (RaF) in potable waters
in Illinois. *In* "The Natural Radiation Environment" (J. A. S. Adams and W. M.
Lowder, eds.), pp. 227–237. Univ. of Chicago Press, Chicago.

Holtzman, R. B. (1980). Normal dietary levels of radium-226, radium-228, lead-210,
and polonium-210 for men. *In* " The Natural Radiation Environment III" (T. F.
Gesell and W. M. Lowder, eds.), U.S. Atomic Energy Commission CONF-780422,
Vol. 2, pp. 755–782. NTIS, Springfield, Virginia.

Holtzman, R. B., and Ilcewicz, F. H. (1966). Lead-210 and polonium-210 in tissues
of cigarette smokers. *Science* **153,** 1259–1260.

Horan, J. R., and Gammill, W. P. (1963). The health physics aspects of the SL-1
accident. *Health Phys.* **9,** 177–186.

Hosker, R. P. (1982). Methods of estimating wake flow and effluent dispersion near
simple blocklike buildings. U.S. Nuclear Regulatory Commission Report NUREG/
CR-252. NTIS, Springfield, Virginia.

Hubner, K. F., and Fry, S. A., eds. (1980). "The Medical Basis for Radiation Accident
Preparedness." Elsevier/North-Holland, New York.

Hueper, W. C. (1942). "Occupational Tumors and Allied Diseases." Thomas, Springfield, Illinois.

Hultqvist, B. (1956). Studies on naturally occurring ionizing radiation. *K. Sven.
Vetenskapsakad., Handl.* **4**(Suppl.).

Hunkin, G. G. (1980). Solution mining economics. *In* "Uranium Resources Technology," pp. 153–172. Seminar III. Colorado School of Mines, Boulder.

Hunt, G. J. (1991). Public radiation exposure due to radioactive waste discharges
into the Irish Sea. *Radiat. Protect. Dosim.* **36,** 205–210.

Hunter, H. F., and Ballou, N. E. (1951). Fission product decay rates. *Nucleonics*
9, C2.

IAEA (1966). Manual on environmental monitoring in normal operations. IAEA
Safety Series No. 16. International Atomic Energy Agency, Vienna and UNIPUB,
Lanham, Maryland.

IAEA (1967). Assessment of airborne radioactivity, proceedings of instrumentation
and techniques for assessment of airborne radioactivity in nuclear operations.
International Atomic Energy Agency, Vienna and UNIPUB, Lanham, Maryland.

IAEA (1973). Regulations for the safe transportation of radioactive materials. International Atomic Energy Agency, Vienna and UNIPUB, Lanham, Maryland.

IAEA (1975). The Oklo phenomenon. Report STI/PUB/405. International Atomic
Energy Agency, Vienna and UNIPUB, Lanham, Maryland.

IAEA (1978). Monitoring of airborne and liquid radioactive releases from nuclear
facilities to the environment. IAEA Safety Series No. 46. International Atomic
Energy Agency, Vienna and UNIPUB, Lanham, Maryland.

IAEA (1985). Press release 6 March, 1985. International Atomic Energy Agency, Vienna.

IAEA (1986). Summary report on the post-accident review meeting on the Chernobyl accident. IAEA Safety Series No. 75-INSAG-1. International Atomic Energy Agency, Vienna and UNIPUB, Lanham, Maryland.

IAEA (1988). The radiological accident in Goiania. International Atomic Energy Agency, Vienna and UNIPUB, Lanham, Maryland.

IAEA (1989). Measurement of radionuclides in food and the environment—A guidebook. IAEA Technical Report Series No. 295. International Atomic Energy Agency, Vienna and UNIPUB, Lanham, Maryland.

IAEA (1990). The international nuclear event scale. *Bulletin of the International Atomic Energy Agency.* **April.**

IAEA (1991a). The international Chernobyl project. International Atomic Energy Agency, Vienna and UNIPUB, Lanham, Maryland.

IAEA (1991b). Airborne gamma-ray spectrometer surveying. IAEA Technical Report Series No. 323. International Atomic Energy Agency, Vienna and UNIPUB, Lanham, Maryland.

IAEA (1992). Effects of ionizing radiation on plant and animals at levels implied by current radiation protection standards. IAEA Technical Report Series No. 332. International Atomic Energy Agency, Vienna and UNIPUB, Lanham, Maryland.

IAEA (1993). Report on radioactive waste disposal. IAEA Technical Report Series No. 349. International Atomic Energy Agency, Vienna and UNIPUB, Lanham, Maryland.

IAEA (1996). One decade after Chernobyl: Summing up the consequences of the accident. Proceedings of the conference held April 8–12, 1996. International Atomic Energy Agency, Vienna and UNIPUB, Lanham, Maryland.

Ibrahim, S. A., and Whicker, F. W. (1992a). Environmental behavior of thorium. *Health Phys.* **62,**(Suppl. 6), S48.

Ibrahim, S. A., and Whicker, F. W. (1992b). Comparative plant uptake and environmental behavior of U-series radionuclides at a uranium mine-mill. *J. Radioanal. Nucl. Chem.* **156,**(2), 253–267.

Ibrahim, S. A., Wrenn, M. E., Singh, N. P., Cohen, N., and Saccomano, G. (1993). Thorium populations in human tissues from two U.S. populations. *Health Phys.* **44**(Suppl. 1), 213–220.

Ibrahim, S. A., Schierman, M. J., and Whicker, F. W. (1996). Comparative distribution of ^{241}Am and ^{239,240}Pu in soils around the Rocky Flats Environmental Technology Site. *Health Phys.* **70,** 520–526.

ICRP (1960). Report of Committee II on permissible dose for internal radiation (with bibliography for biological, mathematical and physical data). *Health Phys.* **3** (Special Volume, 380 pp.).

ICRP (1955). Recommendations of the International Commission on Radiological Protection. *In* "British Journal of Radiology," Suppl. No. 6. British Institute of Radiology, London, UK.

ICRP (1965). "Principles of Environmental Protection Related to the Handling of Radioactive Materials." International Commission on Radiological Protection, Vienna.

ICRP (1966). (Task Group on Lung Dynamics). Deposition and retention models for internal dosimetry of the human respiratory tract. *Health Phys.* **12,** 173–207.

ICRP (1973). "Alkaline Earth Metabolism in Adult Man," ICRP Publ. No. 20. Pergamon, Oxford.

ICRP (1975). "Reference Man: Anatomical, Physiological and Metabolic Characteristics," ICRP Publ. No. 23. Pergamon, Oxford.

ICRP (1977). "Recommendations of the ICRP," ICRP Publ. No. 26, Ann. of the ICRP, Vol. 1(13). Pergamon, Oxford.

ICRP (1979–1988). "Limits for Intakes of Radionuclides by Workers," ICRP Publ. No. 30, Parts 1–4. Pergamon, Oxford.

ICRP (1980). "Biological Effects of Inhaled Radionuclides," ICRP Publ. No. 31, Ann. of the ICRP, Vol. 4 (1–2). Pergamon, Oxford.

ICRP (1983). "Cost–Benefit Analysis in the Optimization of Radiation Protection," ICRP Publ. No. 37, Ann. of the ICRP, Vol. 10(2–3). Pergamon, Oxford.

ICRP (1984). "Principles of Monitoring for the Radiation Protection of the Public," ICRP Publ. No. 43, Ann. of the ICRP, Vol. 15(1). Pergamon, Oxford.

ICRP (1985). "Radiation Protection Principles for Disposal of Solid Radioactive Waste," ICRP Publ. No. 46, Ann. of the ICRP, Vol. 15 (4). Pergamon, Oxford.

ICRP (1986). "The Metabolism of Plutonium and Related Elements," ICRP Publ. No. 48, Ann. of the ICRP, Vol. 16 (2–3). Pergamon, Oxford.

ICRP (1990). "Age-Dependent Doses to Members of the Public from Intake of Radionuclides: Part 1, "ICRP, Publ. No. 56, Ann. of the ICRP, Vol. 20 (2). Pergamon, Oxford.

ICRP (1991). "1990 Recommendations of the ICRP," ICRP Publ. No. 60, Ann. of the ICRP, Vol. 21 (1–3). Pergamon, Oxford.

ICRP (1994a). "Age-Dependent Doses to Members of the Public from Intake of Radionuclides: Part 2, "ICRP Publ. No. 67, Ann. of the ICRP, Vol, 23 (3–4). Pergamon, Oxford.

ICRP (1994b). Human Respiratory Tract Model for Radiological Protection, "ICRP Publ. No. 66, Ann. of the ICRP, Vol. 24 (1–3). Pergamon, Oxford.

ICRP (1995). "Age-Dependent Doses to Members of the Public from Intake of Radionuclides: Part 3," ICRP Publ. No. 69, Ann. of the ICRP, Vol. 25 (1). Pergamon, Oxford.

ICRU (1994). Gamma-ray spectrometry in the environment. Report 53. International Commission on Radiation Units and Measurement, Bethesda, Maryland.

International Standards Organization (1975). General principles for sampling airborne radioactive materials. ISO Report 2889–1975. International Standards Organization, Geneva, Switzlerland.

Intersociety Committee for a Manual of Methods for Ambient Air Sampling and Analysis (1972). Methods of air sampling and analysis. American Public Health Association, Washington, D.C.

Iranzo, E., and Richmond, C. R. (1987). Plutonium contamination twenty years after the nuclear weapons accident in Spain. Oak Ridge National Laboratory, Oak Ridge, Tennessee.

Iranzo, E., Salvador, S., and Iranzo, C. E. (1987). Air concentrations of [239]Pu and [240]Pu and potential radiation doses to persons living near Pu-contaminated areas in Palomares, Spain. *Health Phys.* **52,** 453–461.

Iranzo, E., Espinosa, A., and Iranzo, C. E. (1988). Evaluation of remedial actions taken in agricultural area contaminated by transuranides. 4th International Symposium of Radioecology, Cadarache, France.

Iranzo, E., Rivas, P., Mingarro, E., Marin, C., Espinosa, A., and Iranzo, C. E. (1991). Distribution and migration of soils of an accidentally contaminated environment. *Radiochim. Acta* **52/53** (Pt. 1), 249–256.

Iranzo, E., Espinosa, A., Bellido, A., and Gutierrez, J. (1992). Summary report on the 26 year Palomares surveillance program. Centro de Investigationes Energeticas Medioambientales y Tecnologicas (CIEMAT) Madrid, Spain.

Ishihara, T., and Kumatori, T. (1983). Cytogenetic follow-up studies in Japanese fishermen exposed to fallout radiation. *In* "Radiation-Induced Chromosome Damage in Man," pp. 475–490. Alan R. Liss, New York.

Ishimaru, T., Ichimaru, M., Mikami, M., Yamada, Y., and Tomonaga, Y. (1982). Distribution of onset of leukemia among atomic bomb survivors in the leukemia registry by dose, Horoshima and Nagasaki, 1946–75. Technical Report RERF TR 12–81. Radiation Effects Research Foundation, Hiroshima, Japan.

Ivanova, N. P. (1995). Population doses in Russia from plutonium fallout following the Chernobyl accident. *Radiat. Protect. Dosim.* **58,** 255–260.

Jablon, S., and Kato, H. (1970). Childhood cancer in relation to prenatal exposure to atomic-bomb radiation. *Lancet,* 1000–1003.

Jacobs, G. K., and Spalding, B. P. (1988). *In situ* vitrification demonstration for the stabilization of buried wastes at Oak Ridge National Laboratory. *Nucl. Chem. Waste Manage.* **8,** 249–259.

Janke, R. C., Janke, R. J., Davis, M. J., Habegger, L. J. (1992). RI/FS work plan development for the process area at the U.S. DOE Fernald site. *Health Phys.* **62**(Suppl. 6), S36–S37.

Japan Society for the Promotion of Science (1956). "Research in the Effects and Influences of the Nuclear Bomb Test Explosions." Japan Society for the Promotion of Science, Tokyo.

Jaworowski, Z. (1967). "Stable and Radioactive Lead in Environment and Human Body." Nuclear Energy Information Center, Warsaw.

Jaworowski, Z., Ilkiewicz, J., Kownacka, L., and Wlodek, S. (1975). Artificial sources of natural radionuclides in environment. *In* "The Natural Radiation Environment II" (J. A. S. Adams, W. M. Lowder, and T. F. Gesell, eds.), CONF-71, p. 809. U.S. Department of Energy, Washington, D.C.

Jenne, E. A. (1968). Controls on Mn, Fe, Co, Ni, Cu, and Zn concentrations in soils and water: The significant role of hydrous Mn and Fe oxides. *Adv. Chem. Ser. V* **73,** 337–389.

Jinks, S. M. (1975). An investigation of the factors influencing radiocesium concentrations of fish inhabiting natural aquatic ecosystems. Ph.D. dissertation, New York University, New York.

Jinks, S. M., and Wrenn, M. E. (1976). Radiocesium transport in the Hudson River estuary. *In* "Environmental Toxicity of Aquatic Radionuclides: Models and Mechanisms" (H. W. Miller and J. N. Stannard, eds.). Ann Arbor Science Publ., Ann Arbor, Michigan.

Jirka, G. H., Findikakis, A. N., Onishi, Y., and Ryan, P. J. (1983). Transport of radionuclides in surface waters. *In* "Radiological Assessment" (J. E. Till and H. R. Meyer, eds.), Report NUREG/CR-3332, Chap. 3. U.S. Nuclear Regulatory Commission, Washington, D.C.

Johnson, J. E. (1978). Smoke detectors containing radioactive materials. *In* "Radioactivity in Consumer Products" (A. A. Moghissi, P. Paras, M. W. Carter, and R. F. Barker, eds.), Report NUREG/CP-0001, pp. 434–440. U.S. Nuclear Regulatory Commission, Washington, D.C.

Johnson, R. H., Jr., Bernhardt, D. E., Nelson, N. S., and Calley, H. W., Jr. (1973). Assessment of potential radiological health effects from radon in natural gas. Report EPA-520/1-73-004. U.S. Environmental Protection Agency, Washington, D.C.

Johnson, W. B., and Bailey, P. G. (1983). Study of randon daughter concentrations in Polk and Hillsborough counties. *Health Phys.* **45,** 432–434.

Jonassen, N., and McLaughlin, J. P. (1988). Removal of radon and radon progeny from indoor air. *In* "Radon and Its Decay Products in Indoor Air" (W. W. Nazaroff and A. V. Nero, eds.), pp. 435–458. Wiley, New York.

Jones, C. G. (1994). Ir-192 misadministration and contribution of build-up factors to public dose. *Health Phys.* **66** (Suppl. 6), S11–S12.

Jones, S. R., Fulker, M. J., McKeever, J., and Stewert, T. H. (1991). Aspects of population exposure consequent on discharges of radionuclides to the environment from the nuclear reprocessing plant at Sellafield at Cumbria. *Radiat. Protect. Dosim.* **36,** 199–204.

Jose, D. E. (1985). U.S. court practice concerning compensation for alleged radiation injuries. Paper delivered at Edison Electric Institute–Health Physics Conference, September 5, 1985.

Kahn, B., and Rosson, R. (1993). Radon in groundwater. *Health Phys.* **64,**(Suppl. 6), S46.

Kaplan, S. I. (1971). HTGR safety. *Nucl. Saf.* **12,** 438–447.

Karas, J. S., and Stanbury, J. B. (1965). Fatal radiation syndrome from an accidental nuclear excursion. *N. Engl. J. Med.* **272,** 755–761.

Kathren, R. L. (1984). "Radioactivity in the Environment." Harwood Academic Pub. New York.

Kathren, R., Masse, F., Mossman, K., Roessler, G., and Schaiger, K. (1993a). Scientific and public issues committee statement: Radiation dose limits for the general public. *Health Physics Society Newsletter* **21**(5), 5–6.

Kathren, R., Masse, F., Mossman, K., Roessler, G., and Schaiger, K. (1993b). SPI committee responds to Puskin. *Health Physics Society Newsletter* **21**(8), 5–6.

Kathren, R. L., Denham, D. H., and Salmon, K., Eds. (1993c). Environmental health physics. *In* "Proceedings of the 26th Midyear Topical Meeting of the Health Physics Society." Research Enterprises Publishing Segment, Richland, Washington.

Kauranen, P., and Miettinen, J. K. (1969). [210]Po and ^{210}Pb in the arctic food chain and the natural radiation exposure of Lapps. *Health Phys.* **16,** 287–296.

Keith, L. H., ed. (1988). "Principles of Environmental Sampling." American Chemical Society, Washington, D.C.

Kelly, J. J., ed. (1980). Effluent and environmental surveillance. ASTM Special Technical Publication 698. American Society for Testing and Materials, Philadelphia, Pennsylvania.

Kemeny, J. G. (1979). President's commission on the accident at Three Mile Island. Report of Task Force on Public Health and Safety. U.S. Government Printing Office, Washington, D.C.

Kennedy, W. E., Parkhurst, M. A., Aaberg R. L., Rhoades, K. C., Hill, R. L., and Martin , J. B. (1992). Evaluation of exposure pathways to man from disposal of radioactive materials in sanitary sewer systems. Report PNL-7892 and NUREG/CR-5814. Pacific Northwest Laboratory, Richland, Washington.

Kerber, R. A., Till, J. E., Simon, S. L., Lyon, J. L., Thomas, D. C., Preston-Martin, S., Rallison, M. L., Lloyd, R. D., and Stevens, W. (1993). A cohort study of thyroid disease in relation to fallout from nuclear weapons testing *JAMA, J. Am. Med. Assoc.* **270,** 2076–2082.

Kerr, R. A. (1993). Ocean in a machine starts looking like the real thing. *Science* **260,** 32–33.

Khodakovsky, I. K. (1994). Radionuclide sources of arctic contamination. *Arctic Research of the United States* **8**(Spring), 262–265.

King, L. J., and McCarley, W. T. (1961). Plutonium release incident of november 20. Report ORNL-2989. Oak Ridge National Laboratory, Oak Ridge, Tennessee

King, P. T., Michel, J., and Moore, W. S. (1982). Ground water geochemistry of Ra-228, Ra-226 and Rn-222. *Geochim. Cosmochim. Acta* **46,** 1173–1182.

Kinne, O. ed. (1970). "Marine Ecology, Vol. I, Environmental Factors." Wiley–Interscience, New York.

Kirchmann, R., Darcheville, M., and Koch, G. (1980). Accumulation of radium-226 from phosphate fertilizers in cultivated soils and transfer to crops. *In* "The Natural Radiation Environment III" (T. F. Gesell and W. M. Lowder, eds.), CONF-78042, Vol. 2, pp. 1667–1672. U.S. Department of Energy, Washington, D.C.

Kitto, M. E. (1992). Radioactivity in size-separated municipal incinerator ashes. *Health Phys.* **62,** 529–536.

Kleinman, M. T. (1971). The stratospheric inventory of Pu-238. Report HASL-245. U.S. Atomic Energy Commission, New York.

Klement, A. W. (1959). A review of potential radionuclides produced in weapons detonations. Report WASH-1024. U.S. Atomic Energy Commission, Washington, D.C.

Klement, A. W. (1965). Radioactive fallout phenomena and mechanisms. *Health Phys.* **11,** 1265–1274.

Klemic, G., Shobe, J., Gesell, T., and Shebell, P. (1995). Results of the tenth international intercomparison of environmental dosimeters. *Radiat. Protect. Dosim.* **58,** 133–142.

Klevin, P. B., Weinstein, M. S., and Harris, W. B. (1956). Ground level contamination from stack effluents. *Am. Ind. Hyg. Assoc. Q.* **17,** 189–192.

Klusek, C. S. (1984a). Sr-90 in the U.S. diet, 1982. EML Report 429. U.S. Department of Energy, New York.

Klusek, C. S. (1984b). Sr-90 in human bone in the U.S., 1982. EML Report 435. U.S. Department of Energy, New York.

Klusek, C. S. (1987). Strontium-90 in food and bone from fallout. *J. Environ. Qual.* **16,** 195–199.

Knapp, A. H. (1961). The effect of deposition rate and cumulative soil level on the concentrations of strontium-90 in U.S. milk and food supplies. Report TID-13945. U.S. Atomic Energy Commission, Washington, D.C.

Knapp, A. H. (1963). Iodine-131 in fresh milk and human thyroids following a single deposition of nuclear test fallout. Report TID-19266. U.S. Atomic Energy Commission, Washington, D.C.

Kneale, G. W., and Stewart, A. M. (1976). Mantel-Haenszel analysis of Oxford data. I. Independent effects of several birth factors including fetal irradiation. *J. Natl. Cancer Inst.* **56,** 879–883.

Knief, R. A. (1981). "Nuclear Energy Technology." McGraw–Hill, New York.

Knief, R. A. (1992). "Theory and Technology of Commercial Nuclear Power." Taylor & Francis–Hemisphere, Washington D.C.

Kobayashi, R., and Nagai, I. (1956). Cooperation by the United States in the radio-chemcial analyses. *In* "Research in the Effects and Influences of the Nuclear Bomb Test Explosions II," pp. 1435–1445. Japan Society for the Promotion of Science, Tokyo.

Kocher, D. C. (1991). Perspective on the historical development of radiation standards. *Health Phys.* **61,** 519–527.

Kocher, D. C., Sjoreen, A. L., and Bard, C. S. (1983). Uncertainties in geological disposal of high level wastes: Ground water transport of radionuclides and radiological consequences. Report 5838. Oak Ridge National Laboratory, Oak Ridge, Tennessee.

Kochupillai, N., Verma, I. C., Grewal, M. S., and Ramalingaswami, V. (1976). Down's syndrome and related abnormalities in an area of high background radiation in coastal Kerala. *Nature (London)* **262,** 60–61.

Koczy, F. F. (1960). The distribution of elements in the sea. *In* "Proceedings, Disposal of Radioactive Wastes, 1959. "International Atomic Energy Agency, Vienna.

Kohman, T. (1959). Sources of radiation. *In* "Radiation Hygiene Handbook" (H. Blatz, ed.), Sect. 6. McGraw–Hill, New York.

Kolb, W., and Schmier, H. (1978). Building material induced radiation exposure of the population. *In* "Radioactivity in Consumer Products" (A. A. Moghissi, P. Paras, M. W. Carter, and R. F. Barker, eds.). Report NUREG/CP-0001, pp. 344–349. U.S. Nuclear Regulatory Commission, Washington, D.C.

Konoplya, E. F., and Rolevich, I. V. (1996). "The Chernobyl Catastrophe. Consequences in the Republic of Belarus." Academy of Sciences of Belarus, Minsk.

Koren, G., and Klein, N. (1991). Bias against negative studies in newspaper stories of medical research *JAMA, J. Am. Med. Assoc.* **226,** 1824–1826.

Koshurnikova, N. (1996). The risk of cancer among nuclear workers at the "Mayak" production association: Preliminary results of an epidemiological study. *In* "Proceedings of the 32nd Annual Meeting of the NCRP, April 3–4, 1996," in press. National Council on Radiation Protection and Measurements, Bethesda, Maryland.

Kouts, H., and Long, J. (1973). Tritium production in nuclear reactors. *In* "Tritium" (A. A. Moghissi and M. W. Carter, eds.), pp. 38–45. Messenger Graphics, Phoenix, Arizona.

Kraner, H. W., Schroeder, G. L., and Evans, R. D. (1964). Measurements of the effects of atmospheric variables on radon-222 flux and soil-gas concentrations. *In* "The Natural Radiation Environment" (J. A. S. Adams and W. L. Lowder, eds.), pp. 191–215. Univ. of Chicago Press, Chicago.

Krauskopf, K. B. (1986). Thorium and rare earth elements as analogs for actinide elements. *Chem. Geol.* **55,** 323–335.

Krauskopf, K. B. (1988). "Radioactive Waste Disposal and Geology." Chapman & Hall, London.

Krey, P. W. (1967). Atmospheric burnup of a plutonium-238 generator. *Science* **158,** 769–771.

Krey, P. W. (1976). Remote plutonium contamination and total inventories from Rocky Flats. *Health Phys.* **30,** 209–214.

Krey, P. W., and Hardy, E. P. (1970). Plutonium in soil around the Rocky Flats Plant. Report HASL-235. U.S. Atomic Energy Commission, New York.

Krey, P. W., and Krajewski, B. (1970). Comparison of atmospheric transport model calculations with observations of radioactive debris. *J. Geophys. Res.* **75,** 2901–2908.

Krey, P. W., Kleinman, M. T., and Krajewski, B. T. (1970). Sr-90, Zr-95 and Pu-238 stratospheric inventories 1967–1969. Report HASL-227, pp. 39–69. U.S. Atomic Energy Commission, New York.

Krieger, H. L., and Burmann, F. J. (1969). Effective half-time of [85]Sr, and [134]Cs for a contaminated pasture. *Health Phys.* **17,** 811–824.

Krishnaswami, S., Graunstein, W. C., Turekian, K. K., and Dowd, J. J. (1982). Radium, thorium, and radioactive lead isotopes in groundwater: Application to the *in situ* determination of adsorption–desorption rate constants and retardation factors. *J. Water Res. Res.* **18,** 1633–1675.

Kulp, J. L., and Schulert, A. R. (1962). Sr-90 in man V. *Science* **136,** 619.

Kumatori, T., Ishihara, T., Hirashima, K., Sugiyama, H., Ishii, S., and Miyoshi, K. (1980). Follow-up studies over a 25-year period on the Japanese fishermen exposed to radioactive fallout in 1954. *In* "The Medical Basis for Radiation Accident Preparedness" (K. F. Hubner and S. A. Fry, eds.), pp. 33–54. Elsevier/North-Holland, New York.

Laiche, T. P., and Scott, L. M. A. (1991). Radiological evaluation of phosphogypsum. *Health Phys.* **60,** 691–693.

Lamarsh, J. R. (1966). "Nuclear Reactor Theory." Addison–Wesley, Reading, Massachusetts.

Lamarsh, J. R. (1975). "Introduction to Nuclear Engineering." Addison–Wesley, Reading, Massachusetts.

Lamarsh, J. R. (1983). "Introduction to Nuclear Engineering." Addison–Wesley, Reading, Massachusetts.

Lamont, L. (1965). "Day of Trinity." Atheneum, New York.

Landa, E. (1987). Buried treasure to buried waste: The rise and fall of the radium industry. *Colo. Sch. Mines Q.* **82**(2), 77 pages.

Langroo, M. K., Wise, K. N., Duggleby, J. C., and Kotler, L. H. (1991). A nationwide survey of Rn-222 and gamma radiation levels in Australian homes. *Health Phys.* **61,** 753–761.

Laurer, G. R., and Eisenbud, M. (1963). Low-level *in vivo* measurement of iodine-131 in humans. *Health Phys.* **9**, 401–405.

Laurer, G. R., and Eisenbud, M. (1968). *In vivo* measurements of nuclides emitting soft penetrating radiations. *In* "Proceedings of the Symposium on Diagnosis and Treatment of Deposited Radionuclides," p. 189. Excerpta Medical Foundation, Amsterdam.

Lawrence, E. P., Wanty, R. B., and Nyberg, P. (1992). Contribution of Rn-222 in domestic water supplies to Rn-222 in indoor air in Colorado homes. *Health Phys.* **62**, 171–177.

League of Red Cross and Red Crescent Societies (1990). Report on assessment mission to the areas affected by the Chernobyl disaster, U.S.S.R. League of Red Cross and Red Crescent Societies, Geneva.

Lederer, C. M., and Shirley, V. S., eds. (1978). "Table of Isotopes," 7th Ed. Wiley, New York.

Lee, R. E., Jr. (1974). Measuring particulate matter in air. *In* "Instrumentation for Monitoring Air Quality," ASTM Publication STP-555. American Society for Testing and Materials, Philadelphia, Pennsylvania.

Lei, W., Linsalata, P., Penna Franca, E., and Eisenbud, M. (1986). Distribution and mobilization of cerium, lanthanum and neodymium in the Morro do Ferro basin, Brazil *Chem. Geol.* **55**, 313–322.

Leifer, R., Juzdan, Z. R., and Larsen, R. (1984). The high altitude sampling program: Radioactivity in the stratosphere. EML Report 434. U.S. Department of Energy, New York.

Leipunsky, O. I. (1957). Radioactive hazards from clean hydrogen bomb and fission atomic bomb explosions. *At. Energ. [Sov. At. Energy (Engl. Transl.)]* **3**, 530.

Lengemann, F. W. (1966). Predicting the total projected intake of radioiodine from milk by man. I. The situation where no countermeasures are taken. *Health Phys.* **12**, 825–835.

Lentsch, J. W., Kneip, T. J., Wrenn, M. E., Howells, G. P., and Eisenbud, M. (1972). Stable manganese and Mn-54 distributions in the physical and biological components of the Hudson River estuary. *In* "Proceedings of the Third National Symposium on Radioecology, "CONF-710501-P2, pp. 752–768. U.S. Atomic Energy Commission, Washington, D.C.

Lentsch, J. W., Babad, H., and Kirch, N. W. (1995). Operational experience in mitigating flammable gas releases from Hanford Site Tank 241-SY-101. Westinghouse Hanford Company, Richland Washington.

Lessard, E. T., Miltenberger, R. P., Cohn, S. H., Musolino, S. V., and Conard, R. A. (1984). Protracted exposure to fallout: The Rongelap and Utirik experience. *Health Phys.* **46**, 511–527.

Lettner, H., and Steinhäusler, F. (1988). Radon exhalation of waste gypsum recycled as building material. *Radiat. Protect. Dosim.* **24**(1–4), 415–417.

Lewellen, W. S., Sykes, R. I., and Parker, S. F. (1985). Comparison of the 1981 INEL dispersion data with results from a number of different models. Report NUREG/CR-4159, ARAP 505. U.S. Nuclear Regulatory Commission, Washington, D.C.

Lewellen, W. S., Sykes, R. I., Cerasoli, C. P., and Kornegay, F. C. (1987). Comparison of the 1982 SEADEX dispersion data with results from a number of different models. Report NUREG/CR-4820. U.S. Nuclear Regulatory Commission, Washington, D.C.

Lewin, J. (1977). The Russian approach to nuclear reactor safety. *Nucl. Saf.* **18**, 438–450.

Lewis, E. B. (1959). Statement. *In* "Fallout from Nuclear Weapons Tests," Vol. 2, pp. 1552–1554. Hearing before Joint Committee on Atomic Energy, Special Subcommittee on Radiation. U.S. Government Printing Office, Washington, D.C.

Lewis, H. W., Budnitz, R. J. Kouts, H. J. C. Loewenstein, W. B. Rowe, W. D., von Hippel., F., and Zachariasen, F. (1979). Ad hoc risk assessment review group. Report NUREG/CR-0400. NTIS, Springfield, Virginia.

Lewis, W. L. (1955). "Arthritis and Radioactivity." Christopher Publ. House, Boston.

Libby, W. F. (1952). "Radiocarbon Dating." Univ. of Chicago Press, Chicago.

Libby, W. F. (1958). Paper presented before Swiss Academy of Medicine, Lausanne, Switzerland.

Lindell, B. (1985). "Concept of Collective Dose in Radiological Protection." Organization for Economic Cooperation and Development, Paris.

Linsalata, P. (1984). Sources, distribution and mobility of plutonium and radiocesium in soils, sediments and water of the Hudson River and watershed. Ph.D. dissertation, New York University, New York.

Linsalata, P. (1994). Uranium and thorium decay series radionuclides in human and animal foodchains—A review. *J. Environ. Qual.* **23**, 633–642.

Linsalata, P., Simpson, H. J., Olsen, C. R., Cohen, N., and Trier, R. M. (1985). Plutonium and radiocesium in the water column of the Hudson River estuary. *Environ. Geol. Water Sci.* **7**, 193–204.

Linsalata, P., Cohen, N., and Wrenn, M. E. (1986). Sources, behavior and transport of ^{137}Cs in sediments and water of the Hudson River estuary. Unpublished manuscript, New York University Medical Center, New York.

Linsalata, P., Penna Franca, E., Campos, M. J., Lobao, N., Ballad, R., Lei, W., Ford, H., Morse, R. S., and Eisenbud, M. (1987). Radium, thorium and the light rare earth elements in soils and vegetables grown in an area of high natural radioactivity. *In* "Environmental Research for Actinide Elements" (J. E. Pinder III, ed.), Conf.-841142 (DE86008713). NTIS, Springfield, Virginia.

Litaor, M. I., Thompson, M. L., Barth, G. R. and Molzer, P. C. (1994). Plutonium-239 + 240 in soils east of Rocky Flats, Colorado. *J. Environ. Qual.* **23**, 1231–1239.

Little, C., ed. (1996). Special issue on dose reconstruction. *Health Phys.* **71**(4) 419–601.

Little, C. A. (1983). Development of computer codes for radiological assessments. *In* "Radiological Assessment" (J. E. Till and H. R. Meyer, eds.), Report NUREG/CR-3332, Chap. 13. U.S. Nuclear Regulatory Commission, Washington, D.C.

Little, C. A., and Miller, C. W. (1979). The uncertainty associated with selected environmental transport models. Report ORNL-5528. Oak Ridge National Laboratory, Oak Ridge, Tennessee.

Little, J. B., Radford, E. P., McCombs, H. L., and Hunt, V. R. (1965). Distribution of polonium in the pulmonary tissues of cigarette smokers. *N. Engl. J. Med.* **173**, 1343.

Little, M. P., Charles, M. W., and Wakeford, R. (1995). A review of the risks of leukemia in relation to parental pre-conception exposure to radiation. *Health Phys.* **68**, 299–310.

Littlefield, L., and Lushbaugh, C. C. (1990). Cytogenetic dosimetry for radiation accidents. *In* "The Medical Basis for Radiation Accident Preparedness II" (R. C. Ricks and S. A. Fry, eds.), pp. 461–479. Elsevier, New York.

Lively, R. S., and Krafthefer, B. C. (1995). Radon-222 variations in Mystery Cave, Minnesota. *Health Phys.* **68,** 590–594.

Lockhart, L. B. (1958). Atmospheric radioactivity studies at U.S. Naval Research Laboratory. Report 5249. U.S. Naval Research Laboratory, Washington, D.C.

Lackhart, L. B., Jr. (1964). Radioactivity of the radon-222 and radon-220 series in the air at ground level. *In* "The Natural Radiation Environment I" (J. A. S. Adams and W. M. Lowder, eds.), pp. 331–344. Univ. of Chicago Press, Chicago.

Lodge, J. P., Bien, G. S., and Suess, H. E. (1960). The carbon-14 content of urban airborne particulate matter. *Int. J. Air Pollut.* **2,** 309–312.

Logsdon, J. E., and Chissler, R. I. (1970). Radioactive waste discharges to the environment from nuclear power facilities. Report BRH/DER 70-2. Bur. Radiol. Health, Environ. Radiat. Div., Washington, D.C.

Look Magazine (1960). A sequel to atomic tragedy. April 12, 1960.

Lorenz, E. (1944). Radioactivity and lung cancer; A critical review in miners of Schneeberg and Joachimstahl. *J. Natl. Cancer Inst.* **5,** 1–15.

Lough, S. A., Hamada, G. H., and Comar, C. L. (1960). Secretion of dietary strontium-90 and calcium in human milk. *Proc. Soc. Exp. Biol. Med.* **104,** 194–198.

Lourelro, C. O., Abriola, L. M., Martin, J. E., and Sextro, R. G. (1990). Three-dimensional simulation of radon transport into houses with basements under constant negative pressure. *Environ. Sci. Technol.* **24,** 1338–1348.

Lowder, W. M., and Solon, L. R. (1956). Background radiation. Report NYO-4712. U.S. Atomic Energy Commission, Washington, D.C.

Lowman, F. G., Rice, T. R., and Richards, F. A. (1971). "Radioactivity in the Marine Environment," pp. 161–199. National Academy of Sciences, Washington, D.C.

Lubenau, J. O., and Yusko, J. G. (1995). Radioactive materials in recycled metals. *Health Phys.* **68,** 440–451.

Lubin, J. H. (1994). Invited commentary: Lung cancer and exposure to radon. *Am. J. Epidemiol.* **140,** 323–332.

Lubin, J., You-Lin, Q., Taylor, P., Shu-Xiang, Y., Schatzkin, A., Bao-Lin, M., Jian-Yu, R., Xiang-Zhen, X., and Li, J. (1990). Quantitative evaluation of the radon and lung cancer association in a case control study of Chinese tin miners. *Cancer Res.* **50,** 174–180.

Lubin, J. H., Boice, J. D., Edling, C., Hornung, R. W., Howe, G. R., Kunz, E., Kusiak, R. A., Morrison, H. I., Radford, E. P., Samet, J. M., Tirmarche, M., Woodward, A., Yao, S. X., and Pierce, D. A. (1995). Lung cancer in radon-exposed miners and estimation of risk from indoor exposure. *J. Natl. Cancer Inst.* **87**(11), 817–826.

Lucas, H. F., Jr. (1982). Ra-226 and Ra-228 in drinking water. Paper presented at the 27th Annual Meeting of the Health Physics Society, Las Vegas, Nevada.

Luetzelschwab, J. W., and Googins, S. W. (1984). Radioactivity released from burning gas lantern mantles. *Health Phys.* **46,** 873–881.

Lushbaugh, C. C. (1990). Radiation accidents worldwide: An additional decade of experience. *In* "The Medical Basis for Radiation Accident Preparedness II" (R. C. Ricks and S. A. Fry, eds.), pp. xxxv–xxxix. Elsevier, New York.

Lushbaugh, C. C., Fry, S. A., Hubner, K. F., and Ricks, R. D. (1980). Total-body irradiation: A historical review and follow-up. *In* "The Medical Basis for Radiation Accident Preparedness" (K. F. Hubner and S. A. Fry, eds.), pp. 3–15. Elsevier/North-Holland, New York.

Lutze, W., and Ewing, R. C., eds. (1988). "Radioactive Waste Forms for the Future." North-Holland Physics, Amsterdam.

Lyman, R. C. (1993). The EPA role in the federal radiological emergency response plan. *Health Phys.* **64**(Suppl. 6), S68.

McClure, J. D., and Cashwell, C. E. (1992). Transportation accidents/incidents involving radioactive materials (1971–1991). *In* "Proceedings of the Tenth Annual Symposium on Packaging and Transportation of Radioactive Material," pp. 1076–1082. Yokohama City, Japan.

McDowell-Boyer, L. M. (1979). Estimated radiation doses from thorium and daughters contained in thoriated welding electrodes. NRC Report NUREG/CR-1039. U.S. Nuclear Regulatory Commission, Washington, D.C.

McDowell-Boyer, L. M., and O'Donnell, F. R. (1978). Radiation dose estimates from timepieces containing tritium or ^{147}Pm in radioluminescent paints. Report ORNL/NUREG/TM-150. Oak Ridge National Laboratory, Oak Ridge, Tennessee.

McDowell-Boyer, L. M., Watson, A. P., and Travis, C. C. (1980). A review of parameters describing food-chain transport of lead-210 and radium-226. *Nucl. Saf.* **21**, 486–495.

McEachern, P., Myers, W. G., and White, F. A. (1971). Uranium concentrations in surface air at rural and urban localities within New York state. *Environ. Sci. Technol.* **5**, 700–703.

Machta, L., and List, R. J. (1959). Analysis of stratospheric strontium-90 measurements. *In* "Fallout from Nuclear Weapons Tests," Vol. I, pp. 741–762. Hearings before Joint Committee on Atomic Energy, Special Subcommittee on Radiation. U.S. Government Printing Office, Washington, D. C.

Machta, L., List, R. J., and Hubert, L. F. (1956). World-wide travel of atomic debris. *Science* **124**, 474–477.

McLeod, K. W., Adriano, D. C., Boni, A. L., Corey, J. C., Horton, J. H., Paine, D., and Pinder III, J. E. (1980). Influence of a nuclear fuel chemical separation facility on the plutonium contents of a wheat crop. *J. Environ. Qual.* **9**, 306–315.

MacMahon, B. (1962). Prenatal x-ray exposure and childhood cancer. *J. Natl. Cancer Inst.* **28**, 1173–1191.

Magno, P. J., Groulx, P. R., and Apidianakis, J. C. (1970). Lead-210 in air and total diets in the United States during 1966. *Health Phys.* **18**, 383–388.

Mahar, E. F., Rudich, S. N., and Moeller, D. W. (1987). Effective removal of airborne radon-222 decay products inside buildings. *Health Phys.* **53**, 351–356.

Mangino, J. J., Steele, J. M., and Burnhardt, K. C. (1995). Environmental monitoring and disposal of radioactive waste from U.S. naval nuclear-powered ships and support facilities, 1994. Report NT-95-1. Naval Nuclear Propulsion Program, Washington, D.C.

Mann, L. J., and Cecil, D. (1990). Tritium in ground water at the Idaho National Engineering Laboratory, Idaho. Water Resources Investigations Report 90-4090. U.S. Geological Survey, Idaho Falls, Idaho.

Marcinowski, F., and White, S. W. (1993). EPA's map of radon zones. *Health Phys.* **64** (Suppl. 6), S47.

Marcinowski, F., Lucas, R. M., and Yeager, W. M. (1994). National and regional distributions of airborne radon concentrations in U.S. homes. *Health Phys.* **66,** 699–706.

Markham, O. D., and Halford, D. K. (1982). Radionuclides in mourning doves near a nuclear facility complex in southeastern Idaho. *Wilson Bulletin* **94,** 185–197.

Markham, O. D., Halford, D. K., Rope, S. K., and Kuzo, G. B. (1988). Plutonium, Am, Cm and Sr in ducks maintained on radioactive leaching ponds in southeast Idaho. *Health Phys.* **55,** 517–524.

Marley, W. G., and Fry, T. M. (1956). Radiological hazards from an escape of fission products and the implications in power reactor location. *In* "Proceedings of the UN Conference on Peaceful Uses of Atomic Energy." United Nations, New York.

Marsden, E., and Collins, M. A. (1963). Particle activity and free radicals from tobacco. *Nature (London)* **98,** 962.

Marshall, N. (1983). "Nuclear Power Technology," Vol. 1. Oxford Univ. Press (Clarendon), Oxford.

Martell, E. A. (1974). Radioactivity of tobacco trichomes and insoluble cigarette smoke particles. *Nature (London)* **249,** 215–217.

Martland, H. S. (1925). Some unrecognized dangers in the use and handling of radioactive substances. *Proc. N. Y. Pathological Soc.* **25,** Nos. 6–8.

Martland, H. S. (1951). "Collection of Reprints on Radium Poisoning, 1925–1939." U.S. Atomic Energy Commission, Oak Ridge, Tennessee.

Mason, B. (1960). "Principles of Geochemistry." Wiley, New York.

Mason, B. (1982). "Principles of Geochemistry," 4th Ed. Wiley, New York.

Mauchline, J., and Templeton, W. L. (1964). Artificial and natural radioisotopes in the marine environment. *In* "Annual Review of Oceanography and Marine Biology" (H. Barnes, ed.), Vol. 2, pp. 229–279. Allen & Unwin, London.

May, M. M. (1994). Nuclear weapons supply and demand. *Am. Sci.* **82,** 526–537.

Mayneord, W. V., Radley, J. M., and Turner, R. C. (1958). "The Alpha-Ray Activity of Humans and Their Environment." United Nations, New York.

Mayneord, W. V., Turner, R. C., and Radley, J. M. (1960). Alpha activity of certain botanical materials. *Nature (London)* **187,** 208.

Mays, C. W. (1978). Endosteal dose to thorotrast patients. *Health Phys.* **35,** 123–125.

Mays, C. W., Spiess, H., and Gerspack, A. (1978). Skeletal effects following Ra-224 injections into humans. *Health Phys.* **35,** 83–90.

Means, J. L., Crerar, D. A., Borcsik, M. P., and Duguid, J. O. (1978). Adsorption of Co and selected actinides by Mn and Fe oxides in soils and sediments. *Geochim. Cosmochim. Acta* **42,** 1763–1773.

Media Institute (1979). "Television Evening News Covers Nuclear Energy: A Ten-Year Perspective." Media Institute, Washington D.C.

Medical Research Council (1956). "The Hazards to Man of Nuclear and Allied Radiations." Her Majesty's Stationery Office, London.

Medvedev, Z. A. (1979). "Nuclear Disaster in the Urals." Norton, New York.

Meinke, W. W., and Essig, T. H. (1991a). Offsite dose calculation manual guidance: Standard radiological effluent controls for pressurized water reactors. Report NUREG-1301. U.S. Nuclear Regulatory Commission, Washington, D.C.

Meinke, W. W., and Essig, T. H. (1991b). Offsite dose calculation manual guidance: Standard radiological effluent controls for boiling water reactors. Report NUREG-1302. U.S. Nuclear Regulatory Commission, Washington, D.C.

Melo, D. R., Lipsztein, J. L., de Oliveira, C. A. N., and Bertelli, L. (1994). Cs-137 internal contamination involving a Brazilian accident, and the efficacy of Prussian blue treatment. *Health Phys.* **66**, 245–252.

Menczer, L. F. (1965). Radioactive ceramic glazes. *Radiol. Health Data* **6**, 656.

Meneely, G. R., and Linde, S. M. (1965). "Second Symposium, Whole Body Counting and Effects of Internal Gamma Ray Emitting Isotopes." Thomas, Springfield, Illinois.

Menzel, R. G. (1960). Radioisotopes in soils. *In* "Effects of Amendments on Availability. Radioisotopes in the Biosphere" (R. S. Caldecott and L. A. Snyder, eds.), pp. 37–46. Univ. of Minnesota Press, Minneapolis.

Menzel, R. G. (1964). Competitive uptake by plants of potassium, rubidium, cesium, calcium, strontium and barium from soils. *Soil Sci.* **77**, 419.

Merriam, G. R., and Focht, E. F. (1957). A clinical study of radiation cataracts and the relationship to dose. *Am. J. Roentgenol., Radium Ther., Nucl. Med.* **77**, 759.

Merwin, S. E., and Darwin, R. F. (1991). Determining optimal protective actions for nuclear incidents. *Health Phys.* **61**, 305–316.

Mettler, F. A., Jr., Royal, H. D., Hurley, J. R., Khafagi, F., Sheppard, M. C., Beral, V., Reeves, G., Saenger, E. L., Yokoyama, N., Parshin, V., Griaznova, E. A., Taranenko, M., Chesin, V., and Cheban, A. (1992). Administration of stable iodine to the population around the Chernobyl nuclear power plant. *J. Radiol. Protect.* **12**, 159–165.

Mettler, F. A., Jr., Briggs, J. E., Carchman, R., Altorelli, K. K., Hart, B. L., and Kelsey, C. A. (1993). Use of radiology in U.S. general short term hospitals: 1980–1990. *Radiology* **189**, 377–380.

Meyer, H. R., Till, J. E., Bonner, E. S., Bond, W. D., Morse, L. E., Tennery, V. S., and Yalcintas, M. G. (1979). Radiological impact of thorium mining and milling. *Nucl. Saf.* **20**, 319–330.

Meyer, K. R., Voilleque, P. G., Killough, G. G., Schmidt, D. S., Rope, S. K., Shleien, B., Moore, R. E., Case, M. J., and Till, J. E. (1994). Overview of the Fernald dosimetry reconstruction project. *Health Phys.* **66**(Suppl. 6), S77.

Miettinen, J. K. (1969). Enrichment of radioactivity by arctic ecosystems in Finnish Lapland. *In* "Proceedings of the Second National Symposium on Radioecology, 1969," CONF-670503, pp. 23–31. U.S. Atomic Energy Commission, Washington, D.C.

Miller, C. W. (1984). Atmospheric dispersion and deposition. *In* "Models and Parameters for Environmental Radiological Assessments" (C. W. Miller, ed.), DE81027154(DOE/TIC-11468), DOE Critical Review Series, pp. 11–19. NTIS, Springfield, Virginia.

Miller, C. W., and Hoffman, F. O. (1982). An analysis of reported values of the environmental half-time for radionuclides deposited on the surfaces of vegetation. *In* "Proceedings of the Environmental Migration of Long-Lived Radionuclides." International Atomic Energy Agency, Vienna.

Miller, E. C. (1966). The integrity of reactor pressure vessels. Report No. 15. Nuclear Safety Information., Oak Ridge National Laboratory, Oak Ridge, Tennessee.

Mills, W. A., Flack, D. S., Arsenault, F. J., and Conti, E. F. (1988). "Compendium of Major U.S. Radiation Protection Standards and Guides: Legal and Technical Facts." Oak Ridge Associated Universities, Oak Ridge, Tennessee.

Ministry of Agriculture, Fisheries, and Food (1991). December 11 news release, London.

Minogue, R. B. (1978). NRC's role in regulating consumer products. *In* "Radioactivity in Consumer Products" (A. A. Moghissi, P. Paras, M. W. Carter, and R. F. Barker, eds.), Report NUREG/CP-0001, pp. 11–17. U.S. Nuclear Regulatory Commission, Washington, D.C.

Mistry, K. B., Bharathan, K. G., and Gopal-Ayengar, A. R. (1970). Radioactivity in the diet of population of the Kerala coast including monazite bearing high radiation areas. *Health Phys.* **19,** 353–542.

Miyake, Y., and Saruhashi, K. (1960). Vertical and horizontal mixing rates of radioactive material in the ocean. *In* "Proceedings of the Disposal of Radioactive Wastes, 1959." International Atomic Energy Agency, Vienna.

Mochizuki, Y., Mowafy, R., and Pasternack, B. (1963). Weights of human thyroids in New York City. *Health Phys.* **9,** 1299–1301.

Moeller, D. W., and Fujimoto, K. (1984). Cost evaluation and control measures for indoor radon progeny. *Health Phys.* **46,** 1181–1193.

Moeller, D. W., Rudnick, S. N., and Maher, E. F. (1988). Laboratory and field tests of a hassock fan-ion generator radon decay product removal unit. *Radiat. Protect. Dosim.* **24,** 503–506.

Moghissi, A. A., ed. (1983). Radioactive waste containing transuranic elements. Special issue *Nucl. Chem. Waste Management* **4**(1).

Moghissi, A. A., and Carter, M. W., eds. (1973). "Tritium." Messenger Graphics, Phoenix, Arizona.

Moghissi, A. A., and Lieberman, R. (1970). Tritium body burden of children, 1967–1968. *Radiol. Health Data Rep.* **11,** 227–231.

Moghissi, A. A., Godbee, H. W., Ozker, M. S., and Carter, M. W. (1978a). "Nuclear Power Waste Technology." American Society of Mechanical Engineers, New York.

Moghissi, A. A., Paras, P., Carter, M. W., and Barker, R. F., eds. (1978b). Radioactivity in consumer products. U.S. Nuclear Regulatory Commission Report NUREG/CP-003. NTIS, Springfield, Virginia.

Moghissi, S. A., Herschel, W. G., and Hobart, S. A., eds. (1986). "Radioactive Waste Technology." American Society of Mechanical Engineers, New York.

Mohammadi, H., and Mehdizadeh, S. (1983). Re-identification of ^{232}Th content and relative radioactivity measurements in a number of imported gas mantles. *Health Phys.* **44,** 649–653.

Monson, R. R., and MacMahon, B. (1984). Prenatal X-ray exposure and cancer in children. *In* "Radiation Carcinogenesis: Epidemiology and Biological Significance" (J. D. Boice and J. F. Fraumeni, eds.), pp. 97–105. Raven, New York.

Montgomery, D. M., Kolde, H. E., and Blanchard, R. L. (1977). Radiological measurements at the Maxey Flats radioactive waste burial site—1974 to 1975. Report

EPA-520/5-76/020. Office of Radiation Programs, U.S. Environmental Protection Agency, Cincinnati, Ohio.

Moore, H. E., Poet, S. E., and Martell, E. A. (1972). ^{222}Rn, ^{210}Pb, ^{210}Bi, and ^{210}Po profiles and the aerosol residence times versus altitude. *J. Geophys. Res.* **78,** 7065–7075.

Moore, R. E., Baes III, C. F., McDowell-Boyer, L. M., Watson, A. P., Hoffman, F. O., Pleasant, J. C., and Miller, C. W. (1979). AIRDOS-EPA: A computerized methodology for estimating environmental concentrations and dose to man from airborne releases of radionuclides. EPA Report 520/1-79-009. U.S. Environmental Protection Agency, Washington, D.C. (Originally published as Oak Ridge National Laboratory Report ORNL-5532.)

Morris, R. C. (1993). The implications of lined radioactive waste ponds for waterfowl contamination. *In* "Environmental Health Physics" (R. L. Kathren, D. H. Denham, and K. Salmon, eds.), pp. 147–155. Proceedings of the 26th Midyear Topical Meeting of the Health Physics Society, 24–28 January, 1993, Coeur d'Alene, Idaho. Research Enterprises Publishing Segment, Richland, Washington.

Morse, J. G. (1963). Energy for remote areas. *Science* **139,** 1175–1180.

Morse, R. S., and Welford, G. A. (1971). Dietary intake of ^{210}Pb. *Health Phys.* **21,** 53–55.

Muck, K. (1993). Detection limits and optimisation of search strategies for reactor satellite fragments. *J. Radiol. Protect.* **13**(3), 187–198.

Muller, H. J. (1927) Artificial transmutation of the gene. *Science* **66,** 84–87.

Muller, R. M., and Sprugel, D. J. (1978). Erosional transport and deposition of plutonium and cesium in two small midwestern watersheds. *J. Environ. Qual.* **7,** 171–174.

Musgrave, G. W. (1947). The quantitative evaluation of factors in water erosion—A first approximation. *J. Soil Water Conserv.* **2,** 133–138.

Mussalo-Rauhamaa, Jaakkola, T., Miettinen, J. K., and Laiho, K. (1984). Plutonium in Finnish Lapps—An estimate of the gastrointestinal absorption of plutonium by man based on a comparison of the plutonium content of Lapps and southern Finns. *Health Phys.* **46,** 549–559.

Muth, H., Schraub, A., Aurand, K., and Hantke, H. H. (1957). Measurements of normal radium burdens. *Br. J. Radiol.,* **7** (Suppl.)

Myrloi, M. G., and Wilson, J. G. (1951). On the proton component of the vertical cosmic-ray beam at sea level. *Proc. Phys. Soc., London, Sec. A* **64,** 404.

Nakaoka, A., Takagi, S., Fukushima, M., and Ichikawa, Y. (1985). Evaluation of radiation dose from a coal-fired power plant. *Health Phys.* **48,** 215–220.

Napier, B. A. (1991). Computational model design specification for phase I of the Hanford environmental dose reconstruction project. PNL-7274 HEDR. Pacific Northwest Laboratory, Richland, Washington.

Narayana, Y., Radhakrishna, A. P., Somashekarappa, H. M., Karunakara, N., Balakrishna, K. M., and Siddappa, K. (1993). Distribution and enrichment of radionuclides in the newly discovered high background area in Ullal on the southwest coast of India. *Health Phys.* **69,** 178–186.

NAS/NRC (1956). Pathologic effects of atomic radiation. Publ. 452. NAS/NRC, Washington, D.C.

NAS/NRC (1957a). Disposal of radioactive wastes on land. Publ. 519. NAS/NRC, Washington, D.C.

NAS/NRC (1957b). The effects of atomic radiation on oceanography and fisheries. Publ. 551. NAS/NRC, Washington, D.C.

NAS/NRC (1959). Radioactive waste disposal into Atlantic and Gulf coastal waters. Publ. 655. NAS/NRC, Washington, D.C.

NAS/NRC (1961). Long-term effects of ionizing radiations from external sources. Publ. 849. NAS/NRC, Washington, D.C.

NAS/NRC (1962). Disposal of low-level radioactive waste into Pacific coastal waters. Publ. 985. NAS/NRC, Washington, D.C.

NAS/NRC (1963). Damage to livestock from radioactive fallout in event of nuclear war. Publ. 1078. NAS/NRC Washington, D.C.

NAS/NRC (1964). Civil defense, project harbor summary report. Publ. 1237. NAS/NRC, Washington, D.C.

NAS/NRC (1969). "Civil Defense, Little Harbor Report." A report to the Atomic Energy Commission by a committee of the National Academy of Sciences. NAS, Washington, D.C.

NAS/NRC (1970). "Disposal of Solid Radioactive Wastes in Bedded Salt Deposits." NAS, Washington, D.C.

NAS/NRC (1978). "Radioactive Wastes at the Hanford Reservation: A Technical Review." National Academy Press, Washington, D.C.

NAS/NRC (1980). "The Effects on Populations of Exposure to Low Levels of Ionizing Radiation." National Academy Press, Washington, D.C.

NAS/NRC (1981). "Radioactive Waste Management at the Savannah River Plant: A Technical Review." National Academy Press, Washington, D.C.

NAS/NRC (1982). "Outlook for Science and Technology: The Next Five Years." Freeman, San Francisco, California.

NAS/NRC (1983). "A Study of the Isolation System for Geologic Disposal of Radioactive Wastes." National Academy Press, Washington, D.C.

NAS/NRC (1984). "Assigned Share for Radiation as a Cause of Cancer: Review of Radioepidemiologic Tables Assigning Probabilities of Causation. Final Report, Board of Radiation Effects Research, Committee on Radioepidemiologic Tables." National Academy Press, Washington, D.C.

NAS/NRC (1985a). "The Management of Radioactive Waste at the Oak Ridge National Laboratory: A Technical Review." National Academy Press, Washington, D.C.

NAS/NRC (1985b). "Mortality of Nuclear Weapons Test Participants." National Academy Press, Washington, D.C.

NAS/NRC (1986). "Scientific Basis for Risk Assessment and Management of Uranium Mill Tailings." National Academy Press, Washington, D.C.

NAS/NRC (1988). "Health Risks of Radon and Other Internally Deposited Alpha Emitters." National Academy Press, Washington, D.C.

NAS/NRC (1989). "The Nuclear Weapons Complex." National Academy Press, Washington, D.C.

NAS/NRC (1990). "Health Effects of Exposure to Low Levels of Ionizing Radiation." National Academy Press, Washington, D.C.

NAS/NRC (1991). "Policy Implications of Greenhouse Warming." National Academy Press, Washington, D.C.

NAS/NRC (1992). "Nuclear Power: Technical and Institutional Options for the Future." National Academy Press, Washington, D.C.

NAS/NRC (1994a). "Health Effects of Exposure to Radon: Time for Reassessment? Report of the Committee on Health Effects of Exposure to Radon." National Academy Press, Washington, D.C.

NAS/NRC (1994b). "Building Consensus through Risk Assessment and Management of the Department of Energy's Environmental Remediation Program." National Academy Press, Washington, D.C.

NAS/NRC (1995). "Radiation Dose Reconstruction for Epidemiologic Uses." National Academy Press, Washington, D.C.

NAS/NRC (1996). "Nuclear Wastes: Technologies for Separations and Transmutation." National Academy Press, Washington, D.C.

National Academy of Engineering (1972). "Engineering for the Resolution of the Energy–Environment Dilemma. Committee on Power Plant Siting." National Academy of Sciences, Washington, D.C.

National Advisory Committee on Oceans and Atmosphere (1984). Nuclear waste management and the use of the sea. A Special Report to The President and The Congress. Washington, D.C.

National Institute of Radiation Hygiene (Denmark) (1987). Natural radiation in Danish dwellings. National Institute of Radiation Hygiene, Copenhagen, Denmark. [cited in Damkjaer, A., and Korsbech, U. (1988). A search for correlation between local geology and indoor radon concentration. *Radiat. Protect. Dosim.* **24,** 51–54.

National Radiological Protection Board (1984). Assessment of radiation exposure to members of the public in West Cumbria as a result of discharges from BNFL Sellafield. Report R170. NRPB, London.

National Safety Council (1994). "Accident Facts." National Safety Council, Chicago.

Nazaroff, W. W., and Nero, A. V., eds. (1988). "Radon and Its Decay Products in Indoor Air." Wiley, New York.

Nazaroff, W. W., Doyle, S. M., Nero, A. V., and Sextro, R. G. (1987). Potable water as a source of airborne Rn-222 in U.S. dwellings: A review and assessment. *Health Phys.* **52,** 281–295.

Nazaroff, W. W., Doyle, S. M., Nero, A. V., and Sextro, R. G. (1988). Radon entry via potable water. *In* "Radon and Its Decay Products in Indoor Air" (W. W. Nazaroff and A. V. Nero, eds.), pp. 131–157. Wiley, New York.

NCRP (1941). Safe handling of radioactive luminous compounds. NCRP Report No. 5. National Bureau of Standards (U.S.), Handbook 2. National Council on Radiation Protection and Measurements, Bethesda, Maryland.

NCRP (1954). Permissible dose from external sources of ionizing radiation. NCRP Report No. 17. National Bureau of Standards (U.S.), Handbook. 59. National Council on Radiation Protection and Measurements, Bethesda, Maryland.

NCRP (1957). Maximum permissible exposure to man. Addendum to: Permissible dose from external sources of ionizing radiation. NCRP Report No. 17, 1954. National Bureau of Standards (U.S.), Handbook 59. National Council on Radiation Protection and Measurements, Bethesda, Maryland.

NCRP (1975a). Alpha-emitting particles in lungs. Report 46. National Council on Radiation Protection and Measurements, Bethesda, Maryland.

NCRP (1975b). Krypton-85 in the atmosphere—Significance and control technology. Report 44. National Council on Radiation Protection and Measurements, Bethesda, Maryland.

NCRP (1975c). Natural background radiation in the United States. Report 45. National Council on Radiation Protection and Measurements, Bethesda, Maryland.

NCRP (1976). Environmental radiation measurements. Report 50. National Council on Radiation Protection and Measurements, Bethesda, Maryland.

NCRP (1977a). Cesium-137 from the environment to man: Metabolism and dose. Report 52. National Council on Radiation Protection and Measurements, Bethesda, Maryland.

NCRP (1977b). Protection of the thyroid gland in the event of releases of radioiodine. Report 55. National Council on Radiation Protection and Measurements, Bethesda, Maryland.

NCRP (1977c). Radiation exposure from consumer products and miscellaneous sources. Report 56. National Council on Radiation Protection and Measurements, Bethesda, Maryland.

NCRP (1979). Tritium in the environment. Report 62. National Council on Radiation Protection and Measurements, Bethesda, Maryland.

NCRP (1980). Krypton-85 in the atmosphere—With specific references to the public health significance of the proposed controlled release at Three Mile Island. Commentary No. 1. National Council on Radiation Protection and Measurements, Bethesda, Maryland.

NCRP (1982). Preliminary evaluation of criteria for the disposal of transuranic contaminated waste. Report of Task Group on Criteria for the Disposal of Transuranic (TRU) Contaminated Waste. National Council on Radiation Protection and Measurements, Bethesda, Maryland.

NCRP (1983). Iodine-129: Evaluation of releases from nuclear power generation. Report 75. National Council on Radiation Protection and Measurements, Bethesda, Maryland.

NCRP (1984a). Radiological assessment: Predicting the transport, bioaccumulation, and uptake by man of radionuclides released to the environment. Report 76. National Council on Radiation Protection and Measurements, Bethesda, Maryland.

NCRP (1984b). Exposures from the uranium series with emphasis on radon and its daughters. NCRP Report 77. National Council on Radiation Protection and Measurements, Bethesda, Maryland.

NCRP (1984c). Evaluation of occupational and environmental exposures to radon and radon daughters in the United States. Report 78. National Council on Radiation Protection and Measurements, Bethesda, Maryland.

NCRP (1985a). Induction of thyroid cancer by ionizing radiation. Report 80. National Council on Radiation Protection and Measurements, Bethesda, Maryland.

NCRP (1985b). Carbon-14 in the environment. Report 81. National Council on Radiation Protection and Measurements, Bethesda, Maryland.

NCRP (1986a). A handbook of radioactivity measurement procedure. Report 58, 2nd Ed. National Council on Radiation Protection and Measurements, Bethesda, Maryland.

NCRP (1986b). Neptunium: Radiation protection guidelines. Report 90. National Council on Radiation Protection and Measurements, Bethesda, Maryland.

NCRP (1987a). Exposure of the population of the United States and Canada from natural background radiation. Report 94. National Council on Radiation Protection and Measurements, Bethesda, Maryland.

NCRP (1987b). Ionizing radiation exposure of the population of the United States. Report 93. National Council on Radiation Protection and Measurements, Bethesda, Maryland.

NCRP (1987c). Public radiation exposure from nuclear power generation in the United States. Report 92. National Council on Radiation Protection and Measurements, Bethesda, Maryland.

NCRP (1987d). Radiation exposure of the U.S. population from consumer products and miscellaneous sources. Report 95. National Council on Radiation Protection and Measurements. Bethesda, Maryland.

NCRP (1987e). Recommendations on limits for exposure to ionizing radiation. Report 91. National Council on Radiation Protection and Measurements, Bethesda, Maryland.

NCRP (1988). Measurement of radon and radon daughters in air. Report 97. National Council on Radiation Protection and Measurements, Bethesda, Maryland.

NCRP (1989a). Control of radon in houses. Report 103. National Council on Radiation Protection and Measurements, Bethesda, Maryland.

NCRP (1989b). Exposure of the U.S. population from diagnostic medical radiation. Report 100. National Council on Radiation Protection and Measurements, Bethesda, Maryland.

NCRP (1989c). Guidance on radiation received in space activities. Report 98. National Council on Radiation Protection and Measurements, Bethesda, Maryland.

NCRP (1989d). Screening techniques for determining compliance with environmental standards. Commentary 3. National Council on Radiation Protection and Measurements, Bethesda, Maryland.

NCRP (1989e). Limit for exposure to "hot particles" on the skin. Report 106. National Council on Radiation Protection and Measurements, Bethesda, Maryland.

NCRP (1991a) Effects of ionizing radiation on aquatic organisms. Report 109. National Council on Radiation Protection and Measurements, Bethesda, Maryland.

NCRP (1991b). Some aspects of strontium radiobiology. Report 110. National Council on Radiation Protection and Measurements, Bethesda, Maryland.

NCRP (1991c). Health and ecological implications of radioactively contaminated environments. Proceedings of the 26th Annual Meeting of the National Council on Radiation Protection and Measurements, Bethesda, Maryland.

NCRP (1993a). Limitation of exposure to ionizing radiation. Report 116. National Council on Radiation Protection and Measurements, Bethesda, Maryland.

NCRP (1993b). Radiation protection in the mineral extraction industry. Report 118. National Council on Radiation Protection and Measurements, Bethesda, Maryland.

NCRP (1993c). Risk estimates for radiation protection. Report 115. National Council on Radiation Protection and Measurements, Bethesda, Maryland.

NCRP (1995). Radiation exposure and high altitude flight. Commentary 12. National Council on Radiation Protection and Measurements, Bethesda, Maryland.

NCRP (1996a). Screening models for releases of radionuclides to atmosphere, surface water, and ground. Report 123 I. National Council on Radiation Protection and Measurements, Bethesda, Maryland.

NCRP (1996b). Screening models for releases of radionuclides to atmosphere, surface water, and ground—Work sheets. Report 123 II. National Council on Radiation Protection and Measurements, Bethesda, Maryland.

NCRP (1996c). Sources and magnitude of occupational and public exposures from nuclear medicine procedures. Report 124. National Council on Radiation Protection and Measurements, Bethesda, Maryland.

Neel, J. V., Schull, W. J., Awa, A. A., Satoh, C., Kato, H., Otake, M., and Yoshimoto, Y. (1990). The children exposed to atomic bombs: Estimates of the genetic doubling dose of radiation for humans. *Am. J. Hum. Genet.* **46,** 1053–1071.

Nefzger, M. D., Miller, R. J., and Fujino, T. (1968). Eye findings in atomic bomb survivors of Hiroshima and Nagasaki: 1963–1964. *Am. J. Epidemiol.* **89,** 129.

Neill, R. H. (1978). The role of the Bureau of Radiological Health in controlling radioactivity in consumer products. *In* "Radioactivity in Consumer Products" (A. A. Moghissi, P. Paras, M. W. Carter, and R. F. Barker, eds.), Report NUREG/CP-0001, pp. 38–39. U.S. Nuclear Regulatory Commission, Washington, D.C.

Nero, A. V., Jr. (1985). What we know about indoor radon. Testimony prepared for hearings on radon contamination: Risk assessment and mitigation research, held by the Committee on Science and Technology, U.S. House of Representatives (October 10, 1985).

Nero, A. V., Jr., Schwehr, M. B., Nazaroff, W. W., and Revzan, K. L. (1986). Distribution of airborne radon-222 concentrations in U.S. houses. *Science* **234,** 992–997.

Nero, A. V., Jr. (1988). Radon and its decay products in indoor air: An overview. *In* "Radon and Its Decay Products in Indoor Air" (W. W. Nazaroff and A. V. Nero, eds.), pp. 1–56. Wiley, New York.

Nero, A. V., Jr., Gadgil, A. J., Nazaroff, W. W., and Revzan, K. L. (1990). Indoor radon and decay products: Concentrations, causes and control strategies. DOE Technical Report DOE/ER-0480P. NTIS, Springfield, Virginia.

Neuberger, J. S. (1992). Residential radon exposure and lung cancer: An overview of ongoing studies. *Health Phys.* **63,** 503–509.

Nevessi, A., and Schell, W. R. (1975). Distribution of plutonium and americium in Bikini Atoll lagoon. *Health Phys.* **28,** 539–547.

New Jersey Department of Environmental Protection (1987). Technical background information report for the soil blending program. NJDEP, Trenton.

NYAM (1983). Resolution concerning the disposal in New York City of biomedical wastes containing de minimis levels of radioactivity. New York Academy of Medicine. Committee on Public Health, New York.

New York Academy of Sciences (1981). The Three Mile Island nuclear accident: Lessons and implications. *Ann. N.Y. Acad. Sci.* **365.**

New York State (1959). Protection from radioactive fallout, special task force report to Governor Nelson A. Rockefeller.

New York State Low Level Waste Group (1983). New York state must act to assure uninterrupted disposal capability for its low-level radioactive waste. Position Paper. New York State Low-Level Waste Group, Albany, NY.

Newell, R. W. (1971). The global circulation of atmospheric pollutants. *Sci. Am.* **224,** 32–47.

Ng, Y. C. (1982). A review of transfer factors for assessing the dose from radionuclides in agricultural products. *Nucl. Saf.* **23,** 57–71.

Ng, Y. C., Colsher, C. S., and Thompson, S. E. (1982). Soil to plant concentration factors for dose assessment. NRC Report NUREG/CR-2975. U.S. Nuclear Regulatory Commission, Washington, D.C.

Nies, A. (1990). State of the art of modeling repository performance in salt domes. *In* "Safety Assessment of Radioactive Waste Repositories" (proceedings of a symposium held in Paris). Organization for Economic Cooperation and Development, Paris.

Nikepelov, B. V., Romanov, G., Buldakov, L., Babaev, N., Kholina, Y., and Mikerin, E. (1989). Accident in the southern Urals on 29 September 1957. Report to the International Atomic Energy Agency, July 28. INFCIRC/368. IAEA, Vienna.

Nikepelov, B. V., Lizlov, A. F., and Koshinurkova, N. A. (1990). Experience with the first nuclear installation. *Piroda* **February.**

Nishita, H., Romney, E. M., and Larson, K. H. (1961). Uptake of radioactive fission products by crop plants. *J. Agric Food Chem.* **9,** 101.

Nishiwaki, Y., Kawai, H., Hyono, A., Kondo, M., Goshi, N., Nishiwaki, J., Yamatera, H., Kudo, I., Mori, M., Azuma, T., Tsumori, K., and Nagayama, O. (1956). Studies on the radioactivity of Bikini ash. *In* "Research in the Effects and Influences of the Nuclear Bomb Test Explosions," p. 262. Japan Society for the Promotion of Science, Tokyo.

NOAA (1979). "Assimilative Capacity of U.S. Coastal Waters for Pollutants" (E. D. Goldberg, ed.). U.S. Dept. of Commerce, National Oceanic and Atmospheric Administration, Washington, D.C.

Nowina-Konopka, M. (1993). Radiological hazard from coal-fired power plants in Poland. *Radiat. Protect. Dosim.* **46,** 171–180.

NRC (1974). Measurement of radionuclides in the environment, sampling and analysis of plutonium in soil. Regulatory Guide 4.5. U.S. Nuclear Regulatory Commission, Washington, D.C.

NRC (1975). Reactor safety study: An assessment of accident risks in U.S. commercial nuclear power plants, executive summary. WASH-1400(NUREG 75/014). U.S. Nuclear Regulatory Commission, Washington, D.C.

NRC (1977). Calculation of annual doses to man from routine releases of reactor effluents for the purpose of evaluating compliance with 10 CFR Part 50, Appendix I, Regulatory Guide 1.109. U.S. Nuclear Regulatory Commission, Washington, D.C.

NRC (1979a). Radiological effluent technical specifications for BWR's Report NUREG-0473. U.S. Nuclear Regulatory Commission, Washington, D.C.

NRC (1979b). Quality assurance for radiological monitoring programs (normal operations)—effluent streams and the environment. Regulatory Guide 4.15. U.S. Government Printing Office, Washington, D.C.

NRC (1980a). Radiological effluent and environmental monitoring at uranium mills. Regulatory Guide 4.14. U.S. Government Printing Office, Washington, D.C.

NRC (1980b). Final generic environmental impact statement on uranium milling. Report NUREG 0706. U.S. Nuclear Regulatory Commission, Washington, D.C.

NRC (1981). Standards for protection against radiation. Disposal of specific waste. Title 10 CFR 20.306. U.S. Nuclear Regulatory Commission, Washington, D.C.

NRC (1983). Radiological effluent technical specifications for PWR's Report NUREG-0472, Rev. 3. U.S. Nuclear Regulatory Commission, Washington, D.C.

NRC (1986). Report on the accident at the Chernobyl nuclear power station. Report NUREG 1250. U.S. Nuclear Regulatory Commission, Washington, D.C.

NRC (1993). ALARA levels for effluents from materials facilities. NRC Regulatory Guide 8.37. U.S. Nuclear Regulatory Commission, Washington, D.C.

NRC (1994). Code of Federal Regulations, Title 10, Part 72.3. U.S. Government Printing Office, Washington, D.C.

NRC (1995). Occupational radiation exposure at commercial nuclear power reactors and other facilities, 1994. NUREG-0713, Vol. 27. U.S. Nuclear Regulatory Commission, Washington, D.C.

Nuclear Defense Agency (1975). "Palomares Summary Report." Nuclear Defense Agency, Washington, D.C.

Nuclear Energy Agency (1993). "The Cost of High-Level Waste Disposal in Geological Repositories." Organization for Economic Cooperation and Development, Paris.

Nuclear Energy Insight (1995a). February, p. 3.

Nuclear Energy Insight (1995b). For sale: U.S. Enrichment Corporation. August.

Nuclear Energy Institute (1994). Environmental impact statement gives green light to planned enrichment plant. *Nuclear Energy INFO,* October.

Nuclear Energy Institute (1995). *Energy Information Digest,* February.

Nuclear Energy Institute (1996). News release, March 1996. Nuclear Energy Institute, Washington, D.C.

Nuclear Power Oversight Committee (1989). "Nuclear Energy for the Future: What We Must Do." Available from the Nuclear Management and Resources Council, Washington, D.C.

Nuclear Safety Advisory Center (1981). "Analysis of the Three Mile Island Unit 2 Accident." Electric Power Research Institute, Palo Alto, California.

Nuclear Waste Policy Act (1983). Nuclear Waste Policy Act of 1982, Public Law 97–425, U.S.C. 10101–10226. U.S. Government Printing Office, Washington, D.C.

Nuclear Waste Policy Amendments Act (1987). Amendments to the Nuclear Waste Policy Act of 1982, Public Law 100–203, December 22, 1987, 100th Congress. Title V, pp. 236–266. U.S. Government Printing Office, Washington, D.C.

Nugis, V. Y., and Konchalovskii, M. V. (1993). Biological dosimetry, diagnosis and treatment of bone marrow syndrome in victims of the Chernobyl accident. In "The Chernobyl Papers" (S. E. Merwin and H. I. Balanov, eds.), Vol. 1, pp. 349–384. Research Enterprises, Richland, Washington.

Nydal, R. (1968). Further investigation in the transfer of radiocarbon in nature. J. Geophys. Res. **73**, 3617–3635.

Oakley, D. T. (1972). Natural radiation exposure in the U.S. Report ORD/SID 7201. U.S. Environmental Protection Agency, Washington, D.C.

O'Brien, K., and McLaughlin, J. E. (1972). The radiation dose to man from galactic cosmic rays. Health Phys. **22**, 225–232.

O'Brien, K., Friedberg, W., Duke, F. E., Snyder, L., Darden, Jr., E. B., and Sauer, H. H. (1992). The exposure of aircraft crews to radiations of extraterrestrial origin. Radiat. Protect. Dosim. **45**, 145–162.

OECD (1974). "Monitoring of Radioactive Effluents." Organization for Economic Cooperation and Development, Paris.

OECD (1980). "'Radiological Significance and Management of ^{3}H, ^{14}C, ^{85}Kr, and ^{129}I Arising from the Nuclear Fuel Cycle." Organization for Economic Cooperation and Development, Paris.

OECD (1995). "Chernobyl Ten Years On." Organization for Economic Cooperation and Development, Paris.

O'Donnell, F. R. (1978). Assessment of radiation doses from radioactive materials in consumer products. In "Radioactivity in Consumer Products" (A. A. Moghissi, P. Paras, M. W. Carter, and R. F. Barker, eds.), Report NUREG/CP-0001, pp. 241–252. U.S. Nuclear Regulatory Commission, Washington, D.C.

Office of Technology Assessment (1979). "The Effects of Nuclear War." U.S. Government Printing Office, Washington, D.C.

Office of the President of the Russian Federation (1993). "Facts and Problems Related to Radioactive Waste Disposal in Seas Adjacent to the Territory of the Russian Federation." Moscow.

Office of the Surgeon General (1970). "Remedial Action Recommendations of the Surgeon General for Exposures to Mill Tailings Used in Building Construction." U.S. Government Printing Office, Washington, D.C.

Okrent, D. (1981). "Nuclear Reactor Safety." Univ. of Wisconsin Press, Madison.

Okubo, T. (1990). Radium in the oceans and seas. The environmental behavior of radium. IAEA Technical Report Series 310 (2 vols.). International Atomic Energy Agency, Vienna and UNIPUB, Lanham, Maryland.

Oliveira, A. R., Hunt, J., Valverde, N., Brandao-Mello, C., and Farina, R. (1991). Medical and related aspects of the Goiania accident: An overview. Health Phys. **60**, 17–24.

Onishi, Y., Seine, R. L., Arnold, E. M., Cowan, C. E., and Thompson, F. L. (1981). Critical review: Radionuclide transport, sediment transport, and water quality mathematical monitoring and radionuclide adsorption/desorption mechanisms. Report NUREG/CR-1322. Pacific Northwest Laboratory, Richland, Washington.

Ophel, I. L., and Judd, J. M. (1966). Effects of internally deposited radionuclides on the thermal tolerance of fish. In "Proceedings on the Radioactive Wastes into Seas, Oceans and Surface Waters," pp. 825–833. International Atomic Energy Agency, Vienna.

Optical Manufacturers Association (1975). "Ophthalmic Glass Radiological Standards." Optical Manufacturers Association, Arlington, Virginia.

ORNL (1970). Siting of fuel reprocessing plants and waste management facilities. Oak Ridge National Laboratory Report ORNL-4451. U.S. Atomic Energy Commission, Washington, D.C.

ORNL (1980). Spent fuel and waste inventories and projections. Report ORO-778 prepared by Oak Ridge National Laboratory for DOE, Office of Nuclear Waste Management. NTIS, Springfield, Virginia.

ORNL (1995). Integrated data base—1994: U.S. spent fuel and radioactive waste inventories, projections and characteristics. Report DOE/RW-0006, Rev. 11. Office of Scientific and Technical Information, Oak Ridge, Tennessee.

Oxenberg, T. P., and Davis, L. S. (1993). Construction of catchboxes at an army test center to enhance recovery of depleted uranium (DU) projectiles and limit the spread of contamination. *Health Phys.* **64** (Suppl. 6), S62.

Palmer, H. E., and Beasley, T. M. (1965). Iron-55 in humans and their food. *Science* **149,** 431–432.

Paredes, C. H., Kessler, W. V., Landolt, R. R., Ziemer, P. L., and Paustenbach, D. J. (1987). Radionuclide content of and Rn-222 emanation from building materials made from phosphate industry waste products. *Health Phys.* **53,** 23–29.

Parikh, S. R., Belden, R. D., and Cook, K. E. (1995). Pilot-scale soil washing treatability test to investigate the removal of radionuclides from contaminated soils in the 100 Area of the Hanford Site. *In* Proceedings of the Symposium "Environmental Restoration 95," August 13–17 in Denver, Colorado. U.S. Department of Energy Office of Environmental Restoration, Washington, D.C.

Parker, F. L., Schmidt, G. D., Cottrell, W. B., and Mann, L. A. (1961). Dispersion of radiocontaminants in an estuary. *Health Phys.* **6,** 66–85.

Parker, F. L., Churchill, M. A., Andrew, R. W., Frederick, B. J., Carrigan, P. H., Jr., Cragwall, J. S., Jr., Jones, S. L., Struxness, E. G., and Morton, R. J. (1966). Dilution, dispersion and mass transport of radionuclides in the Clinch and Tennessee rivers. *In* "Proceedings on the Disposal of Radioactive Wastes into Seas, Oceans and Surface Waters, 1966," pp. 33–55. International Atomic Energy Agency, Vienna.

Parker, H. M., and Healy, J. W. (1956). Effects of an explosion of a nuclear reactor. *In* "Proceedings of the International Conference on Peaceful Uses of Atomic Energy," p. 482. United Nations, New York.

Parker, H. M., and Norwood, W. D. (1946). H. I. Section report for October 1946. H. W. 7-5301. General Electric Company, Hanford Works, Richland, Washington.

Parker, L., Belsky, J. L., Yamamoto, T., Kawamoto, S., and Keehn, R. J. (1974). Thyroid carcinoma after exposure to atomic radiation. *Ann. Int. Med.* **80,** 600–604.

Parks, B. S. (1992). User's guide for CAP88-PC version 1.0. Las Vegas Facility Report 402-B-001. U.S. Environmental Protection Agency, Las Vegas, Nevada.

Parsly, L. F. (1971). Removal of elemental iodine from reactor containment atmospheres by spraying. Report ORNL-4623. Oak Ridge National Laboratory, Oak Ridge, Tennessee.

Party, A., Wilkerson, A., and Gershey, E. L. (1989). Need for broad, generic BRC rulings for biomedical low-level radioactive waste. *Radiat. Protect. Manage.* **6**(6), 45–51.

Paschoa, A., Tranjan Filho, A., and Rosenthal, J. (1993). Revisiting Goiania: Toward a final repository for radioactive wastes. *IAEA Bulletin,* **35,** 28–31.

Pearce, D. W., Lindroth, C. E., Nelson, J. L., and Ames, L. L. (1960). A review of radioactive waste disposal to the ground at Hanford. *In* "Proceedings on the Disposal of Radioactive Wastes, 1959." International Atomic Energy Agency, Vienna.

Pearson, J. E. (1967). Natural environmental radioactivity from radon-222. U.S. Public Health Service Publication 999-RH-26. National Center for Radiological Health, Rockville, Maryland.

Pelletieri, M. W., and Welles, B. W. (1985). History of nuclear materials transportation and packaging research and development sponsored by the U.S. federal government. Report SAND 85-7153. Sandia National Laboratories, Albuquerque, New Mexico.

Pendleton, R. C., and Hanson, W. C. (1958). Absorption of cesium-137 by components of an aquatic community. *In* "Proceedings of the Second International Conference on Peaceful Uses of Atomic Energy," Vol. 18, pp. 419–422. United Nations, New York.

Pendleton, R. C., Lloyd, R. D., and Mays, C. W. (1963). Iodine-131 in Utah during July and August, 1962. *Science* **141,** 640–642.

Pendleton, R. C., Mays, C. W., Lloyd, R. D., and Brooks, A. L. (1964). Differential accumulation of ^{131}I from local fallout in people and milk. *In* "Proceedings of the Hanford Symposium on Biology of Radioiodine, 1964," p. 72. Pergamon, Oxford.

Pendleton, R. C., Ways, C. W., Lloyd, R. D., and Church, B. W. (1965). A trophic level effect on Cs-137 concentration. *Health Phys.* **11,** 1503–1510.

Penna Franca, E., Fiszman, M., Lobao, N., Costa Ribeiro, C., Trindade, H., Dos Santos, P. L., and Batista, D. (1968). Radioactivity of Brazil nuts. *Health Phys.* **14,** 95–99.

Penna Franca, E., Fiszman, M., Lobao, N., Trindade, H., Costa Ribeiro, C., and Santos, P. L. (1970). Radioactivity in the diet in high background areas in Brazil. *Health Phys.* **19,** 657–662.

Pennders, R. M. J., Koster, H. W., and Lembrechts, J. F. (1992). Characteristics of Po-210 and Pb-210 in effluents from phosphate-producing industries: A first orientation. *Radiat. Protect. Dosim.* **45**(Suppl. 1–4), 737–740.

Perkins, R. W., and Nielsen, J. M. (1965). Cosmic-ray produced radionuclides in the environment. Health Phys. **11,** 1297–1304.

Pershagen, G., Axelson, O., Clavensjö, B., Damber, L., Enflo, A., Lagarde, F., Mellander, H., Svartengren, M., Swedjemark, G.-A., and Akerblom, G. (1993). Radon in dwellings and cancer. A country-wide epidemiological investigation. [in Swedish] IMM Report 2/93. Institutet för Miljomedicin, Karolinska Institutet, Stockholm, Sweden.

Pershagen, G., Akerblom, G., Axelson, O., Clavensjö, B., Damber, L., and Desai, G. (1994). Residential radon exposure and lung cancer in Sweden. *N. Engl. J. Med.* **330,** 159–164.

Persson, B. R. (1972). Radiolead (^{210}Pb), polonium (^{210}Po) and stable lead in the lichen, reindeer, and man. *In* "The Natural Radiation Environment II" (J. A. S. Adams, W. M. Lowder, and T. F. Gesell, eds.), U.S. Department of Energy CONF-720805-P2, pp. 347–367. NTIS, Springfield, Virginia.

Pertsov, L. A. (1964). "The Natural Radioactivity of the Biosphere." Atomizdat, Moscow. (Translated by Israel Program for Scientific Translations, Jerusalem, 1967.)

Peterson, H. T., Jr. (1983). Terrestrial and aquatic food-chain pathways. *In* "Radiological Assessment" (J. E. Till and H. R. Meyer, eds.), Report NUREG/CR-3332, Chap. 5. U.S. Nuclear Regulatory Commission, Washington, D.C.

Peterson, H. T., Jr. (1984). Regulatory implications of radiation dose–effects relationships. Health Phys. **47,** 345–359.

Petterssen, S. (1958). "Introduction to Meteorology," 2nd Ed. McGraw–Hill, New York.

Petterssen, S. (1968). "Introduction to Meteorology," 3rd Ed. McGraw–Hill, New York.

Petukhova, E. V., and Knizhnikov, V. A. (1969). Dietary intake of Sr-90 and Ca-137. Publ. A/AC.82/G/L.1245. United Nations, New York.

Philip, P. C., Jayaraman, S., and Pfister, J. (1984). Environmental impact of incineration of low-level radioactive wastes generated by a large teaching medical institution. *Health Phys.* **46,** 1123–1126.

Pickering, R. J., Carrigan, P. H., Jr., Tamura, T., Abee, H. H., Beverage, J. W., and Andrew, R. W., Jr. (1966). Radioactivity in bottom sediment of the Clinch and Tennessee rivers. *In* "Proceedings on the Disposal of Radioactive Wastes into Seas, Oceans and Surface Waters, 1966," pp. 57–88. International Atomic Energy Agency, Vienna.

Pierce, D. A., Shimuzy, Y., Preston, D. L., Vaeth, M., and Mabuchi, K. (1996). Studies of the mortality of A-bomb survivors, Report 12, Part 1. Cancer: 1950–1990. *Radiat. Res.* in press.

Piersanti, E. G., Macri, M. A., Nannini, D., Pona, C., Sgattoni, L., and Di Luzio, S. (1993). Control on contaminated liquid wastes from a laboratory of nuclear medicine. *Phys. Med.* **9**(Suppl. 1), 298–301.

Pimpl, M., and Schuttelkopf, H. (1981). Transport of plutonium, americium, and curium from soils into plants by root uptake. *Nucl. Saf.* **22,** 214–225.

Pohl-Rüling, J. (1993). The scientific development of a former gold mine near Badgastein, Austria, to the therapeutic facility "Thermal Gallery." *Environ. Int.* **19,** 455–465.

Pohl-Rüling, J., and Fischer, P. (1979). The dose–effect relationship of chromosome aberrations to alpha and gamma radioactivity. *Radiat. Res.* **80,** 61–81.

Pohl-Rüling, J., and Scheminzky, F. (1954). Das Konzentrationsverhaltnis Blut/Luft bei der Radon-inhalationund die Radon-aufnahme in den Menschliehen Korper in Radioaktiven Thermalstollen von Badgastein/Bockstein. *Strahlentherapie* **95,** 267.

Porter, C. R., Phillips, C. R., Carter, M. W., and Kahn, B. (1967). The cause of relatively high Cs-137 concentrations in Tampa, Florida, milk. *In* "Proceedings of the International Symposium on Radioecological Concentration Processes, 1966," pp. 95–101. Pergamon, Oxford.

Powell, R. W., and Puls, R. W. (1993). Passive sampling of ground water monitoring wells without purging: Multilevel well chemistry and tracer disappearance. *J. Contam. Hydrol.* **12**, 51–77.

Prantl, F. A., Tracy, B. L., and Quinn, M. J. (1980). Health significance of the radioactive contamination of soils and plants in Port Hope, Ontario. Paper presented at 1st Annual Conf. Canadian Radiat. Prot. Assn. (Montreal, May 1980).

Preston, A., and Jeffries, D. F. (1969). The I.C.R.P. critical group concept in relation to the Windscale discharges. *Health Phys.* **16**, 33–46.

Prichard, H. M., and Gesell, T. F. (1983). Radon-222 in municipal water supplies in the central United States. *Health Phys.* **45**, 991–993.

Prichard, H. M., Gesell, T. F., and Davis, E. (1981). Iodine levels in sludge and treated municipal wastewaters near a large medical complex. *Am. J. Public Health* **71**, 47–52.

Pritchard, D. W. (1960). The application of existing oceanographic knowledge to the problem of radioactive waste disposal into the sea. *In* "Proceedings on the Disposal of Radioactive Wastes, 1959." International Atomic Energy Agency, Vienna.

Pritchard, D. W. (1967). What is an estuary: Physical viewpoint. *In* "Estuaries," Publ. 83. American Association for the Advancement of Science, Washington, D.C.

Pritchard, D. W., Reid, R. O., Okubo, A., and Carter, H. H. (1971). Physical processes of water movement and mixing. *In* "Radioactivity in the Marine Environment," pp. 90–136. National Academy of Sciences, Washington, D.C.

Pugh, C. E. (1992). Aging and safety in reactor pressure vessels. Oak Ridge National Laboratory Review, No. 2. Oak Ridge National Laboratory, Oak Ridge, Tennessee.

Puskin, J. S. (1993). The value of a day of life. *Health Physics Society Newsletter* **21**(8), 5.

Raabe, O. G. (1984). Comparison of the carcinogenicity of radium and bone-seeking actinides. *Health Phys.* **46**, 1241–1258.

Raabe, O. G. (1994). Three-dimensional models of risk from internally deposited radionuclides. *In* "Internal Radiation Dosimetry" (O. G. Raabe, ed.), pp. 633–656. Medical Physics Publ., Madison, Wisconsin.

RAC (1996). "Radiation Doses and Risk to Residents from FMPC Operations from 1951–1988." Draft Report, Task 6. Radiological Assessments Corporation, Neeses, South Carolina.

Radford, E. P., Jr., and Hunt, V. R. (1964). Polonium-210: A volatile radioelement in cigarettes. *Science* **143**, 247–249.

Radhakrishna, A. P., Somashekarappa, H. M., Narayana, Y., and Siddappa, K. (1993). A new natural background radiation area on the southwest coast of India. *Health Phys.* **65**, 390–395.

Rahola, T., and Miettinen, J. K. (1973). Accumulation of ^{137}Cs in Finnish Lapps. *Arch. Environ. Health* **26**, 6769.

Rajewsky, B., and Stahlhofen, W. (1966). ^{210}Po activity in the lungs of cigarette smokers. *Nature (London)* **209,** 1312–1313.

Ramsdell, J. V. (1991). Atmospheric transport and dispersion modeling for the Hanford Environmental Dose Reconstruction Project. Report PNL-7198-HEDR. Pacific Northwest Laboratory, Richland, Washington.

Ramsden, D. (1969). The measurement of plutonium-239 *in vivo. Health Phys.* **16,** 145–154.

Randerson, D., ed. (1984). Atmospheric science and power production. Report DOE/TIC-27601. U.S. Department of Energy, Washington, D.C.

Rankama, K. (1954). "Isotope Geology." McGraw–Hill, New York.

Rankama, K., and Sahama, T. G. (1950). "Geochemistry." Univ. of Chicago Press, Chicago.

Ray, S. S., and Parker, F. G. (1977). Characterization of fly ash from coal-fired power plants. Report EPA-600-7-77-010. Office of Energy, Minerals and Industry, U.S. Environmental Protection Agency, Washington, D.C.

Reid, D. G., Sackett, W. M., and Spaulding, R. F. (1977). Uranium and radium in livestock feed supplements. *Health Phys.* **32,** 535–540.

Reid, G. K., and Wood, R. D. (1976). "Ecology of Inland Waters and Estuaries," 2nd Ed. Van Nostrand, New York.

Reid, G. W., Lassovszky, P., and Hathaway, S. (1985). Treatment, waste management and cost for removal of radioactivity from drinking water. *Health Phys.* **48,** 671–694.

Reinig, W. C., ed. (1970). "Environmental Surveillance in the Vicinity of Nuclear Facilities." Thomas, Springfield, IL.

Reiter, E. R. (1974). The role of the general circulation of the atmosphere in radioactive debris transport. Report COO-1340-38. U.S. Atomic Energy Commission, Washington, D.C.

Reitz, G., Schnuer, K., and Shaw, K. (1993). Workshop on radiation exposure of civil aircrew. *Radiat. Protect. Dosim.* **48,** 3.

Revelle, R., and Schaefer, M. B. (1957). General considerations concerning the ocean as a receptacle for artificially radioactive materials. Publ. 551. National Academy of Sciences, Washington, D.C.

Revelle, R., Folsom, T. R., Goldberg, E. D., and Isaacs, J. D. (1956). Nuclear science and oceanography. *In* "Proceedings of the International Conference on Peaceful Uses of Atomic Energy," p. 177. United Nations, New York.

Ricks, R. C., and Fry, S. A., eds. (1990). "The Medical Basis for Radiation Accident Preparedness II—Clinical Experience and Followup Since 1979." Elsevier, New York.

Ritchie, J. C., and McHenry, J. R. (1973). Vertical distribution of fallout ^{137}Cs in cultivated soils. *Radiat. Data Rep.* **14,** 727–728.

Ritchie, J. C., Clebsch, E. E. C., and Rudolph, W. K. (1970). Distribution of fallout and natural gamma radionuclides in litter, humus and surface mineral soil layers under natural vegetation in the Great Smoky Mountains, North Carolina–Tennessee. *Health Phys.* **18,** 479–489.

Roberts, H., Jr., and Menzel, R. G. (1961). Availability of exchangeable and nonexchangeable strontium-90 to plants. *J. Agric. Food Chem.* **9,** 95.

Robinson, A. R., and Marietta, M. G. (1985). Research, progress and the mark a box model for physical, biological and chemical transports. Report SAND 84-0646, UC-70. Sandia National Laboratories, Albuquerque, New Mexico.

Robinson, E. W. (1968). The use of radium in consumer products. Report MORP 68-5. U.S. Public Health Service, Washington, D.C.

Robison, W. L., Conrado, C. L., and Bogen, K. T. (1994). An updated dose assessment for Rongelap Island. Report UCRL-LR-107036. Lawrence Livermore Laboratory, Livermore, California.

Roche-Farmer, L. (1980). Study of alternative methods for the management of liquid scintillation counting wastes. Report NUREG-0656. Div. Tech. Information, U.S. Nuclear Regulatory Commission, Washington, D.C.

Roesch, W. C., ed. (1987). "U.S.–Japan Joint Reassessment of Atomic Bomb Radiation Dosimetry in Hiroshima and Nagasaki." Radiation Effects Research Foundation, Hiroshima, Japan.

Roessler, C. E., Smith, Z. A., Bolch, W. E., and Prince, J. R. (1979). Uranium and radium-226 in Florida phosphate materials. Health Phys. **37,** 269–277.

Roessler, C. E., Kautz, R., Bolch, W. E., Jr., and Wethington, J. A., Jr. (1980). The effect of mining and land reclamation on the radiological characteristics of the terrestrial environment of Florida's phosphate regions. In "The Natural Radiation Environment III" (T. F. Gesell and W. M. Lowder, eds.), CONF-780422, Vol. 2, pp. 1476–1493. U.S. Department of Energy, Washington, D.C.

Roessler, C. E., Roessler, G. S., and Bolch, W. E. (1983). Indoor radon progeny exposure in the phosphate mining region: A review. Health Phys. **45,** 389–396.

Rogers, V. C. (1991). Disposal alternatives for oil and gas NORM. Health Phys. **60**(Suppl. 2), S38.

Rogovin, M., and Frampton, G. J. (1980). A sequence of physical events. Three Mile Island, Vol. II, Part 2, pp. 309–340. Report to the Commissioners and the Public. Special Inquiry Group, U.S. Nuclear Regulatory Commission, Washington, D.C.

Romney, E. M., Wallace, A., Schulz, R. K., and Dunway, P. (1982). Plant root uptake of ^{239,240}Pu and ^{241}Am from soils containing aged fallout materials. In "Environmental Migration of Long-Lived Radionuclides," IAEA-SM-257/83. International Atomic Energy Agency, Vienna.

Ron, E., Lubin, J. H., Mabuchi, K., Modan, B., Pottern, L. M., Schneider, A. B., Tucker, M. A., and Boice, J. D. (1995). Thyroid cancer after exposure to external radiation: A pooled analysis of seven studies. Radiat. Res. **141,** 259–277.

Rosenstein, M. (1976). Organ doses in diagnostic radiology. U.S. Department of Health Education and Welfare Publication FDA 76-8030. U.S. Government Printing Office, Washington, D.C.

Rosenthal, J. J., de Almeida, C. E., and Mondancer, A. H. (1991). The radiological accident in Goiania: The initial remedial actions. Health Phys. **60,** 7–13.

Roser, F. X., and Cullen, T. L. (1964). External radiation measurements in high background regions of Brazil. In "The Natural Radiation Environment" (J. A. S. Adams and W. M. Lowder, eds.), pp. 825–836. Univ. of Chicago Press, Chicago.

Roser, F. X., Kegel, G., and Cullen, T. L. (1964). Radiogeology of some high-background areas of Brazil. In "The Natural Radiation Environment" (J. A. S.

Adams and W. M. Lowder, eds.), pp. 855–872. Univ. of Chicago Press, Chicago.

Rothman, S., and Lichter, S. R. (1985). Elites in conflict: Nuclear energy, ideology, and the perception of risk. *J. Contemp. Stud.* **VIII**, 23–44.

Rowland, R. E. (1994). Radium in humans: A review of U.S. studies. Report ANL-ER3. Argonne National Laboratory, Argonne, Illinois.

Russell, R. S. (1965). Interception and retention of airborne material on plants. *Health Phys.* **11**, 1305–1315.

Russell, R. S. (1966). "Radioactivity and Human Diet." Pergamon, Oxford.

Russell, R. S., and Bruce, R. S. (1969). Environmental contamination with fallout from nuclear weapons. *In* "Proceedings on Environmental Contamination by Radioactive Material, 1969." International Atomic Energy Agency, Vienna.

Russell, W. L. (1968). Recent studies on the genetic effects of radiation in mice. *Pediatrics* **41**, 223–230.

Ryan, M. T. (1981). Radiological impacts of uranium recovery in the phosphate industry. *Nucl. Saf.* **22**, 70–76.

Sabbarese, C., De Martino, S., Signorini, C., Gialanela, G., Roca, V., Baldassini, P. G., Cotellessa, G., and Sciocchetti, G. A. (1993). Survey of indoor Rn-222 in the Campania region. *Radiat. Protect. Dosim.* **48**, 257–263.

Samet, J., Stolwijk, J., and Rose, S. (1991). Summary: International workshop on residential Rn epidemiology. *Health Phys.* **60**, 223–227.

Schaefer, H. J. (1971). Radiation exposure in air travel. *Science* **173**, 780–783.

Scheminzky, F. (1961). 25 Jahre Baderforschung in Gastein mit Verzeichnis der Wissenschaftlichen Veroffenlichungen bis 1960. *Sonderabdruck Badgasteiner Bladeblatt* Nos. 35 and 36.

Schiager, K. J. (1974). Analysis of radiation exposures on or near uranium mill tailings piles. *Radiat. Data Rep.* **15**, 411–425.

Schiager, K. J. (1986). Disposal of uranium mill tailings. *In* Proceeding of the Twenty-First Annual Meeting of the NCRP." National Council on Radiation Protection and Measurements, Bethesda, Maryland.

Schlenker, R. A. (1982). Risk estimates for bone. *In* "Critical Issues in Setting Radiation Dose Limits," pp. 153–163. Proc. 17th Annual Meeting of NCRP. National Council on Radiation Protection and Measurements, Bethesda, Maryland.

Schmitz, J., and Fritsche, R. (1992). Radon impact at underground workplaces in western Germany. *Radiat. Protect. Dosim.* **45**(Suppl. 1–4), 193–195.

Schreckhise, R. G., and Cline, J. F. (1980a). Uptake and distribution of ^{232}U in peas and barley. *Health Phys.* **38**, 341–343.

Schreckhise, R. G., and Cline, J. F. (1980b). Comparative uptake and distribution of Pu, Am, Cm and Np in four plant species. *Health Phys.* **38**, 817–824.

Schulz, R. K. (1965). Soil chemistry of radionuclides. *Health Phys.* **11**, 1317–1324.

Schwartz, F. G., and Strider, P. V. (1995). Management of Pit 9—Highlights of accomplishments and lessons learned to date. *In* Proceedings of the Symposium "Environmental Restoration 95," August 13–17 in Denver, Colorado. U.S. Department of Energy Office of Environmental Restoration, Washington, D.C.

Scott, A. G. (1988). Controlling indoor exposures. *In* "Radon and Its Decay Products in Indoor Air" (W. W. Nazaroff and A. V. Nero, eds.), pp. 407–434. Wiley, New York.

Scott, H. L. (1992). Initial studies on levels of indoor radon at U.S. Department of Energy facilities. *Health Phys.* **62**(Suppl. 6), S59.

Scott, N. S. (1897). X-ray injuries. *Am. X-Ray J.* **1,** 57–65.

Scott, R. L., Jr. (1971). Fuel-melting incident at the Fermi reactor on Oct. 5, 1966. *Nucl. Saf.* **12,** 123–134.

Seaborg, G. T. (1958). "The Transuranic Elements." Addison–Wesley, Reading, Massachusetts.

Seaborg, G. T. (1981). "Kennedy, Khrushchev, and the Test Ban." Univ. of Calif. Press, Berkeley, California.

Seaborg, G. T., and Bloom, J. (1970). Fast breeder reactors. *Sci. Am.* **223**(5), 13–21.

Seelentag, W., and Schmier, H. (1963). Radiation exposure from luminous watch dials. *Radiol. Health Data* **4,** 209–213.

Sehmel, G. A. (1979). Particle and gas dry deposition: A review. *Atmos. Environ.* **14,** 983–1011.

Sehmel, G. A. (1984). Deposition and resuspension. *In* "Atmospheric Science and Power Production" (R. Randerson, ed.), Report DOE/TIC-27601, pp. 533–583. U.S. Department of Energy, Washington, D.C.

Semenov, B. A., and Oi, N. (1993). Nuclear fuel cycles: Adjusting to the new realities. *IAEA Bulletin* **35**(No. 3), 2–13.

Shafer, C. K. (1959). Testimony before Joint Committee on Atomic Energy. Hearings on Biological and Environmental Effects of Nuclear War. Washington, D.C.

Shapiro, P. S., and Sorg, T. J. (1988). Reduction of radon from household water supplies. *Radiat. Protect. Dosim.* **24**(1–4), 523–525.

Shapley, D. (1971). Rocky Flats: Credibility gap widens on plutonium plant safety. *Science* **174,** 569–571.

Shearer, S. D., Jr., and Lee, G. F. (1964). Leachability of radium-226 from uranium mill solids and river sediments. *Health Phys.* **10,** 217–227.

Shearer, S. D., Jr., and Sill, C. W. (1969). Evaluation of atmospheric radon in the vicinity of uranium mill tailings. *Health Phys.* **17,** 77–88.

Sheets, R. W., Thompson, C. C., and Petefish, H. M. (1995). Thorium in collectible glassware. Paper presented at the Sixth International Symposium on the Natural Radiation Environment, June 5–9, 1995, Montreal, Quebec, Canada.

Sheppard, M. I. (1980). The environmental behavior of radium. Report AECL-6796. Atomic Energy of Canada Ltd., Whiteshell.

Shipman, T. L., ed. (1961). Acute radiation death resulting from an accidental nuclear critical excursion. *J. Occup. Med.* **3**(No. 3), Special Suppl.

Shleien, B., Glavin, T. P., and Friend, A. G. (1965). Particle size fractionation of airborne gamma-emitting radionuclides by graded filters. *Science* **147,** 290–292.

Shleien, B., Cochran, J. A., and Magno, P. J. (1970). Strontium-90, strontium-89, plutonium-239, and plutonium-238 concentrations in ground-level air, 1964–1969. *Environ. Sci. Technol.* **4,** 598–602.

Shleien, B., Tucker, T., and Johnson, D. W. (1978). The mean active bone marrow dose to the adult population of the United States from diagnostic radiology. *Health Phys.* **34,** 587–601.

Shlyakhter, A., and Wilson, R. (1992). Chernobyl and glasnost: The effects of secrecy on health and safety. *Environment* **34,** 25–30.

Shore, R. E. (1992). Issues and epidemiological evidence regarding radiation-induced thyroid cancer. *Radiat. Res.* **131,** 98–111.

Shore, R. E., Woodward, E. D., and Hempelmann, L. H. (1984). Radiation-induced thyroid cancer. *In* "Radiation Carcinogenesis: Epidemiology and Biological Significance" (J. D. Boice and J. F. Fraumeni, eds.), pp. 131–138. Raven, New York.

Shore, R. E., Hemplemann, L. H., and Woodward, E. D. (1986). Carcinogenic effects of radiation on the human thyroid gland. *In* "Radiation Carcinogenesis" (A. C. Upton, *et al.,* eds.), pp. 293–309. Elsevier, New York.

Simon, S. I., Till, J. E., Lloyd, R. D., Kerber, R. L., Thomas, D. C., Preston-Martin, S., and Stevens, W. (1995). The Utah leukemia study: Dosimetry methodology and results. *Health Phys.* **68,** 460–471.

Simon, S. L., and Graham, J. (1994). Radiological survey of northern Rongelap Atoll. Republic of the Marshall Islands, Majuro.

Simon, S. L., and Graham, J. (1995). Findings of the nationwide radiological study. Republic of the Marshall Islands, Majuro.

Simon, S. L., Barron, A. M., Graham, J. C., and Duffy, S. (1993). An overview of the Marshall Islands nationwide radiological survey. *In* "Proceedings of the Twenty-Sixth Midyear Topical Meeting of the Health Physics Society." Research Enterprises Publ. Segment, Richland, Washington.

Simpson, H. J., Linsalata, P., Olsen, C. R., Cohen, N., and Trier, R. M. (1986). Transport of fallout and reactor radionuclides in the drainage basin of the Hudson River estuary. *In* "Environmental Research for Actinide Elements" (J. E. Pinder III, ed.), CONF 941142 (DE86008713). NTIS, U.S. Department of Energy, Washington, D.C.

Simpson, R. E., and Shuman, F. G. D. (1978). The use of uranium ceramic tableware. *In* "Radioactivity in Consumer Products" (A. A. Moghissi, P. Paras, M. W. Carter, and R. F. Barker, eds.), Report NUREG/CP-0001, pp. 470–474. U.S. Nuclear Regulatory Commission, Washington, D.C.

Sinclair, W. K. (1993). Science, radiation protection and the NCRP. The Lauriston S. Taylor Lecture of the National Council on Radiation Protection and Measurements, Bethesda, Maryland.

Sjöblom, K.-L., and Linsley, G. (1994). Sea disposal of radioactive waste: The London Convention (1972). *IAEA Bulletin* **36**(2), 12–16.

Slinn, W. G. (1984). Precipitation scavenging. *In* "Atmospheric Science and Power Production" (D. Randerson, ed.), Report DOE/TIC-27601. U.S. Department of Energy, Washington, D.C.

Smith, B. M., Grune, W. N., Higgins, F. B., Jr., and Terrill, J. G., Jr. (1961). Natural radioactivity in ground water supplies in Maine and New Hampshire. *J. Am. Water Works Assoc.* **53,** 75.

Smith, F. A., and Dzuiba, S. P. (1949). Preliminary observations of the uranium content of photographic materials. Unpublished memo, University of Rochester, Rochester, New York.

Smith, P. G., and Doll, R. (1982). Mortality of patients with ankylosing spondylitis after a simple treatment course with X rays. *Br. Med. J.* **284,** 449–460.

Soffer, L., Burson, S. B., Ferrell, C. M., Lee, R. Y., and Ridgely, J. N. (1992). Accident source terms for light-water nuclear power plants. Draft for comment. NUREG 1465. U.S. Nuclear Regulatory Commission, Washington, D.C.

Soldat, J. K. (1963). The relationship between I-131 concentrations in various environmental samples. *Health Phys.* **9,** 1167–1171.

Soldat, J. K. (1965). Environmental evaluation of an acute release of ^{131}I to the atmosphere. *Health Phys.* **11,** 1009–1015.

Solon, L. R., Lowder, W. M., Zila, A., LeVine, H. D., Blatz, H., and Eisenbud, M. (1958). External environmental radiation measurements in the United States. *Science* **127,** 1183–1184.

Sowinski, J. (1996). Personal communication.

Spalding, R. F., and Sackett, W. (1972). Uranium in runoff from the Gulf of Mexico distributive province. Anomalous concentrations. *Science* **175,** 629–631.

Spiers, F. W. (1966). Dose to bone from strontium-90: Implications for the setting of maximum permissible body burden. *Radiat. Res.* **28,** 624–642.

Spiers, F. W. (1968). "Radioisotopes in the Human Body: Physical and Biological Aspects." Academic Press, Orlando, Florida.

Spitsyn, V. I., and Balukova, V. D. (1979). The scientific basis for, and experience with, underground storage of liquid radioactive wastes in the U.S.S.R. *In* "Scientific Basis for Nuclear Waste Management" (G. J. McCarthy, ed.), p. 237. Plenum, New York.

Spitsyn, V. I., *et al.* (1958). A study of the migration of radioelements in soils. *In* "Proceedings of the UN Conference on Peaceful Uses of Atomic Energy," p. 2207. United Nations, New York.

Spitsyn, V. I., *et al.* (1960). Sorption regularities in behavior of fission product elements during filtration of their solutions through ground. *In* "Proceedings on Disposal of Radioactive Wastes, 1959." International Atomic Energy Agency, Vienna.

Stallings, Hon. R. A. (1994). Personal communication.

Stanley, R. E., and Moghissi, A. R., eds. (1974). Noble gases—Proceedings of a symposium held in Las Vegas, Nevada, August, 1972. Energy Research and Development Administration Report CONF-730915. U.S. Government Printing Office, Washington, D.C.

Stannard, J. N. (1988). "Radioactivity and Health: A History," DOE/RL/01830-T59. Battelle Press, Richland, Washington.

Start, G. E., and Wendell, L. L. (1974). Regional effluent dispersion calculations considering spatial and temporal meteorological variables. NOAA Technical Memorandum ERL ARL-44. National Oceanic and Atmospheric Administration, Air Resources Laboratory, Idaho Falls, Idaho.

Start, G. E., Cate, J. H., Sagendorf, J. F., Ackerman, G. R., Dickson, C. R., Nukari, N. H., and Thorngren, L. G. (1985). Idaho field experiment 1981. NRC Report NUREG/CR-3488, Vol. 3. U.S. Nuclear Regulatory Commission, Washington, D.C.

Stein, J. L. (1971). Evaluation of the sampling frequency of the pasteurized milk network. *Radiol. Health Data Rep.* **12,** 451–455.

Steinberg, E., and Glendenin, L. (1956). "Proceedings of the UN Conference on the Peaceful Uses of Atomic Energy," p. 614. United Nations, New York.

Steinhäusler, F. A. (1975). Long-term measurements of Rn-222, Rn-220, Pb-214 and Pb-212 concentrations in the air of private and public buildings and their dependence on meteorological parameters. *Health Phys.* **29,** 705–713.

Steinhäusler, F., Hoffmann, W., Daschill, F., and Reubel, B. (1988). Chernobyl and its radiological and socio-economic consequences for the province of Salzburg, Austria. *Environ. Int.* **14,** 91–112.

Steinhäusler, F., Landa, E. R., Rundo, J., Stebbings, J. H., and Moghissi, A. A., eds. (1993). Special issue: History of radium, thorium and related nuclides in industry and medicine. Proceedings of the International Workshop, Badgastein, Austria, October 1–3, 1991. *Environ. Int.* **19**(5).

Stevens, D. L., Jr. (1963). "A Brief History of Radium." Bur. Radiol. Health, U.S. Public Health Service, Washington, D.C.

Stevens, W., Thomas, D. C., Lyon, J. L., Till, J. E., Kerber, R. A., Simon, S. L., Lloyd, R. D., Eghany, N. A., and Martin, S. P. (1990). Leukemia in Utah and radioactive fallout from the Nevada Test Site. *JAMA, J. Am. Med. Assoc.* **264,** 585–591.

Stevens, W., Till, J. E., Thomas, D. C., Lyon, J. L., Kerber, R. A., Preston-Martin, S., Simon, S. L., Rallison, M. A., and Lloyd, R. D. (1992). Assessment of leukemia and thyroid disease in relation to fallout in Utah. University of Utah, Salt Lake City.

Stewart, A., and Kneale, G. W. (1970). Radiation dose effects in relation to obstetric X-rays and childhood cancers. *Lancet,* 1185–1188.

Stewart, N. G., *et al.* (1957). World-wide deposition of long-lived fission products from nuclear test explosions. Report MP/R2354. United Kingdom Atomic Energy Authority, London, UK.

Stidley, C. A., and Samet, J. (1993). A review of ecologic studies of lung cancer and indoor air. *Health Phys.* **65,** 234–251.

Stigall, G. E., Fowler, T. W., and Krieger, H. L. (1971). [131]I discharges from an operating boiling water reactor nuclear power station. *Health Phys.* **20,** 593–599.

Stoller, S. M., and Richards, R. B. (1961). "Reactor Handbook," Vol. 2. Wiley–Interscience, New York.

Stonier, T. (1964). "Nuclear Disaster." World Publ., Cleveland, Ohio.

Stovall, J. E. (1989). HP implications of ionization type smoke detectors. *Health Physics Society Newsletter* **17**(4), 10–11.

Stowasser, W. F. (1977). "Phosphate Rock—1975 Mineral Yearbook." U.S. Bureau of Mines, Washington, D.C.

Suess, H. E. (1958). The radioactivity of the atmosphere and hydrosphere. *Annu. Rev. Nucl. Sci.* **8,** 243.

Summerlin, J., and Prichard, H. M. (1985). Radiological health implications of lead-210 and polonium-210 accumulations in LPG refineries. *Am. Ind. Hyg. Assoc. Q.* **46,** 202–205.

Sundaram, K. (1977). Down's syndrome in Kerala. *Nature (London)* **267,** 728.

Sunta, C. M., David, M., Abani, M. C., Basu, A. S., and Nambi, K. S. V. (1982). Analysis of dosimetry data of high natural radioactivity areas of southwest coast of India. *In* "The Natural Radiation Environment IV" (K. G. Vohra, U. S. Mishra, K. C. Pillai, and S. Sadasivan, eds.), Wiley Eastern, Bombay/New Delhi.

Sutcliffe, W. B., Condit, R. H., Mansfield, W. G., Meyers, D. S., Layton, D. W., and Murphy, P. W. (1995). A perspective on the dangers of plutonium. Report UCRL-JC-118825. Lawrence Livermore National Laboratory, Livermore, California.

Sutton, O. G. (1953). "Micrometeorology." McGraw–Hill, New York.

Sverdup, H. V., Johnson, M. W., and Fleming, R. H. (1963). "The Oceans: Their Physics, Chemisty and General Biology." Prentice–Hall, Englewood Cliffs, New Jersey.

Swindle, D. W. (1990). Remedial action measures. In "Proceedings of the Twenty-Sixth Annual Meeting of the NCRP, Health and Ecological Implications of Radioactively Contaminated Sites" (C. R. Richmond, ed.), pp. 119–139. National Council on Radiation Protection and Measurements, Bethesda, Maryland.

Tabor, W. H. (1963). Operating experience of the Oak Ridge research reactor through 1962. *Nucl. Saf.* **5,** 116–123

Tait, J. H. (1983). Uranium enrichment. In "Nuclear Power Technology" (W. Marshall, ed.), Vol. 2 Oxford Univ. Press (Clarendon), Oxford.

Tajima, E. (1956). A summary of the investigations of the fallout from nuclear detonations. In "Research in the Effects and Influences of the Nuclear Bomb Test Explosions," p. 419. Japan Society for the Promotion of Science, Tokyo.

Tajima, E., and Doke, T. (1956). Airborne radioactivity. *Science* **123,** 211–214.

Tanner, A. (1964). Radon migration in the ground: A review. In "The Natural Radiation Environment" (J. A. S. Adams and W. M. Lowder, eds.), pp. 161–190. Univ. of Chicago Press, Chicago.

Tanner, A. B. (1980). Radon migration in the ground: A supplementary review. In "The Natural Radiation Environment III" (T. F. Gesell and W. M. Lowder eds.), CONF-780422, Vol. 1, pp. 5–56. U.S. Department of Energy, Washington, D.C.

Tanner, A. B. (1986). Geological factors that influence radon availability. In "Indoor Radon, Proceedings of the APCA International Specialty Conference, Philadelphia, Pennsylvania. February 24–26, 1986," Publication SP-54, 1–12. Air Pollution Control Association, Pittsburgh, Pennsylvania.

Tanner, A. B. (1989). The source of radon in houses. In "Radon" (N. H. Harley, ed.), pp. 159–168. In "Proceedings of the Twenty-Fourth Annual Meeting of the National Council on Radiation Protection and Measurements," NCRP Proceedings 10. National Council on Radiation Protection and Measurements, Bethesda, Maryland.

Tanner, A. B. (1992). Bibliography of radon in the outdoor environment and selected references on gas mobility in the ground. Open-file Report 92-351. U.S. Geological Survey Open File Reports Section, Denver, Colorado.

Taylor, L. S. (1971). "Radiation Protection Standards." Chem. Rubber Publ. Co., Cleveland, Ohio.

Taylor, L. S. (1981). The development of radiation protection standards (1925–1940). *Health Phys.* **41,** 227–232.

Templeton, W. L., Nakatani, R. E., and Held, E. E. (1971). Radiation effects. In "Radioactivity in the Marine Environment," pp. 223–239. National Academy of Sciences, Washington, D.C.

Tengs, T., Adams, M., Pliskin, J., Safran, D., Siegel, J., Weinstein, M., and Graham, J. D. (1995). Five hundred life-saving interventions and their cost-effectiveness. *Risk Analysis* **15,** 369–390.

Thomas, R. G. (1992). Dose–effect parameters from the Argonne radium dial painter study. *Health Phys.* **62**(Suppl. 6), S18.

Thompson, D. L. (1978). Recommendations on the use of uranium in porcelain teeth. *In* "Radioactivity in Consumer Products" (A. A. Moghissi, P. Paras, M. W. Carter, and R. F. Barker, eds.), Report NUREG/CP-0001, pp. 475–478. U.S. Nuclear Regulatory Commission, Washington, D.C.

Thompson, R. C. (1982). Neptunium—The neglected actinide: A review of the environmental and biological literature. *Radiat. Res.* **90,** 1–32.

Thompson, T. J., and Beckerley, J. G. (1964). "The Technology of Nuclear Reactor Safety," Vol. 1 MIT Press, Cambridge, Massachusetts.

Tichler, J., Norden, K., and Congemi, J. (1989). Radioactive materials released from nuclear power plants. Annual Report 1989. Report NUREG/CR-2907 (BNL-NUREG-51581), Vol. 10. Brookhaven National Laboratory, Upton, New York.

Till, J. E., and Meyer, H. R., eds. (1983). Radiological assessment: A textbook on environmental dose analysis. Report NUREG/CR-3332. U.S. Nuclear Regulatory Commission, Washington, D.C.

Till, J. E., Simon, S. L., Kerber, R., Lloyd, R. D., Stevens, W., Thomas, D. C., Lyon, J. L., and Preston-Martin, S. (1995). The Utah thyroid cohort study: Analysis of the dosimetry results. *Health Phys.* **68,** 472–483.

Time, October 31 1988.

Tobias, C. A., and Todd, P. (1974). "Space Radiation Biology and Related Topics." Academic Press, Orlando, Florida.

Totter, J. R., and MacPherson, H. G. (1981). Do childhood cancers result from prenatal X rays? *Health Phys.* **40,** 511–524.

Totter, J. R., Zelle, M. R., and Hollister, H. (1958). Hazard to man of carbon-14. *Science* **128,** 1490–1495.

Trabalka, J. R., and Garten, C. T., Jr. (1983). Behavior of the long-lived synthetic elements and their natural analogs in food chains. *Adv. Radiat. Biol.* **10,** 39–104.

Trabalka, J. R., Eyman, L. D., and Auerbach, S. I. (1980). Analysis of the 1957–1958 Soviet nuclear accident. *Science* **209,** 345–353.

Tracy, B. L., Prantl, F. A., and Quinn, J. M. (1984). Health impact of radioactive debris from the satellite Cosmos 954. *Health Phys.* **47,** 225–233.

Trapeznikov, A. V., Pozolotina, V. N., Chebotina, M. Y., Chukanov, V. N., Trapeznikova, V. N., Kulikov, N. V., Nielsen, S. P., and Aarkrog, A. (1993). Radioactive contamination of the Techa River, the Urals. *Health Phys.* **65,** 481–488.

Trevathan, M. S., and Price, K. R. (1985). Ambient krypton-85 sampling at Hanford. *In* "Proceedings of the Fifth DOE Environmental Protection Information Meeting," CONF-841187. Vol. 2, pp. 655–662. U.S. Department of Energy, Washington, D.C.

Triffet, T. (1959). Basic properties and effects of fallout. Biological and Environmental Effects of Nuclear War. Hearings before the Joint Committee on Atomic Energy. U.S. Government Printing Office, Washington, D.C.

Trunk, A. D., and Trunk, E. V. (1981). Three Mile Island: A resident's perspective. The Three Mile Island nuclear accident: Lessons and implications. *Ann. N.Y. Acad. Sci.* **365,** 175–185.

Tsivoglou, E. C., *et al.* (1960a). Estimating human radiation exposure on the Animas River. *J. Am. Water Works Assoc.* **52,** 1271.

Tsivoglou, E. C., Stein, M., and Towne, W. S. (1960b). Control of radioactive pollution of the Animas River. *J. Water Pollut. Control Fed.* **32,** 262.

Tsuzuki, M. (1955). The experience concerning radioactive damage of Japanese fishermen by Bikini fallout. *Muench. Med. Wochenschr.* **97,** 988.

Turk, B. H., Brown, J. T., Geisling-Sobotka, F., Grimsrud, D. T., Harrison, J., Koonce, J. F., and Revzan, K. L. (1986). Indoor air quality and ventilation measurements in 38 Pacific Northwest commercial buildings. Report LBL-21453. Lawrence Berkeley Laboratory, Berkeley, California.

Turner, D. B. (1970). "Workbook of Atmospheric Dispersion Estimates." U.S. Environmental Protection Agency, Washington, D.C.

Underhill, D. K., and Muller-Kahle, E. (1993). World uranium supply and demand: The changing market. *IAEA Bulletin* **35,** 8–13.

Unger, W. E., Browder, F. N., and Mann, S. (1971). Nuclear safety in American radiochemical processing plants. *Nucl. Saf.* **23,** 234.

United Kingdom Agricultural Research Council (1961). Strontium-90 in milk and agricultural materials in the United Kingdom 1950–1960. Report No. 4. U.K. Agricultural Research Council, London, UK.

United Kingdom Atomic Energy Office (1957). "Accident at Windscale No. 1 Pile on October 10, 1957." Her Majesty's Stationery Office, London.

United Kingdom Atomic Energy Office (1958). "Final Report on the Windscale Accident." Her Majesty's Stationery Office, London.

United Nations (1956). "Proceedings of the 1955 UN Conference on the Peaceful Uses of Atomic Energy," 17 vols. United Nations, New York/Geneva.

United Nations (1981). Report of the working group on the use of nuclear power in outer space. Document A/AC.105/287. United Nations, New York.

U.S. Council for Energy Awareness (1992). Strategic plan for building new nuclear power plants. Summary of progress, November 1991–November 1992. U.S. Council for Energy Awareness, Washington, D.C.

U.S. Department of Agriculture (1957). "Soils. The 1957 Yearbook of Agriculture." U.S. Department of Agriculture, Washington, D.C.

U.S. Government Printing Office (1959). Report of fallout prediction panel. Hearings before Joint Committee on Atomic Energy on Fallout from Nuclear Weapons Tests. Washington, D.C.

U.S. Public Health Service (1971). State and federal control of health hazards from radioactive materials other than materials regulated by the Atomic Energy Act of 1954. U.S. Public Health Srvice, Washington, D.C.

U.S. Weather Bureau (1955). "Meteorology and Atomic Energy." U.S. Government Printing Office, Washington, D.C.

UNSCEAR (1958–1994). A series of reports concerning the sources, effects and risks of ionizing radiation. United Nations Scientific Committee on the Effects of Atomic Radiation reports to the General Assembly of the United Nations with annexes. United Nations, New York.

Upton, A. C. (1977). Radiobiological effects of low doses: Implications for radiological protection. *Radiat. Res.* **71,** 51–74.

Upton, A. C., Chase, H. B., Hekhius, G. L., Mole, R. H., Newcombe, H. B., Robertson, J. S., Schaeffer, H. J., Snyder, W. S., Sondhaus, C., and Wallace, R. (1966). Radiobiological aspects of the supersonic transport. A report of the ICRP Task Group on the Biological Effects of High Energy Radiations. *Health Phys.* **12,** 209–226.

Uranium Institute (1989). "Uranium Market Issues 1989–1990." Uranium Institute, London.

USA Today (1995). Fewer fire deaths. August 19, 1995, p. 3A.

Uzinov, I., Steinhäusler, F., and Pohl, E. (1981). Carcinogenic risk of exposure to radon daughters. *Health Phys.* **41,** 807–813.

Van Middlesworth, L. (1954). Radioactivity in animal thyroids from various areas. *Nucleonics* **12,** 56.

Vennart, J. (1981). Limits for intakes of radionuclides by workers: ICRP Publication 30. *Health Phys.* **40,** 477–484.

Vetter, R. J. (1992). Incineration of biomedical low-level radioactive waste. Editorial. *Health Phys.* **62,** 121.

Villforth, J. C. (1964). Problems in radium control. *Public Health Rep.* **79,** 337–342.

Villforth, J. C., Robinson, E. W., and Wold, G. J. (1969). A review of radium incidents in the U.S.A. *In* "Proceedings of Handling Radiation Accidents." International Atomic Energy Agency, Vienna.

Voigt, G., and Paretzke, H. G. (1992). Cs-137 intake with cafeteria food after the Chernobyl accident. *Health Phys.* **63,** 574–575.

Voilleque, P. G., and Gesell, T. F., eds. (1990). Evaluation of environmental radiation exposures from nuclear testing in Nevada: A symposium. *Health Phys.* 59, Number 5.

Volchok, H. L. (1970). Worldwide deposition of ^{90}Sr through 1969. Report HASL-227. U.S. Atomic Energy Commission, New York.

Volchok, H. L., and Kleinman, M. T. (1971). Worldwide deposition of Sr-90 through 1970. Report HASL-243. U.S. Atomic Energy Commission, New York.

von Hevesey, G. (1966). Radioactive tracers and their application. *Isot. Radiat. Technol.* **4,** 9–12.

Wadleigh C. H. (1957). Growth of plants. *In* "Soils," pp. 38–48. U.S. Department of Agriculture, Washington, D.C.

Wald, N. (1980). The Three Mile Island accident in 1979: The state response: *In* "The Medical Basis for Radiation Accident Preparedness" (K. F. Hubner and S. A. Fry, eds.), pp. 491–500. Elsevier/North-Holland, New York.

Wald, N., Thoma, G. E., Jr., and Broun, G., Jr. (1962). Hematologic manifestations of radiation exposure in man. *Prog. Hematol.* **3.**

Wallace, R. (1973). Comparison of calculated cosmic ray doses to a person flying in subsonic and supersonic aircraft. Report. LBL-150, p. 21. Lawrence Berkeley Laboratory, University of California, Berkeley, California. (Also U.S. Atomic Energy Rep. TID-4500 R 61.)

Walton, G. N. (1961). Fission and fission products. *In* "Atomic Energy Waste, Its Nature, Use, and Disposal" (E. Glueckauf, ed.), Chap. 1. Wiley–Interscience, New York.

Watson, J. E., Jr. (1980). Upgrading environmental data. Report EPA 520/1-80-012. U.S. Environmental Protection Agency, Washington, D.C.

Watson, J. E. Jr., and Mitsch, B. F. (1987). Ground-water concentrations for Ra-226 and Rn-222 in North Carolina phosphate lands. *Health Phys.* **52,** 361–365.

Watters, R. L., and Hansen, W. R. (1970). The hazards implication of the transfer of unsupported ^{210}Po from alkaline soil to plants. *Health Phys.* **18,** 409–413.

Watters, R. L., Edington, D. N., Hakonson, T. E., Hanson, W. C., Smith, M. H., Whicker, F. W., and Wildung, R. E. (1980). Synthesis of the research literature. *In* "Transuranic Elements in the Environment" (W. C. Hanson, ed.), Report DOE/TIC-22800. Technical Information Center/U.S. Department of Energy, Washington, D.C.

Webb, J. H. (1949). The fogging of photographic film by radioactive contaminants in cardboard packaging materials. *Phys. Rev.* **76,** 375–380.

Webb, J. W., and Voorhees, L. D. (1984). Revegetation of uranium mill tailings sites. *Nucl. Saf.* **25,** 668–675.

Weber, A. H., and Garrett, A. J. (1985). Proceedings of the DOE/AMS air pollution model evaluation workshop. DOE Report DP—1701-1, Vols. 1–3. E. I. Dupont de Nemours & Co., Savannah River Laboratory, Aiken, South Carolina.

Weems, S. J., Layman, W. G., and Haga, P. B. (1970). The ice-condenser reactor containment system. *Nucl. Saf.* **11,** 215–222.

Wei, L., ed. (1980). Health survey in high background radiation areas in China. Report by High Background Radiation Research Group, China. *Science* **209,** 877–880.

Wei, L., Zha, Y., Tao, Z., He, W., Chen, D., and Yuan, Y. (1990). Epidemiological investigation of radiological effects in high background area of Yangjiang, China. *J. Radiat. Res.* **31,** 119–136.

Weinberg, A. M., and Wigner, E. P. (1958). "The Physical Theory of Neutron Chain Reactors." Univ. of Chicago Press, Chicago.

Weiss, E. S., Rallison, M. L., London, W. T., and Thompson, G. D. C. (1971). Thyroid nodularity in southwestern Utah school children exposed to fallout radiation. *Am. J. Public Health* **61,** 241–249.

Welford, G. A., and Sutton, D. (1957). Determination of the uranium content of the National Bureau of Standards iron and steel chemical standards. Report NYOO-4755. U.S. Atomic Energy Commission, Washington, D.C.

Wennerberg, L. S. (1991). NORM contamination of oil and natural gas: An emerging regulatory problem. *Health Phys.* **60**(Suppl. 2), S37.

Wennerberg, L. S. (1992). Regulatory control of NORM contamination: Options and constraints. *Health Phys.* **62**(Suppl. 6), S20.

Wenslawski, F. A., and North, N. S., Jr. (1979). Response to a widespread, unauthorized dispersal of radioactive waste in the public domain. *In* "Proceedings of the Health Physics Society Twelfth Midyear Topical Symposium, Low-Level Radioactive Waste Management," EPA 520/3-79-002. Office of Radiation Programs, U.S. Environmental Protection Agency, Washington, D.C.

Wenzel, D. R. (1994). The radiological safety analysis computer program (RSAC-5) user's manual. Report WINCO-1123. Westinghouse Idaho Nuclear Company, Idaho Falls, Idaho.

Wheelright, J. (1995). Atomic overreaction. *The Atlantic Monthly* **April.**

Whicker, F. W. (1995). Environmental pathway analysis in dose reconstruction. Manuscript submitted for the Proceedings of the Thirty-first Annual Meeting of the National Council on Radiation Protection and Measurements, Arlington, Virginia, April 12–13, 1995 (to be published by the National Council on Radiation Protection and Measurements, Bethesda, Maryland).

Whicker, F. W., and Ibrahim, S. A. (1994). "Plutonium and Americium in the Environs of Rocky Flats: Spatial Distribution, Environmental Transport, and Human Exposure." Colorado State University, Fort Collins.

Whicker, F. W., and Kirchner, T. B. (1987). PATHWAY: A dynamic food-chain model to predict radionuclide ingestion after fallout deposition. *Health Phys.* **52,** 717–737.

Whicker, F. W., and Schultz, V. I. (1982). "Radioecology: Nuclear Energy and the Environment." CRC Press, Boca Raton, Florida.

Whicker, F. W., Kirchner, T. B., Breshears, D. D., and Otis, M. D. (1990). Estimation of radionuclide ingestion: The "PATHWAY" food chain model. *Health Phys.* **59,** 645–657.

White, S. B., Bergsten, J. W., Alexander B. V., Rodman, N. F., and Phillips, J. L. (1992). Indoor Rn-222 concentrations in a probability sample of 43,000 houses across 30 States. *Health Phys.* **62,** 41–50.

Whitehead, D. C. (1984). The distribution and transformation of iodine in the environment. *Environ. Int.* **10,** 321–339.

WHO (1964). "Prevention of Cancer." Report of a WHO Expert Committee. Technical Report Series No. 276. World Health Organization, Geneva.

WHO (1986). Report of the working group on assessment of radiation dose commitment in Europe due to the Chernobyl accident. World Health Organization Regional Office for Europe, Copenhagen.

WHO (1993). Indoor air quality: A risk-based approach to health criteria for radon indoors. Document BUR/ICP/CEH 108(S). World Health Organization Regional Office for Europe, Copenhagen.

WHO (1995). Report on the pilot phase of the international programme on the health effects of the Chernobyl accident. World Health Organization, Geneva.

Wick, R. R. and Gossner, W. (1983). Late effects of ^{224}Ra treated ankylosing spondylitis patients. *Health Phys.* **44**(Suppl. 1), 187–192.

Widner, T. E. (1994). Oak Ridge phase I health studies: A project overview. *Health Phys.* **66**(Suppl. 6), S98–S99.

Wilkening, M. (1952). Natural radioactivity as a tracer in the sorting of aerosols according to mobility. *Rev. Sci. Instrum.* **23,** 13.

Wilkening, M. (1964). Radon-daughter ions in the atmosphere. *In* "The Natural Radiation Environment" (J. A. S. Adams and W. M. Lowder, eds.), pp. 359–368. Univ. of Chicago Press, Chicago.

Wilkening, M. (1982). Radon in atmospheric studies: A review. *In* "The Natural Radiation Environment IV" (K. G. Vohra, U. C. Mishra, K. C. Pillai, and S. Sadasivan, eds.), pp. 565–574. Wiley Eastern, Bombay/New Delhi.

Wilkening, M. H., and Watkins, D. E. (1976). Air exchange and Rn-222 concentrations in the Carlsbad Caverns. *Health Phys.* **31,** 139–145.

Williams, A. R. (1982). Biological uptake and transfer of ^{226}Ra: A review. *In* "Proceedings of Environmental Migration of Long-Lived Radionuclides," IAEA-SM-257/92. International Atomic Energy Agency, Vienna.

Wilson, A. J., and Scott, L. M. (1992). Characterization of radioactive petroleum piping scale with an evaluation of land contamination. *Health Phys.* **63,** 681–685.

Wilson, D. W., Ward, G. M., and Johnson, J. E. (1969). A quantitative model of the transport of Cs-137 from fallout to milk. *In* "Proceedings on Environmental Contamination by Radioactive Material, 1969." International Atomic Energy Agency, Vienna.

Wilson, J. W. (1981). Solar radiation monitoring for high altitude aircraft. *Health Phys.* **41,** 607–617.

Wilson, M. L., Gauthier, J. H., Barnard, R. W., Barr, G. E., Dockery, H. A., Dunn, E., Eaton, R. R., Guerin, D. C., Lu, N., Martinez, M. J., Nilson, R. Rautman, C. A., Robey, T. H., Ross, B., Ryder, E. E., Schenker, A., Shannon, S. A., Skinner, L. H., Halsey, W. G., Gansemer, L. C., Lewis, L. C., Lamont, A. D., Triay, I. R., Meijer, A., and Morris, D. E. (1994). Total-system performance assessment for Yucca Mountain—SNL second iteration (TSPA-1993). Report SAND 93-2675. Sandia National Laboratories, Albuquerque, New Mexico.

Wischmeier, W. H., and Smith, D. O. (1978). "Predicting Rainfall Erosion Losses—A Guide to Conservation Planning," USDA Handbook No. 537. U.S. Government Printing Office, Washington, D.C.

Wittels, M. C. (1966). Stored energy in graphite and other reactor materials. *Nucl. Saf.* **8,** 134.

Wodrich, D. (1994). Historical perspective of Handford's radioactive waste. Unpublished set of visual aids. Westinghouse Hanford Company, Richmond, Washington.

Wolff, T. A. (1984). "Transportation of Nuclear Materials," SAND 84-0062, Sandia National Laboratories, Albuquerque, New Mexico.

Wood, W. W. (1981). Guidelines for collection and field analysis of ground-water samples for selected unstable constituents. *In* "United States Geological Survey Techniques of Water Resources Investigations," Book 1, Chap. D2. U.S. Geological Survey, Washington, D.C.

Woods, S. E., Young, S. C., Abernathy, S. E., and Chambers, D. G. (1993). The disposal of naturally occurring radioactive materials through hydraulic fracturing. *Health Phys.* **64**(Suppl. 6), S52.

Wrenn, M. E. (1968). The dosimetry of ^{55}Fe. *In* "Proceedings of the First Congressional Radiation Protection, 1968," pp. 843–850. Pergamon, Oxford.

Wrenn, M. E., and Cohen, N. (1967). Iron-55 from nuclear fallout in the blood of adults. Dosimetric implications and development of a model to predict levels in blood. *Health Phys.* **13,** 1075–1082.

Wrenn, M. E., and Cohen, N. (1979). Dosimetric and risk/benefit implications of americium-241 in smoke detectors disposed of in normal wastes. *In* "Proceedings of the Health Physics Society Twelfth Midyear Topical Symposium 1979," EPA 520/3-79-002. U.S. Environmental Protection Agency, Washington, D.C.

Wrenn, M. E., Mowafy, R., and Laurer, G. R. (1964). ^{95}Zr–^{95}Nb in human lungs from fallout. *Health Phys.* **10,** 1051–1058.

Wrenn, M. E., Lentsch, J. W, Eisenbud, M., Lauer, J., and Howells, G. P. (1972). Radiocesium distribution in water, sediment, and biota in the Hudson River estuary from 1964 through 1970. *In* "Proceedings of the Third National Symposium on Radioecology," CONF-710501-P2, pp. 752–768. U.S. Atomic Energy Commission, Washington, D.C.

Wrenn, M. E., Singh, N. P. Cohen, N., Ibrahim, S. A., and Saccomanno, G. (1981). Thorium in human tissues. NUREG/CR-1227. NTIS, Springfield, Virginia.

Wrenn, M. E., Singh, N. P., and Xue, Y. H. (1994). Urinary excretion of ^{239}Pu by the general population: Measurement techniques and results. *Radiat. Protect. Dosim.* **53,** 81–84.

WSSRAP (1995). "Weldon Spring Site Remedial Action Project." WSSRAP Community Relations Department, St. Charles, Missouri.

Wyckoff H. O. (1980). From "quantity of radiation" and "dose" to "exposure" and "absorbed dose"—An historical review. NCRP Taylor Lecture No. 4. *In* "Proceedings of the Sixteenth Annual Meeting, Quantitative Risk in Standards Setting." National Council on Radiation Protection and Measurements, Bethesda, Maryland.

Wyerman, T. A., Farnsworth, R. K., and Stewart, G. L. (1970). Tritium in streams in the United States, 1961–1968. *Radiol. Health Data Rep.* **11,** 421–439.

Xuan, X. Z., Lubin, J. H., Yang, L. F., Wang, J. Z., Yan, F. L., and Blot, W. J. (1993). A cohort study in southern China of workers exposed to radon and radon decay products. *Health Phys.* **64,** 120–131.

Yamagata, N., and Yamagata, T. (1960). The concentration of cesium-137 in human tissues and organs. *National Institute of Public Health (Japan)* **9,** 72.

Yarborough, K. A. (1980). Radon- and thoron-produced radiation in National Park Service caves. *In* "The Natural Radiation Environment III" (T. F. Gesell and W. M. Lowder, eds.), CONF-780422, Vol. 2, pp. 1371–1395. U.S. Department of Energy, Washington, D.C.

Zach, R., and Mayoh, K. R. (1984). Soil ingestion by cattle: A neglected pathway. *Health Phys.* **46,** 426–431.

Zapp, F. C. (1969). Testing of containment systems used with light-water-cooled power reactors. *Nucl. Saf.* **10,** 308–315.

Zelle, M. R. (1960). Radioisotopes and the genetic mechanism: Mutagenic aspects. *In* "Radioisotopes in the Biosphere," pp. 160–180. Univ. of Minnesota Press, Minneapolis.

Index

A

Accidental releases, diffusion from, 82
Accidents
  aircraft carrying nuclear weapons, 429
  CASTLE fallout, March 1954, 380
  Chernobyl, 409; *see also* Chernobyl accident
  Cosmos 954, reentry to atmosphere, 401
  deaths due to, in the United States nuclear energy program, 525
  Goiania, 435
  Houston incident, 393
  Juarez, 434
  Marshall Islands, 379–386
  Oak Ridge plutonium release, 395
  Palomares, 429
  reactor accidents, *see* Nuclear reactors
  record of government agencies, 524
  Rocky Flats, plutonium fire, 489
  SL-1 reactor explosion, 398
  SNAP 9A, abortive reentry, 399
  Three Mile Island Unit 2, 402
  during transportation of radioactive materials, 330
  Urals, 1957 explosion, 427
  Windscale reactor, 387
Actinide elements, 109; *see also* specific nuclides
Actinium series, 139
Acute radiation effects, *see* Radiation, biological effects of
Advisory Committee on Reactor Safeguards, 264
AEC, *see* Atomic Energy Commission
Aerosols, *see* Particulates
Air-cooled reactors, 231, *see also* Windscale, liquid wastes from
Airline employees as radiation workers, 187
ALARA, 47
  application to nuclear plant effluents, 242
Alpha radiation, relative effectiveness of, 26

Americium, 323
  association with plutonium, 96
  decay from ^{241}Pu, 311
  in fallout from 1954 Bikini test, 381
  flux to ocean in Marshall Islands, 123
  properties of, 568
  residue from nuclear weapon
      accident in Greenland, 433
  in sewage sludge, 324
  in smoke detectors, 323
  in transuranic waste, 350
Ankylosing spondylitis, treatment by
      radiation, 23, 25
Annual limit of intake, 56
Aplastic anemia, 5
Aquatic ecosystems, *see also*
      Estuaries, diffusion in; Oceans,
      characteristics of; Rivers
  aquatic organisms
      bioaccumulation in, 128–131
      effect of temperature on, 115–116
      radiation effects on, 117
  behavior of suspended solids in, 118,
      131
  concentration factors in, 128–131
  K_d, effects of, 94, 119
  mixing characteristics, 118
  sediments, role of, 118
Aquifers, 113–115
Arctic region
  lichens, elevated ^{137}Cs in, 107, 307
  remediation of Project Chariot, 520
  Soviet nuclear tests in, 169
  waste disposal in, 355, 428
Army Stationary Low Power Reactor,
      see SL-1 accident
Atmosphere, *see also* Diffusion;
      Fallout, from nuclear explosions;
      specific nuclides
  behavior of aerosols in, 86
  comparison of diffusion models, 81
  composition of, 64
  diffusion in, 73, 78
  diffusion models, 78
  diurnal cooling, 68

  friction layer, 69
  lapse rate, 67
  properties of, 63
  residence time of dust, 87
  structure of, 65
  temperature inversion, 67–69
  transport mechanisms, 68–81
  turbulent diffusion, 65
  vertical temperature gradient, 65,
      66, 69
Atomic Energy Act, 203, 219, 479
Atomic Energy Commission, 7
Atomic energy industry, 5
  safety record of, 379, 522–524
Atomic Safety and Licensing Board,
      264

B

Background radiation, *see* Radiation;
      Radioactivity
Bad Gastein, 191, 193
Beta radiation, skin injury from, 383,
      416, 525
Bikini, *see* Marshall Islands
Black sands, 193
Blood, effects of radiation on, 15; *see
      also* Leukemia
Bombs, nuclear and thermonuclear,
      see Nuclear explosions
Bone, cancer of, 24; *see also* Lead-210;
      Polonium-210; Radium-224;
      Radium-226; Radium-228;
      Strontium-90
Boron, use as burnable poison, 226
Brazil, *see also* Morro do Ferro;
      Guarapari
  radioactivity of, 195
Brazil nuts, radium in, 147
Building materials, radioactivity of,
      174–175
  in Brazil, 194
  source of radon, 158
  in Taiwan, 433
  use of uranium mill tailings in, 208,
      508
Burnable poison, 226

C

Calcium
  available, 102
  dietary sources of, 300–302
  effect of mineral additives, 104
  influence on strontium uptake, 103
  similarity of strontium to, 102
  in soils, 92–93, 97
Cancer, radiation-induced
  in bone, 24
  from Chernobyl, 421
  in general population, 535–537
  in Japan, 23
  in lung, 3, 26, 29–32
  in thyroid, 32–33, 385, 420
  in uranium miners, 29, 203
Carbon-14
  atmospheric content of, 181, 376
  concentration in biological material, 181
  dose from, 181, 377
  properties of, 549
  sources of, 181, 376
Cataracts, 34
Cation exchange in soils, 93
Centrifuges for isotope separation, 6
Ceramics, radioactivity of, 319
CERCLA, see Comprehensive Environmental Response, Compensation and Liability Act
Cesium-134, properties of, 558
  use as an indicator of reactor emissions, 243
Cesium-137
  anomalous transfer to man, 106
  behavior in soils, 104
  from Chernobyl, 415–422
  concentrations in reindeer and lichens, 106, 107
  contamination from Goiania accident, 435
  deposition of, 104
  effect of potassium deficiency in soils, 104, 307
  in fallout, 306

foliar deposition on, 104, 306
  in food, 105–106, 307–308
  human measurements of, 106
  pangola grass, effect on uptake by milk, 106
  properties of, 558
  Prussian Blue for decorporation, 436
  relationship to ^{90}Sr, 306
  in wood ash, 107
Chariot project, remediation of, 520
Chemical reprocessing of irradiated fuel, 213–215
Chernobyl accident, 409
  acute radiation effects, 416
  circumstances of the accident, 412
  construction of the sarcophagus, 416
  contamination levels and dosimetry
    beyond 30 km, 418
    on site, 416
    within 30 km, 417
  delayed effects, 417
    leukemia, 423
    thyroid cancer, 33, 420
  description of the RBMK, 410
  economic cost, 424
  effects on general health, 421
  evacuation, 417, 422
  land contamination, 422
  on-site cleanup, 416
  skin beta burns, 416
China
  atmospheric nuclear tests, 269
  lung cancer in tin mine, 29
  studies in monazite areas, 23, 196
Chromosomes
  abnormal in peripheral blood, 36
  radiation effects on, 194
Cigarettes, radioactivity in, 170
CIS, see Commonwealth of Independent States
Clay, adsorption of radionuclides on, 92
Coal, radioactivity of, 175; see also Fossil fuels, radioactivity of
Coastal waters, mixing properties of, 125

Cobalt-60
    accident in Juarez, 434
    contamination of buildings in
        Taiwan, 433
    hot particles containing, 227
    properties of, 553
Columbia River, *see* Rivers
Commonwealth of Independent States,
    425
    accidents other than Chernobyl, 425
    development of nuclear program,
        426
    Mayak, 426
    waste disposal in Arctic, 355, 428
Comprehensive Environmental
    Response, Compensation and
    Liability Act, 496
Concentration factors in fresh water
    and marine environment, 130
Congeners, 97
Containment, reactor, 258
Cosmic radiation, 185
Critical organ, 43
Criticality, 225
    prompt, 227
Crookes tubes, as source of X-rays, 11
Cross section, 224

D

Darcy's Law, 113
Decontamination following plutonium
    release, 396
De minimis dose, 539
Default values, 79, 97, 443
Denver Radium Site, 502
Department of Transportation, 61
Depleted uranium projectiles, 320
Deposition velocity, 84
Derived working limits for Windscale
    liquid wastes, 354
Detriment, the concept of, 37
Diffusion, 69, 72
    from accidental release, 82
    in coastal waters and estuaries, 125
    from continuous point source, 70
    during inversions, 67

in oceans, 120
in rivers, 125
tropospheric and stratospheric, 85
Disposal of wastes, *see* Waste
    management
Distribution coefficient, *see* K_d
Dose, *see also* specific nuclides and
    sources
    absorbed, 42, 44
    biological indicators of, 416
    collective, 35, 54, 243
    effective dose, 42
    effective dose equivalent, 42
    equivalent, 42
    equivalent dose, 42
    limits, 43–49
        basic standards, 47
        early concepts of, 40
        evolution of, 41
        limits for occupational
            exposure, 40
        monetary criteria for, 57
        organizations involved in, 58
        public, 46
        quantities and units for, 42
        risk basis of, 45, 53
        specific environmental sources, 49
    rate reduction factor, 20, 30
    reconstruction, 475
        basic elements of, 475–478
        current examples of,
            Fernald, 492–493
            Hanford, 484–489
            Rocky Flats, 489–491
        early examples of, 470
        uncertainties in, 476
Dust, *see* Particulates

E

e-folding time, 226
Ecology, 8
Effective stack height, 74
Electrical energy, demand for, 220
Elemental phosphorus plants, 50, 172
Environmental Protection Agency, 59

Environmental radioactivity, *see also*
     Radiation; Radioactivity; specific
     sources
  public attitudes toward, 530
Environmental surveillance, 438
  aquatic, 458
  atmospheric, 452–455
  basic principles, 440
  drinking water, 462
  emergency, 467
  food, 463
  gamma radiation, 446–449
  groundwater, 459
  human measurements, 466
  objectives, 439
  operational phase, 443
  preoperational phase, 442
  requirements, 438
  specific nuclides, 456
  station design, 445
  surface deposition, 449
Erythema dose, 40
Estuaries, diffusion in, 118
  role of sediments, 118
Evans, Robley D., 12, 21, 25, 41
External sources of natural radiation,
     182

F

Failla, G., quantitation of erythema
     dose, 40
Fallout, from nuclear explosions, *see
     also* Nuclear explosions; specific
     nuclides
  gamma radiation from wartime, 282,
     286–288
  induced activities in, 279
  ingestion, 289, 298–308
  inhalation, 314
  livestock, effects on, 291
  particle size of, 266, 280
  partitioning of debris from, 280, 295
  patterns of, 282–288
  residence time in atmosphere, 87, 296
  short-term effects, 285–289
  worldwide, 291–294

FAO, *see* Food and Agricultural
     Organization
Fast reactors, 231
FDA, *see* Food and Drug
     Administration
Federal Radiation Council, 48, 59
Fernald, 211
  dose reconstruction, 492–493
  remediation, 511–512
Fertilizer, radioactivity of, 172
Final Safety Analysis Report, 264
Fission, 223
  products
     from reactors, inventory of, 234,
        247
     yields of in nuclear weapons, 277
  spontaneous, 135
Foliar deposition, 90
Food and Agricultural Organization, 49,
     62, 424
Food and Drug Administration, 60
Food chains, transport of radionuclides
     in, 89, 101, 289, 298–308
Foods, radioactivity of, 141, *see also*
     specific nuclides and sources
Formerly Utilized Sites Remedial
     Action Program, 210, 505–508
Fossil fuels, radioactivity of, 175
Fuel elements
  construction of, 212
  escape rate coefficients from, 239
  reprocessing, 212–215
Fukuru Maru, fallout on, 381
  effects on fishermen, 381, 383–385
FUSRAP, *see* Formerly Utilized Sites
     Remedial Action Program

G

Gas-cooled reactors, 231
Gaseous diffusion plants, 6, 203,
     211–212
Gastrointestinal tract, *see* Radiation,
     biological effects of
Genetic effects, 18, 35
  absence of, in Japanese studies, 36
Genetically significant dose, 35

Glass, radioactivity in, 319
Goiania, Brazil, abandoned ^{137}Cs
    gamma source, 435
Grand Junction, Colorado, misuse of
    uranium mill tailings, 208
    cleanup of tailings, 508
Gray (unit), 42
Groundwater, properties of, 113
    effects of K_d on, 114
Growth and development, radiation
    effects on, 37
Grübbe, E. H., early experiences with
    X rays, 11
Guarapari, Brazil, abnormal radiation
    levels, 194

H

Hanford Nuclear Reservation, transport
    of wastes through soil, 95
    dose assessment for, 484–489
    high-level waste storage tanks, 347
    release of radioactive iodine from,
        215, 487–488
    remediation of, 516
    use of Columbia River water for
        irrigation, 131
Hospitals, incineration of wastes from,
    339
    release of radionuclides from, 323, 442
Hot channel factor, 228
Hot particles, 227
Houston incident of 1957, 393
Hudson River, 133, 118, 131, 243
Hydraulic gradient, 113
Hydrogen-3, see Tritium

I

IAEA, see International Atomic Energy
    Agency
ICRP, see International Commission on
    Radiological Protection
Idaho National Engineering Laboratory,
    contamination of aquifer, 114
    SL-1 accident, 398
    radionuclide uptake by water fowl,
        131
    remediation of, 519

Incineration of low-level wastes, 339
Incinerator ash, 179
India, elevated radiation levels in
    Kerala from monazite, 195
    nuclear explosion in, 269
Indoor radon, see Radon and decay
    products
Induced radionuclides
    by cosmic rays, 85
    in nuclear reactors, 235
    from nuclear weapons, 280
Inhalation of radioactive dust, see
    Respiratory system, deposition
    and clearance of inhaled
    particulates
Internal emitters, see specific nuclides
International Atomic Energy Agency
    nuclear accident event scale,
        256–257
    waste categorization system, 335
International Commission on
    Radiological Protection, 39
    formation of, 39
    limits for exposure, 43–45
International Labor Organization, 61
Iodine-129, residence time in soil, 96
    from Chernobyl, 419
    properties of, 556
    worldwide collective dose from, 377
Iodine-131
    at Chernobyl, 415
    deposition velocity of, 108
    effects on Marshallese children, 386
    in food chains, 108
    fresh milk as source of, 108
    from fuel reprocessing, 215, 487–488
    measurement in milk and thyroids,
        312
    from nuclear weapons, 311
    from power reactors, 241, 251
    properties of, 557
    from reactor accidents, 251
    in sewage, 323, 442
    at Three Mile Island, 406
    thyroid blocking, use of potassium
        iodide for, 109

thyroid cancer from, 32
thyroid uptake from, 109
Utah children, doses to thyroids of, 312–313
in weapons fallout, 311–313
at Windscale, 254
Iridium-192, *see* Accidents, Houston incident
Iron-55 and iron-59
properties of, 551, 552
in reactor coolant, 235
in weapons fallout, 313

J

Jamaica, high cesium content of milk in, 104–105
Japan, studies of atomic bomb survivors, 19–24, 32–34
Japanese fishermen, *see* Fukuru Maru, fallout on
Joachimsthal, 13
Juarez, accident involving ^{60}Co teletherapy unit, 434

K

K_d, 94, 119
Kerala, *see* India, elevated radiation levels in Kerala from monazite
Krypton-85
atmospheric accumulation of, 374
from fuel reprocessing, 217
properties of, 554
Kuroshio current, 123

L

Lead-210, 167
in atmosphere, 154, 168
content in man, 169
dose from, 169
in human diet, 168–169
in lung of smokers, 169
properties of, 559
sources of, 169
Lethal radiation dose, 14

Leukemia, 23
among children irradiated in utero, 24
clusters near nuclear facilities, 24
ratios to solid tumors, 24
risk coefficients for, 24
Light-water-cooled and moderated reactors, *see* Nuclear reactors
Limits, *see* Dose, limits
Linear and linear-quadratic hypotheses, 18–19
Liquefied petroleum gas, radon in, 167
Low-level waste, *see* Waste management
Lower plants and animals, radiation effects on, 88
Lucky Dragon, *see* Fukuru Maru, fallout on
Luminous paints, 5, 41, 318
Lung, *see also* Respiratory system, deposition and clearance of inhaled particulates
cancer
in Chinese tin mine, 29
in Eastern European miners, 29
in fluorospar miners, 29
studies in relation to indoor radon, 31
in U.S. uranium miners, 29

M

Manganese, and ^{54}Mn, 93, 118, 133, 235
Marshall Islands
cost of cleanup, 386
dose to inhabitants, 382, 386
effects on inhabitants, 384–387
fallout, March 1954, 379
radiation limits, 387
radionuclides released to ocean, 123
weapons testing in, 268, 281, 296, 313
Martland, Harrison, 25
Maximum permissible dose, *see* Dose, limits
Maximum permissible exposure to radionuclides, *see* Annual Limit of Intake

Mechanical harvesting, effect on
    resuspension, 100
Medical facilities, *see* Hospitals,
    incineration of wastes from
MESODIF, *see* Meso-scale Diffusion
    Model
Meso-scale Diffusion Model, 80
Mesothorium, *see* Radium-228
Milk contamination, *see* specific
    nuclides
Mill tailings piles, *see* Uranium mines
    and mills
Mineral waters, radioactivity of, 191
Mixed waste, *see* Waste management
Models, use in risk assessment, 78, 474
    need for validation, 81
    uncertainties in use of, 476
Monazite, 193; *see also* Brazil; China;
    India, elevated radiation levels in
    Kerala from monazite
Monitoring methods, *see*
    Environmental surveillance
Montclair, New Jersey, radium site,
    503
Morro do Ferro, 195
    dose to indigenous rodents, 195
    mobilization rates of transuranic
        actinide elements, 365–366
    presence of thorium and rare earths,
        110
Muller, H. J., 12
Municipalities, regulatory function
    of, 61

N

National Council on Radiation
    Protection and Measurements, 39
    evolution of standards, 44
    first recommendation of maximum
        permissible dose, 40
    formation of, 39
National Emission Standards for
    Hazardous Air Pollutants, for
    radionuclides, 49
Natural analogues for radioactive
    wastes, *see* Morro do Ferro; Oklo,
    natural reactor at

Natural gas, radon in, 167
Natural radioactivity, *see* Radioactivity
Natural reactor, Oklo, 197
NCRP, *see* National Council on
    Radiation Protection and
    Measurements
Neptunium
    behavior in soil, 112
    concentration ratios, 112
    importance for high-level
        repositories, 361
    metabolic behavior, 112
    properties of, 566
NESHAP, for radionuclides, *see*
    National Emission Standards for
    Hazardous Air Pollutants, for
    radionuclides
Neutron
    cross section, 224
    generation time, 226
Neutrons
    delayed, prompt, 227
    production in fission, 224
New York, New York
    natural radionuclides in diet of, 141,
        148
    Sr-90 deposition, 298
    Sr-90 in food of, 303
    U in air of, 138
NRC, *see* Nuclear Regulatory
    Commission
Nuclear explosions
    activation products from, 279–280
    C-14 from, 308
    countries involved, 268
    cloud height, 268
    contamination of
        land, by, 281–290
        photographic products, 271–272
    effects of
        precipitation, 300
        wartime, 285–291
    fallout
        doses, methods of estimating, 282
        in Marshall Islands, 379

patterns from, 283–294
   dosage calculations for, 282
   partitioning of, 266, 295
  shelters, 290
fireball, 267, 276
fission products from, 276–279
  fractionation of, 280
fusion in, 275
height of burst, influence of,
   267–268, 281
induced activities, 279–280
local fallout from, 281–290
physical aspects, 271–280
radionuclides from, 279
strontium-90 produced, 270
test-ban treaties, 270
thermonuclear reactions, 275
worldwide fallout from, 291–295
yields of tests, 269
Nuclear power for generation of
  electricity, 2, 219
Nuclear reactors, 219
  accidents, 244, 402, 409
   assumptions as to source terms,
    250
   Chernobyl, *see* Chernobyl accident
   consequences of, 246, 250
   involving core damage, 246, 250
   loss of coolant, 245
   radioactive releases in event of,
    246, 250
   Rasmussen study, 252, 253
   Three Mile Island, 251, 402
   types of, 244–255
  advanced designs, 260
  burnable poisons, use of, 226, 239
  cladding failure, 227
  containment structures, 258
  coolant purification, 239
  criticality, 224, 227
  decommissioning, 265
  departure from nucleate boiling, 228
  electricity production by, 221
  engineered safeguards, 255, 258
  exothermic reactions, 222
  final safety analysis report, 264

fission product inventory, 233, 236,
  248
fuel, types of, 222
gaseous wastes from, 240, 242
history of, 219, 221
hot channel factor, 228
licensing procedures, 264
materials of construction, 227
neutron generation time, 226
neutron multiplication factor,
  223–224
physical principles, 222–228
population exposures from, due to
  accidents, 263, *see also*
      Chernobyl accident,
      contamination levels and
      dosimetry; Three Mile Island
      Unit 2
  routine operation, 241
prompt critical condition, 227
radioactive effluents from,
  compared to coal burning power
    plants, 241
  gaseous, 240
  liquid, 236
reactivity, 225
reactor period, 226
regulatory procedures, 264
site selection criteria, 261
tritium production by, 236
types of, 229–232
wastes from, *see* Waste management
Nuclear Regulatory Commission, 59
Nuclear war, 285
Nuclear weapons, *see* Nuclear
  explosions

O

Oak Ridge National Laboratory
  plutonium accident, 395
Observed ratio, 102–104
Occupational exposure
  limitation of, 40–46
  national experience, 524
Oceans, characteristics of, 120
  currents, 121
  diffusion coefficients of, 124

Oceans (*continued*)
  diffusion in, 123
  Kuroshio current, 121, 123
  mixing characteristics of, 121,
       124–126
  pycnocline, 121
  thermocline, 121
  vertical mixing, 120, 124
  waste disposal in, 336, 354, 355, 428
Oil and gas industry, *see* Petroleum
       industry, natural radionuclides in
Oklo, natural reactor at, 197

P

Palomares, 429–432
Particulates
  deposition
    in lung, 26
    on surfaces, 82–85, 98–101
    velocity, 84
  diffusion of, 82
  residence time, in atmosphere, 87
  resuspension of, 82–85, 100
  settling rates, in air, 83, 84
  washout factor, 87
Petroleum industry, natural
       radionuclides in, 167, 179, 180
Phosphorus industry, natural
       radionuclides in, 50, 172
Plants
  root function, 98
  uptake of radionuclides, *see* specific
       nuclides
Plücker, observation of X-ray
       fluorescence, 11
Plutonium-238
  in fallout, 310
  properties of, 566
  use in RTGs and SNAP units,
       325–327
Plutonium-239, accidental release at
       Oak Ridge, 395
  behavior in soil, 112
  concentration ratios, 112
  dose from in fallout, 314
  in fallout, 310
  naturally occurring, 135

  at Palomares, *see* Palomares
  production rate in fission reactors,
       109
  properties of, 567
  at Rocky Flats, 489–491
  thorium as an analogue for, 110
  uptake by plants, 110, 112
Pollutants, dispersion in aquatic
       systems, 118
Polonium-210, 167
  in cigarette smoke, 169
  concentration by lichens, 170
  correlation with Cs-137, 170
  emissions from elemental
       phosphorous plants, 50, 171
  ingrowth in vegetation, 170
  Laplanders, exposure to, 170
  in lungs of smokers, 169
  properties of, 560
Population exposure, 528; *see also*
       Radiation, sources of exposure to
Porosity, 113
Potassium-40, 135
  dose from, 171, 183–184
  in fertilizers, 171
  human body content of, 171
  properties of, 550
  in seawater, 171
  in soils, 140
Power reactors, *see* Nuclear reactors
Practical threshold, 21
Preliminary Safety Analysis Report,
       264
Probability of causation, 22
Protective action guides, 424
Public concerns about environmental
       radioactivity, 530
Public perception of nuclear power, 530

Q

Quality factor, 42

R

Radiation
  biological effects of
    acute radiation syndrome, 13–14

bone cancer, 12, 21, 24–26, 41
cataracts, 12, 34
central nervous system effects, 15
chromosome mutations, 36
early history of, 10
effects on skin, 11, 12, 40, 383
epilation, 15
gastrointestinal tract effects, 15
genetic effects, 12, 18–19, 35
growth and development, 37
latency, 21, 26, 34
leukemia, 23–24
lung cancer, 26
probability of causation, 22
threshold dose, 18–19
sources of exposure to
medical, 323, 527, 528
natural, 182–184, 200
nuclear power, 241–243, 527–529
occupational, 324–327
standards for, see Dose, limits;
International Commission on
Radiological Protection;
National Council on Radiation
Protection and Measurements
summary of all sources, 527–529
various other sources, 332–333
weapons testing, see Nuclear
explosions
Radioactive dust, see also Particulates
mean residence time, in
stratosphere, 86–87, 296
troposphere, 86–87, 292
Radioactive wastes, see Waste
management
Radioactivity; see also specific sources
abnormal levels of, 191–197
atmosphere, see specific nuclides
Brazil nuts, 147
ceramics, 319
early history of, 2
fossil fuels, 175
glass, 319
houses, 174–175

liquefied petroleum gas, 167
mineral springs, 191
monazite sands, 193
natural, 134–200; see also specific
nuclides
natural gas, 167
petroleum industry, 167, 179, 180
phosphate fertilizers, 141, 172
phosphorus industry, 50, 172
rocks and soils, 140
spontaneous fission, 135
technologically enhanced natural,
190
terrestrial, dose from, 182
water, 143–144, 146
wood ash, 107
Radioisotope thermal generator, see
SNAP program
Radiological assessment, 471
Radionuclides, see specific listings
Radium-224, 152
bone cancer from, 25
medical uses of, 25
volatized from lantern mantles, 321
Radium-226, 142
bone cancer from, 41
Brazil nuts, 147
concentrations in rocks and soil, 142
early use of, 3, 12
food, 111, 146, 147
in human tissues, 147, 149
leaching from tailing piles, 111, 208
luminous dial painting, 318
maximum permissible body
burden, 41
mineral springs, 191
mobilization rate from soil, 111
nostrums, 25
properties of, 561
Radium-228, 151
cancer caused by, 25
concentration ratio for, 110
properties of, 562
in water, 143

Radon and decay products, 152
concentration in general
atmosphere, 153
correlation of decay products with
concentration of inert dust, 156
depletion of decay products during
storms, 155
drinking water, 157, 159
emanation rate, 153, 205–207
groundwater, 157
lung cancer from, 3, 13, 29–32
lung dose from, 28, 156, 200
mill tailings, 207
mines, 28–29, 50, 203
properties of, 560
decay products, 254, 156, 168
soil gas, 152–153
working level, definition of, 28
Radon-220, *see* Thoron and decay
products
Rain splash, 98
Rare earths, mobilization rates, 95
use as analogues for actinide
elements, 110
Reactivity (of nuclear reactors), 225
Reactors, *see* Nuclear reactors
Reference man, 56
Regulations and radiation standards,
39–62
Relative biological effectiveness, 42
Rem (radiation unit), 42
Remediation, 494, 521
Chariot site, 520
costs, 512–513
Denver Radium Site, 502
Fernald site, 511–512
FUSRAP sites, 505
Hanford Nuclear Reservation, 516
Idaho National Engineering
Laboratory, Pit Nine, 519
mill tailings, 508
Montclair, New Jersey, 503
Oak Ridge Reservation, 517
process of, 498–500
risk assessment for, 497–498

Rocky Flats, 518
Savannah River, 513
sites, locations of, 505
standards for, 495–497
technologies for, 500–502
Weldon Spring site, 509–511
Rep (historical radiation unit), 42
Reprocessing of nuclear fuel, 212
radioactive emissions from, 216
Respiratory system, deposition and
clearance of inhaled
particulates, 26
dose from
inhaled radioactive dust, 28
radon series, 28
physiological parameters in dose
calculations, 28
Resuspension, 100, 82–85
Risk
coefficients, 20, 37
de minimis, 539
difference between real and
perceived, 530
Rivers
Animas, 207
Delaware, 126
Clinch, 132
Colorado, 207
Columbia, 126
Hudson, 115
Miami, 483
mixing studies of, 126
San Miguel, 207
Susquehanna, 407
Techa, 426
Tennessee, 518
Rocky Flats
atmospheric dispersion code for, 80
dose reconstruction for, 489–491
fires at, 489
plutonium in soil of, 112
remediation of, 518
Roentgen, adoption of, 40
Rongelap, *see* Marshall Islands, fallout
in March 1954

RTG, *see* Radioisotope thermal
    generator; SNAP program
Rubidium-87, 172

S

Safety record of nuclear program,
    occupational, 524
Salt beds, as a high-level waste
    repository, 357, 371
Samarium-149, effect on reactivity, 225
Satellite nuclear auxiliary power units,
    see SNAP program
Savannah River plant, remediation of,
    513
Schneeberg, 13
Scott, N. S., early compilation of
    radiation injuries, 12
Sediments, *see* Aquatic ecosystems;
    Estuaries, diffusion in
Sellafield, *see* Windscale, liquid wastes
    from
Sievert (unit), 42
Site selection for nuclear reactors, 261
Skin, beta burns to, 383, 416, 525
SL-1 accident, 398
SNAP program, 325–328
    abortive reentry of SNAP 9A, 399
Soils, 90
    behavior of radionuclides in, 93
    cation adsorption in, 92
    composition of, 90
    erosion of, 101
    exchange capacity of, 92
    horizons, 90
    role of
        clays, 90
        leaf litter, 96
Solar flares, 187–190
Somatic effects, 22
Source term, *see* Nuclear reactors,
    accidents; Dose, reconstruction
Soviet Union, *see* Commonwealth of
    Independent States
Space programs, 190
Spent fuel, 212
    civilian use, from, 352
    costs of management, 347

defense production, from, 349
options for management, 352
regulation of, 59
reprocessing. 212
storage of, temporary, 214, 352, 373
transportation, 212–213, 330–333
as waste, 335, 352
Yucca Mountain, proposed repository
    for, 366
Stack gases
    behavior of plumes from, 68–70
    dispersion of, 70
    effect of
        temperature gradient on, 69, 71
        effect of terrain on dispersion
            of, 76
    effective height, 74
    effects of buildings on, 76
    maximum downwind
        concentration, 73
    total exposure to, 73
Standard man, *see* Reference man
States, regulatory function of, 61
Stratosphere, 85
    dust in, 86–87, 296
    residence time in, 86, 296
Strontium-89
    in fallout, 306
    in nuclear reactors, 241
    properties of, 554
Strontium-90
    behavior in soil, 95
    dependence of fallout on
        precipitation, 300
    deposition of, global, 293
    dietary sources of, 301, 304
    dose to bone from, 305
    in fallout, 298
    field loss, rate of, 101
    human bone, 305
    influence of calcium content of soil,
        103, 104
    inventory of, 297
    leaching rate from soil, 96
    milk, 303

Strontium-90 (*continued*)
  observed ratios, 102, 103
  properties of, 555
  relative importance of soil uptake
    and foliar deposition, 99
  removal during lactation, 103
  stratospheric inventory of, 297
Strontium–calcium ratio, 102
Strontium unit, 103
Supersonic flight, 188
Surface water systems, 115
Surveillance programs, *see*
    Environmental surveillance

T

Taiwan, ⁶⁰Co contamination of
    buildings, 433
Techa River, 426
Terrain-responsive atmospheric
    code, 80
Terrestrial radiation, 183–184; *see also*
    Radioactivity, natural
Thermocline, 121
Thorium-230
  concentration ratios for, 111
  in high-level waste, 360
  in human tissue, 151
  properties of, 562
  in raffinate, 511
  in uranium mill tailings, 205–208
Thorium-232, 135, 149–151, 217
  analogue for Pu-239, 110
  in Brazil, 194–195
  concentration in rocks, 140, 143
  decay series, 139
  gas mantles, 217
  in human tissues, 152
  mobilization rate, 95, 111
  in monazite, 193
  Morro do Ferro, 195
  nuclear fuel, potential source of,
    201, 217
  properties of, 563
  uptake from soil, 111
  welding electrodes, 321
Thorium series, 139

Thoron and decay products, 150,
    152
  diffusion of, 150
  dose to rodents from, 195
Three Mile Island Unit 2, 402
Threshold effects, *see* Radiation,
    biological effects of
Thyroid, *see also* Iodine-131
  blockage of radioiodine uptake by
    KI, 34, 109, 416, 418, 425
  cancer
    from diagnostic procedures, 32
    in Japanese survivors, 33
    in Marshallese, 33, 385
    result of Chernobyl accident, 33,
      420
Title 10 Code of Federal
    Regulations, 59
Tobacco, *see* Cigarettes,
    radioactivity in
TRAC, *see* Terrain-responsive
    atmospheric code
Transportation of radioactive material,
    328–333
  of spent fuel, 212–213, 330–333
Transuranic elements
  contamination by, 512
  natural analogues for, 365
  naturally occurring, 135
  waste, *see* Waste management,
    transuranic wastes
Tritium, 182
  content of surface waters, 182
  dose from, 182, 373–376
  forecast accumulation of,
    373–376
  natural production rate, 182
  production in nuclear explosions,
    313, 375
  properties of, 548
  sources of, 182
Tropopause, 66, 86, 267
Troposphere, 66, 86–87; *see also*
    Atmosphere, residence time of
    dust

Troy, New York, unusual rainout, 273, 292

U

United Nations Scientific Committee on the Effects of Atomic Radiation, 8
United States Army, Vicksburg Waterways Experiment Station, 126
United States Navy, reactor-powered vessels, 219
UNSCEAR, *see* United Nations Scientific Committee on the Effects of Atomic Radiation
Ural Mountains, 1957 accident in, 427
Uranium, 135
  atmospheric concentration of, 138
  ceramic glazes, 319
  concentration in rocks and soils, 140
  decay series, 138
  dietary intake of, 141
  fuel manufacture, 212
  human bone, 142
  industry, Chapter 7
  isotopes of, 135
  isotopic enrichment, 211
  mineral deposits, 202
  mines and mills, 13, 29, 203–211
    refining, 210
    remediation of tailings, 497, 508
    tailings from, 205–208
  occurrence of, 137, 202
  phosphate fertilizers, 111
  phosphate rock, 111
  photographic emulsions, 138
  projectiles, 320
  refining, 209
  reserves of, 202
  series, 138
  tap water, 141
Uranium-235, properties of, 564
Uranium-238, properties of, 565
  projectiles, 320

W

WASH-740 (1957 reactor safety study), 250
WASH-1400 (1975 reactor safety study), 252
Waste Isolation Pilot Plant, repository for transuranic waste, 371–372
Waste management; *see also* Nuclear reactors; specific sources
  high-level wastes, 345
    amounts of, 346, 349, 351–352
    civilian, 352
    defense-related, 345
    options for disposal, 353
      basalt, 358
      geological cavities, 356
      granite, 359
      on-site solidification, 353
      salt, bedded, 357
      sea, 354
      sea bed, 355
      tuff, 358
      Yucca Mountain proposed repository, 366–371
    risk assessment, 359
    site selection criteria, 357, 363
  low-level wastes, 236
    incineration of, 339
    Low-Level Radioactive Waste Policy Act, 342
    near-surface land burial, 340
    NRC regulatory requirements, 338–340
    U.S. disposal sites, 340
  mixed wastes, 335, 344
  transuranic wastes, 350
  Waste Isolation Pilot Plant, 371
Water sampling for radioactivity, *see* Environmental surveillance
Weapons testing, *see* Nuclear explosions
Weighting factors, for
  common radiations, 45
  tissues, 46
Weldon Springs, 509

Welsbach Gas Mantle, 3

Windscale, liquid wastes from, 130, 354
   reactor accident, 387–393
      dose to public from, 392
      lessons learned from, 392
      neutrons effects on graphite, 388,
         389
      nuclides released during, 389
      survey procedures, 389

WIPP, *see* Waste Isolation Pilot Plant,
   repository for transuranic waste

Wood ash, ^{137}Cs in, 107

Working Level, definition of, 28
   dose per Working Level Month, 29

World Health Organization, 62

X

Xenon-135, effects on reactivity,
   225

Y

Yucca Mountain proposed high-level
   waste repository, 366–371

Z

Zinc-65 properties of, 553

Zirconium-95 in fallout, 289, 314
   lung dose from, 314

ISBN 0-12-235154-1